Variation and Evolution in Plants

NUMBER XVI OF THE

COLUMBIA BIOLOGICAL SERIES

Edited at Columbia University

Variation and Evolution in Plants

BY G. LEDYARD STEBBINS, JR.

Columbia University Press
NEW YORK · 1950

Manufactured in the United States of America

To My Father and Mother

Columbia
BIOLOGICAL SERIES

EDITED AT COLUMBIA UNIVERSITY

Preface

THE LAST twenty years have been a turning point in the history of man's knowledge and thinking about organic evolution. Great advances have been made in the fields of genetics, cytology, and the statistical study of populations, as well as in the more traditional descriptive fields of systematics and morphology. These coupled with increasing co-operation and interchange of ideas between scientists with different training and background have made possible a far broader conception of the processes of evolution as a whole than any which was open to evolutionists of even a generation ago. There is no need now for seeking hidden causes of evolutionary diversification or evolutionary progress, except in regard to certain specific processes, like mutation. Instead, the attention of scientists has been focused on evolution as a series of problems in dynamics. The direction and speed of the evolution of any group of organisms at any given time is the resultant of the interaction of a series of reasonably well-known factors and processes, both hereditary and environmental. The task of the evolutionist, therefore, is to seek out and evaluate all these factors and processes in respect to as many different groups of organisms as possible, and from the specific information thus acquired to construct such generalizations and hypotheses as he can. This requires the broadest possible knowledge of biology, which, if it cannot be acquired through direct contact with original research, must be built up vicariously through communication with biologists in different fields.

The present book is intended as a progress report on this synthetic approach to evolution as it applies to the plant kingdom, and in particular to the seed plants. It does not intend to offer any new hypotheses, except on certain limited phases of plant evolution. On the other hand, some hypotheses, speculations, and generalizations are offered whenever they seem to serve as stimuli for further research. No attempt has been made to give a final answer to any of the major problems confronting evolutionists, but the information and ideas are presented here in the hope that they will help to open the way toward a deeper understanding of evolutionary problems and more fruitful research in the direction of their solution.

Nevertheless, certain concepts of the nature of evolution seem to be now so well founded that they are taken as major premises on which the arguments of this book as a whole are based. The first of these is that evolution must be considered on three levels; first, that of individual variation within an interbreeding population or within a single colony of an asexual organism; second, that of the distribution and frequency of variants within a system of actually or potentially interbreeding populations, that is, the variation pattern on the population level within a species; and, third, the separation and divergence of populations or population systems as a result of the building up of isolating mechanisms, or the origin of species and consequently of separate evolutionary lines. If terms are desired, these could be called individual variation, microevolution, and macroevolution. The dominant evolutionary processes are different at each of these levels. Individual variation is dominated by gene mutation and genetic recombination, microevolution by natural selection, and macroevolution by a combination of the effects of selection and of the development of isolating mechanisms, chiefly of a genetic-physiological nature. The second major premise is that at all three levels evolution has progressed chiefly by the accumulation of small changes, each with a relatively slight effect, rather than by single great jumps. The third and final one is that the speed and the direction of these changes have not been constant in any one line, as might be expected if they had been predetermined for that line. On the contrary, evolution appears as a sort of progressive opportunism, which in both direction and speed is regulated at any one time by the genetic constitution of the population and the environmental influences acting upon it at that time. Progressive evolution, when it has taken place, has been caused either by progressive alterations of the environment, by the action of selection in canalizing the genetic potentialities of the evolving population, or by both of these factors. This type of evolution has been much less prominent in plants than in animals and therefore has received comparatively little attention in the present book. In brief, evolution is here visualized as primarily the resultant of the interaction of environmental variation and the genetic variability occurring in the evolving population.

Much of this book is based on a series of Jesup Lectures delivered at Columbia University in October and November, 1946. Parts of the manuscript were read by Edgar Anderson, Ernest B. Babcock, W. S. Boyle, F. G. Brieger, Spencer W. Brown, David G. Catcheside, Ralph W. Chaney, Jens Clausen, Roy E. Clausen, Ralph E. Cleland, Lincoln Constance, Pierre Dansereau, Everett R. Dempster, Th. Dobzhansky, Ralph Emerson, Carl Epling, Rudolph Florin, Adriance S. Foster, Edward Garber, H. Bentley Glass, Ake Gustafsson, E. Heitz, M. J. Heuts, W. Horowitz, James A. Jenkins, David D. Keck, I. Michael Lerner, H. E. McMinn, Herbert L. Mason, George Papenfuss, Harold H. Smith, Leon A. Snyder, A. H. Sturtevant, O. Tedin, Juan I. Valencia, Mogens Westergaard, Thomas W. Whitaker, and Ira L. Wiggins. I am much indebted to all of these men for valuable suggestions. In particular, Drs. Babcock, Dobzhansky, Anderson, and Epling have discussed these problems extensively with me both before and during the writing of this book, and have given much needed assistance and encouragement at all times. Naturally, however, the complete responsibility for the factual statements and the theoretical conclusions presented here is mine, except where other authors are cited.

G. Ledyard Stebbins, Jr.

Berkeley, California
April, 1949

Contents

Figures

Tables

Variation and Evolution in Plants

CHAPTER I

Description and Analysis of Variation Patterns

ORGANIC EVOLUTION has through the ages produced a multitude of kinds of living things. Biologists now consider this to be a statement of fact, not a theory. Therefore, our first understanding of evolution must obviously come from an intimate knowledge of the array of diverse living organisms which it has produced. Two cardinal facts stand out about this diversity. First, by and large, no two individuals are exactly alike. Second, organisms can be grouped into "kinds"; races, species, genera, families, and higher categories. These categories can be recognized because groups of individuals resemble each other more than they do members of other groups. The hierarchy of categories is a multidimensional pattern of variation in nature, and the gaps or discontinuities in this pattern give reality to the various categories.

The general outlines of this variation pattern are familiar to everyone. The word "oak" means to us one of a group of trees which have certain characteristics in common: heavy bark; hard wood; wide branching; leaves of various shapes; and acorns in the fall. We distinguish between kinds or species of oaks, such as the red oak, the white oak, the live oak, and the blue oak, by recognizing still other similarities and differences in groups of characteristics: shape of tree; persistence or fall of the leaves in winter; size, shape, and lobing of the leaves; shape and manner of growth of the acorns and acorn cups. When we learn to recognize different kinds of oaks or any other type of plant or animal, we are analyzing nature's pattern of variation in terms of its individual dimensions, that is, the characters by means of which the various categories are distinguished.

EXPLORING AND CHARTING PATTERNS OF VARIATION

The evolutionist, therefore, must be familiar with the science of classification, or systematics. The fund of information built up by systematic botanists and zoologists during the past three hundred years is the first source of his factual data. Nevertheless many systematists, even in modern times, have not adopted a point of view or a method of attack on their problems which is acceptable to the scientist interested primarily in evolution. The reason for this lies partly in the complexity of nature's pattern of variation. Not only are there hundreds of thousands of different kinds or species of plants and animals; in addition the individual differences by which we could recognize and distinguish between them, even if we confined our attention to a small group of related species, are numbered by the score. Our minds are well equipped to compare two or three objects on the basis of a number of characteristics, or several individuals on the basis of one or two characteristics, such as size, color, or general form. But in order to analyze fully nature's pattern of variation we must compare thousands of individuals on the basis of scores of characteristics.

If the reader wishes to grasp the magnitude of this task, let him try to form an accurate mental classification of a roomful of twenty or thirty people on the basis of even five or six individual characteristics, such as sex, height, stoutness, color of hair, color of eyes, and prominence of nose. If he starts by classifying all the individuals in the room on the basis of two characteristics, such as sex and color of hair, he can quickly arrive at a good generalization, such as that the group consists of about one half men and one half women, and two thirds of the men have light hair, while only one fourth of the women are blondes. But as he adds the three or four other characteristics, both the difficulty of his task and the complexity of his generalizations increase manyfold. And if he starts his classification by considering all six characteristics, he can quickly single out a few striking individuals, such as the short, stout man with light hair, blue eyes, and a snub nose, or the tall brunette with large brown eyes and delicate pointed nose. But as with the other approach, his difficulties increase out of all proportion to the number of individuals added to the classification.

If we classify a group of individuals on the basis of only two characteristics, we can easily form a diagram on paper or in our mind, representing each individual by a dot with a definite position on a field of variation, so that the clustering of the dots represents the pattern of similarities and differences formed by the group, as in Fig. 38 (p. 410). If we use three characteristics, we can make three-dimensional diagrams, which are relatively easy to picture in our minds, but harder to represent on paper. Kern and Alper (1945) have attempted to express variation in a four-dimensional diagram, with somewhat dubious success. A true diagram of the variation pattern in any group, however, would have to contain as many dimensions as there are character differences in the group. Diagrams in eight, nine, or ten dimensions can neither be imagined nor drawn.

It is obvious, therefore, that the entire pattern of variation is so vast and multidimensional that we could not understand it fully even if all the facts about it were available to us. The great problem involved in learning about the variation pattern is not so much the acquisition of additional factual knowledge, but rather the selection of facts to be emphasized, and their proper organization. Most systematists, so that they may recognize quickly and easily a large number of species, solve this problem by a series of short cuts not very different from those which all of us employ for identifying and classifying the objects about us. They focus their attention on a few easily recognized diagnostic or "key" characteristics. The other similarities and differences between individuals and groups are not ignored, but usually they are not carefully observed and evaluated. The effect of this method on the minds of some systematists is to give them a picture of the variation pattern which is accurate and precise in respect to certain characteristics found in the great majority of the individuals of each group, but which is indefinite in regard to its non-diagnostic characteristics, as well as to the nature, frequency, and distribution of those occasional individuals which deviate from the normal combination of characters.

For those systematists whose principal aim is to identify specimens and to place them into a series of species and genera which can be characterized by a set of keys and descriptions, the recognition and use of a few diagnostic key characters is perhaps the most efficient procedure. But the systematist who wishes to find in the contemporary pattern of nature a clue to the past must consider such diagnostic characters as only the beginning of his study. He must do three things which systematists interested primarily in identification would consider superfluous. First, he must seek as many additional characters as he can. He must be interested not only in those characteristics which are relatively certain clues to the identity of a species but also in those which are usually but not always different in two or more related species, and even the characteristics which show about the same range of variability in each species or subspecies, but in which this variability possesses certain regularities in relation to geographical or ecological factors. Thus, in his study of variation within *Asclepias tuberosa,* Woodson (1947) made a careful and illuminating analysis of leaf size, even though he was reasonably certain that the three subspecies under investigation did not differ significantly from each other in this respect. Second, the systematist interested in evolution must find ways of estimating variation quantitatively, both in respect to single characters and to the relationship between characters. For this purpose, the accepted methods of statistics and biometry are ideally the best; but, as will be pointed out later in this chapter, the evolutionary systematist cannot always obtain factual data suitable for such statistical treatment. He can, however, resort to a number of relatively crude diagrams and indices which provide a greater degree of precision and communicability than the intuitive judgment relied on by many systematists. Third, he can never obtain a reliable understanding of evolution in his group unless he is able to analyze at least partly the factors responsible for its variation pattern. In particular, he needs to know how much of the variation between individuals is due to environmental causes rather than genetic causes, as well as something of the different cytological and genetic processes which have been operating in the group. As stated clearly by Turrill (1936a, 1938a, 1940), the modern botanist who wishes to employ systematics as a tool for studying evolution must be thoroughly grounded in genetics, cytology, and ecology, and must integrate the evidence from all these fields in approaching his problem.

The need for an amplified taxonomy for dealing with problems of evolution has been recognized by the better systematists ever since the acceptance of Darwin's theory. But in recent years the rise of genetics, which deals with the relationships between groups of organisms in an entirely different way, has opened up new opportunities to the student of systematics and evolution, and therefore has brought to the fore the need for synthesizing and combining the approach to evolutionary problems from these two disciplines. This has produced discussions of "The New Systematics" (Huxley 1940), of "Alpha and Omega Taxonomy" (Turrill 1938b), and of proposed new disciplines such as "experimental taxonomy" (Clausen, Keck, and Hiesey 1940), "biosystematy" (Camp and Gilly 1943), and "genonomy" (Epling 1943). All these proposals are too new to be properly evaluated at present. With descriptive taxonomy of the traditional sort as his foundation, the evolutionist can explore each of the proposed new methods, both descriptive and analytical, and decide which of them is best suited to the problem which he has at hand. Their ultimate integration and evaluation must be carried out over a period of many years by the cooperation of scientists with different points of view and training who can nevertheless understand each other.

The reader might ask at this point why a book which is intended to discuss the principles and dynamics of plant evolution should devote a major part of its first chapter to a description of methods of investigation. The answer is that our impressions of all scientific phenomena are largely the result of the methods which we use to investigate them, and this is particularly true of such complex phenomena as the variation pattern of living organisms in nature. This book is written chiefly for those who are carrying out or intend to pursue research connected with evolution, a subject which covers a large proportion of the field of biology. For these readers, the writer wishes to repeat a bit of advice which has often been given to scientists. Your understanding of general principles and hypotheses will be sound only if you select the best methods for obtaining facts relating to these hypotheses and are fully aware at all times of the weaknesses as well as the strong points of the methods which you have adopted. This point of view should be held not only in relation to your own research but also in your judgment of the work of others.

OLD AND NEW MORPHOLOGICAL CHARACTERS IN THE HIGHER PLANTS

The classification of the vascular plants has been based traditionally on the characteristics of gross external morphology. Among these characteristics there is usually an ample number which serve the purpose of delimiting species and genera, and in most groups the systematist interested only in identification need not go beyond them. On the other hand, evolutionary studies must make use of all the significant character differences, and these include many features of internal anatomy, as well as of the individual cells. For this reason, one of the chief ways in which systematic studies are being amplified for evolutionary purposes is through the increased use of anatomical, histological, and cytological characteristics.

The most outstanding example of the use of anatomy and histology in conjunction with gross morphology and distribution to elucidate the evolutionary phylogeny of a group is the long series of cooperative studies which have been carried on at Harvard University on the woody members of the Ranales (Bailey and Nast 1943, 1945, 1948, Nast and Bailey 1945, 1946, Smith 1945, 1946, Swamy 1949; see these for additional references). This study of the most primitive living angiosperms has up to the present brought forth convincing evidence to support the following conclusions. The group includes a high proportion of species which on the basis of all characteristics must be placed not only in monotypic genera but even monogeneric or digeneric families. They are obviously relict types of which the close relatives have long since become extinct. Furthermore these families, as well as the genera within one of the largest of them, the Winteraceae, are not progressive stages in either one or several separate lines of adaptive radiation. Their interrelationships are more nearly reticulate, with most of the species specialized in one or more characteristics, but primitive in others. The significance of this conclusion in relation to the chromosome numbers found in the primitive angiosperms is discussed in Chapter IX (p. 363). Furthermore, both the stamens and the carpels of these primitive groups are clearly seen to be modified sporophylls. Finally, distributional studies show that the genera and species are at present strongly concentrated in eastern Asia and Australasia, and at least one family, the Winteraceae, may have radiated from the latter center (Smith 1945). This family was dispersed through the Antarctic regions, while others, such as Cercidiphyllaceae, the Trochodendraceae, and probably the Magnoliaceae, were Holarctic in distribution. The Winteraceae do not support Wegener's hypothesis of drifting continents. The concrete results obtained from these studies show that similar investigations will contribute greatly to our knowledge of relationships and evolutionary tendencies in the higher plants. The rapid technique described by Bailey and Nast (1943) of making cleared preparations for anatomical study from herbarium specimens has made such investigations practical on a much larger scale than previously. Stebbins (1940b) and Babcock (1947), using a somewhat cruder method, found that the evidence from comparative anatomy of the flower contributed greatly to an understanding of relationships and phylogeny in *Crepis* and its relatives.

Histological characters have been found useful in determining relationships and probable phylogeny by a number of workers. Stebbins (1940b) found that various species and genera of the tribe Cichorieae, family Compositae, are characterized by the presence or absence of sclerenchymatous tissue at the summit of the ovary and by the type of crystal, whether simple or aggregate, found in the ovary wall. Avdulov (1931) showed that in the family Gramineae the evidence on generic and tribal relationships obtained from cytology is much more in accord with that obtained by Harz from a study of the types of starch grains and by Duval-Jouve and others from the distribution of the chlorophyll-bearing parenchyma of the leaf than it is with the evidence from gross morphology which has traditionally been used to subdivide the family. In addition Prat (1932, 1936) has shown that the histology of the leaf and sheath epidermis of grasses likewise provides evidence which agrees with that from the starch grains, chlorophyll tissue, chromosomes, and geographic distribution, while Reeder (1946) has obtained additional concurrent evidence from a study of the embryos of some genera. In the highly important grass family, therefore, the present evidence indicates that histological and cytological characteristics provide more certain clues to the true affinity of genera than do those of the gross external morphology of the inflorescence (including spikelets and flowering scales) which are the basis for the traditional systems of classification of the family. The use of epidermal characteristics, particularly trichomes, can undoubtedly be expanded. These have been for many years considered of value by students of the oaks and of the mustard family (Cruciferae), while Rollins (1944) has found that in the genus *Parthenium* of the family Compositae two species, *P. argentatum* and *P. incanum*, differ so strikingly from each other in the shape of their epidermal trichomes that the nature of these structures provides the most reliable clue to the identity of natural hybrids and hybrid derivatives involving them. Finally, Foster (1945, 1946) has shown that the sclereids of the leaf tissue provide a valuable clue to racial differentiation in *Trochodendron aralioides* and are highly characteristic of species and species groups in the genus *Mouriria* of the Melastomaceae.

Pollen grains in many instances provide valuable additional characteristics for determining relationships, as has been shown by a number of investigators, particularly Wodehouse (1935). In some instances they may provide additional evidence for phylogenetic trends, as in certain primitive genera of the Compositae, tribe Cichorieae (Stebbins 1940b). A most ingenious and promising use of pollen grains for investigating phylogeny was worked out by Covas and Schnack (1945). They found that within a group of related species, such as the genus *Glandularia* of the family Verbenaceae, the size of the pollen grain is closely related to the distance which the pollen tube must traverse from the stigma to the ovules. The longer the style, the larger is the pollen grain; which is an obvious adaptation in providing the pollen tube with sufficient food to reach its destination. Nevertheless, although the ratio (size of pollen grain/length of stigma) is constant for the different species of a genus, and sometimes for the genera of a family, it varies greatly from one family to another. Furthermore, those families which on morphological grounds are usually considered primitive most often have a high ratio, indicating unusually large pollen grains, while the more advanced families, such as the Verbenaceae, Labiateae, and Scrophulariaceae, tend to have smaller ones. Covas and Schnack have suggested that this reduction in the relative size of the pollen grains is a phylogenetic tendency associated with the development of more efficient metabolic processes in the more advanced families, which makes possible the securing of energy for the growth of the pollen tube on the basis of a relatively small amount of stored food. This tendency, if true, is so fundamental that it needs further exploration from the standpoint of comparative physiology as well as of comparative morphology.

CYTOLOGICAL, SEROLOGICAL, AND DISTRIBUTIONAL CHARACTERS

The cytological characteristics which have been most widely used in recent years in connection with systematic studies are those of the chromosomes. A full account of this work is presented in Chapter XII. In the present discussion, however, the fact should be noted that the chromosomes, since they are the bearers of the hereditary factors, are in a class apart from other structures. Some chromosomal studies, particularly those which merely describe the number and external morphology of the somatic chromosomes, do no more than provide the systematist with additional morphological characteristics. But studies of polyploidy in conjunction with external morphology as well as geographical and ecological distribution can be analytical rather than descriptive to the extent that they may make possible the formulation of hypotheses as to the particular evolutionary processes which have been at work in the group concerned. These hypotheses, furthermore, can be tested experimentally, as discussed in Chapter IX. Even more in the nature of analysis are studies of chromosome behavior at meiosis, particularly in artificial and putative natural interspecific hybrids, as discussed in Chapters VI and VII. Chromosomes, therefore, can provide merely additional systematic characters or they may afford the best means which we have of analyzing many of the cytogenetic processes which are going on during the evolution of a group.

An additional characteristic for comparing species and families which is based on biochemistry rather than on morphology or histology, but which nevertheless is descriptive rather than analytical in nature, is provided by serodiagnostic studies of differences in proteins. Such studies were carried out on a large scale by Mez and his associates (Mez and Siegenspeck 1926) and resulted in the formation of an entire phylogenetic tree. Mez's results have been accepted by some, but severely criticized by other botanists. Some attempts to repeat the work have produced results largely in agreement with those of Mez, but other such attempts have not. The entire subject has been reviewed and discussed in detail by Chester (1937). The technique, which involves the injection of plant proteins into an experimental animal and the detection of the production of antibodies by that animal, is so unfamiliar to most botanists that they cannot evaluate it. According to Chester, the interpretation of the serological results is only partly objective and may in some instances be obscured by precipitation reactions not related to the antigenic properties of the plant sera. The techniques involved can be handled only by experienced biochemists and so are out of the reach of most botanists. Furthermore, the interpretation of the results to determine phylogenetic relationships requires the assumption that changes in protein specificity always proceed at the same rate in relation to other types of evolutionary change, or at least in relation to the amount of genetic and cytological change which is taking place. For such assumptions there is as yet no evidence, either positive or negative.

Since protein specificity is undoubtedly one of the important foundations of the differences between species and other groups of organisms, we can safely assume that studies of this specificity by serodiagnosis or any other method will eventually shed much light on such differences. In animals, the work of Irwin, Cole, and their associates on species and racial crosses in pigeons (see Dobzhansky 1941, pp. 85–87) has shown the value of such studies in analyzing the differences between closely related species. Investigations of this nature on similar groups of plant species would undoubtedly be very rewarding. In the opinion of the present writer, they would be a necessary prelude to the much more difficult task of analyzing the interrelationships between families and orders on this basis.

Finally, descriptive systematics is being amplified by additional data on the distribution and ecological relationships of groups of organisms. Distributional studies have always been an essential part of taxonomy, but in recent years they have become increasingly thorough and illuminating. Modern collectors are paying more attention than ever to careful notes on the environment of the plants which they are gathering, so that many herbarium labels are miniature essays on ecology. The inclusion of distributional maps of each species discussed is becoming standard practice for monographers, and the correlation of these distributions with the geological history of the regions concerned is an accepted procedure. The importance of such information to students of evolution is obvious. These approaches are discussed further in Chapter XIV, and have been thoroughly reviewed from the evolutionary point of view by Cain (1944).

QUANTITATIVE METHODS IN DESCRIPTIVE SYSTEMATICS

For the systematist interested in evolution, the study of additional morphological characteristics, even such fundamental ones as those of the chromosomes and geographic distribution, is only the beginning of his amplification of taxonomy over the minimal knowledge needed for identification. The next step in descriptive systematics is to obtain as complete a quantitative picture as possible of the variation within species, both in respect to individual characters and to the relationship between different characteristics. Emphasis must be placed not on the similarity between the individuals of a species in respect to certain diagnostic "key characters," but on the fact that in any sexually reproducing, cross-fertilized species no two individuals or populations are exactly alike. Furthermore, we must realize that some of this variation within species is more or less at random, while variation in other characteristics follows regular geographic patterns, which may or may not be associated with the pattern of variation between species.

As was stated at the beginning of this chapter, the pattern of variation which exists in any widespread species is so complex and multidimensional that it cannot be analyzed in its entirety. On the basis of preliminary exploration the investigator must decide what parts of this pattern are likely to provide the most significant information on the evolution of the group concerned, and he must then select the methods which will enable him to obtain this information as efficiently as possible. Three types of approach have been most frequently used for this purpose. One is the intensive study of one or two separate characters, particularly those which show geographic regularities of distribution, or clines (Huxley 1938, 1939, Gregor 1939). The second is the study of the interrelationships of morphological characteristics, through diagrams, numerical indices, or statistical methods. The third is the intensive study of the greatest possible number of both morphological and physiological characteristics of a relatively small series of samples of the species population, as is now the standard practice in the field of experimental taxonomy. These methods will be discussed in turn.

The advantage of studying variation in respect to single characteristics is that by this means large numbers of individuals can be examined and the samples treated diagrammatically as well as statistically with relative ease. When we realize that the total population of most species is numbered in hundreds of thousands or more often in millions of individuals, and that consequently a sample consisting even of thousands of specimens can rarely be more than a few percent or even a fraction of a percent of the total, we see that this advantage is not inconsiderable. This type of study, however, requires great care both in selecting characteristics to be studied and in obtaining the sample of the species population or populations to be investigated. The characters should be those which, as a result of previous systematic studies, are known to be significant in separating species or subspecies, or those which are known to have or are suspected of having regularities of geographic distribution which might be connected with the adaptive qualities and the evolutionary history of the species concerned. The selection of characters easy to observe and measure, such as flower color (Epling and Dobzhansky 1942), possesses obvious advantages in respect to efficiency, but is not essential. Woodson (1947) obtained valuable information about variation within *Asclepias tuberosa* by studying two angles of rather complex derivation, one indicating the degree of attenuation of the apex of the leaf blade and the other, the shape of the leaf base.

The sample of any species population most readily available to a systematist is that found in the larger herbaria. Although this will yield valuable information, it rarely is wholly suitable for quantitative studies of variation within species. As Woodson (1947) has pointed out, herbarium specimens are never a random sample of the species population, since they have been collected in great numbers near the centers where universities and botanical gardens are located, as well as in national parks and other points of interest, while many intervening areas have been sampled little or not at all. Furthermore, most herbarium specimens consist of only one or a few individuals from any locality and therefore cannot give a picture of the range of variation in a population, a type of information which is of the utmost importance in interpreting some of the genetic processes which may be taking place. The investigator will therefore be forced to make for himself additional collections designed to obtain a more even sampling of the range of his species and to show something of the variability at any one locality.

For the latter purpose, Anderson and Turrill (1935) and Anderson (1941, 1943) have advocated making mass collections in which a few critical parts of 25 to 50 or more individuals from each locality are pressed and kept together as a unit. The term local population samples, adopted by Woodson (1947), seems to the present writer more descriptive and appropriate. The statistical work of Gregor, Davey, and Lang (1936) on *Plantago* has served to emphasize the importance of securing a comparable part of each plant. Lewis (1947) has shown that in *Delphinium variegatum* the variation on the same individual plant from year to year in respect to the number of lobes per leaf may be greater than that found in an entire population sample. On this ground, he concluded that data obtained from population samples alone may give a very distorted picture of variation, a picture less accurate than that provided by the average series of herbarium specimens. There is little doubt that studies of a small number of population samples will give an accurate picture of a variation pattern only when combined with parallel studies of a good herbarium collection of the usual type, as well as with observations of the amount of variation induced by the environment, made either on cultivated representatives or on the same wild colony during several successive years.

Once the collections have been made and the individual data recorded, the data must be presented in such a fashion that they may be understood and evaluated as readily as possible, with a minimum amount of personal bias. For simple measurements, the well-known statistical constants, namely, the mean, the standard deviation, the coefficient of variability, and the chi-square test for the significance of differences between groups, should be used whenever appropriate. Erickson (1943, 1945), in his study of

Clematis fremontii var. *riehlii,* has shown how the variation pattern of one relatively restricted entity can be recorded almost in its entirety by the use of such statistical methods, and has on this basis been able to demonstrate regularities of geographic distribution of individual characters within an area as small as 400 square miles in habitats which are essentially uniform ecologically. Data like these are essential for an understanding of such evolutionary factors as the rate with which genes may be spread through populations and the effects of isolation on populations of different size. The significance of such problems is discussed by Dobzhansky (1941) and in Chapter IV, below.

Statistics are clearer to most readers when accompanied by diagrams. Simple histograms or curves effectively illustrate the frequency distribution of the variants within one sample with respect to a single character, while two-dimensional "scatter diagrams" like those presented in Fig. 38 (p. 410) are equally effective for showing the relationship between two characteristics in a sample. The geographic variation of a single character, or of the frequency in the populations of one or more distinct morphological types, can be shown on a map in a number of different ways, as has been done by Dobzhansky (1924), Fassett (1941, 1942), Miller (1941), McClintock and Epling (1946), Woodson (1947), and others. None of these methods has been used often enough to permit judgments of which is the most generally effective. At present, the investigator should be aware of their advantages and weaknesses, as well as of the possibility of devising improvements over all of them.

For studying the more complex problem of the interrelationship between characteristics in the variation pattern various methods have been devised, the suitability of which depends upon what the investigator wishes to learn about the populations concerned. Of the numerous problems which center about the interrelationship between characteristics, those most frequently met and most likely to be clarified by quantitative methods are the following. First, within any species population, what characteristics are correlated with each other, either with respect to the variation at any locality or to the geographic distribution of the variants, or both? Second, can two species populations which are known to be separated by barriers of physiological or genetic isolation (see Chapter VI), but which cannot be absolutely distinguished on the basis of any single characteristic, be distinguished on the basis of a combination of characteristics? Third, how can we estimate the relative amount of difference or similarity between a series of three or more different categories (species, genera, or families) on the basis of several unrelated morphological characteristics? Fourth, how can we estimate the degree of distinctness or the amount of intergradation between two populations on the basis of several characteristics?

The first problem, that of correlation, is relatively simple if only two characteristics are involved. Statistically, it can be solved by the correlation coefficient; and it may be represented diagrammatically by two-dimensional scatter diagrams. When three or more varying characteristics are involved, however, the problem becomes much more complex. Multiple correlations form a difficult statistical problem, and diagrams in three or more dimensions are difficult to construct and even more difficult to interpret. Here the best solution is to use visual methods, in which the particular characteristics to be studied are presented as simply and directly as possible, with the irrelevant characteristics either not illustrated at all or made inconspicuous.

Photographs of a series of leaves or flowers of the different individuals to be compared are often highly successful, particularly if care is taken in the arrangement and lighting of the specimens to be photographed, so that the observer can compare as easily as possible the particular characteristics to which his attention should be drawn. Examples of this technique may be seen in the work of Anderson and Whitaker (1934) on *Uvularia,* of Anderson (1936b) on *Iris,* of Clausen, Keck, and Hiesey (1940) on *Potentilla* and other genera, of Erickson (1945) on *Clematis,* and many others. When the plants or plant parts to be studied differ from each other in a large number of characteristics that are difficult to define or measure, such photographs provide the most satisfactory means of demonstrating with a minimum of personal bias the variation in these characteristics.

Another way of representing several different characteristics so as to permit easy comparison is the use of simple line diagrams, or ideographs. The floral diagrams of classical taxonomy are good examples of these, since they illustrate simply and effectively the number of sepals, petals, stamens, and carpels possessed by a family or genus, as well as whether these parts are free or united, and similar traits. As Anderson and others have shown in a number of studies, such diagrams may be adapted in an endless number of different ways, depending on the ingenuity of the investigator. They may represent the shape of a tree (Fassett 1943), the branching and internode pattern of an herb (Anderson and Whitaker 1934), the shapes and relative sizes of the sepals and petals of a flower (Anderson 1936b), or even a complex of characters taken from the leaves, inflorescence, calyx, corolla, and seeds (Epling 1944; see Fig. 1). Anderson (1946) has shown that the

FIG. 1. Ideograph showing the average measurements of the important diagnostic characteristics of a subspecies of the genus *Lepechinia* (Labiatae), constructed from a series of measurements of the various parts. From Epling 1944, by permission of the University of California Press.

correlation between the shape and size of various structures among the variant individuals of a population can be shown by superimposing ideographs representing the shape upon a graph in which the ordinates and abscissae represent certain size measurements (Fig. 2). Diagrams of this type are termed pictorialized scatter diagrams (Anderson 1949).

The second question — that of distinguishing entities on the basis of a combination of characters — was answered in the affirmative by Anderson and Whitaker (1934) in their study of *Uvularia grandiflora* and *U. perfoliata.* These two closely related species, which both occur together in the forests of the eastern

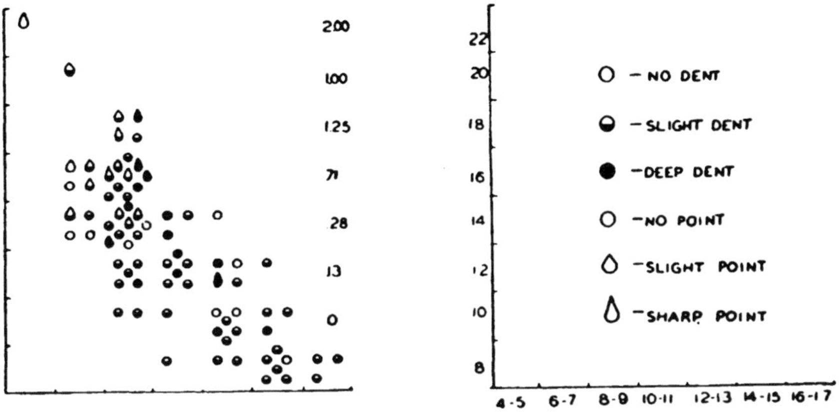

FIG. 2. At left, diagram showing the mean values in respect to four different characteristics of a series of collections of maize from Mexico. The position of each symbol on the vertical scale represents the mean number of rows of kernels per ear for that collection; its position on the horizontal scale represents the mean width of the kernel. At right, explanation of the symbols. From Anderson 1946.

United States, cannot be absolutely distinguished from each other on the basis of any single diagnostic characteristic. There are, however, at least 14 characteristics of the stems, leaves, and flowers by which a large proportion of the plants of one species can be distinguished from those of the other, and if a particular individual is examined for all of these, it can be placed in one or the other species with a reasonable degree of certainty. The reason for this is that the characters under consideration vary independently within each species, so that any individual of *U. grandiflora* which resembles *U. perfoliata* in a character such as the number of nodes below the lowest branch of the stem is typical of its own species in all or nearly all of the other characteristics, such as number of flowers, leaf shape, flower size, the character of the perianth segments, and the length of the style.

Anderson and Whitaker devised a crude numerical index by which each individual could be rated according to several characteristics with less danger of personal bias than that resulting from purely subjective judgments. This index can be used only if all the characteristics selected have about the same range of variation in the two species and if they all vary in the same direction. Nevertheless, they obtained by this method a series of values for *U. grandiflora* which was entirely distinct from the range of values for the same index in *U. perfoliata,* although in every single character from which the index was compounded the ranges of values for the two species were found to overlap. Fisher (1936) demonstrated statistically the soundness of the postulate that two entities can be separated on the basis of a combination of characteristics even when no single characteristic entering into this combination is sufficient, and he devised a more precise, but statistically much more complex, index. This is termed the linear discriminant function, and is discussed further by Hoel (1947).

The third problem, that of deciding which of several entities are more similar to and which are more remote from each other, is basic to the task of grouping species into genera, as well as to the delimitation of higher categories. The most common procedure of systematists when faced with this problem is either to use an intuitive judgment based on long experience with the group concerned or else to follow a few well-marked and presumably fundamental diagnostic or "key" characteristics. The first method has the disadvantage of being almost impossible for other systematists to interpret or to repeat, while the latter inevitably leads to a certain amount of artificiality. Anderson and Abbe (1934) have devised a method of estimating the "aggregate difference" for this purpose, which serves to systematize and standardize the subjective judgments of the monographer. The same purpose is served by representing the values for each of the groups concerned on a polygonal graph, as recommended by Hutchinson (1936) and Davidson (1947). This method has the advantage of showing not only which entities are the most similar and which the most different but also the particular characters which are most alike, and to a certain degree the amount of correlation between characteristics.

The fourth problem, that of obtaining and systematizing evidence on intergradation between subspecies or, more usually, species, was attacked by Clausen (1922) on *Viola arvensis* and *V. tricolor* and by Raunkiaer (1925) on *Crataegus monogyna* and *C. oxyacantha* by tabulating the characteristics of a number of individuals in various natural populations. This method is long, cumbersome, and difficult to follow. Anderson (1936c) has devised a much simpler and neater method of systematizing judgments on possible natural hybridization. This he has called the hybrid index. It is discussed with illustrations in Chapter VII.

The third type of quantitative approach to the study of the

variation pattern is well exemplified by the monographs of Clausen, Keck, and Hiesey (1940, 1945a) on experimental studies of the nature of species. These combine quantitative descriptive accounts of both external morphology and physiology with analysis of the environmental and genetic basis of the variation observed. They are therefore discussed in the next section, under analytical methods for studying variation.

The fact is obvious that each of these three approaches has its advantages and limitations. The first permits a relatively complete degree of sampling, but is obviously inadequate to explain the nature of variation in the organism as a whole. The study of the interrelationship between characteristics is more satisfactory in this respect, but it still provides only a part of the information needed for an interpretation of variation in terms of evolution, and it involves highly complex problems of methodology, which at present can be solved only on the basis of subjective judgments, however systematized. Furthermore, no descriptive study of variation in phenotypes is satisfactory unless something is known about how much the environment contributes to this phenotypic variation and how much of it is actually due to genotypic variation. This point will be discussed more fully in the next section and in Chapter III. On the other hand, complete descriptive and experimental studies of variation are so laborious and time-consuming that they can never be carried out on more than a small sample of the species population and therefore cannot be relied on as the only source, or even the principal source, of our knowledge about variation. Ideally, all three of these approaches should be made on the same groups, and the results should be integrated and compared before evolutionary interpretations are made.

ANALYZING THE VARIATION PATTERN

The final and most essential stage in the study of variation, that of analysis, should be begun while observational, descriptive studies are still in progress. Four different methods of analysis are commonly used in the higher plants. The first of these methods is transplantation of different genetic types into a uniform habitat, so that the effects of differences in the environment on the phenotype are eliminated, and the genotypes of different individuals, varieties, and species may be compared directly. The earliest experiments of this sort were the classic ones of Gaston

Bonnier, but they were carried out under such poor conditions of cultivation that their results and the inferences drawn from them are completely unreliable (Hiesey 1940). On the other hand, Turesson (1922a,b), Gregor (1938a, 1939), Gregor, Davey, and Lang (1936), Clausen, Keck, and Hiesey (1940), Turrill (1940), and others have shown by careful transplant experiments that in all the species studied plants adapted to different habitats are usually different genetically, so that they remain different in appearance even when grown side by side. This is the basis of the ecotype concept of differentiation within species, which will be discussed in the next chapter. Constant-environment gardens are now standard procedure in the experimental analysis of species, as well as in the application of this work to the practical task of reforestation, soil conservation, and revegetation of depleted stock ranges.

A refinement of the transplant technique has been developed with great success by Clausen, Keck, and Hiesey (1940), as well as by Turrill (1940) and his associates. This consists of dividing a single large perennial plant into several parts and growing these clonal divisions under different environmental conditions. By this method important physiological differences between closely related types have been discovered. Some genotypes have a constitution which renders them very plastic, so that their phenotype may be greatly altered by the environment, while others are much more rigid and can be modified little or not at all. These authors, as well as Anderson (1929), working with *Aster anomalus,* and Brainerd and Peitersen (1920), working with *Rubus,* showed that the vegetative parts of the plant can be modified much more by the environment than the flowers and fruits, thus confirming experimentally the common belief of taxonomists that the latter organs are generally more reliable as diagnostic criteria to separate species and other categories. Turrill (1936a) and his coworkers, using similar methods, have also shown that some species, like *Centaurea nemoralis,* are relatively little modified by the environment, while others, like *Plantago major,* are extremely plastic.

A probable solution to a problem which has puzzled systematists for some time is provided by Goodwin's (1941, 1944) uniform garden studies of *Solidago sempervirens.* This species of goldenrod, which is found along the seashore of the Atlantic coast from Newfoundland south to Florida, flowers in August in the northern end of its range and progressively later as one goes southward, so that in Florida it flowers in November. This situation is typical of wide-ranging, fall-blooming plants and is at least in part adaptive, since if the plants bloomed late in the northern regions they would be killed back by frosts before maturing their seeds. Early blooming in the south is a probable disadvantage also, because of the hot weather prevailing in the summer months. Goodwin found that when plants collected respectively at Ipswich, Massachusetts, Ocean City and Point Lookout, Maryland, and Fort Myers, Florida, were grown under uniform conditions at Rochester, New York, those from Massachusetts flowered earliest, those from Maryland next, and the Florida plants latest. The Florida plants could, however, be made to flower relatively early by covering them with hoods of heavy black cloth from late afternoon until early morning, thereby subjecting them artificially to day lengths of nine to ten hours, according to the well-known method of Garner and Allard. It is evident, therefore, that the three races studied differ genetically in physiological factors affecting their reaction to photoperiodism, or day length. This leads to the hypothesis that *S. sempervirens* and other fall-blooming species consist of a large number of different genetic types, each with a photoperiodic reaction which adjusts it to a time of blooming most favorable for its particular habitat.

A second and related method of analysis is the progeny test. For this purpose, seed of single representative individuals of a natural population are gathered and planted under uniform conditions. This provides valuable evidence on the degree of homozygosity or heterozygosity of the plants in question, and therefore of the amount of variation which they can produce by segregation and recombination, without the occurrence of new mutations. If the species is predominantly self-pollinated, the progeny grown under uniform, optimum conditions will be very much alike. The progeny of a single plant of a cross-pollinated species will, on the other hand, be very variable, but this variability will be limited by the size of the population and the conditions under which it is growing. For instance, Hiesey, Clausen, and Keck (1942) found that plants of *Achillea borealis* from a large colony (San Gregorio, California) growing under favorable conditions produced very variable offspring and were apparently highly heterozygous. On the other hand, progeny from individuals growing in a small population on an exposed coastal bluff (Bodega, California) were much more nearly uniform. This indicates a much more severe action of selection in the latter locality (Fig. 3).

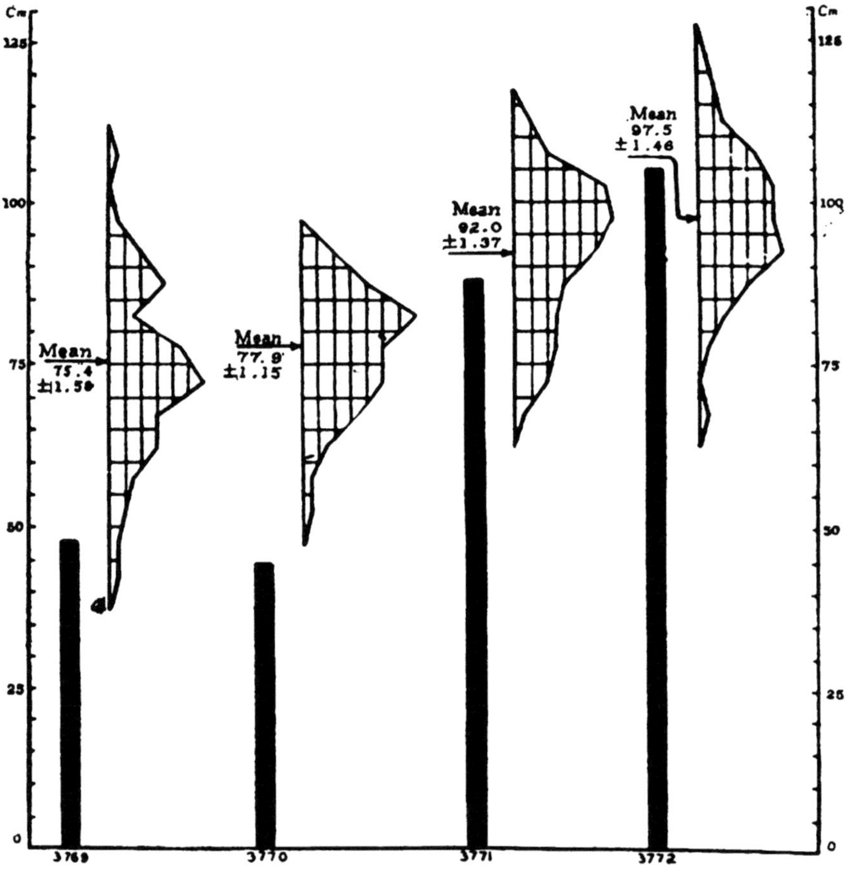

Fig. 3. Diagram showing the variability in the progeny of four different plants of *Achillea borealis* from central California. Black columns show the height of the parent plants in their native habitat; frequency curves, the variation in height of their offspring at Stanford. From Hiesey, Clausen, and Keck 1942.

The progeny test is particularly useful in testing the hybrid nature of individuals which appear to be intermediate between two species or subspecies. A well-known example of the use of this method is the work of Anderson (1936a) on *Apocynum*. In this case a taxonomist, Dr. R. E. Woodson, separated a series of specimens into three groups, identified respectively as *A. androsaemifolium, A. cannabinum,* and *A. medium,* the latter of which he suspected of being a collection of F_1 hybrids between the two former, or of F_2 and backcross types derived from them. The geneticist Anderson then grew progeny from each of these plants and sent flowering and fruiting specimens to Woodson for identification, but did not reveal their parentage. All the progeny of typical *A. androsaemifolium* and *A. cannabinum* were identified as belonging to their own species, but among those of *A. medium* were individuals identified as *A. androsaemifolium* and *A. cannabinum,* as well as those referable to *A. medium.* There is no doubt, therefore, that in respect to the characters considered diagnostic by taxonomists, *A. medium* segregates in the direction of one or both of its putative parents, and its hybrid nature is very likely.

In long-lived woody plants, in which testing putative hybrids by artificial crossing of the supposed parents is particularly difficult because of the long time involved, progeny tests of seedlings are often of great value. In the genus *Quercus,* for instance, the hybrid nature of *Q. heterophylla,* which is intermediate between the very distinct species *Q. phellos* and *Q. borealis maxima,* was verified by this method (MacDougal 1907). The writer has used it successfully to verify the occurrence of hybridization between the California blue oak (*Q. douglasii*) and the Oregon oak (*Q. garryana*) first noted by Epling (Dobzhansky 1941, p. 259). Seedlings of intermediate trees collected with their parents in a grove west of Kenwood, Sonoma County, California, were grown beside progenies of typical *Q. douglasii* and *Q. garryana.* The latter are relatively uniform and quite distinct from each other, but the seedlings from intermediate trees included segregates strongly resembling both parents, giving good evidence of their hybrid nature. In a similar experiment designed to test the hybrid nature of a population of aberrant trees resembling the black oak, *Q. kelloggii,* the seedlings closely resembled those from a typical representative of that species, so that the aberrant character of the parental trees must be ascribed to causes other than hybridization. Other examples of the use of the progeny test for demonstrating segregation in hybrid trees are those of Henry (1910), Allard (1932), and Yarnell (1933).

The ultimate analysis of both species differences and species barriers is obtained by artificial hybridization, accompanied by careful cytological and genetical studies of the F_1 hybrids and their progeny, if any are produced. Unfortunately, this method is possible in only a limited number of groups which are easily raised and in which the plants are not too large or the generations too long. Furthermore, as should become evident from material to be presented in later chapters, the relation between visible, external differences between species and the extent and nature of their genetic and cytological differences varies so greatly from one group to another that one rarely can make inferences about those groups which are impossible to grow in cultivation on the basis of results obtained with those which can be cultivated.

We should be forced to conclude, therefore, that in a large majority of plant genera a clear picture of species interrelationships can never be obtained, were it not for the fact that nature often makes experiments for us. Spontaneous hybrids between plant species are by no means uncommon, and in just those types in which they are most needed, namely, shrubs and trees, they are particularly large, long-lived, and easy to recognize. Nevertheless, no plant can be considered a hybrid after mere casual inspection. Hybrids must be distinguished from plants representing a primitive, intermediate species or subspecies, from aberrant types which may segregate from a population near the periphery of its geographic range, and from members of different species which have converged because of recent mutation. This requires careful analysis, of which the most important steps are as follows. Whenever two related species occur in the same territory, this region should be carefully searched for intermediate individuals. If none are found, this is in itself evidence that the two species are separated by some efficient isolating mechanism, or a system of them. As described in Chapter VI, many types of isolating mechanisms exist, and while some of them, like seasonal isolation, are easy to detect in the field, most of them are not. Plants appearing at first glance to be intermediate between two species should be carefully compared, character by character, with both parents. Since most morphological differences between species of plants depend on multiple factors rather than single genes and show relatively little dominance, the hybrid can be expected to be intermediate between its parents in nearly every character. Consequently, the larger the number of characters in which a particular plant can be found to be intermediate between its putative parents, the greater is the probability of its hybrid origin. Often the microscopic appearance of certain specialized cells is particularly diagnostic, as in the leaf trichomes of *Parthenium argentatum* and *P. incanum* (Rollins 1944). For evaluating systematically a large number of such differences the numerical hybrid index (described in Chapter VII) is of great value.

The morphological examination of the putative hybrid must be followed by as thorough a cytological and genetic analysis of it as is possible. A study of chromosome behavior at meiosis will be decisive if the parents differ in the number and structure of their chromosomes. But even if meiosis is normal, the fertility of both pollen and seed should be studied. If the plant is completely healthy and growing in company with its putative parents, then the discovery of a high proportion of well-filled pollen grains in the latter and 50 percent or more of empty ones in the suspected plant is strong evidence that this plant is a hybrid between two forms that are partly isolated from each other genetically. Low seed set is a much more hazardous criterion, since seed sterility can be caused by self-incompatibility or "self-sterility" and is much more subject to environmental influences than is abortion of the pollen. Finally, if morphological evidence suggests intermediacy in all or a majority of characters and the plant is fertile enough to produce seed, then a progeny test will give the crucial criterion of genetic heterozygosity for the characters which separate the species. If one or two hybrids of a particular combination have been subjected to such a careful analysis, then others of this combination can be detected with relative ease, and the collective hybrid indices of populations expressed graphically gives a rough but reasonably reliable estimate of the amount of gene interchange taking place between the two parental species populations. This is valuable both in determining the systematic position of the populations concerned and in leading toward an analysis of the isolating mechanisms which have promoted their divergence.

This completes our brief survey of the combined use of descriptive and analytical methods in studying the interrelationships of a group of species. It must be clearly understood not only that both these approaches are essential to an understanding of the evolutionary processes at work within a group of organisms but that they also require diligent, careful work and a meticulous attention to accuracy of detail which can never be overlooked. It is just as misleading and reprehensible to speculate on the degree of relationship between species or on their phylogeny on the basis of subjective examination of a series of herbarium specimens as it is to speculate on chromosome numbers and polyploidy when no counts have been obtained or to revise generic, species, or varietal names on the basis of cytological data alone, without careful studies of the traditional systematic nature. Combined attacks of this sort are often most successful when carried out by a group of cooperators, but in this case it is essential that the systematists, the cytologists, and the geneticists be not only sympathetic with the aims and methods of their coworkers in the other fields, but in addition it is necessary for them to have a good understanding of the scope and the limitations of these methods. Too often the cytologist looks to the systematist merely to give him a reliable, unchanging name for the form on which he is working, while the systematist in turn hopes that the cytologist will give him a precise tag, which he can add to his list of diagnostic key characters, thereby enabling him to make infallible diagnoses of his species. Unfortunately, however, the more thorough study of nature's variation pattern reveals that fickle Dame Nature has very different ends in view from that of making neat hierarchies of species and genera which naturalists can file away tidily in cabinets with the least possible trouble.

The relation between the various approaches to the study of variation may be illustrated by the following analogy. The pattern of variation among organisms may be likened to the topography of the earth's surface. The races, species, and genera are the various eminences, mountain peaks, and mountain ranges, while the gaps between them are the canyons, the gorges, and the valleys. The study of any part of the earth's surface has three stages. First come the explorers, who in a general way and with the crude methods at their disposal find the approximate location, height, and breadth of the mountain peaks, the depth and direction of the major valleys, and the trend of the mountain ranges. When the mountains of the region are similar in height, are separated by broad valleys, and are grouped into ranges that run straight and parallel to each other, the job of the explorers is easy. But if the topography is very rough and irregular, they must leave many purely geographical problems unsolved. The next step is the surveying of this partly known region. Exact distances and heights are obtained by triangulation, meridians and bench marks are established, and contours expressing the exact steepness and direction of the slopes of each mountain are established by means of careful observation with appropriate instruments. The third stage, that of geological study, must either follow or partly accompany the surveying process. Not until the geology is known can anything be really understood about the past history of the region in question.

The descriptive phases of systematics and cytology are analogous to exploration and surveying. They are essential preliminaries to all further study of evolution, but in themselves can never do more than suggest lines of attack and provide tentative working hypotheses for analytical studies. The simpler its evolutionary history, the more easily can the "topography" of a genus (that is, its species and subspecies) be recognized by descriptive methods. But even in these simple groups, the degree to which we can consider that we understand the evolution of a group depends to a large extent upon how many of our hypotheses and predictions are verified by experimental analysis.

Moreover, in difficult groups having a confused "topography" resulting from a complex evolutionary history, descriptive methods alone are never sufficient, even to provide a basis for recognizing biologically valid species or subspecies. Some descriptive systematists, dealing with the flora of a well-known region, often turn to the difficult or critical groups, such as *Rubus, Rosa, Lupinus, Antennaria, Hieracium,* and *Poa,* hoping that their intuition, insight, and creative ability will enable them to improve on their predecessors who have studied the same or similar series of herbarium specimens. In this they are wrong. For various reasons, which will be discussed in detail in Chapters VIII-XI, these groups are intrinsically complex in their variation patterns, and repeated studies by the traditional descriptive methods can scarcely serve to lessen the confusion in our knowledge about them. Such genera can be understood only with the aid of analytical techniques, supplemented by quantitative descriptive studies.

On the other hand, the student of evolution must not place undue emphasis on the problems presented by these critical groups. Most of them are difficult because of the presence of polyploidy, apomixis, structural hybridity, or some other complicating factor. From the broader viewpoint of the evolution of all living organisms, these are not generally distributed and significant evolutionary processes, but are more like excrescences or rococo decorations on the fundamental framework of the variation pattern. It is a basic principle of science that we can learn much about normal phenomena from studies of the abnormal, but it is equally true that we must know thoroughly the familiar, everyday situations. The basic facts about variation and evolution are most likely to be learned from studies of simple, diploid species, like the studies of Gregor (1938a, 1939) on *Plantago maritima,* of Clausen, Keck, and Hiesey (1940) on *Potentilla glandulosa,* of Epling (1944) on the genus *Lepechinia,* of Erickson (1943, 1945) on *Clematis fremontii* var. *riehlii,* and of Woodson (1947) on *Asclepias tuberosa.*

SOME PRINCIPLES OF VARIATION

Before systematic units or their patterns of variation can be studied intelligently, certain basic principles about their nature must be clearly understood. These are as follows.

First, all taxonomic entities, including varieties or subspecies and species as well as genera and higher categories, are not simple units, but complex systems of populations. This fact, though obvious to anyone who really knows species as they grow in nature, is often overlooked both by systematists and geneticists whose experience has been confined to herbarium specimens or garden cultures. Some systematists have based their philosophy on the superficially simple statement, "if two things are different, they should be described as different species." The fatal weakness of this philosophy is that these workers never form, even in their own minds, a clear conception of what they mean by the "things" or by "different." Usually the "thing" is conceived as a carefully selected type specimen plus a series of other specimens or vaguely remembered living plants which share with the type a series of easily recognized diagnostic characters. And the "different thing" is another group of specimens which for one reason or another do not fit the key characters established for the first "thing." Once the variation pattern in almost any group of species is understood, the fact becomes clear that any two individuals, populations, or series of populations of a sexually reproducing, cross-fertilized organism are somewhat different from each other, and that if the systematist is to produce a classification which will help him understand the evolution of his group, he must find out as nearly as possible which differences are biologically significant and what is the nature of the groups which they separate.

Second, any population will be constant for some characters and highly variable for others. In genetic terms, this means that in any population some genetic factors are present in 100 percent of the individuals, while the gene frequency for other factors may have any value down to a fraction of a percent. Conversely, any character or group of characters may be constant in one population and variable in other populations of the same species or subspecies.

From these two principles follows the corollary that populations must be compared in terms of differences in the frequency of qualitative, "presence or absence," characters and in the modal or mean values of quantitative characters. A second corollary is that all differences between populations are multidimensional; that is, they exist in several different characters, and the degree of difference is never the same in any two characters. Populations, therefore, should be compared in respect to an aggregate of differences, and it is for this reason that the graphs and indices described earlier in this chapter are valuable.

The third principle, and the most important one for the systematist, is that when series of populations are compared, they may often be separated into two or more groups on the basis of discontinuities in certain characters. In terms of "presence or absence" characters, this means that the frequency of the character is very low in some populations and very high in others, and that intermediate frequencies do not exist. For instance, in one group of populations the percentage of individuals with hairy sepals may vary from 0 to 20 percent, while in another, 80 to 100 percent of the individuals may have hairy sepals, and populations containing intermediate percentages of hairy-sepaled individuals may be rare or absent. In quantitative characters, a discontinuity between two groups of populations will consist of the absence or scarcity of certain intermediate values between the contrasting series of mean or modal values found in the two groups. Taxonomic categories are based on these discontinuities between series of populations. No matter how great and numerous are the differences between certain individual plants, they cannot be segregated into biologically significant subspecies, species, or other categories unless the populations to which they belong are separated as wholes by at least partial discontinuities in several characters.

TAXONOMIC CATEGORIES AND THEIR GENETIC SIGNIFICANCE

The integration of systematics and genetics, which is essential for a true understanding of the variation pattern in nature, is possible only if workers in both fields have a full understanding of and a reasonable amount of agreement on the terms used by both themselves and their colleagues to describe the phenomena of variation. This cannot be accomplished by discarding or redefining the terms now used in either discipline or by inventing new terms. The terminology of both disciplines has grown up in response to the need for describing phenomena in a certain way, and therefore both sets of terms have their own value. They should be retained as parallel series, each fitted for describing certain natural phenomena on the basis of a particular type of evidence. The following discussion will attempt to give as nearly as possible the genetic significance of the most useful taxonomic terms, and vice versa.

The fact that no two populations are exactly alike would make possible the recognition of a limitless number of categories below the rank of species. Practically, the number and nature of the categories which are to be recognized depends largely on convenience, that is, how many such categories give us a clear picture of the variation in the group concerned. The experience of most plant systematists, as well as their zoological colleagues, has been that the recognition of infraspecific units of several degrees of rank, such as subspecies, variety, subvariety, and form, produces more confusion than order. Units of one rank, termed subspecies by all zoologists and many contemporary botanists, are enough to express the great majority of the biologically significant infraspecific variation that can be comprehended by anyone not a specialist in the group. The English term variety, as well as its Latin equivalent *varietas,* is often used in nearly the same sense. But these terms have been used in a number of different ways by systematists, horticulturists, and others, so that they lack the precise connotation that is usually attached to the term subspecies. The subspecies or geographic variety is a series of populations having certain morphological and physiological characteristics in common, inhabiting a geographic subdivision of the range of the species or a series of similar ecological habitats, and differing in several characteristics from typical members of other subspecies, although connected with one or more of them by a series of intergrading forms. The only difference between this definition and that accepted by most zoologists is that in plants, which are more closely tied to their habitats than animals, two subspecies of the same species are more likely to coexist over the same territory, but are likely to be at least partly isolated from each other by habitat preferences. This is well illustrated by *Potentilla glandulosa,* which is discussed in Chapter II.

In applying the subspecies concept, the systematist cannot adhere too rigidly to certain diagnostic "key characters." After all, the characteristics which enable a plant to grow in a particular region or habitat are chiefly intangible ones, such as root growth, seasonal rhythm, transpiration rate, photosynthesizing ability, preference for certain soil types, and so forth. Some of the diagnostic characters, such as leaf size, height, and pubescence, contribute to this adaptation, but others do not, and are merely associated with the adaptive complex for one reason or another, as discussed in Chapter IV. Moreover, since different subspecies may interbreed and exchange genes, and since mutants simulating a distantly related subspecies may occasionally become established, individuals may occur which have all or nearly all of the adaptive characteristics of one subspecies, but certain diagnostic characteristics of another. Goldschmidt (1940, pp. 54–55) has given a graphic account, based on actual experience, of the errors which a systematist made when confronted with such individuals. For this reason, the assigning of individuals to a subspecies should always be done on the basis of as broad a knowledge as possible, and the locality and habitat in which the plant grew should be given as much if not more consideration than any particular morphological characteristic which the plant might possess.

In addition to the subspecies, another infraspecific unit, the form or forma, is used by some plant systematists. As a category, this has no genetic or evolutionary significance, since it consists of all of those individuals which possess in common some conspicuous aberration from the norm of the species or subspecies. Typical examples are white-flowered forms of a species which normally has colored flowers, and laciniate-leaved forms of a species with entire-margined leaves. The recognition and listing of such artificial categories in manuals and local floras is useful to botanists working in a limited area and with limited facilities, as it points out to them the relatively minor significance of plants which might otherwise be given undue attention because of their conspicuousness. Furthermore, the population geneticist may find in such conspicuous forms, many of which differ from their normal relatives by a single genetic factor, valuable material for a study of gene frequency or even of mutation rates in nature.

The *species,* the fundamental unit of the systematist, has been subjected to more arguments over its proper definition than has any other biological term. In the opinion of the writer, however, the concepts held by those systematists and other biologists concerning the nature of this entity which are expressed by Dobzhansky (1941), Mayr (1942), and Huxley (1940) and his collaborators in the same volume are notable, not for their divergence, but for their essential similarity. All agree that species must consist of systems of populations that are separated from each other by complete or at least sharp discontinuities in the variation pattern, and that these discontinuities must have a genetic basis. That is, they must reflect the existence of isolating mechanisms which greatly hinder or completely prevent the transfer of genes from one system of populations to another. Further consideration of these mechanisms and of the nature of species will be left to Chapter VI, in which the species problem will be taken up in detail.

In discussions of species and infraspecific units, reference must often be made to the fact that they occur in different geographic regions or in the same one. For this reason, Mayr (1942) has established the term *sympatric* for systematic units of which the geographic ranges coincide or overlap, and *allopatric* for those which do not occur together. More recently Mayr (1947) has recognized the difficulties of applying these terms to sedentary species and to those which occupy radically different but adjacent habitats. Nevertheless, they are very useful in general discussions of the relationship between geographic distribution and species formation and so will be employed for this purpose in the present volume.

The categories higher than the species, that is, genus, family, order, class, and phylum, are impossible to describe or to define except in highly subjective terms. This does not mean that they are purely artificial aggregates without biological meaning. On the other hand, the systematist sometimes finds that species fall naturally into clusters which have many characteristics in common and share only a few of these with the most nearly related cluster. In the conifers, for example, most botanists are in agreement over the boundaries of such genera as *Pinus, Picea, Abies, Tsuga,* and most of the others. This is because any species of *Pinus* resembles all other pines much more closely than it resembles any species of spruce, fir, or hemlock, and the same is true for the other genera mentioned. This situation has undoubtedly resulted from the fact that the conifers are a very ancient group and that forms which previously connected the genera have become extinct long ago. In fact, it is likely that most families in which the genera are well defined have suffered the extinction of many species, and further that most boundaries between neighboring genera represent gaps left by species which have perished. If this fact is kept in mind, then the search for natural boundaries to genera has some meaning to the evolutionist and is not entirely a matter of convenience. The recognition of partial discontinuities between genera and of a small number of species frankly transitional between two genera is probably necessary in complex families like the Gramineae and the Compositae. But in any case the boundaries between genera must be drawn on the basis of the largest possible number of characters rather than one or two conspicuous "key characters," as is now the case in several groups. From the genetic point of view, the separation into different genera of species which intercross and form partially fertile hybrids is particularly unnatural, so that genera like *Festuca* and *Lolium*, *Aegilops* and *Triticum*, *Elymus* and *Sitanion*, and *Zea* and *Euchlaena* in the Gramineae, as well as *Laelia Cattleya*, and their relatives in the Orchidaceae, should certainly be united. But this does not mean that species which cannot be intercrossed necessarily belong in different genera. If hybrids can be obtained it is possible to determine the degree of relationship of the parental species. But failure to obtain hybrids may be due to any one of a number of causes, as discussed in Chapter VI, and may have no connection with the degree of genetic relationship between the parental species.

SOME GENETIC TERMS AND THEIR TAXONOMIC SIGNIFICANCE

On the genetic side, the terminology begins with individuals and other units within a population, and on this level was first codified by Johannsen (1926).[1] In the first place, each individual organism has a genotype and a phenotype. The genotype is the sum total of all the genes present in the individual. For any particular locus on a chromosome, the genotype may be homozygous, that is, it may possess identical genes or alleles on the two homologous chromosomes, which are the corresponding chromosomes derived from opposite parents. Or if the two allelic genes at a locus are different, then the individual is heterozygous for that gene pair. The number of genes for which an individual is heterozygous determines the degree of its homo- or heterozygosity.

The most important genetic factor affecting the dynamics of evolution in populations is the degree of homo- or heterozygosity of the individuals composing them. Evidence for this statement is discussed fully in Chapters IV and V. In preparation for this discussion, however, the reader should remind himself of these elementary genetic principles. All the first-generation progeny of a cross between two completely homozygous individuals have

[1] The reference is to the third and final edition of Johannsen's well-known text, since here the final word on the terms is given. The original use of most of them, however, dates from the first edition of this text, published in 1909.

exactly similar genotypes, no matter how different are the two parents. But depending on the degree of heterozygosity of one or both parents of any mating, whether within a population, between varieties, or between species, the genotypes of the offspring will differ from each other to a greater or lesser degree. Homozygous parents produce uniform progeny; heterozygous parents, diverse progeny. Over a period of generations, the maintenance of genetic variability within a population depends either on the constant existence of heterozygosity or on the occasional crossing between homozygotes bearing different genotypes.

The *biotype* consists of all the individuals having the same genotype. Since even a small amount of cross-fertilization will produce heterozygosity and differences between genotypes for at least a few of the hundreds of genes present in any organism, the biotype in cross-fertilized organisms usually consists of a single individual. But in self-fertilizing plants, the individuals may become completely homozygous and produce by selfing a progeny of individuals all with the same genotype, and therefore belonging to the same biotype. The offspring of a single homozygous individual is called a *pure line,* which therefore represents the appearance of the same biotype in successive generations. Pure lines can exist only in self-fertilizing organisms, and they are maintained in nature only if natural selection constantly purifies the line of all new mutations.

When asexual reproduction occurs, either through vegetative means or through seed produced by apomixis, as discussed in Chapter X, several individuals of the same biotype may be produced even if the parent is heterozygous. All asexually produced offspring of an original individual are collectively termed a *clone.* Members of a clone and individuals of a biotype resemble each other in that both consist of many individuals, perhaps hundreds or thousands, having the same genotype and therefore appearing identically with each other when grown under the same conditions. Both clones and biotypes have been classified by some systematists and geneticists as Jordanons, isoreagents, microspecies, or even species. But such terms are more misleading than helpful if they are used for both asexual clones and sexual biotypes, since the genetic behavior of these two entities is entirely different. All the individuals of the same homozygous biotype produce genetically uniform progenies, unless crossing occurs with individuals of another line. But most clones are heterozygous, because asexual reproduction is most common in groups where cross-fertilization prevails. Hence, when members of a clone do occasionally produce offspring through the sexual process, these are usually very diverse and different from the progeny of any other member of the same clone.

A practical illustration of this point is familiar to gardeners and farmers. If one sows the seed of any plant of a well-selected, purified variety of wheat, barley, tomato, or snapdragon, the seedlings are alike and resemble the offspring of any other plant belonging to that variety. But cultivated varieties of apples, peaches, potatoes, dahlias, or lilies never come true from seed. The latter varieties are heterozygous clones and therefore produce very diverse offspring by the sexual process. Their constancy is maintained entirely by asexual reproduction. Further discussion of these points, with specific examples, is presented in Chapters V and X.

The collective term most widely used in evolutionary genetics is the *population*. This term has no single precise meaning and never denotes a unit which is analogous to any of the systematic categories. That is, whatever the scope of a population, the individuals composing it are grouped together not because they look alike or because they have any characteristics in common, but because they bear a certain temporal and spatial relationship to each other. In sexual organisms, this permits them to intercross. It is true that the members of the same population often resemble each other more than they resemble members of different populations, but this is because they constantly exchange genes and are subjected to similar selective agencies and is not implicit in the concept of the population. In sexually reproducing organisms, therefore, a population may be defined as a group of individuals among which a larger or smaller amount of interbreeding and gene exchange can occur.

This admittedly elastic definition obviously can include groups of individuals of almost any size. Furthermore, Wright (1940a,b, 1943) has emphasized the fact that the effective size of the population, which determines the degree to which its members can interbreed and share each other's genes, is much less than its apparent size. There are two reasons for this. In the first place, organisms are most likely to mate with, or to receive pollen from, individuals adjacent to them in the population, so that individuals occurring at opposite ends of a large population may be isolated from each other nearly as effectively as if they occurred in separate populations. Secondly, many populations fluctuate greatly in size from generation to generation, and in such populations the effective breeding size is, according to Wright's mathematical calculations, near the minimum size reached.

In large populations divergent evolution of segments of the same population, due to isolation by distance, has been shown by Wright to be at least a theoretical possibility. Wright has, furthermore, attempted to define subdivisions of the population of such a size that this isolation by distance does not occur or is at least a negligible factor. The numerous variables involved here require complex mathematical treatment which cannot be effectively summarized; for further details the reader is referred to the original papers of Wright (1943, 1946).

A series of population units which can be plotted and visualized is illustrated by Erickson's (1945) analysis of the distribution of *Clematis fremontii* var. *riehlii* in the limestone glades of Missouri (Fig. 4). He found that within the local populations or colonies of this species occupying a single glade, groups of plants numbering up to several hundred, which he termed aggregates, were the smallest ones that could be demonstrated to be homogenous within themselves in respect to leaf shape, and possibly differentiated from other such groups. On the other hand, his calculations suggested that differentiation might be effective only between groups of plants numbering about two thousand.

Further discussion of the problem of population size in relation to gene exchange and divergent evolution will be deferred to Chapter IV, after the discussion of selection. The present account, however, should serve to emphasize the difficulties involved in determining the size of the constant N, or effective population size, in any natural population (cf. Dobzhansky 1941, pp. 164 ff.). These difficulties indicate that we must know much more about the actual rate at which genes can spread through populations before we can apply the mathematical models constructed by the theoretical evolutionists like Wright and Fisher to many actual situations in nature.

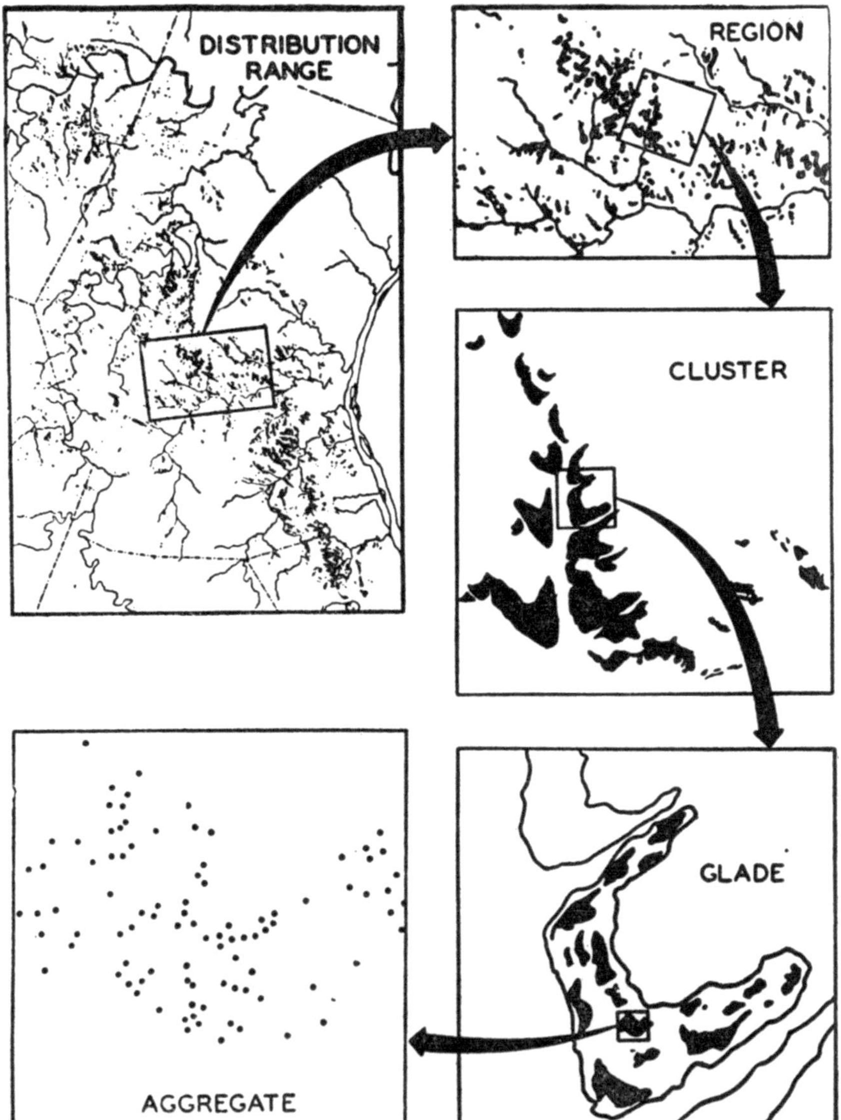

FIG. 4. Diagram illustrating the organization of the distribution range of *Clematis fremontii* var. *riehlii* into a series of subdivisions. From Erickson 1945.

The concepts embraced by the terms ecotype, ecospecies, and cenospecies (Turesson 1922a, Clausen, Keck, and Hiesey 1939, 1940) are both systematic and ecological, as well as genetic. They can be best defined and characterized in connection with accounts either of the nature of variation within species or of the species problem. Discussion of them is therefore deferred until following chapters.

The methods and concepts discussed in this chapter are the principal tools with which the evolutionist can work at present. The aim of the subsequent chapters will be to describe and to interpret as far as possible the results which up to this time have been obtained by these methods and to suggest further avenues of approach to problems of evolution.

CHAPTER II

Examples of Variation Patterns within Species and Genera

BEFORE DISCUSSING the individual factors responsible for evolution, it seems desirable to describe some of the patterns of variation which are the outcome of these processes. These exist on two different levels; first, that within the species and, second, that involving the different species of a genus or larger grouping. As is discussed in greater detail in Chapter VI, the distinction between these two levels is that of whether the pattern is essentially continuous or only partly discontinuous, owing to the more or less free interchange of genes between the various individuals or populations constituting the pattern, or whether the presence of isolating mechanisms preventing or greatly restricting mating and gene interchange has produced a number of sharply discontinuous and distinct populations or population systems.

THE ECOTYPE CONCEPT

The type of variation within species which is most important in evolution is that showing certain regularities, particularly in connection with adaptation to ecological conditions. For this reason, major emphasis has been placed in recent years on the concept of the *ecotype*. This term was originally defined by Turesson (1922a) as "the product arising as a result of the genotypical response of an ecospecies or species to a particular habitat." Turesson, in a long series of publications (1922a, 1922b, 1925, 1927, 1931a, 1936, etc.; see Clausen, Keck, and Hiesey 1940 or Hiesey 1940 for complete list), has described ecotypes in a large number of wide-ranging Eurasian species, mostly perennial herbs. He has emphasized (1936) the fact that differentiation into ecotypes is much more likely to be found in common, widespread species than in rare, local, or endemic ones. In this country, the studies of Clausen, Keck, and Hiesey (1940, 1947) have shown a similar condition in several species of western North America, while various other workers have demonstrated intraspecific genetic variation correlated with habitat differences. Groups of biotypes like those which Turesson has recognized as ecotypes undoubtedly exist in most wide-ranging plant species. Two questions arise in connection with the ecotype concept. First, to what extent are the different biotypes of a species grouped into partly discontinuous aggregates which may be recognized as distinct ecotypes, and to what extent do they form a continuous series? Second, what is the relation between the ecotype concept in plants and the concept of *polytypic species,* or *Rassenkreise,* as it has been developed by modern zoological systematists like Rensch (1939), Mayr (1942), and Miller (1941)? These two questions will be considered in turn.

Although both Turesson (1936) and to a lesser extent Clausen, Keck, and Hiesey (1940) have tended to emphasize the distinctness of ecotypes, other authors have found difficulty in recognizing well-marked groups of genetic variants because of the presence of a more or less continuous series of morphologically and ecologically intermediate populations. Engler (1913), Burger (1941), and Langlet (1936) showed the presence of much variation of a continuous type within *Pinus sylvestris* of Europe, although a slight discontinuity in the variation pattern in northern Scandinavia permits the recognition of a separate ecotype or subspecies for the pines of Lapland, as suggested by Turesson (1936) on the basis of Langlet's preliminary and incomplete data. Gregor, Davey, and Lang (1936) found that in *Plantago maritima* "there are . . . many quantitative characters which vary continuously within populations. The ranges of these in different populations nearly always overlap, and even if they do not, a series could be arranged so that there could be continuous variation throughout." Faegri (1937) has pointed out that the apparent distinctness of ecotypes in many of the species studied by Turesson results from comparison of biotypes taken from a relatively small number of widely separated localities. The same comment may be made about many of the examples given by Clausen, Keck, and Hiesey

(1940), as was suggested by Turrill (1942b). The validity of Turrill's criticism is evident from a comparison of the discussion of *Achillea* given in the above-mentioned work with the later, more complete study by the authors of the same group (Clausen, Keck, and Hiesey 1948). There is no doubt that in plants, as in animals, many species may be divided into races or groups of genetic types which are adapted to the different ecological conditions found in different parts of their ranges, and that these subdivisions are separated from each other by partial discontinuities in the variation pattern. But in addition, many widespread species possess a considerable amount of ecotypic, that is, directly adaptive, genetic variation which because of its continuous nature does not permit the recognition of distinct ecotypes.

ECOTYPIC AND CLINAL VARIATION

For this reason, much of the variation within certain species is best portrayed by the use of the "auxiliary taxonomic principle" defined by Huxley (1938, 1939) as the *cline,* or character gradient. Clines are probably common in plant species, but the ordinary methods of systematics, which deal with combinations of characters and are aimed at detecting character correlations and discontinuities, are not likely to reveal them.

Among the best examples of clines within plant species are those described by Langlet (1936) in *Pinus sylvestris* for genetic variation in chlorophyll content, length of mature leaves, hardiness, and rapidity of shoot development in the spring. Clausen, Keck, and Hiesey (1948a) found within the ecotypes of *Achillea lanulosa* and *A. borealis* clines for height of plant when grown and compared under uniform cultural conditions. In *A. lanulosa*, the tallest genetic types were from the lowest altitudes, and the decrease in height was more or less continuous with increasing altitude (Fig. 5). Olmstead (1944) found clinal variation in vigor and in reaction to photoperiodism in strains of side-oats grama grass (*Bouteloua curtipendula*) obtained from different latitudes in the Great Plains. Clinal trends were found by Böcher (1943, 1944) in *Plantago lanceolata* and *Veronica officinalis.* It is likely that most species with a continuous range that includes more than one latitudinal or altitudinal climatic belt will be found to possess clines for the "physiological" characteristics

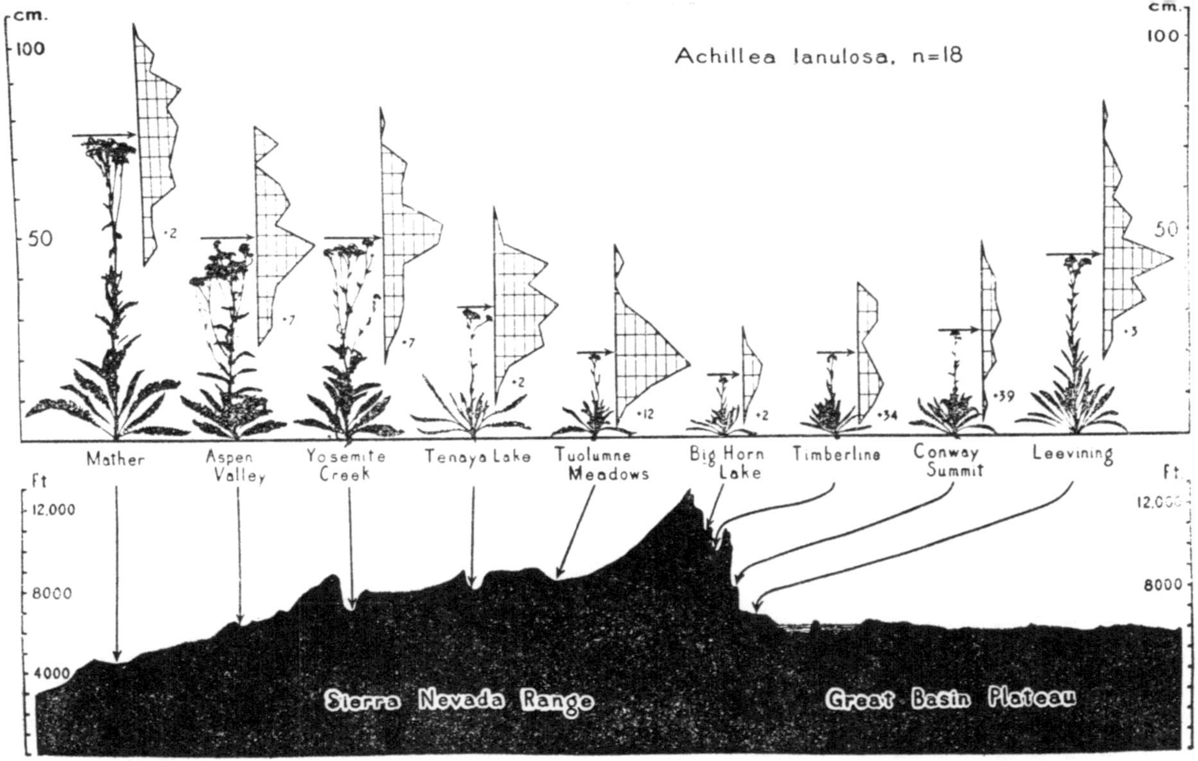

FIG. 5. Representatives of populations of *Achillea lanulosa* as grown in a uniform garden at Stanford. These originated in the localities shown in the profile below of a transect across east central California at approximately 38° N. latitude.

The plants are herbarium specimens, each representing a population of approximately 60 individuals. The frequency diagrams show variation in height within each population: the horizontal lines separate class intervals of 5 cm according to the marginal scale, and the distance between vertical lines represents two individuals. The numbers to the right of some frequency diagrams indicate the nonflowering plants. The specimens represent plants of average height, and the arrows point to mean heights. From Clausen, Keck, and Hiesey 1948.

adapting them to the conditions prevailing in the different parts of their range.

Clines have also been reported for the frequency of occurrence in a population of a single qualitative character, such as pubescence. Fassett (1942) found that the pubescent form (var. *hypomalaca*) of *Diervilla lonicera* in the region north of Lake Huron shows a cline ranging from 80 percent frequency at Espanola, Ontario, to 0 percent at Callander, about 100 miles farther east. The data of McClintock and Epling (1946) indicate that in *Teucrium canadense* there is a cline for the frequency of plants with glandular hairs on the calyx ranging from a high of 100 percent in the Rocky Mountains southeast to 0 percent in southern Florida.

Gregor (1939), in his studies of *Plantago maritima,* recognized two types of clines, topoclines and ecoclines. The former are similar to the geographic or regional clines described above for other genera and include variation both in quantitative characters, such as the length/width index of bracts and sepals, and in the frequency of such characters as pubescence and purple spots on the leaves. Ecoclines, on the other hand, are clines related to ecological gradients within a restricted area (see also Gregor 1944). They occur in characteristics such as density of spike, habit of growth, and length of scape, which vary genetically in the same region according to whether the plants are growing under the poor drainage conditions of a salt marsh or waterlogged coastal mud or under the better drainage conditions prevailing higher up and on maritime rocks and cliffs. The extreme forms which are the ends of these ecoclines constitute edaphic ecotypes (see p. 52).

Treatment of intraspecific variation by clines has advantages as well as disadvantages over the more usual method of dealing in terms of the character combinations which distinguish races or subspecies. One advantage is that it makes possible the analysis of the individual characters of these combinations and is the first step toward the causal analysis of these differences in terms of selection or any other factor. It also focuses attention on the continuous variation in quantitative characteristics which is present in many wide-ranging species and is of great importance in their adaptation to the environment, but which because of its very continuity is difficult to use as a basis for classification. One disadvantage of the use of clines is that each individual cline can be recognized only after analysis of extensive samples from many localities. This makes it impossible to describe all the clines existing in a species, so that the method of necessity places particular emphasis on certain individual characters, which may or may not be the most important ones from the standpoint of the biology and evolution of the species as a whole. Furthermore, since selection in populations is based on character combinations (see Chapter IV), too much emphasis on gradients in single characters may serve to confuse rather than clarify the whole picture of variation.

The relation of clines to subspecies depends on the character of the clines. If they are continuous over long distances, and if the extreme types are relatively limited in distribution, as with the clines in *Pinus sylvestris,* then they cannot form the basis for classification. But if a number of clines run parallel and are partly discontinuous, with a steep gradient in some regions and a moderate one or a constant level in others, then the different levels of variation in the characters forming the clines may constitute part of the basis of local races or subspecies. These are the intergroup clines of Huxley (1939).

Various species undoubtedly differ from each other in the distinctness of their ecotypes, depending on the regions they occupy and the size of their population units in terms of cross-breeding and seed dispersal. In species occupying an area like the eastern United States, which is comparatively uniform in many climatic characteristics and where a single set of factors, such as temperature and length of growing season, varies gradually and continuously, continuous or clinal ecotypic variation will be particularly prevalent. On the other hand, diversity and discontinuity of the available habitats will promote the differentiation of more distinct, easily recognized groups of biotypes within the species, and therefore of more distinct ecotypes. In addition, as Turesson (1936) has pointed out, species with obligate cross-pollination, particularly those like pines and other wind-pollinated trees of temperate regions, in which the pollen may be carried through the air for many miles, are most likely to show continuous genetic variation. In them the interbreeding population is relatively large, and the seedlings which survive are selected from a great store of genetic variants. The chances are therefore particularly favorable that the survivors will be closely adapted to their environment and will reflect more or less exactly the continuity or discontinuity of the environmental gradient. On the other hand, species in which pollination is nearly always between neighboring individuals, and particularly those with a relatively high proportion of self-pollination, will tend to be more uniform within colonies of closely adjacent plants and to show more differences between colonies. In them, therefore, distinct ecotypes are likely to be recognized with greater ease. Although much more experimental work is needed before any really conclusive generalizations can be made, the present evidence justifies the following tentative answer to the first question posed above. Intraspecific variation of an ecotypic (that is, strictly adaptive) nature is found in nearly all species with a wide ecological distribution; but the ease with which distinct ecotypes can be recognized probably varies greatly from one species to another and depends on various factors, both external and internal, which promote continuity or partial discontinuity in the variation pattern.

From the foregoing discussion, the fact should be clear that clines and ecotypes are not mutually exclusive concepts, but merely express different ways of approaching the same problem. Clinal variation may occur in the characters which determine the nature of adaptation, and therefore form the basis of the ecotypes, or it can also be found in characters of no apparent adaptive value. Correspondingly, ecotypic variation may consist of a series of clines, running either parallel to each other or in opposite directions, or it may have such well-marked discontinuities or be of such an irregular type that constant gradients are not apparent. As Gregor (1939) has emphasized, both these approaches are valuable aids to an understanding of the variation within species.

ECOTYPES AND SUBSPECIES

There are two differences between the ecotype concept and that of the polytypic species, or *Rassenkreise*. In the first place, subspecies are based primarily on recognizable differences, while ecotypes are distinguished primarily by their reaction to the environment, and may or may not possess well-marked morphological differences which enable them to be recognized in the field. Clausen, Keck, and Hiesey consider that (1939, p. 106) "limits of subspecies (a morphologic term) correspond to the limits of one or a group of several ecotypes (an experimental term)." And on the same page they state specifically that "it is sometimes necessary to include more than one ecotype in a subspecies." The relation between ecotypes and subspecies is expressed briefly as follows (Clausen, Keck, and Hiesey 1940, p. 33), "we consider a morphologically distinguishable ecotype the basis of a subspecies." A specific example of this relationship may be cited in their discussion of *Potentilla glandulosa* subsp. *nevadensis,* as follows (1940, p. 43): "It consists of two ecotypes: one is a dwarf, early-flowering alpine that occurs above 2600 m., while the other is subalpine and may be distinguished in garden cultures by its taller stature and later flowering. Since the differences between the two are not sufficiently distinct to be recognized in the field or in the herbarium with certainty, the two ecotypes are here included in one subspecies." In this example the genetic differences between the two ecotypes of this subspecies, stature and time of flowering, are clearly among the characteristics most strongly affected by the environment, so that the inability to recognize ecotypes in the field or the herbarium is due to the masking of genetic differences by environmental modification. This effect can be eliminated only by growing the two ecotypes under uniform conditions. This example is probably typical of those in which a single subspecies contains more than one ecotype. It permits the generalization that the frequency of such situations in plants probably results from the fact that the most important genetic adaptive characteristics are often paralleled by environmental modifications. Since environmental modification in animals is much less frequent than in plants, and only rarely masks racial differences, zoologists do not need a purely genetic-ecological term in addition to a primarily systematic one.

The second difference between the concept of ecotypes and that of subspecies is that the ecotype is primarily an ecological and adaptational concept, while the subspecies is primarily a morphological, geographical, and historical one. Gregor (1944) graphically describes the origin of ecotypes as follows, "The habitat environment is likened to a sieve which sorts out from among the constituents of a population those genotypes which are best fitted to survive." Ecotypes correspond with subspecies in so far as different geographical regions possess different ecological conditions. But the presence of two or more ecotypes of the same species is expected in a single geographic region wherever several ecological habitats available to the species occur. Thus, within a relatively small section of southern Sweden, Turesson (1922b) described four ecotypes of *Hieracium umbellatum;* one typical of shifting sand dunes, one of arenaceous fields, one of sea cliffs, and one of interior woodlands. Furthermore, since the ecotype is produced by the direct selective action of the environment on a heterozygous population, the same ecotype may originate independently in different localities. This is also shown by Turesson's studies of *Hieracium umbellatum.* Figure 6 shows the distribution of the different coast types of this species in the province of Scania, in southern Sweden. Turesson points out that the dune ecotype found at Torekov (upper left of map) has leaves which in many respects resemble those of the cliff and inland types occurring in the same region and are very different from those of the dune ecotype found at Sandhammar (lower right). The same is true of the cliff types found respectively at Stenshufvud on the east coast and at Kullen or Hofs Hallar on the west coast. Each of these cliff and dune races seems to have originated independently from the more widespread inland ecotype of *H. umbellatum.* The "cliff ecotype" and the "dune ecotype" are therefore aggregates of races which have originated independently in response to the same selective forces. The cliff ecotype and the salt marsh ecotype described by Gregor (1939) in *Plantago maritima* have undoubtedly arisen repeatedly. The subspecies, on the other hand, is usually conceived of as a group of populations with a common origin and a more or less integral geographic distribution, which has acquired its distinctive morphological characteristics partly through the influence of similar environmental factors, but also to a large extent through partial isolation from other subspecies. Lawrence (1945) found that in *Deschampsia caespitosa* "the evolution and distribution of its taxonomic variants appear to be entirely independent of any ecotypic adaptation." In this species, different subspecies possess similar ecotypes.

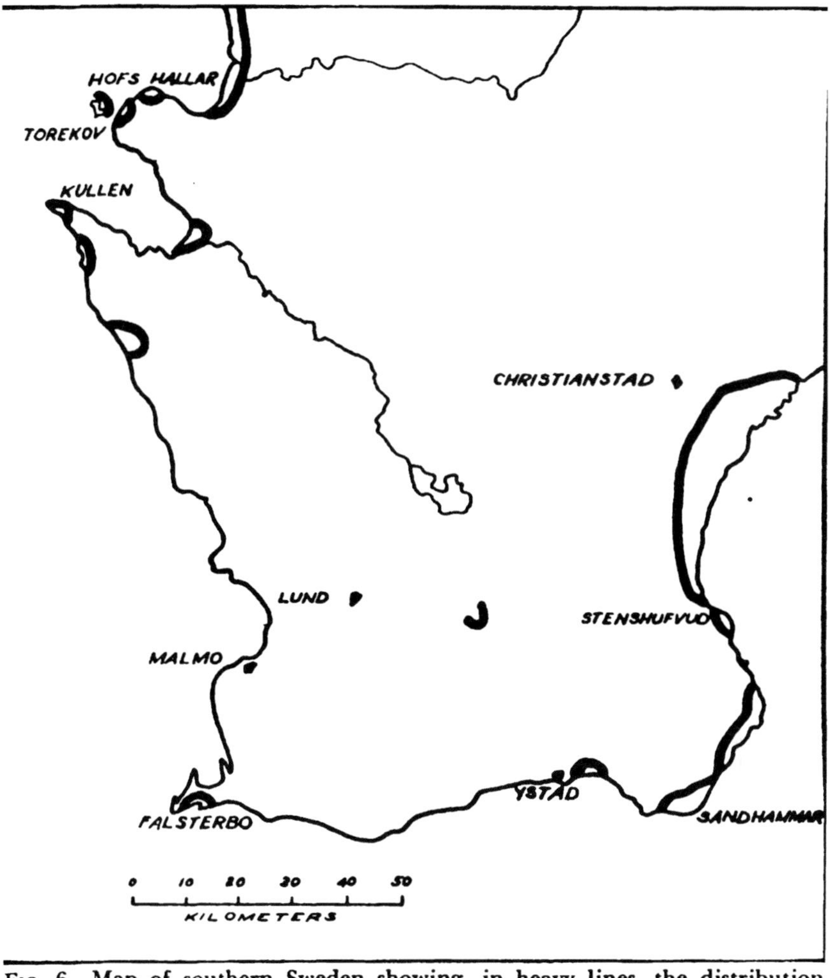

Fig. 6. Map of southern Sweden showing, in heavy lines, the distribution of the maritime ecotype of *Hieracium umbellatum*. Simplified, from Turesson 1922b.

The ecotype and subspecies have been interpreted by some authors in a relatively broad sense and by others in a relatively narrow one, but this has been true to a considerably greater degree in regard to the ecotype. This is because the concept of what constitutes an ecological difference is largely relative. As Turrill (1946) points out, the ecotypes of Turesson include both regional groupings and small local groupings of biotypes, while in the work of Clausen, Keck, and Hiesey emphasis is placed on regional or climatic adaptation. Sinskaia (1931a) and Gregor (1939) have restricted the term ecotype to local ecological variants within a particular geographic region and use another term — climatype (Sinskaia) and topotype (Gregor) — for the geographic-ecological units corresponding to the usual subspecies. In other publications (1931, 1942, 1944), Gregor recognized the geographic groupings as a different sort of ecotype and proposed a system of nomenclature for subspecific units which took into account both regional and ecological groupings and the presence of continuous variation in addition to discontinuous variation. Lawrence (1945) presented another classification of ecotypes, based on the factors responsible for their segregation, as follows:

1. Climatic ecotype (Turesson 1925)
   Synonym: climatype (Sinskaia 1928, 1931a)
2. Edaphic ecotype (Sinskaia 1928; Gregor 1942)
3. Biotic ecotype (Sinskaia 1931a)
   a. Synecotype (Sinskaia 1931a)
   b. Agroecotype (Gregor 1938b)
4. Geographic ecotype
   Synonyms: seclusion type (Turesson 1927)
   geoecotype (Gregor 1931)

Some parts of this classification may prove difficult to apply. The distinction between geographic and climatic ecotypes is impossible in most cases because different geographical regions almost always have different climates, and the process of establishment of a species in any region usually includes both selection by climatic factors and the effects of "chance introduction into mechanically isolated areas," as postulated by Gregor for the origin of geoecotypes. As recognized by Sinskaia, the biotic ecotype may be considered a special type of edaphic ecotype adapted to the type of plant competition existing under human cultivation. Like all other problems of classification, that of the infraspecific categories will be solved only by impartial discussion among all scientists interested in the subject and by mutual agreement as to what constitutes the most convenient and efficient method of expressing the actual biological situation.

VARIATION ON THE LEVEL OF THE SPECIES AND GENUS

In the case of variation patterns involving species — that is, populations or population systems separated from each other by physiological or genetic isolating mechanisms which prevent or greatly restrict interbreeding and the exchange of genes—the nature of the pattern is largely determined by the degree to which these barriers are developed and the size and diversity of the populations which they isolate. For this reason, some plant geneticists have adopted a series of terms characterizing species or species groups according to the degree of development of the isolating mechanisms separating them from other groups. These terms are the *ecospecies,* the *cenospecies,* and the *comparium.* The first two terms, both developed by Turesson (1922b), are defined on the basis of ease of crossing and the fertility of the hybrids of the F_1 and later generations. The ecospecies is a system of populations or ecotypes "so related that they are able to exchange genes freely without loss of fertility or vigor in the offspring" (Clausen, Keck, and Hiesey 1945a, p. vi). This unit corresponds most closely to the usual taxonomic species (Clausen, Keck, and Hiesey 1939). It is discussed more fully in Chapter VI. The cenospecies consists of "all the ecospecies so related that they may exchange genes among themselves to a limited extent through hybridization" (Clausen, Keck, and Hiesey 1945a, p. vi). This means that the various ecospecies composing a cenospecies can hybridize to a limited extent and form at least partially fertile hybrids, but crossing between members of different cenospecies either is unsuccessful or yields completely sterile hybrids. The cenospecies may correspond exactly with the ecospecies, but often it consists of a section or subgenus, as in *Pinus, Quercus,* and *Ceanothus,* or it may comprise a whole genus, as in *Aquilegia.* The comparium (Danser 1929) is a more inclusive and strictly genetical term and includes (Clausen, Keck, and Hiesey 1945a, p. vi) "all the cenospecies between which hybridization is possible, either directly or through intermediaries." Like the cenospecies, it may in isolated instances consist of only one ecospecies, but usually it approaches the size of a genus, and in some plant groups, such as the grasses and the orchids, it may include a whole series of recognized genera.

Although the terms and concepts which have already been discussed and the factors of evolution which will be reviewed in the remaining chapters of the present volume all apply to the great majority of plant groups, nevertheless different interrelationships between evolutionary factors have produced in various genera and families a great diversity of variation patterns. These patterns may differ not only between closely related genera but even between different sections of the same genus. Part of this diversity is due to the sporadic or frequent occurrence of such phenomena as self-fertilization, polyploidy, apomixis, and structural hybridity, as is brought out in Chapters V and VIII–XI. Nevertheless, even groups of which all the members are diploid, sexual, and cross-fertilizing may differ greatly from each other in the degree to which isolating barriers are developed and the amount of genetic and ecotypic diversity contained within the species. In some genera, the species are genetically homogeneous, have relatively restricted ranges, and have no close relatives. At the other extreme are genera in which all the species are polytypic in the sense of Mayr (1942, Chaps. V and VI); that is, they consist of a greater or lesser number of partially discontinuous, morphologically recognizable subspecies or ecotypes which replace each other geographically. Most genera have variation patterns somewhere intermediate between these extremes. Camp and Gilly (1943) have attempted a classification of the types of species found in plants, but their terminology has proved cumbersome and not always soundly based on genetic phenomena, so that it has not been generally accepted. Nevertheless, as knowledge of the genetic bases of variation patterns increases, the adoption of a system like that of Camp and Gilly may prove valuable in orienting the thinking of botanists on problems of evolution; their system may be considered premature rather than altogether inappropriate. In the remainder of this chapter, a series of examples will be presented which is intended to give an idea of the diversity of genetic patterns on the diploid level.

PATTERNS IN THE FAMILY RANUNCULACEAE

In the Ranales, the most primitive order of angiosperms, three genera have been studied by experimental as well as by descriptive systematic methods. These are *Paeonia, Delphinium,* and *Aquilegia.* In *Paeonia* the two New World species are complicated by structural hybridity and are discussed in Chapter XI. In the Old World there are ten to twelve diploid species plus a number of polyploids which will not be discussed here (Stebbins 1939, Stern 1946). Four of these diploid species, *P. suffruticosa, P. emodi, P. tenuifolia,* and *P. cretica,* are fairly certainly homogeneous in nature, and two others, *P. albiflora* and *P. intermedia,* are probably so. *P. delavayi* is very polymorphic in nature, but the evidence from herbarium specimens does not indicate any geographic segregation of its variants. Careful field study may nevertheless reveal that it is a polytypic species. The three remaining species groups, *P. anomala* (including *P. veitchi, P. woodwardii,* and *P. beresowskii*), *P. obovata* (including *P. japonica*), and the *P. daurica-mlokosewitschii-broteri-cambessedesii* series, are characteristic polytypic species or species groups. Experimental evidence on most of these has shown that they are separated from each other by strongly developed genetic barriers (Saunders and Stebbins 1938, Stebbins 1938a). It is possible that some of the units of the last-mentioned complex are also genetically isolated from each other. In *Paeonia,* therefore, homogeneous and polytypic species occur with about equal frequency.

The genus *Delphinium* has been treated in two monographs, one of which (Ewan 1945) explored thoroughly the North American species. In addition, the chromosomes are known of all the species native to California, and many of these have been further analyzed by means of transplants, progeny tests, and interspecific hybridization (Epling and Lewis 1946 and unpublished). These are all diploid, except that three, *D. hanseni, D. variegatum,* and *D. gypsophilum,* contain related tetraploid forms. Ewan described subspecies in 23 of the 79 species recognized by him, and further field studies, as well as genetic analyses, may show that some of the species which he recognizes are no more than subspecies. A group of "species" which probably form a single polytypic species (Epling, oral communication) is that consisting of *D. decorum, D. patens, D. menziesii, D. nuttallianum,* and their relatives, including 13 species placed by Ewan in his "tuberiform series." These forms replace each other geographically, and Ewan cites examples of intermediate and intergrading populations.

In another species group are found seven species which have been extensively observed in the field and intercrossed, namely, *D. variegatum, D. hesperium, D. hanseni, D. gypsophilum, D. parryi,* and *D. amabile.* These simulate the subspecies of a polytypic species in that they in general replace each other geographically or ecologically, and they are obviously closely interrelated. But they all overlap with one or more of their relatives in a considerable proportion of their ranges, and in these regions Drs. Mehlquist, Epling, Ewan, Lewis, and the writer have repeatedly and independently seen two or three species growing side by side with no signs of intergradation. The isolating mechanisms are, however, of a somewhat puzzling nature, and they will be discussed further in Chapter VI. Five of these species are themselves polytypic, with two to four subspecies each. Two California species, *D. uliginosum* and *D. purpusii,* are sharply distinct relict endemics. In *Delphinium,* as in *Paeonia,* the pattern is that of closely related homogeneous species more or less sharply isolated from each other, as well as of polytypic species, the two types occurring with about equal frequency.

Aquilegia has become a classic example of a genus in which isolating barriers between species are weak or absent. The latest monograph of the genus, that of Munz (1946), lists 67 species, distributed through temperate Eurasia and North America. They are almost entirely allopatric; in a few places two species grow in the same region, but rarely, if ever, more than two, if we except regions which because of their great topographic relief include more than one climatic zone. All the species have the same vegetative habit and essentially similar leaves, and the differences between them in stamens and seed follicles are relatively slight. The species are based almost entirely on differences in the size, shape, proportions, and color of the sepals and petals. It has long been known among gardeners that they hybridize freely and that a large proportion of F_1 hybrids, even between species that are relatively remotely related to each other, are highly fertile. Natural hybridization between two such species, *A. formosa* and *A. flavescens,* has been recorded by Anderson (1931) and apparently is not uncommon.

The genetic barriers between some of the species have been studied by Skalinska (1928a,b). She found that the F_1 hybrid between the European *A. vulgaris* and the rather remotely related *A. chrysantha* of western North America has about 50 percent normal pollen and 20 percent normal seed setting, regardless of which way the cross is made. On the other hand, if the Japanese species, *A. flabellata nana,* a close relative of *A. vulgaris,* is crossed with *A. chrysantha,* the F_1 is highly sterile and resembles the mother when *A. flabellata nana* is the pistillate parent, and furthermore it tends to produce matroclinous segregates in the F_2 generation. But the reciprocal hybrid, with *A. chrysantha* as the pistillate parent, is normal and rather highly fertile (Skalinska 1929, 1935). On the other hand, if *A. flabellata nana* is crossed with another North American species, *A. californica,* the F_1 hybrid is normal and has almost normal fertility no matter in which direction the cross is made (Skalinska 1929). In a later paper (Skalinska 1931) the author reported matroclinous behavior and high sterility in *A. flabellata nana* × *A. truncata.* Since according to Munz (1946) *A. californica* is merely a synonym of *A. truncata,* the evidence of Skalinska would suggest that in this group of *Aquilegia* species the sterility and the genetic behavior of the F_1 hybrid differs according to the strain of the parental species used in the cross. At any rate, there are no strong genetic barriers to gene exchange between the Old World species centering about *A. vulgaris* and their morphologically very different relatives in North America.

Evidence of a different type of genetic isolating mechanism between species of *Aquilegia* was found by Anderson and Schafer (1933). They grew a plant of *A. vulgaris* which was homozygous for two recessive genetic factors in a garden beside plants of the same species which had the corresponding dominants, and in addition plants of three other rather distantly related species, *A. pyrenaica, A. skinneri,* and *A. caerulea.* Seed harvested from open pollination gave 16 percent of offspring which were the result of outcrossing to other plants of *A. vulgaris,* but no interspecific hybrids were found.

It is evident, therefore, that while many of the described species of *Aquilegia* are only subspecies from the standpoint of their genetic isolation from each other, nevertheless the whole genus cannot be viewed as one large polytypic species. Munz, however, was unable to divide the genus into natural species groups even on the basis of the most careful study of herbarium specimens and living plants in the garden, and he pointed out great divergences between previous groupings of the species. The true boundaries of the species, therefore, cannot be determined by a study of the specimens alone, but will become apparent only after a careful and systematic series of hybridizations. However, we may now safely accept the statement of Clausen, Keck, and Hiesey (1945a) that "*Aquilegia* is one huge cenospecies composed of only a few ecospecies. Probably most of the recognized 'species' are merely morphologically distinguishable ecotypes or subspecies." The largest of these ecospecies is undoubtedly the European *A. vulgaris.* Twenty-nine of the 67 species recognized by Munz have been recognized by at least one other botanist as varieties or subspecies of *A. vulgaris,* and 12 more are obviously closely related to these. *A. vulgaris,* therefore, is probably a polytypic species which in its complexity rivals *Peromyscus maniculatus* among the vertebrates. Among the North American Aquilegias, the red-flowered series, consisting of the eastern *Aquilegia canadensis* and the western *A. formosa* and their relatives, is probably another rather complex polytypic species. The suggestion of Clausen, Keck, and Hiesey, that *Aquilegia* "possibly represents a youthful stage experienced by many other, now mature genera" is reasonable if the reference is confined to the evolution of the interspecific barriers themselves. But *Aquilegia* is probably not a youthful genus in terms of chronological time. It is distributed chiefly in the old floristic communities of the mesophytic forests of the north temperate regions (see Chapter X). Furthermore, it belongs to the primitive family Ranunculaceae and in some respects is primitive even for that family. In *Aquilegia,* therefore, the development of complex polytypic species has been accompanied by much geographic isolation of segments of its populations, along with great variability of certain organs of the flower. On the other hand, the rest of the plant, as well as the chromosomal apparatus, has remained comparatively stable.

VARIATION PATTERNS IN THE GENUS *Potentilla*

The second family of angiosperms to be considered, the Rosaceae, is noted for the frequency of polyploidy, both within and between genera, as will be discussed in Chapter VIII. Phylogenetically it is usually placed in an intermediate position. The genus that will be considered here, *Potentilla,* contains at least three different types of variation pattern. Most of the genus consists of one or a few large polyploid or agamic complexes and will be discussed in Chapters VIII–X. The subgenus *Drymocallis*, however, consists entirely of diploid species, with the somatic number of 14 chromosomes. Its most widespread and complex species is *P. glandulosa*, which contains at least 11 subspecies (Clausen, Keck, and Hiesey 1940). These are adapted to regions as widely diverse as the warm-temperate, semiarid coast of Southern California and the arctic-alpine meadows of the high Sierra Nevada. Diagnostic characters for distinguishing the subspecies are found in the stolons, leaves, stems, inflorescences, sepals, petals, and seeds, and 16 such characters are tabulated for the 11 subspecies by Clausen, Keck, and Hiesey. In addition, transplant experiments revealed great physiological differences in rhythm of growth and other characteristics, and an important biological distinction is that some of the subspecies are self-incompatible and normally cross-pollinated, while others are self-compatible and largely self-fertilized.

Within *P. glandulosa*, therefore, we find all the types of morphological and physiological differences that often separate good, distinct species. If the extreme forms, such as subspp. *typica* and *nevadensis*, should become completely isolated from each other geographically and should develop genetic sterility barriers, they would become amply distinct species even without any further divergence in morphological and physiological characteristics. What keeps the species a single unit is the fact that every subspecies is at some locality or localities in contact with at least one other subspecies, and where these contacts occur, hybrid swarms are regularly found.

Two other species of the subgenus *Drymocallis* occur in the United States, *P. arguta*, which has two subspecies, and *P. fissa*, which is confined to high altitudes in the central Rocky Mountains and is homogeneous. Intermediates between these two and *P. glandulosa* are reported from herbarium specimens, but apparently have not been observed in the field. In the Medicine Bow Mountains *P. fissa* and *P. glandulosa* subsp. *glabrata* occur sympatrically, but no information is available on the extent and manner of their isolation from each other in that region. The uncertainty in regard to these species, which belong to one of the groups best known from the experimental point of view, is clear evidence of how much work is necessary before the true nature of species even in a relatively simple group can be made clear.

Since *Potentilla glandulosa* is one of the best-known species of flowering plants in North America from the standpoint of experimental taxonomy, a further examination of the variation pattern within this species is desirable. Its four best-known subspecies are *typica,* most characteristic of coastal California, and subspp. *reflexa, hanseni,* and *nevadensis,* all best known from the Sierra Nevada. Subspecies *hanseni,* when compared with subspecies as they are usually understood in animals, has two unusual characteristics. In the first place, it is included entirely within the geographic range of subsp. *reflexa.* The two subspecies are, however, ecologically isolated from each other, subsp. *hanseni* occurring in moist meadows and subsp. *reflexa* on dry slopes. Where these two habitats come together, as at Aspen Valley in Yosemite National Park, hybrid swarms are found, but elsewhere the subspecies maintain themselves as fairly distinct. This separation is aided by the fact that subsp. *reflexa* is self-pollinating and therefore has a relatively small chance of receiving pollen from subsp. *hanseni.* These two subspecies, therefore, represent edaphic ecotypes and differ from the strictly geographic subspecies commonly recognized in animals.

The second peculiarity about subsp. *hanseni* is that in morphological characteristics and ecological preferences it is intermediate between subsp. *reflexa* and subsp. *nevadensis.* Furthermore, all the moist meadows which it occupies have resulted from the filling up of postglacial lakes and are therefore one of the most recent habitats in the region. The suggestion of Clausen, Keck, and Hiesey (1940, p. 44) that subsp. *hanseni* has resulted from hybridization between subspp. *nevadensis* and *reflexa,* with subsequent stabilization of a group of segregates adapted to meadow conditions, is very plausible. It may be true that this type of origin of new subspecies in plants is not infrequent.

The other seven subspecies of *Potentilla glandulosa* are much less well known, but the fact that four subspecies occur together in another region, namely, the Siskiyou mountain area of northwestern California, suggests that here also edaphic factors rather than climatic or geographic factors are maintaining the isolation between subspecies. The subgenus *Drymocallis* contains, in addition to the three American species mentioned, six in Eurasia. One of these, *P. rupestris,* is widespread, while the other five are endemic to mountain areas in southwestern Asia, from Asia Minor and the Crimea to the Himalaya. The interrelationships between these species are relatively poorly known, but available evidence indicates that some of them may be subspecies of *P. rupestris.* Attempts to produce hybrids between *P. rupestris* and *P. arguta* have failed, suggesting that the American and the Eurasian species of the subgenus *Drymocallis* are well isolated from each other and may represent different cenospecies.

The subgenus *Drymocallis* of *Potentilla,* therefore, contains two widespread, polytypic species, one in each hemisphere of the Holarctic region, and several less complex species about equally distributed in each hemisphere.

THE GENUS *Quercus*

The next variation pattern to be considered is that of the oaks, a typical genus of angiospermous trees of the north temperate regions. According to present systematic treatments the genus *Quercus,* of the family Fagaceae, contains between 250 and 300 species, but it is very likely that many of these actually are subspecies of a smaller number of polytypic species. Cytologically all species of *Quercus* investigated, as well as all other genera of the Fagaceae, have the same chromosome number, $n = 12$, and only slight differences exist between them in size and in morphology of the chromosomes (Hoeg 1929, Ghimpu 1929, 1930, Jaretzky 1930, H. J. Sax 1930, Natividade 1937a,b, Duffield 1940). Natural hybrids between generally recognized species are not uncommon, and both cytological and genetic studies indicate that most are fertile (Sax 1930, Yarnell 1933, Wolf 1938, 1944).

The usual treatments in the various regional floras do not describe geographic varieties or subspecies within most of the oak species, although *Q. borealis,* the common red oak of the eastern United States, and three species of live oaks of California, *Q. wislizenii, Q. agrifolia,* and *Q. dumosa,* are commonly recognized to be polytypic. Nevertheless, the experience of the writer and of most other observers with whom he has spoken indicates that a much larger number of species will be found to possess geographic variation when their intraspecific variation patterns are studied more carefully. Furthermore, hybrid swarms are often found in regions where the ranges of markedly different allopatric species overlap. A conspicuous example of this is *Quercus garryana* and *Q. douglasii,* mentioned in Chapter I. Furthermore, similar hybridization apparently occurs at the southern end of the range of *Q. douglasii,* where it overlaps with *Q. dumosa* and its var., *turbinella.* There is thus good evidence of gene interchange between a chain of forms ranging from a large tree with broad, deciduous, lobed leaves to a shrub with very small, coriaceous, evergreen, unlobed leaves; and the range of differences in acorn cups and other reproductive structures within this series of forms is equally great. This situation is further complicated by the fact that the valley oak of California, *Q. lobata,* which is the counterpart of *Q. garryana* both in habitat and in the character of the tree, shows relatively little evidence of hybridization in localities occupied by these two species. Furthermore, *Q. lobata* and *Q. douglasii,* although they occur sympatrically and often in mixed stands throughout most of their ranges, also show little evidence of intergradation with each other. In the oaks of California, therefore, the magnitude of the visible differences between species seems to bear no direct relationship to the amount of intergradation that can occur between them.

A relationship of a somewhat different type is shown by *Q. marilandica* and *Q. ilicifolia,* two species of the subgenus *Erythrobalanus* (black oak) in the eastern United States. These species occur together chiefly on the coastal plain of New Jersey, Long Island, and Staten Island, New York, but elsewhere are usually well separated from each other geographically or ecologically. Both species are relatively uniform throughout their ranges, as is evidenced from the fact that a single population sample of *Q. marilandica* from Cliffwood, New Jersey, showed a range of variation in six important diagnostic characteristics equal to that found in a series of specimens representing the entire geographic range of the species from Pennsylvania to Kansas and Arkansas (Stebbins, Matzke, and Epling 1947). In New Jersey, the two species occur sympatrically over a large part of the pine-barren area and in general remain quite distinct from each other. But shrubs intermediate between them do occur and were described as natural hybrids more than fifty years ago (Davis 1892). Study of a population sample from one of these localities by Stebbins, Matzke, and Epling (1947), with the aid of Anderson's hybrid index method (see Chapter VII), revealed that eight out of the sample of 86 specimens were sufficiently intermediate between the two species so that they could be considered F_1 hybrids or their derivatives, while about 35 percent of the specimens showed some evidence of admixture of genes derived from the two species. Most of them were introgressive types of *Q. marilandica,* containing a few genes from *Q. ilicifolia.*

The variation pattern in *Quercus* is further complicated by the fact that species may occur together in many regions without showing any signs of intergradation, but in other areas these same species may be connected by both first-generation hybrids and other intergrading types which are the result of segregation or backcrossing. *Q. douglasii* and *Q. lobata,* mentioned above, are an example of this. Although they are distinct nearly everywhere, a series of hybrid types under the name × *Q. jolonensis* has been described from a small area in central California. In the eastern United States, where in many regions as many as ten or eleven species of oaks may occur sympatrically, examples of this type are more numerous. Two very distinct species of this region are *Q. imbricaria,* the shingle oak, which has elliptic entire-margined leaves, and *Q. velutina,* the black oak, which has much broader, deeply lobed leaves and differs greatly in acorn cups, bark, and other characteristics. The easily recognizable hybrid, which bears the name × *Q. leana,* has often been noted as an occasional isolated tree, and in some regions considerable numbers of hybrid and hybrid derivative trees have been seen (W. H. Camp, oral communication). *Quercus coccinea,* the scarlet oak, resembles *Q. velutina* more closely than does *Q. imbricaria,* and although the two species are sufficiently distinct over much of their range, in some areas, such as the Atlantic coastal plain of Long Island and southeastern Massachusetts, intermediate types are almost as frequent as trees typical of one or the other species. Furthermore, in the coastal plain of New Jersey and probably elsewhere the populations of *Q. velutina* contain genes from *Q. marilandica,* the blackjack oak. It is likely, therefore, that most of the geographic and other variability of *Q. velutina* has resulted from hybridization and introgression of genes from other species, as is discussed in Chapter VII. Turning to the white oaks, an even more marked degree of intergradation has been noted by

Hampton (Trelease 1924) and by E. Anderson (oral communication) between *Q. bicolor,* the swamp white oak, and *Q. macrocarpa,* the bur oak, both of the central United States. It is remarkable that these two species, although they occur sympatrically over most of their geographic ranges and inhabit rather similar ecological sites, should be able to intergrade so freely and nevertheless retain their identity.

In Europe, the species of *Quercus,* though extremely variable, cannot be segregated into well-marked geographical races or subspecies (Ascherson and Graebner 1913, pp. 445–544). Of the 15 species recognized by Ascherson and Graebner, at least 12 occur sympatrically in parts of the Balkan Peninsula. The three most variable species are the closely interrelated group, *Q. robur, Q. sessilis (Q. sessiliflora),* and *Q. lanuginosa.* Hybrids between all three of these species have been reported as frequent by numerous botanists from many parts of Europe, and there is some evidence that triple hybrids occur, involving all three species. Furthermore, forms intermediate between the hybrids and their parental species are frequent in many places, so that many forms have been considered by some specialists to be of hybrid derivation and by others as merely aberrant forms of the parental species. Salisbury (1940) has given an account of the ecological relationships between *Q. robur* and *Q. sessiliflora* in Hertfordshire, England. In this region, the soil preferences of the two species are different, and hybrid trees are frequent along the zone where two types of soil meet. The same hybrid was studied in Denmark by Hoeg (1929), who noted also trees intermediate between the hybrid and its parents. Studies of chromosome behavior in meiosis showed perfectly regular, normal conditions in the parental species, but in several trees of an intermediate type Hoeg found irregularities typical of interspecific hybrids, such as lagging univalent chromosomes and tetrads with extra nuclei.

One interesting feature of the progeny of both artificial and natural oak hybrids is that in respect to vegetative characteristics they usually segregate so sharply that even among a relatively small number of individuals the parental types can be recovered (MacDougal 1907, Ness 1927, Coker and Totten 1934, Allard 1942, Wolf 1938, 1944, Yarnell 1933, and Stebbins, unpublished). This is in striking contrast to the behavior of interracial and interspecific hybrids in most other plant groups, in which the number of genetic factors controlling the differences between them is so large that it is relatively difficult to recover the parental types (Müntzing 1930a, Goodwin 1937, Winge 1938, Jenkins 1939, etc.). The evidence available suggests that the number of genes by which species of oaks differ from each other is considerably smaller than it is in the case of most other plant groups.

Palmer (1948) has presented a complete list of the hybrid oaks of the United States, compiled from the extensive references in the literature, but based also on his own experience with them in nature over a period of many years. He has pointed out that although hybrids may be expected between almost any two species belonging to the same section, nevertheless they always form a very small percentage of the total population. They are most often found near the edge of the geographic range of a species, where it is rare and related species are more common. He has concluded that this rarity is due, not to the difficulty of producing interspecific hybrids, but to the small chance which the hybrids have of becoming established. This conclusion is borne out by the results of Pjatnitzky (1946a,b) from experiments on artificial cross-pollination of several different species. He obtained in every case a rather low percentage of acorns set, but this was also true in pollinations between members of the same species. A noteworthy fact is that some intersectional crosses, such as *Q. macranthera* × *robur* and *Q. macranthera* × *alba*, gave higher yields of acorns and seedlings than did the maternal parents when pollinated with their own species. Pjatnitzky was even able to obtain a type of hybrid not known at all in nature, namely, one between the subgenera *Lepidobalanus* and *Erythrobalanus* (*Q. robur* × *borealis maxima* and *Q. macranthera* × *borealis maxima*). These seedlings, though vigorous, have apparently not yet reached maturity, and no data are available on their fertility.

In *Quercus*, therefore, we have a rather large number of species, many of them polytypic, which are often capable of exchanging genes with other, morphologically distinct populations occurring in the same region and in similar habitats. These sympatric populations have always been kept apart as distinct species, but some of them, such as *Q. bicolor* and *Q. macrocarpa*, behave genetically more like subspecies. The correct systematic treatment of such a variation pattern is a difficult problem, and it may be remarked that other genera, for instance, *Ceanothus* (McMinn 1944), and probably *Vitis*, are essentially similar to *Quercus*. The great evolutionary possibilities of such situations will be discussed in Chapter VII.

PATTERNS IN THE FAMILY COMPOSITAE

In the great family of Compositae, three genera will be discussed, all of which have recently been studied with the use of quantitative and analytical methods. The first is the genus *Wyethia*, monographed by Weber (1946). This consists of 14 species, all deep-rooted and large-flowered perennials, native to the arid and semiarid regions of the western United States. The four species counted have the basic haploid chromosome number $x = 19$. Only one, *W. scabra*, is recognized as polytypic, but in three others, *W. angustifolia*, *W. ovata*, and *W. arizonica*, the presence of variant forms is noted. In the most primitive of the three subgenera, subg. *Agnorhiza*, the six species are allopatric and markedly different, as well as distinct from each other. In the other two subgenera, *Alarçonia* and *Euwyethia*, each species is wholly or partly sympatric with at least one other of the same subgenus. When two such species occur together, occasional hybrids between them are found. Although these are fertile, they rarely produce enough offspring in nature to form extensive hybrid swarms, and even when two species are completely sympatric, like *W. helianthoides* and *W. amplexicaulis*, they completely retain their identity. Such small groups of hybrids as are found have probably resulted from recent spreading of the species in range lands overgrazed by cattle. *Wyethia*, therefore, presents a pattern of mostly homogeneous, sharply differentiated species which form occasional hybrids when they come in contact with each other. This situation is not uncommon in the higher plants.

The next genus to be discussed, *Layia*, is one of the most thoroughly investigated cytologically and genetically, as well as systematically, of any of the plant kingdom, but only part of these investigations have been published (Clausen, Keck, and Hiesey 1941, 1945a, 1947). It contains 13 diploid and one tetraploid species. Of the diploids five — *L. platyglossa*, *L. chrysanthemoides*, *L. gaillardioides*, *L. pentachaeta*, and *L. glandulosa* — are polytypic, while the others are genetically rather homogeneous and of limited geographic distribution. Interspecific hybridization, which has been attempted in nearly every possible combination, has in many cases been completely unsuccessful and in others has yielded hybrids of various degrees of sterility and reduction in chromosome pairing. The results of these hybridizations are summarized in Fig. 7.

The pattern of species relationships in *Layia* is in some respects

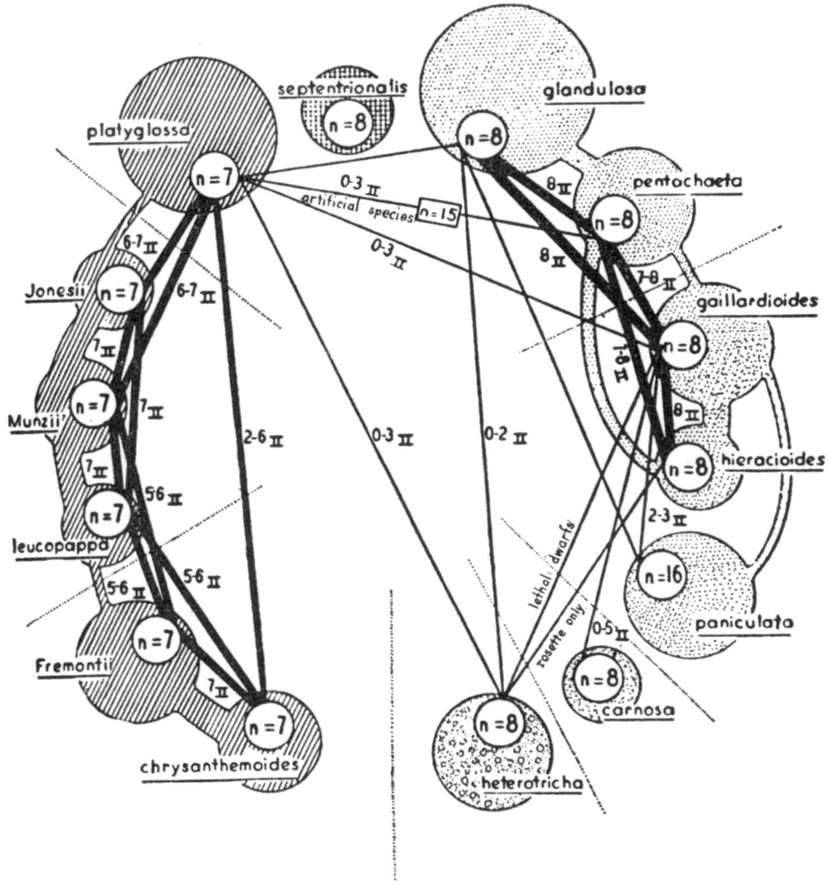

FIG. 7. Relationships in the genus *Layia*. The circles represent species, with chromosome numbers indicated; shaded connections show degrees of genetic affinity, and width of solid black lines represents degree of chromosome pairing in interspecific hybrids. The dotted lines indicate major morphological breaks in the genus. From Clausen, Keck, and Hiesey 1941.

different from that found by Clausen, Keck, and Hiesey (unpublished) in other genera of Madinae, but it is probably similar to that in other genera of annuals which make up a large proportion of the California flora. The presence of closely related species in adjacent areas and occupying habitats known to be geologically recent, along with genetic barriers in various stages of development, suggests that *Layia* is in an active state of relatively rapid evolution.

The genus *Crepis* is the most thoroughly studied of the larger genera of angiosperms. The monumental work of Babcock (1947) on this genus shows a remarkable insight into the genetic evolutionary processes that have been operating in it. With the exception of the polyploid, apomictic species of North America, the bulk of the species are diploid and form an aneuploid series with the numbers $x = 7, 6, 5, 4$, and 3. Of the 196 species recognized by Babcock, 97 are known to be diploid, and 78 are probably so. Of these 175 diploid species, only 18 are recognizably polytypic. The best known of these is *C. foetida* (Babcock and Cave 1938), which has three subspecies, within some of which local varieties and forms have been recognized.

The possibility is remote that any appreciable number of the species of *Crepis* are actually allopatric subspecies of more comprehensive species, as Mayr (1942, p. 122) has suggested may be true of plant genera. In 22 of the 27 sections into which Babcock has divided the genus, the majority of the species are either sympatric or with overlapping distributions, and natural hybrids between them are rare or unknown. Of the five sections with predominantly allopatric species, the one which has been investigated genetically possesses complete barriers of genetic isolation separating its species. The species of the four other sections are morphologically at least as sharply distinct from each other as are species of *Crepis* known to be separated by reproductive isolating mechanisms; hence, it is unlikely that they represent subspecies under any criterion.

Jenkins (1939) has described an example of four allopatric, insular endemic species of *Crepis* which are separated by imperfectly developed barriers of hybrid sterility and might therefore be considered to be species in the making. It is very likely that there are other similar examples in the genus, but both morphological and genetical evidence indicates that these are uncommon. In another example, *C. pulchra, C. palaestina,* and *C. pterothecoides,* all of which occur sympatrically in Palestine and Lebanon, were found to form fertile or only partly sterile hybrids (Babcock 1947, p. 150). These are apparently separated from each other by barriers of ecological or seasonal isolation (see Chapter V).

In *Crepis,* therefore, the great majority of the diploid species, though somewhat variable, are not polytypic in the sense that they consist of well-marked subspecies which replace each other geographically or ecologically. And when studied, they have been found to be separated from each other by more or less well-developed barriers of reproductive isolation, which differ in character according to the group investigated. The processes which have given rise to these barriers, and to the alteration in chromosome number and morphology which is one of the most distinctive features of the genus, will be discussed in later chapters.

THE FAMILY GRAMINEAE

The fifth family to be discussed here, the Gramineae, probably has the highest proportion of polyploid species of any family of angiosperms. The only genus containing more than a few species which is strictly diploid is *Melica.* This genus was studied cytologically and systematically by Boyle (1945), while Joranson (1944) has investigated representative hybrids between closely related species. Of the 18 North American species, all of which have the somatic chromosome number $2n = 18$, only two are polytypic, and these contain only two subspecies each. Three very closely related species, *M. imperfecta, M. torreyana,* and *M. californica,* have been artificially crossed. They form almost completely sterile hybrids, so that gene exchange between them is impossible or nearly so. The systematic treatments of the Eurasian species in such standard floras as those of Hegi, Ascherson and Graebner, and Komarov indicate that the variation pattern in them is similar to that in North America, except that *M. ciliata* and its relatives apparently form a polytypic species with more subspecies than are found in any North American species. The South American Melicas are less well known, but apparently similar. *M. papilionacea* is probably polytypic, but few, if any, of the others are. *Melica,* therefore, in contrast with most other genera of grasses, consists mostly of homogeneous or only slightly variable species which are sharply separated by barriers of genetic isolation.

In another genus, *Bromus,* the subgenus *Bromopsis* consists in western North America mostly of diploid species. Two of these, *B. ciliatus* and *B. orcuttianus,* have at least two subspecies, but the others — *B. kalmii, B. porteri, B. laevipes, B. grandis, B. vulgaris,* and *B. suksdorfii* — appear to be homogeneous. Morphologically these species are rather similar, and for the most part they replace each other geographically except for smaller or larger regions of overlap. They might be taken for well-marked subspecies were it not for the fact that hybrids have been made connecting the majority of them, and all these hybrids are completely or almost completely sterile (Stebbins, unpublished). The variation pattern in this section of *Bromus,* therefore, is similar to that in *Melica,* except that the species are less sharply defined in characteristics of external morphology.

GENERAL CONCLUSIONS

This survey of ten genera of seed plants belonging to five families scattered throughout the system of classification and phylogeny may be summarized as follows. Four of them — *Wyethia, Crepis, Melica,* and *Bromus* subg. *Bromopsis* — contain a majority of homogeneous or variable, but not polytypic, species, but in every case a larger or smaller minority of the species consists of two or more geographically separated subspecies. In *Crepis, Melica, Bromus,* and probably *Wyethia,* genetic diversity exists in the form of variants distinguishable when grown side by side under uniform conditions and occupying somewhat different habitats. These are ecotypes in the original sense of Turesson, but they cannot be recognized as subspecies because they are not distinguished by morphological differences recognizable over a geographic range of any extent. The recognized species, whenever investigated, have been found to be separated by well-marked barriers of reproductive isolation. These barriers are of a diverse nature, but the partial or complete sterility of pollen or seed in F_1 hybrids is the most common type.

Of the remaining six genera three, *Paeonia, Delphinium,* and

Layia, contain polytypic and homogeneous species in about equal proportions. In *Paeonia,* barriers of hybrid sterility between the species are very well developed, but in *Delphinium* they are rather weak. In only three of the ten genera, *Quercus, Aquilegia,* and *Potentilla* subg. *Drymocallis,* are the majority of the species polytypic. But it is noteworthy that the woody genus discussed is of this type. What is known of the woody genera of temperate regions, such as *Acer, Fraxinus, Populus, Vaccinium, Ceanothus,* and *Diplacus,* indicates that the pattern found in *Quercus* is typical of such genera (see Chapter VI). A notable feature of this pattern is not only the fact that the species are mostly wide ranging and polytypic; in addition the good species, even those having ranges that coincide almost completely, often form fertile F_1 hybrids. The isolating mechanisms best developed are of an ecological nature and operate through the failure of most hybrid derivatives to become established.

The generalization emerging from these examples, as well as from the additional ones given in Chapters V to XI, is that we cannot apply uncritically the criteria of species which have been developed in one group to the situations existing in another, particularly if the groups are distantly related to each other and have very different modes of life. This generalization applies with particular force to attempts of zoologists to reinterpret the species which have been recognized by botanists, or of the latter to use their yardsticks of species on animal material. General principles of systematics, as well as of evolution, must be based on as broad a knowledge as possible of different groups of plants and animals.

CHAPTER III

The Basis of Individual Variation

IN THE PREVIOUS chapters we have studied variation on two different levels; first, that between different individuals of a population and, second, the statistical variation between populations and population systems. Although the variation which is responsible for the origin and divergence of evolutionary lines is entirely on the population level, this variation is itself a product of differences between individuals. Heritable individual variations are the basic materials of evolution; the forces acting on populations are the mechanisms which fashion these materials into an orderly, integrated pattern of variation. Obviously, therefore, our understanding of evolution must come from an understanding of differences both between individuals and between populations.

ENVIRONMENTAL MODIFICATION AND ITS EFFECTS

The variation seen between the individuals of any population is based on three factors: environmental modification, genetic recombination, and mutation. Of these, environmental modification is the least important in evolution, although sometimes very conspicuous. As mentioned in Chapter I, the first step in any analysis of natural variation is the performing of transplant experiments by which the effects of environmental differences are largely neutralized. Such experiments have shown that each genotype has its own genetically determined degree of modifiability or plasticity (see Fig. 3). The adult individual we see, therefore, is a phenotype which is always the product of the effect of a given environment on an individual with a particular hereditary background, or genotype. That this axiom of genetics disposes at once of many futile arguments as to the relative importance of heredity and environment or "nature and nurture" has been pointed out by many geneticists, including Dobzhansky (1941, pp. 16–17).

Recent transplant experiments, as exemplified by the work of Clausen, Keck, and Hiesey (1940), have shown that the phenotype can be altered much more profoundly in some characteristics than in others. Most easily modified is the absolute size of the plant and of its separate vegetative parts, that is, roots, stems, or leaves. Hardly less plastic are the amount of elongation of the stem, the number of branches, and the number of leaves, inflorescences, or flowers. The quantity of hairiness or pubescence also is relatively easily modified. On the other hand, many particular characteristics of individual form and pattern can be modified only slightly or not at all by the environment; their appearance in the phenotype is almost entirely the expression of the genotype. Such constant characteristics are in *Potentilla* the pinnate character of the leaves; the type of serration of the leaf margins; the type of pubescence, whether glandular or nonglandular; the shape of the inflorescence; and most of the floral characteristics, such as the size and shape of the sepals, the petals, and the carpels. This relative plasticity of certain vegetative characteristics and constancy of reproductive ones has long been realized by plant systematists; upon it is based the greater emphasis in classification on reproductive characteristics as compared to vegetative characteristics.

The basis of this differential plasticity is probably to be found in the manner of growth of the plant shoot. The easily modified characteristics such as absolute size, elongation, branching, and number of parts are determined by the length of time during which the shoot meristems are actively growing and on the amount of cell elongation which takes place during the later expansion of the various organs of the plant. These processes are relatively easily affected by such external factors as nutrition, water supply, light, and temperature. On the other hand, the fundamental morphological pattern of individual organs, such as bud scales, leaves, and the parts of the flower, is already impressed on their primordia at a very early stage of development (Goebel 1928, 1933, Foster 1935, 1936) when the influence of the external environment is at a minimum. The relative stability of the reproductive structures is due largely to the fact that they are differentiated, not in a simple serial fashion as are the leaves and the branches, but more or less simultaneously or according to a rather complex development pattern (see Chapter XIII). Furthermore, the growth of the primordia after their differentiation is usually less in floral parts than it is in leaves and branches.

Many aquatic or semiaquatic plants are an exception to the usual rule that the shape pattern of organs is little modified by the environment. The difference in outline and internal structure between the submersed and the aerial leaves of such plants as *Myriophyllum verticillatum* has long been used as an example of the extreme modification of phenotypic expression by the environment. Other aquatic types, such as *Sium, Polygonum amphibium,* and various species of *Sagittaria* and *Potamogeton,* are equally capable of modification in their leaves (Fernald 1934). The difference between an aqueous medium and an aerial medium is one of the most effective external forces in modifying the organization and growth of the vegetative meristem and its appendages. In all these aquatic species, however, the reproductive parts are constant, no matter what the nature of the external medium.

In this discussion of the relative plasticity of vegetative as compared to reproductive characteristics, emphasis must be placed on the fact that we are dealing only with the modifiability or plasticity of the individual genotype and phenotype. The term "plasticity" is often used to denote variability between genotypes of a population or species. This, of course, is a totally different phenomenon and, as will be pointed out later in this chapter, is governed by entirely different causes. There is no a priori reason for assuming that genotypic variation between individuals should be greater in vegetative characteristics than in reproductive characteristics. In fact, there are many plant species, particularly among those of garden flowers, in which the variation between individuals and varieties is much greater in the reproductive parts of the plant than it is in the vegetative parts.

Environmental modification is a source of variability which must be kept in mind by the evolutionist because it affects every individual we see in nature. But as a direct factor in evolutionary divergence it is not significant. The Lamarckian concept of evolution through the direct effect of the environment on the individual, and the inheritance of acquired modifications, has in the past seemed even more attractive to explain evolution in the plant kingdom than in animals. As has just been mentioned, the genotypes of plants are relatively easily modified by environmental agencies, such as temperature, moisture, and nutrition, which are simple and ever present. Furthermore, the type of modification induced by the environment frequently simulates that which is produced by genetic factors. For instance, many species of plants possess genotypes which normally promote erect growth, but produce prostrate and spreading phenotypes when the plants are exposed to the extreme conditions of wind and salt spray prevailing at the seacoast. In certain of these species, such as *Succisa pratensis, Atriplex* spp. and *Matricaria inodora,* maritime ecotypes exist in which the prostrate spreading character is determined genetically and is maintained even in garden cultures (Turesson 1922b). Furthermore, the principle established by Weismann of the separation between somatoplasm and germ plasm does not hold for plants as it does for animals. In any perennial plant, the apical meristem produces alternately vegetative and reproductive organs, and any permanent modification of these meristematic cells at any time will affect the gametes they will eventually produce, and thereby the individual's progeny. Theoretically, therefore, the direct alteration of heredity in adaptive response to a changed environment would seem on the surface not impossible.

But the actual facts are that in plants, as in animals, there is no valid experimental evidence to indicate that acquired characters are inherited, and some experiments exist which show that they are not. The extensive transplant experiments of Clausen, Keck, and Hiesey (1940), already discussed many times, have failed to give any evidence of this type of inheritance. An earlier experiment which has the advantage of including a relatively large number of generations is that of Christie and Gran (1926) on cultivated oats.

This does not mean that the hereditary material cannot be altered by the environment. On the contrary, the effect on the germ plasm of such environmental agencies as heat, cold, radiations, and certain chemical substances is well known. But at least in the higher organisms the changes induced by extreme environmental conditions are either at random, like the apparently spontaneous mutations of the germ plasm, or of a specific nature, like the induction of polyploidy by means of heat shocks or the action of colchicine. In any case, they do not cause the progeny of the organism to become better adapted to the environmental agency used.

THE IMPORTANCE OF RECOMBINATION

The second factor in variation, recombination, is the main immediate source of variability for the action of selection and other external forces which direct the evolution of populations. By far the greater part of the genotypic variation in any crossbreeding population is due to the segregation and recombination of genic differences which have existed in it for many generations. This fact is brought into striking relief when one raises under uniform conditions the progeny of any single plant of a self-fertilized species. Such progenies are remarkably uniform, in striking contrast to the variable progeny obtained from a single plant of any cross-pollinated species. New mutations are undoubtedly occurring constantly in most natural populations, but they are so rare that their total effect on genotypic variability in the generation in which they occur is negligible. As Sumner (1942) has pointed out, mutations become material for alteration of the population by selection only after they have existed in it for several generations, and the primary adaptation to any new environment is accomplished by segregating variants of a heterozygous population.

The importance of sexual reproduction and cross-fertilization stems directly from this fact. For this reason, a particular study of the action of different types of sexual reproduction is essential to an understanding of the dynamics of evolution. Since such a study must be made on the basis of the importance for efficient natural selection of each type of reproduction, it will be deferred to Chapter V, following the discussion of selection.

TYPES OF MUTATION AND THEIR SIGNIFICANCE

Although the immediate basis of the variation which makes evolution possible is genetic segregation and recombination, its ultimate source is mutation. This fact is implicit in the definition of mutation, as accepted by most modern geneticists. Good discussions of the history and meaning of this term have been given by Dobzhansky (1941, pp. 19–22) and Mayr (1942, pp. 65–70); they need not be repeated here. A brief and satisfactory definition is given by Mayr (p. 66): "A mutation is a discontinuous chromosomal change with a genetic effect." One should understand clearly, though, that the adjective "chromosomal" refers to chemical changes in a small part of the chromosome, as well as to alterations of its physical structure. A particularly significant fact about this definition is that the chromosomal change rather than the genetic effect is characterized as discontinuous. This emphasizes the presence of mutations having less visible effect on the phenotype than the changes produced by environmental modification or by the genetic effect of segregation and recombination in a typical heterozygous natural population. Such mutations with slight effect, since they do not produce a visible discontinuity in any character, blend into the continuous or fluctuating individual variation normally present in panmictic populations. They contrast sharply with the type of mutations originally described by De Vries, which produced striking and discontinuous variations in the phenotype.

Mutations may be classified either on the basis of the changes taking place in the chromosomes or according to their effect on the phenotype. The former classification is the one most widely used. In it we may recognize four main groups: first, multiplication of the entire chromosome set, or polyploidy; second, the addition or subtraction of one or a few chromosomes of a set; third, gross structural changes of the chromosomes; and fourth, submicroscopic changes, probably including chemical alterations of the chromosomal material. The first of these, polyploidy, is of such special significance that it will be treated in detail in Chapters VIII and IX. The second includes chiefly polysomic or monosomic types which possess an extra whole chromosome or are deficient for one. Such types are partly sterile and genetically unstable, and therefore of relatively little significance in evolution. They are very rarely found in nature. Other types of aneuploid changes in chromosome number occur, but they are usually associated with rearrangements of chromosomal segments. They are discussed in Chapter XII.

The third type of mutation, gross structural alteration of the

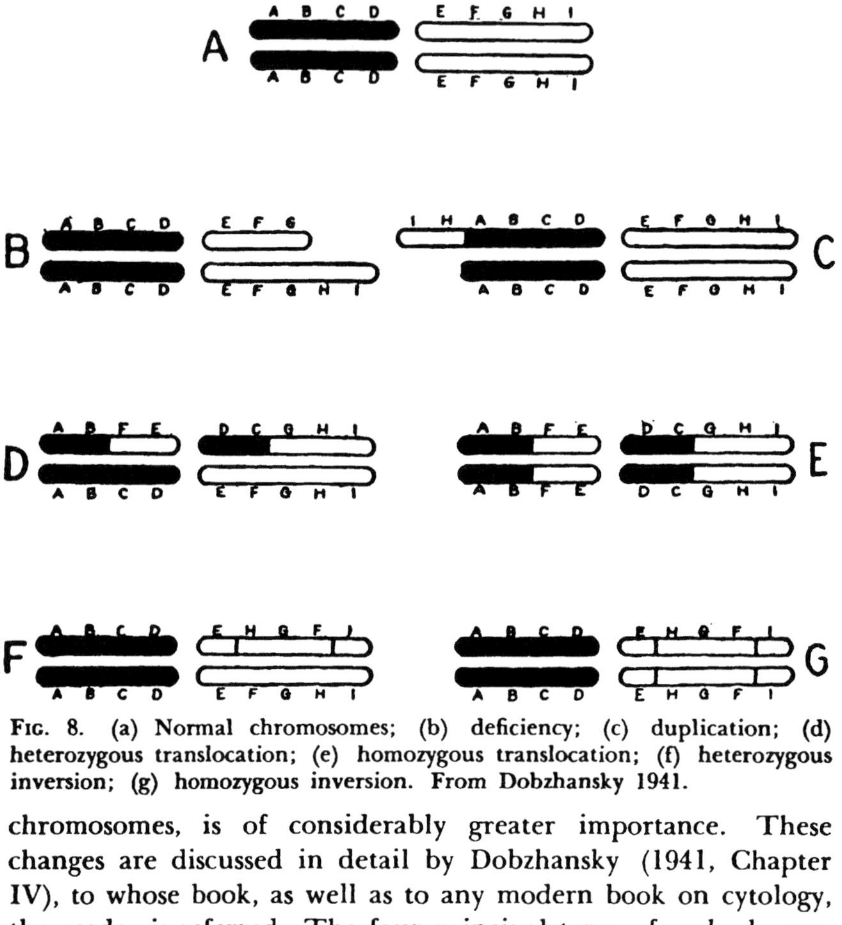

FIG. 8. (a) Normal chromosomes; (b) deficiency; (c) duplication; (d) heterozygous translocation; (e) homozygous translocation; (f) heterozygous inversion; (g) homozygous inversion. From Dobzhansky 1941.

chromosomes, is of considerably greater importance. These changes are discussed in detail by Dobzhansky (1941, Chapter IV), to whose book, as well as to any modern book on cytology, the reader is referred. The four principal types of such changes are deficiencies, duplications, translocations, and inversions. These are diagrammatically shown in Figs. 8–10, taken from Dobzhansky (1941).

Deficiencies are usually lethal or semilethal when homozygous, so that the majority of them are of relatively little importance in evolution. It is possible, however, that a considerable proportion of the genetic changes which have been regarded as point mutations are actually minute deficiencies (McClintock 1938, 1941, 1944). Within the limits of four chromomeres in the short arm of chromosome 5 of maize, McClintock (1941) found six distinct nonallelic mutations. All of these appear to be homozygous minute deficiencies. A similar series was later (1944) produced in the

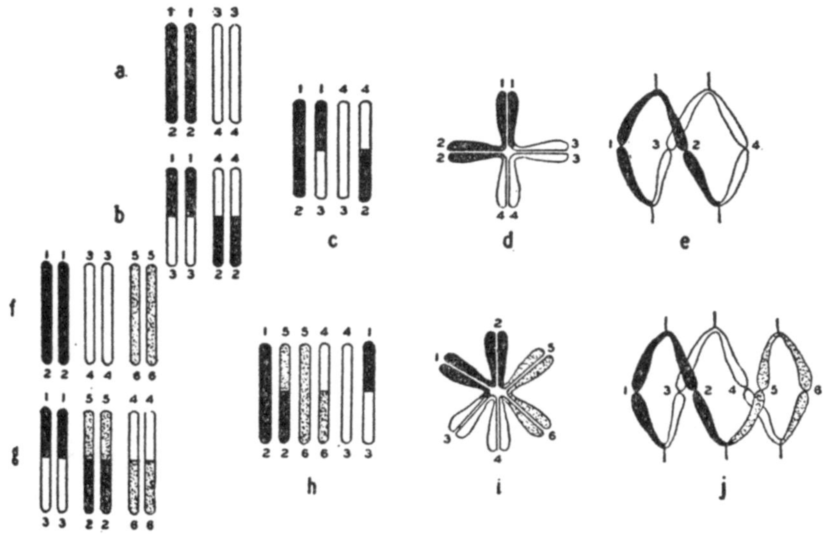

FIG. 9. Translocation between two (b–e) and between three (g–j) chromosomes. Normal chromosomes (a) and (f); (b) and (g) translocation homozygotes; (c) and (h) translocation heterozygotes; (d) and (i) chromosome arrangement at pairing stages; (e) and (j) arrangement of chromosomes at the metaphase of the meiotic division. From Dobzhansky 1941.

short arm of chromosome 9. There is no reason to believe that these regions are more mutable than other parts of the maize complement or that the maize plant is unusual in the type of mutations it can produce. Similar minute deficiencies simulating mutations have been produced by several workers in *Drosophila.* In some groups of organisms, therefore, minute deficiencies may play a role in evolution similar to that of point mutations.

Duplications may be of even greater significance than deficiencies, although as yet not much is known about them. They are most easily detected in the giant chromosomes of the salivary glands of *Drosophila, Sciara,* and other flies, where they appear as "repeats" of certain patterns of arrangement of the dark staining bands. In plants, their spontaneous occurrence can be inferred chiefly from the behavior of chromosomes at meiosis in haploids or in species hybrids of certain types. Thus, in haploid *Oenothera* (Emerson 1929, Catcheside 1932), barley (*Hordeum distichum,* Tometorp 1939), rye (*Secale cereale,* Levan 1942a), *Triticum monococcum* (L. Smith 1946), and *Godetia whitneyi*

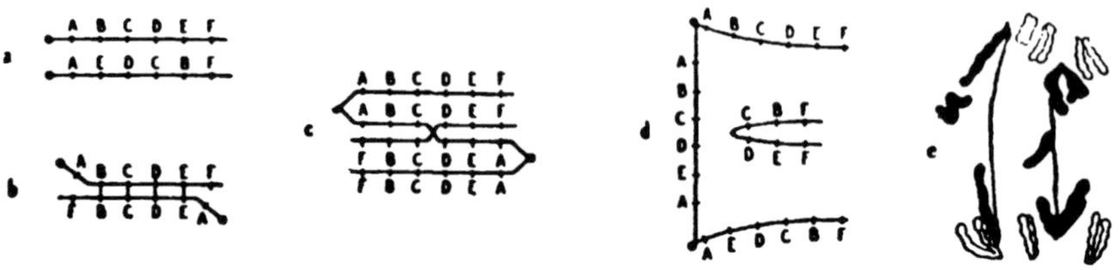

FIG. 10. Crossing over in inversion heterozygotes, leading to the formation of chromosomes with two and with no spindle attachments (d). (e) Chromatin bridges in the hybrid *Lilium martagon* × *L. hansonii*. (After Richardson.) From Dobzhansky 1941.

(Håkansson 1940), there is occasional pairing of chromosomal segments. Although some of this pairing may be the association of nonhomologous, genetically different segments, as suggested by Levan (1942a), nevertheless the work of McClintock (1933) on nonhomologous association in maize suggests that this phenomenon does not occur extensively unless the pairing chromosomes have some homologous segments in common. If these chromosomes are different members of a single haploid set, then this set must contain duplications. In the hybrid between *Bromus catharticus* ($n = 21$) and *B. carinatus* ($n = 28$) the normal pairing at meiosis is 21 pairs and 7 single chromosomes (Stebbins and Tobgy 1944). But in a very few cells two of these univalents will pair with each other in a single small segment, indicating the presence of a duplication.

Dobzhansky (1941, p. 129) and, particularly, Metz (1947) have emphasized the importance of repeats or duplications as the only method, aside from polyploidy, by which the number of genes in the germ plasm of an organism may be increased. Just how essential to evolutionary progress is an increase in the number of genes cannot be decided until more is known about the physicochemical structure of chromosomes and genes and about the nature of mutations. It must be mentioned here, however, that aside from polyploidy there is no evidence of a regular increase in the amount of chromosomal material during the evolution of the land plants. The most archaic types known, the Psilotales and the Ophioglossales, have larger chromosomes than the majority of flowering plants. In the larger subdivision of the angiosperms, the dicotyledons, most of the species with the largest chromosomes are in the most primitive order, the Ranales (*Paeonia*, *Helle-*

borus, Delphinium, Ranunculus, Podophyllum), while the most highly evolved types (Labiatae, Scrophulariaceae, Compositae, and so on) usually have small or medium-sized chromosomes. Since the genes themselves probably constitute only a small fraction of the substance which makes up the chromosomes, we cannot say whether or not differences in the size of the chromosomes are correlated with differences in the number or size of the genes which they contain. Nevertheless, plant cytology provides no evidence in favor of the hypothesis that progressive evolution has been accompanied by a regular increase in the number of gene loci contained on each chromosome. A more likely hypothesis would be that progressive physicochemical diversification of a more or less constant amount of chromosomal substance is responsible for increasing evolutionary specialization. The role of duplications or repeats might be to act as "buffers," offsetting certain deleterious effects of otherwise beneficial mutations.

The two remaining types of gross structural changes, translocations and inversions, are much better known. Translocations are common in a number of plant species, where they are recognized by the presence of rings or chains of chromosomes at meiosis in individuals heterozygous for them (Fig. 9). Typical examples are *Datura stramonium* (Blakeslee, Bergner, and Avery 1937), *Campanula persicifolia* (Gairdner and Darlington 1931, Darlington and Gairdner 1937), *Polemonium reptans* (Clausen 1931a), species of *Tradescantia* (Darlington 1929a, Sax and Anderson 1933, Anderson and Sax 1936), and *Triticum monococcum* (Smith 1936). Further examples are cited by Darlington (1937) and Dobzhansky (1941, pp. 114–115). The accumulation of translocations to give structural heterozygotes with most or all of their chromosomes associated in a system of rings or chains, as in *Oenothera, Rhoeo,* and the North American species of *Paeonia,* is a special phenomenon which will be discussed in Chapter XI.

Inversions of chromosome segments are likewise well known and probably occur in an even larger number of plant species than do translocations. They can readily be detected in the heterozygous condition in species like maize, in which chromosome pairing at mid-prophase (pachytene) of meiosis is easily observed by the presence of a loop, formed by the pairing of the normal with the inverted chromosome segment (McClintock

1933; see diagram, Fig. 7, of Dobzhansky 1941). In most species of plants, however, this stage cannot be easily observed, and inversions are recognized only by the results of chiasma formation and crossing over in the inverted segment. This leads to the formation at meiotic metaphase and anaphase of one chromatid with two spindle fiber attachments and one free, acentric fragment (Fig. 10; see also Darlington 1937, pp. 265–271). Since the fragment always includes the entire inverted segment plus those parts of the exchange chromatids distal to it, the size of the fragment indicates the position (not the size) of the inversion on the chromosome arm. The frequency of bridge-fragment configurations in any species or hybrid obviously depends on three factors: chiasma frequency, number of inversions, and size of inversions.

A considerable list of plant species containing individuals heterozygous for inversions is given by Darlington (1937, pp. 273–274), as well as by Dobzhansky (1941, p. 126). These lists could be greatly extended on the basis of more recent observations, and interspecific hybrids are even more likely to be heterozygous for inversions. The two genera which up to date seem to exceed all others in the occurrence of detectable inversions are *Paris* and *Paeonia.* Geitler (1937, 1938) found that in Tirolean populations of *Paris quadrifolia* every individual is heterozygous for one or more inversions, which are distributed over every chromosome arm in the entire complement. Stebbins (1938a) found a greater or lesser percentage of cells with bridge-fragment configurations in every one of the Old World species of *Paeonia* investigated by him, and this is likewise true of every individual of the two New World species, *P. brownii* and *P. californica* (Stebbins and Ellerton 1939, Walters 1942). In one species, *P. triternata* var. *mlokosewitschii,* 20 percent of the cells contained these configurations. Although the different chromosome arms could not all be distinguished from each other, in all probability the inversions in this species, as in *Paris quadrifolia,* are scattered through the entire chromosomal complement. Analysis of the different bridge-fragment configurations found in a single chromosome arm of an interspecific hybrid, *Paeonia delavayi* var. *lutea* × *P. suffruticosa,* showed that this arm was heterozygous for several inversions, probably of small size. In all likelihood, therefore, the high frequency of bridge-fragment configurations in species of *Paeonia,* and perhaps also in *Paris*, is due to the presence of many small inversions rather than a few large ones. If this is true, then the actual frequency of inversions in many other cross-fertilized plant species may approach or equal that in *Paris* and *Paeonia*. These two genera have among the largest chromosomes in the plant kingdom, so that inverted segments which are relatively short in relation to the length of the chromosome would nevertheless be long when compared to the total length of the chromosomes in other species. It is possible, therefore, that in species with smaller chromosomes inversions as small in proportion to the total length of a chromosome arm as are many of those in *Paris* and *Paeonia* would never form chiasmata when paired with a normal segment. The fact must be emphasized that the evidence from bridge-fragment configurations indicates only the minimum number of inversions for which an organism is heterozygous, and may come far from revealing all of them.

The evidence presented above suggests that translocations and inversions occur frequently in most species of plants. They do not, however, produce any recognizable effect on either the external morphology or the physiological adaptive characteristics of the species. Examination of numerous translocation types produced artificially in maize (Anderson 1935), *Crepis* (Levitzky 1940, Gerassimova 1939), and other plants has failed to show any conspicuous morphological effect of these chromosomal alterations. The same is true of gross structural differences in the chromosomes of wild species. Blakeslee (1929) and Bergner, Satina, and Blakeslee (1933) found that the "prime types" of *Datura*, which differ from each other by single translocations, are morphologically indistinguishable, while differences between individuals are no greater in those species of *Paeonia* which are heterozygous for a large number of inversions than they are in those with relatively few inversions (Stebbins 1938). In *Paeonia californica*, furthermore, some populations are structurally homozygous or nearly so and others contain a great array of different interchange heterozygotes (Walters 1942 and unpublished). These two types of populations are indistinguishable in external morphology in respect to both individual variation and the differences between populations. In plants, the phenomenon of position effect, which has considerable importance in *Drosophila*, has been clearly demonstrated only in *Oenothera* (Catcheside 1939, 1947a). It appears to be so uncommon that it has relatively little importance in evolution. Gross structural alterations of the chromosomes are not the materials that selection uses to fashion the diverse kinds of organisms which are the products of evolution.

The importance of these chromosomal changes, however, lies in an entirely different direction. In most structural heterozygotes there is a certain amount of chromosomal abnormality at meiosis, leading to the production of a certain percentage of inviable gametes. This is relatively small when single translocations or inversions are involved, so that individual structural differences scattered through the chromosome complement cause a relatively slight reduction of fertility. But when individuals differ by groups or complexes of inversions and translocations, particularly if these are closely associated in the same chromosome arm, pairing at meiosis may be considerably disturbed, and even with relatively normal pairing the effects of crossing over and chromosomal segregation will be the production of gametes containing duplications and deficiencies, and therefore inviable. Gross structural changes of the chromosomes, therefore, are the units from which are built up many of the isolating mechanisms separating plant species. These mechanisms will be discussed in greater detail in Chapter VI, in connection with species formation.

From the summary presented above we may conclude that the majority of the morphological and physiological differences important in evolution come about, not through alterations of the number and gross structure of the chromosomes, but through changes on a submicroscopic level; the "point mutations" of classical genetics. This conclusion agrees in general with that of most geneticists, such as Sturtevant and Beadle (1939), and was also reached by Babcock (1942, 1947) as a result of his prolonged and intensive studies of cytogenetics and evolution in the genus *Crepis*. The physicochemical nature of these mutations is not known. That they represent a class of phenomena different from larger structural changes in the chromosomes is suggested by the results of Stadler and Sprague (1936) on the effects of ultraviolet radiation as compared with X rays. The latter produce a relatively high proportion of chromosomal alterations, while ultraviolet radiations produce mostly point mutations. Since point mutations, as Stadler (1932) has pointed out, are merely those alterations in the germ plasm for which a mechanical chromosomal basis cannot be detected, they may actually consist of a whole assemblage of different physical and chemical changes, including rearrangements of the micellae of a colloidal system, of individual molecules, and of various rearrangements of rings, side chains, or even of individual atoms within the molecule. Since among the visible structural alterations of the chromosomes there is no close correlation between the size of the change and the magnitude of its genetic effect, there is no reason for believing that such a correlation exists on a submicroscopic level.

The obvious implication of the above remarks is that the nature of mutation is no longer a problem primarily of evolutionary biology or even of genetics per se, but of nuclear biochemistry. From now on, our new knowledge of mutation will come almost entirely from a better understanding of the physicochemical structure of chromosomes and of the biochemistry of gene action. The rapid advances being made in this field suggest that the evolutionist may not have very long to wait for the vital information he needs in order to understand the ultimate source of evolutionary change.

GENETIC EFFECTS OF MUTATIONS

Nevertheless, our understanding of the role of mutation in evolution will be greatly helped by a summary of the salient facts about the nature of the genetic effects of point mutations and about the rates at which they occur. And in discussing these genetic effects, we must always keep in mind this cardinal fact: the most direct, immediate action of genes is on the processes of development, and genes produce effects on the visible, morphological characters of the mature organism only through their influence on development. Similarly, mutations directly alter processes, and only indirectly, characters. The recent developments in biochemical genetics (Beadle 1945, 1946) have served particularly to emphasize this point.

With these facts in mind, we can classify mutations according to the number of different parts of the organism they affect and also according to the extent and nature of their effect on individual organs. We can find in the different mutations now known all possible combinations of extent and intensity of effect, from those which profoundly affect all parts of the organism to those which have a hardly detectable effect on a single character.

The obvious limit in extent and magnitude of genetic effects is set by the lethal mutations, which alter one or more of the metabolic processes of the cell so radically that death results. Hardly less drastic are those mutations which, in addition to altering profoundly the vegetative parts, change the reproductive structures of the plant to such an extent that they are sterile in the homozygous condition. These produce plants which in the older morphological terminology were called teratological monstrosities. Typical examples are those given by Lamprecht (1945a,b) for *Pisum* and *Phaseolus*. For instance, the *unifoliata* mutant of *Pisum* produces entire or three-parted leaves instead of the pinnately divided type usually found in peas; the sepals and the petals are transformed into leaflike structures bearing rudimentary ovules; and the carpel, although recognizable as such, is by no means normal (Fig. 11, left). In another mutant, *laciniata*, the margins of the leaflets are strongly toothed, and most of them

FIG. 11. The *unifoliata* (left) and the *laciniata* (right) mutants of the garden pea (*Pisum*). In the former diagram a pistilloid sepal (b), the pistil (c), and a pistilloid stamen (d) are shown in addition to a part of the plant; in the latter are shown a part of the plant plus the gynoecium. From Lamprecht 1945.

are replaced by branching tendrils, while the pod is correspondingly modified (Fig. 11, right). Both of these types are completely sterile.

Of considerably greater interest are those mutations which produce a marked effect on many or all of the parts of the plant, but which do not result in weakness or sterility. Some of these are known as single gene differences in natural populations of wild or cultivated plants, where they presumably arose as mutations in the past and have since been designated as distinct species. Two good examples are the *oxyloba* gene of *Malva parviflora* (Kristofferson 1926) and the *sphaerococcum* gene of wheat, *Triticum aestivum* L. (Ellerton 1939, Sears 1947). The former produces a striking change in leaf shape, causing the lobes to be sharply pointed, and effects a similar change in the sepals. The *sphaerococcum* mutation produces a shorter, stiffer stem, a profuse tillering, small spikes or ears, practically hemispherical, inflated glumes, and small, round grains. Ellerton (1939) believed it to consist of a deficiency of a chromosomal segment, but Sears (1947) has demonstrated that this is not true, and has shown furthermore that it is a true recessive, since two *sphaerococcum* genes must be present before the character is expressed.

These two factors are typical of genes whose effect has been termed *pleiotropic,* that is, altering simultaneously several characteristics of the adult phenotype. There is good reason to believe, however, that this pleiotropy is not a peculiar property of certain genes, but results from the fact that certain developmental processes are important in the same way in several different organs, so that genes affecting these processes indirectly affect many characters. A mutant which illustrates this point is the *compacta* gene of *Aquilegia vulgaris* (Anderson and Abbe 1933). This gene affects all parts of the plant. It produces shorter, thicker stems, with the internodes much reduced in length; shorter petioles and rachises of the leaves; and a shorter inflorescence with the flowers on short, stout, erect peduncles. All these effects have been shown by Anderson and Abbe to be the result of precocious secondary thickening of the cell walls throughout the plant.

A series of genes which affect the leaves and also the parts of the flower, as well as those affecting only the latter, is described by

Anderson and DeWinton (1935) in *Primula sinensis* (see Fig. 12). They noted that those genes which affect leaf shape usually have a supplementary effect on the flower. Following the lead of Schultz, in his work on *Drosophila,* they suggest that these multi-

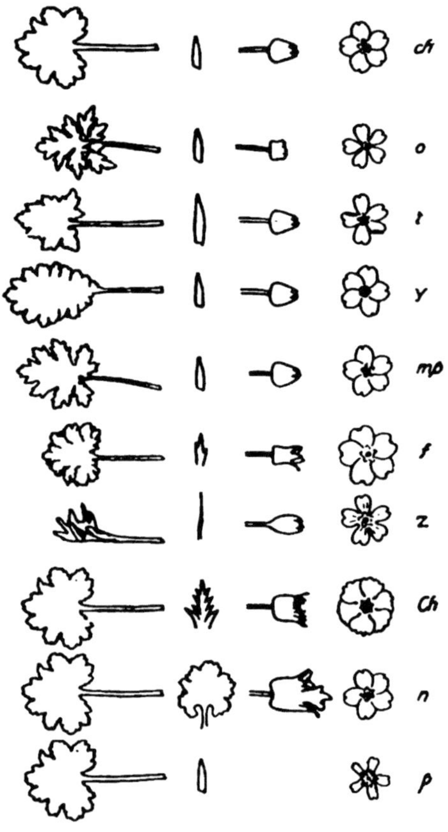

FIG. 12. Leaf, bract, calyx, and corolla of the "wild" type of *Primula sinensis* and of nine mutants affecting leaf and inflorescence shapes. All are shown on a wild type background except the calyx of "nn," which is shown on a ChCh background. From Anderson and DeWinton 1935.

ple effects can be explained on the assumption that "each gene has its own job to do" in terms of affecting certain processes of growth. Crimp (ff), for instance, extends marginal growth on leaves, bracts, sepals, and corolla. Oak (oo) increases the sinuses on both corolla and leaf. In tongue, lateral growth is reduced in leaves, bracts, and corolla lobes. A gene with a similar type of

TABLE 1

SUMMARY OF THE GENETIC COEFFICIENTS DIFFERENTIATING *Nicotiana alata* FROM *N. langsdorffii* [a]

Genetic coefficients	Vegetative phase		Reproductive phase									Other coefficients of which this may be a further expression
							COROLLA					
	Axis	Leaf	Axis	Bract	Pedicel	Calyx	Tube	Throat	Limb	Stamen	Pistil	
(1) Cell size	x	x	x	x	x	x	x	*	x	x	x	(2)
(2) Cell elongation	x	x	x	x		x	*		x		x	
(3) Geotropic response	x	x	x	x	*			?	?			(2)
(4) Leaf-vein angles		x							?			
(5) Plastid color		x			x		x	x	*			
(6) Foliar periphery		x		x		x			*		x	
(7) Foliar base		x		x		x			*			(2)
(8) Pollen color										*		
(9) Time of blooming							x	x	*			(5)
(10) Scent							*					
(11) Inflorescence			*									(2)

x = Organs in which the action of the genetic coefficient is evident.
* = The organ in which it can probably be measured most efficiently.
[a] From Anderson and Ownbey 1939.

effect is the "petioled" gene of *Nicotiana tabacum* (designated as S in Setchell, Goodspeed, and Clausen 1922). This causes the leaves to have true petioles, without wing margins and with narrower, more acute or acuminate blades. In addition, the calyx lobes are narrower and more acute; the corolla lobes are more acuminate; the anthers are longer; and the capsule is narrower. Examination by the writer of developing leaves and flowers has shown that in "petioled" (S) plants the primordia of the leaves, when less than 1 mm long, are already narrower than are those of normal (s) ones. The "petioled" gene, therefore, apparently has the one effect of changing the shape of the primordia, but this affects each part of the plant in a somewhat different way, according to the developmental processes normally taking place in a particular primordium. Genes with effects of this nature are probably responsible for the genetic coefficients described by Anderson and Ownbey (1939) in hybrids between *Nicotiana alata* and *N. langsdorffii*. Eleven such coefficients are listed, each of them affecting two or more different characters of the plant (Table 1). One of the most interesting of these is that affecting cell elongation. This causes *N. alata* to have narrower leaves, longer calyx lobes, a much longer style, a more pointed ovary, and particularly a longer corolla tube than *N. langsdorffii*. Nagel (1939) found that all these differences are due to the fact that *N. alata* inactivates auxin less readily than *N. langsdorffii*.

Genes with relatively restricted effects are also very common in plants. In *Nicotiana tabacum* the gene "broad" (A in Setchell, Goodspeed, and Clausen 1922), like "petioled," affects most strongly the basal part of the leaves, causing the wing margins of the leaves to become broader. But it does not alter appreciably the shape of the blade, and it has no visible effect on any of the floral parts. In contrast to those of "petioled," leaves of "broad" (A) plants are indistinguishable from those of normal or constricted (a) ones until they are well differentiated and about 2 cm long. This gene, therefore, acts on growth processes which take place relatively late in leaf development and have no counterpart in the development of the floral parts.

That there is no fundamental difference between these more "typical" genes and the "pleiotropic" ones described in the preceding paragraph is evident from the extensive studies of Stadler (unpublished) on alleles at the R locus in maize. At this single locus there is a long series of multiple alleles for anthocyanin color. Some of these affect the entire plant, others only the endosperm, and still others various combinations of plant parts, such as the coleoptile, the apex of the leaf sheaths, and the anthers. One could imagine a similar series affecting, not the formation of pigment, but of some growth-promoting substance. In such a series, the different alleles could affect the shape of one or several organs, depending on the developmental conditions under which they produced this growth substance. Thus, the apparent diversity of gene action may be largely a result of the complexity of the developmental processes taking place in any higher plant or animal. The same initial gene-controlled enzymatic process may give very different end results, depending on other processes going on at any particular stage of development.

The final set of genic effects to be considered are those which alter the phenotype very slightly, the so-called "small mutations." These were first described by Baur (1924) in *Antirrhinum majus* and its relatives, where they were estimated to occur at the extraordinarily high rate of one in ten gametes. East (1935b), by doubling the somatic chromosome number of a haploid plant of *Nicotiana rustica,* obtained a completely homozygous diploid. The first selfed generation from this plant was strikingly uniform, but in later generations considerable variability was observed. This was not measured quantitatively, but could not be explained except as a result of the occurrence of several mutations with slight phenotypic effect during the four generations of selfing. Lindstrom (1941) obtained quite a different result from selfing a completely homozygous strain of tomato obtained from a haploid. This line remained completely uniform for nine generations. The frequent occurrence of small mutations in *Drosophila* has been shown by a number of experiments of different types (Dobzhansky 1941, pp. 58–60). Their existence in natural populations of both plants and animals seems likely from the fact that most character differences between races and species show complex segregations in hybrids so that they appear to be governed by multiple factors (Tedin 1925, Müntzing 1930a, East 1935a, Winge 1938, etc.). The building up of a character difference based upon multiple factors, as well as the development of a system of modifiers which alter the expression of genes with large effects, must be due to the occurrence and establishment in the population of many small mutations. The statement of Harland (1934) that "the modifiers really constitute the species" can be interpreted in no other way than that small mutations are the most important ones in building up the differences between species, as Harland showed clearly in a later, more detailed review (1936). The large body of evidence which has now accumulated compels us to agree with East that (1935a, p. 450) "the deviations forming the fundamental material of evolution are the small variations of Darwin."

The examples given above and many others like them answer effectively the question which has often been raised as to whether mutations always produce abnormal or weakening effects or whether some of them may be considered the basic units of progressive evolutionary change. Mutant types like those cited in *Malva, Triticum,* and *Nicotiana* are in every way comparable to the individual character differences which separate natural subspecies and species. To be sure, the origin of these mutants was not actually observed, but the indirect evidence for their manner of origin is very strong. It is, however, true that most of the mutations appearing in artificial cultures of plants or animals, both spontaneously and under the influence of such agencies as X rays, produce phenotypes less viable than the original "wild" type. But as Dobzhansky (1941, p. 26) and many others have pointed out, this result is expected. The gene complement of any species is a complex system composed of many different units interacting in a balanced, harmonious fashion. It can be thought of as a complex machine which runs smoothly only when all its parts bear the correct relationship to each other, both in structure and in function. In such a machine, the number of changes that will improve the working of the whole is small, and the number that will injure or destroy its activity is very large. Furthermore, most modern species have been on the earth for thousands or even millions of generations, and during that time have been exposed to similar types of environment. During this time, therefore, one would expect most of the mutations which might improve the adjustment of the species to these environments to have already occurred several times and to have had plenty of opportunity to become incorporated into the genic complement of the species by natural selection, so that even occasional valuable mutations could be expected only in species exposed to a new type of environment. Under any hypothesis about the role of mutations in evolution we should expect most of these changes to be neutral or harmful to the species and only occasional ones to have even potential importance in evolution.

These hypotheses have recently been confirmed experimentally in a striking fashion. Gustafsson (1941a,b, 1947a) has by X radiation induced a large number of mutations in barley, of which the great majority reduce the viability or fertility of the plant. Among those without clearly deleterious effects, nearly all reduce the total yield of grain, and most of them produce other agriculturally undesirable characteristics. But out of the hundreds of mutations obtained and tested, about ten were found which either increase the yield or produce other valuable characteristics, such as stiffness of straw and earliness. These characteristics might be incorporated by hybridization into established high-yielding varieties. Gustafsson (1947a) mentions in addition the production by Granhall and Levan of a typical "defect" or "loss" mutation, namely, a yellowish-green chlorophyll mutation in flax, which nevertheless gave a higher yield of straw and a better quality of fiber than the variety in which it arose.

Another type of evidence of the occurrence of beneficial mutations was obtained by Dobzhansky and Spassky (1947) in *Drosophila pseudoobscura.* By means of a special technique, they produced strains of flies completely homozygous for certain particular chromosome pairs which were known to carry genes causing a reduction in vigor when in the homozygous condition. Seven strains of this type were isolated and bred for fifty generations. At the end of this period, five of the seven strains showed improvement in viability, and two of them were as viable or more so than flies carrying the corresponding chromosomes in the heterozygous condition. That these striking improvements were caused by new mutations was demonstrated by tests of the individual chromosomes concerned on a neutral genetic background. On the other hand, control stocks which were kept in a balanced heterozygous condition by means of a homologous chromosome containing an inversion and certain marker genes did not improve their viability during the same fifty generations of breeding, and in fact tended to deteriorate somewhat. This shows that if the genotype is artificially made subnormal and then bred in a normal environment, mutations which were being eliminated from the genetically normal population now have a high selective value and can therefore become established. Mutation, therefore, although it is the ultimate source of genetic variability in populations, acts only occasionally and under certain special conditions as a direct active agent in evolutionary change.

The foregoing discussion has been based on the assumption that mutation is at random, at least in respect to adaptation and to the origin of races, species, or phylogenetic trends in evolution.

The contrary assumption, that species, genera, or directed, "orthogenetic" evolutionary lines are produced by means of special types of mutations, has been maintained by a number of workers, in recent times notably Goldschmidt (1940) and Willis (1940) in respect to special types of mutations; and Schaffner (1929, 1930, 1932) and Small (1946) in respect to orthogenetic trends. Criticisms of their opinions have been voiced by Wright (1941a,b), Dobzhansky (1941), Mayr (1942), Simpson (1944, pp. 150–157), and many others. These criticisms are in general valid. The arguments in favor of the importance of "systemic mutations, macromutations" and internally directed orthogenetic series are largely negative. Their proponents are unable to find conditions intermediate between intergrading, interfertile subspecies and sharply distinct species; or between the presence and the absence of some character of phylogenetic importance; or they cannot see the selective basis of certain evolutionary changes; and therefore they assume that these conditions do not exist. But such negative evidence is convincing only if it can also be shown that we know and understand fully all the facts about species formation or about the selective value of certain gene combinations. The cases presented by the authors mentioned above indicate no such omniscience.

There is no positive evidence whatever in favor of the occurrence of internally directed mutational changes which force the evolutionary trend of a line in any particular direction, as is postulated by adherents of the strictly orthogenetic concepts of evolution, such as Schaffner. Repeated observations and experiments on both spontaneous and artificially induced mutations in a great variety of organisms have shown that they are at random, at least in respect to the species differences and evolutionary trends which have occurred in the groups studied. It is highly unlikely that future experiments, at least in the higher animals and plants, will ever produce such positive evidence.

Positive evidence in favor of the importance in species formation of "large" mutations consists chiefly of the demonstration that mutations occur which produce certain character differences important for distinguishing species, genera, or families. Thus Goldschmidt (1940) stresses the importance in *Drosophila* of mutant flies with no wings or with four wings instead of the two ordinarily present. A comparable mutant in plants is the *radialis* mutant of *Antirrhinum majus* (Baur 1924, Stubbe and Wettstein 1941), which transforms the two-lipped, zygomorphic corolla ordinarily found in snapdragons into a structure with radial symmetry and therefore characteristic of more primitive families. Numerous other examples are known, notably those described by Lamprecht (1939, 1945) in *Pisum, Phaseolus,* and other genera, and by Anderson and DeWinton (1935) in *Primula sinensis.*

The problem in connection with all these radical changes in a single character is whether they ever are or can be incorporated into the normal gene complement of a species. Darwin, in the *Origin of Species,* first expressed the opinion that they alter so drastically the relation between the individual and its environment that they must nearly always be a great disadvantage in competition with the normal condition, and most modern evolutionists agree with this opinion. This is undoubtedly true of the *radialis* mutation of *Antirrhinum*: the flowers of this mutation are so constructed that they cannot be visited by insects, and the anthers and stigma are so situated that pollen from the former cannot normally reach the latter, so that the mutation is sterile unless artificially pollinated. On the other hand, Stubbe and Wettstein (1941) have described other mutations in *A. majus* which produce equally drastic changes in the flower, but which are nevertheless fully fertile and viable. The most interesting of these is mut. *transcendens,* which reduces the number of stamens. This effect partly simulates the trends toward reduction in stamen number which have taken place in many genera of the family Scrophulariaceae, to which *Antirrhinum* belongs. Stubbe and Wettstein have suggested that in the evolution of the genera concerned, changes in the organization of the flower may have been caused originally by the establishment of mutations with large effects, and that the existence of multiple-factor inheritance in respect to these characteristics could have been acquired later through the establishment of modifier complexes, buffering or reducing the effect of the original mutations. Experiments to test this hypothesis are urgently needed, but although it may be true in many instances, there are nevertheless many phylogenetic trends in plants which are represented by so many transitional stages among existing species that their progress through the accumulation of mutations with small effects seems most likely. Both "switch genes," with large phenotypic effects, and multiple factors or "polygenes," with small ones, have probably been important in the origin of new evolutionary lines in different plants, and their relative importance has probably varied in different groups. Furthermore, as Stubbe and Wettstein have pointed out, the two types of mutations are not sharply defined categories, but are connected by a continuous series of mutations with intermediate effects. Finally, as will be brought out in Chapter VI, even the most drastic of single mutations cannot possibly be considered as able to produce at a single step a new species or evolutionary line. This requires the establishment not only of complexes of genes affecting external morphology but also of genetically controlled isolating mechanisms.

RATES OF MUTATION

Although new mutations probably serve only rarely as the direct agents of evolutionary change, nevertheless the continuation of evolution obviously depends on the maintenance of a sufficiently high mutation rate. For this reason some knowledge of spontaneous as well as of environmentally controlled mutation rates is essential to an understanding of evolution.

Mutations do not by themselves produce evolution, but the mutation rate may under some conditions be a limiting factor. As has been emphasized by Dobzhansky (1941, pp. 34–42), spontaneous rates of mutation, though nearly always low, vary enormously from one race to another of the same species and from one gene locus to another of the same individual. In *Drosophila melanogaster,* the frequency of lethal mutations may be ten times as great in one strain as it is in another, and there is no reason to believe that the amount of variability in the rate of visible mutation is any less. And the data of Stadler in maize show that spontaneous mutation in such loci as R (color factor) may be more than 500 times as frequent as in other loci, like Wx (waxy). For this reason, mutation rates themselves must be looked upon as racial and specific characteristics subject to the same laws of evolutionary change as are the other morphological and physiological characteristics of the organism.

Since the epoch-making discovery of Muller that mutations can be induced in a high frequency by means of X rays, a very large number of experiments has shown that not only these radiations but also ultraviolet and infrared rays, as well as temperature shocks and some (though surprisingly few) chemical agents, can influence directly or indirectly the mutation rate. But these agents are effective only in very high, sublethal doses, and Muller (1930) has himself pointed out that only a fraction of 1 percent of the mutations occurring in natural populations can be accounted for by the radiation present in nature.

Of more interest to the evolutionist are the demonstrations of the effectiveness of certain chemicals, particularly mustard gas and related compounds, in producing mutations (Auerbach, Robson, and Carr 1947). Although the germ plasm may not be exposed very often to the action of such chemicals coming from outside, some of these mutations may be produced regularly or occasionally as by-products of metabolism within the cell. They may therefore be a cause of mutability produced by internal as well as by external agencies. At any rate, the hypothesis which seems most plausible at present is that mutations under natural conditions are usually caused, not by factors of the external environment, but by internal factors, either the slight physical or chemical instability of the gene or the action upon it of substances produced by the organism in which it is located.

Strong evidence in favor of such a hypothesis is provided by the examples of the increase in mutability of one gene through the effect of others with which it is associated. The best example of this was obtained by Rhoades (1938) in maize. The gene a_1 causes absence of anthocyanin pigment in the endosperm. Normally it is very stable, but when a gene known as dt, lying in an entirely different chromosome from a_1, is replaced by its dominant mutant Dt ("dotted"), this mutant causes the gene a_1 to mutate to its dominant allele A_1, which produces anthocyanin pigment in the cells of the endosperm and in other parts of the plant. The dominant mutant Dt is itself very stable and has no visible effect on the phenotype except to increase the mutability of a_1. A comparable example is that described by Harland (1937) in cotton. Both of the two related species, *Gossypium purpurascens* and *G. hirsutum,* have dominant genes for the occurrence of colored spots at the base of the petals and corresponding recessives for the absence of this spot. If a plant of *G. purpurascens* homozygous for its dominant gene for petal spot (S^pS^p) is crossed with a homozygous spotless (ss) plant of *G. hirsutum*, the resulting hybrid, as expected, has petals with spots on most of its flowers, but evidently there is frequent mutation to the spotless condition. These mutations may occur in the vegetative growing point and produce branches on which all the flowers are spotless, or they may occur as late as the differentiation of the petal tissue itself, so that the spots contain streaks of colorless tissue. Pollen from flowers with spotted petals, when used on spotless plants of *G. hirsutum*, gives the expected 1:1 ratio of mutable heterozygous and spotless homozygous progeny; but when pollen was taken from spotless flowers of the same F_1 plant, the resulting back-cross progeny with spotless *G. hirsutum* were all homozygous and spotless. This test, like a comparable one performed by Rhoades on the Dt mutant, shows that the changes induced are actually gene mutations.

The widespread occurrence of genetic factors affecting mutability has been strikingly demonstrated in the corn smut, *Ustilago zeae*, by Stakman, Kernkamp, King, and Martin (1943). This species includes an enormous number of haploid biotypes, many of which yield typical segregation ratios for visible characteristics when crossed with each other. Some of these biotypes are relatively constant, while others are highly mutable, and all sorts of intermediate conditions exist. In one cross, the four monosporidial haploid lines derived from a single zygote consisted of two which did not produce any sector mutants in 100 different flasks and two others which showed as many as 360 distinct sectors in 89 flasks. The mutants were of various morphological types, suggesting that the mutability factor concerned was a general one, affecting many different gene loci. In another cross between two mutable lines, segregation for mutability was also observed and was found to be linked with a factor determining the presence or absence of asexual sporidia, although recombination types, presumably crossovers, were occasionally found. Appropriate crosses between lines of the same kind with respect to variability and constancy make possible breeding either for increased constancy or for increased mutability.

Further examples of this sort are needed, but the evidence to date suggests that mutation rates in nature are more frequently and more drastically altered by internal agents than they are by external agents. If this is the case, then the hypothesis of Sturtevant (1937) and Shapiro (1938), that mutation rates are themselves controlled by natural selection, becomes very plausible. These authors conclude that since in all well-established and adapted species most mutations will be harmful, selection will through the ages tend to lower the mutation rate in the genes of a species by establishing mutant genes which act as mutation suppressors. The behavior of the petal-spot gene of *Gossypium purpurascens* in the genetic environment of another species, away from many of the genes normally associated with it, supports this hypothesis. Other supporting evidence is provided by Gustafsson's (1947a) studies of spontaneous and induced mutation rates in three different varieties of cultivated barley. The Golden variety, which "represents a very old pure line (isolated before 1900)," has the lowest rate of spontaneous mutation as well as of morphological mutations of various types induced by X rays. The two varieties Maja and Ymer, both of which, although they are now essentially homozygous, are the products of relatively recent hybridization and have higher frequencies of both spontaneous and X ray-induced morphological mutations. One may conclude that the history of Golden barley and of other old varieties of cereals has included many generations of artificial selection for constancy, which has involved in part the establishment of gene combinations reducing the mutation rate. The breaking up of these combinations by intervarietal hybridization is responsible for the increased mutation rates in the newer varieties. Similar evidence in *Drosophila* has been obtained by Berg (1944). She found that small populations of *D. melanogaster* found in a mountain valley near Erivan, Armenia, where conditions are limital for the species and selection pressure must be severe and rigid, had a relatively low mutation rate, while the larger populations occurring in more favorable sites had more rapid mutation rates both for lethal mutations and for morphological mutations at particular loci, such as yellow.

The evidence presented above suggests that the facts about mutations and mutation rates which are most important for studies of evolution are likely to be obtained in the future chiefly from the as yet hardly developed study of comparative genetics. Mutation is not a completely autonomous process, but is integrated with the other genetically controlled physiological processes of the organism. Like them it is under the influence of natural selection and of other forces controlling the frequency of genetic types in populations.

CHAPTER IV

Natural Selection and Variation in Populations

SINCE ITS EXPOSITION in the classic work of Darwin, the theory of natural selection has had a tortuous history. In the half-century following the publication of the *Origin of Species,* this theory was regarded as the cornerstone of evolutionary biology, and nearly all of the research on evolutionary problems involved studies either of adaptation and selection or of the phylogenetic relationships between organisms. The proponents of selection undoubtedly went to unjustified extremes in their attempts to show the all-inclusiveness of this principle. In the early part of the present century, however, the prestige of the selection theory declined until many biologists regarded it not only as a relatively unimportant factor in evolution but in addition as a subject not worthy of study by progressive, serious-minded biologists. The reasons for this decline lay partly in the excesses of speculation committed by the extreme selectionists of the previous generation. But a more important cause was the impact on biological thinking of two fundamental discoveries in genetics. The first of these was the discovery of mutations by De Vries, and the second was the demonstration by Johannsen that artificial selection is ineffective in a pure line of completely homozygous organisms. These two workers and their followers maintained that all true evolutionary progress comes about by sudden mutational steps, which create at once a new race or species. Selection was regarded as a purely negative force, eliminating the new variants which are obviously unfit, but not taking part in the creation of anything new. The prestige of these two geneticists, as well as the rapid strides being made by the science they helped to found, caused their views to be accepted widely and often uncritically, so that the mutation theory was often looked upon as a modern substitute for the largely outmoded hypothesis of Darwin.

This negative attitude toward natural selection has again become reversed during the past twenty years. The prevailing concept of evolution at present is the neo-Darwinian one, in which mutation and selection are looked upon, not as alternative, but as complementary processes, each essential to evolutionary progress and each creative in its own way. This revival of belief in natural selection as a creative force has resulted from four different trends in the development of genetics. These are the following.

In the first place, our increased knowledge of mutation has made necessary a considerable broadening of our conception of this phenomenon as compared to the opinions about it held by De Vries, Johannsen, Bateson, and other early geneticists. Geneticists now realize that the great majority of mutations with a large effect on the phenotype – the only type that could possibly produce the "elementary species" visualized by De Vries – are in the nature of semilethal or at least less viable abnormalities. They are comparable to the "sports," the existence of which Darwin fully recognized, though he rejected them as of little significance in evolution because of their usually inadaptive character. On the other hand, evidence which has accumulated from various sources now indicates that mutations with small effects are very common, occurring perhaps at a greater frequency than the conspicuous type, as discussed in the preceding chapter. Furthermore, as will be brought out in more detail below, the nature of inheritance in crosses between natural races and species indicates that the formation of these groups has involved mutations with a small effect far more often than it has involved the conspicuous ones. The "elementary species" of De Vries is either a figment of the imagination or a phenomenon peculiar to plants with self-pollination and an anomalous cytological condition, like that of *Oenothera,* or some other deviation from normal sexual reproduction and cross-fertilization as found in most organisms.

Second has been the realization that the pure line, which according to Johannsen represents the limit of selection, is approached in nature only in organisms with self-fertilization, and is therefore completely absent in the higher animals and in most of the higher plants. Self-fertilization is relatively frequent in those plants which form the most suitable material for experiments on genetics and plant breeding, namely the annual grains, vegetables, and flowers commonly grown in the garden and field. Strains which approach the condition of a pure line are to be expected in such plants. However, these groups are not typical of the higher plants. Study of the wild relatives of the self-pollinated crop and garden plants shows that most of these relatives and presumable ancestors are frequently or regularly cross-pollinated, often being self-incompatible. In these organisms heterozygosity for a large number of genes is the normal condition in nature, and selection may be carried on in one direction for many generations without exhausting the possibilities inherent in the supply of genetic variation present in a population. This fact has now been well demonstrated by the experiments of Winter (1929) on maize, of Goodale (1942) on mice, and particularly of Payne (1920) and Mather (1941, Mather and Wigan 1942) on *Drosophila*. We can no longer think of mutation as the primary source of directive tendencies in evolution and of selection in the purely negative role of eliminating unfavorable tendencies. On the contrary, the direction of evolution is determined largely by selection acting on the gene fund already present in the population, the component genes of which represent mutations that have occurred many generations ago. New mutations are important chiefly as a means of replenishing the store of variability which is continuously being depleted by selection.

The third line of genetic knowledge which has restored our belief in the creative importance of natural selection has been obtained from studies of the genetics of natural populations. From such studies has come the realization that most differences between natural races and species are inherited, not according to simple Mendelian ratios, but in a manner indicating that they are controlled by multiple factors or polygenes. This evidence is discussed in detail by Dobzhansky (1941, pp. 68–82), Huxley (1942, pp. 62–68), and Mather (1943). Without the action of some guiding force, the individual mutations making up the multiple factor or polygenic series will be combined in a multitude of different ways, of which only a small percentage will produce a noticeable effect on the phenotype. Natural selection, if not the only guiding force, is at least by far the most important agent in producing the regular accumulation of these small genetic changes. To be sure, always acts by eliminating genes and gene combinations. But just as a sculptor creates a statue by removing chips from an amorphous block of marble, so natural selection creates new systems of adaptation to the environment by eliminating all but the favorable gene combinations out of the enormous diversity of random variants which could otherwise exist. Muller (1947) has correctly characterized mutation as a "disrupting, disintegrating tendency" in natural populations and selection as the force which "brings order out of mutation's chaos despite itself."

Strong support for these generalizations is provided by the work of Fisher (1930), Haldane (1932a), Wright (1940a,b), Tschetwerikoff, Dubinin, and others on the genetics of natural populations. This work, which has been ably summarized by Dobzhansky (1941, Chaps. V, VI, X) and Simpson (1944, pp. 48–74), has brought the realization that evolutionists must think of variation on two levels. The lower level, that of the variation between individuals within an interbreeding population, is the direct effect of mutation and gene recombination, but by itself is of little significance in evolution. The variation most important to evolutionary progress is on a higher level, that between populations in space and time, with respect to their gene frequencies. New variability on this higher level does not originate through mutations alone. Only two agencies are known which can produce differences between populations in gene frequency — the chance fixation of random variation and the directive action of natural selection. These forces obey the rules of the physiology of populations, not of individuals. Dr. Sewall Wright (oral communication) has characterized evolution as "the statistical transformation of populations." Such a transformation can be accomplished to the degree represented by existing races, species, and genera of organisms, with their great diversity of adaptive gene combinations, only through the guiding influence of natural selection acting on the raw material of variability provided by gene mutation and recombination.

Even stronger support for the creative role of natural selection is provided by the fourth trend in genetics, namely, the modern amplifications of Darwin's analogy between artificial selection as practiced by plant and animal breeders and the selective action of the natural environment. Breeders now recognize that, with the exception of disease resistance, the desirable qualities toward which they are working usually result from favorable combinations of several different measurable characteristics (Hayes and Immer 1942, Lush 1946, Frankel 1947, Lerner and Dempster 1948). For instance, yield of grain in any cereal crop is the result of such diverse characteristics as number of flowering heads, size of grain, fertility, and resistance to disease. Of a similar complex nature are characteristics like milling and baking qualities in wheat, flavor in tomatoes, apples, or other fruits, and fiber quality in cotton.

The survival or extinction of an individual or race in a natural environment is determined by qualities with a similar complex nature. Resistance to arid or desert conditions, for instance, involves such diverse characteristics as depth of penetration of the root system; extent of transpiration surface of the leaves and shoots; size, number, distribution, and morphological character of the stomata; and various physical and chemical properties of the protoplasm which enable it to withstand desiccation. Winter hardiness, competitive ability, seed fertility, and efficiency of seed dispersal are adaptive qualities of equal importance with a similar complex basis. Individual mutations, therefore, even if they produce entirely new characteristics, are important in evolution chiefly in relation to the other characteristics already present in the population.

Furthermore, the effectiveness of selection depends not only on the nature of gene combinations rather than of individual genes; it is also influenced strongly by the degree to which the phenotype can be modified by the environment. This is the basis of the principle of *heritability,* as Lerner and Dempster (1948) have discussed it in relation to animal breeding. Heritability, as defined by animal breeders, is the degree to which the additive genetic effects of multiple factors determine the phenotype, as compared with parallel effects on the same phenotype caused by environmental modification. The direct action of selection is, after all, on the phenotype. If, therefore, the heritability of a character is very low, then strong adverse selection pressure may fail to reduce its magnitude or frequency in a population, since the individuals which die or produce too few offspring because they exhibit this character may have genotypes entirely similar to or deviating in the opposite direction from those of the surviving individuals. Heritability is particularly important in considerations of selection on the basis of combinations of characteristics which, as was pointed out in the beginning of this chapter, is one of the principal ways that selection works. Lerner and Dempster have presented examples to show that under certain conditions of heritability and genetic correlation of characteristics, a population may be altered with respect to a particular character in the opposite direction to that of the selection pressure which is active. This does not mean, of course, that selection can ever cause a population to become less fit. It merely emphasizes the fact that fitness, and therefore the response to selection, is based on combinations of characteristics, and that the same type of selection pressure does not always alter similar phenotypic characteristics in the same direction.

To students of plant evolution perhaps the greatest significance of the concept of heritability is in connection with the fact brought out in Chapter III, namely, that the degree to which the expression of the phenotype may be modified by the environment differs greatly from one genotype to another. There seems to be good reason for believing that in some species natural selection has favored a high degree of phenotypic plasticity in terms of environmental modification, in spite of the low heritability and consequent inefficiency of selection which this brings about. On the other hand, in other species low phenotypic plasticity and high heritability have been maintained, presumably because of their selective advantage to the evolving population. The study of the distribution of different degrees of phenotypic plasticity and heritability will undoubtedly be one of the important lines of approach to evolutionary problems in the comparative genetics of the future.

EXPERIMENTAL EVIDENCE FOR NATURAL SELECTION

As with most other natural phenomena, the most convincing evidence for natural selection is coming and will continue to come from carefully conducted experiments. Unfortunately, however, the number of these experiments, particularly on the higher plants, is as yet very small. This is not because evolutionists have failed to recognize their importance. To perform them adequately, particularly in relatively large, slow-growing organisms like the seed plants, requires much time, space, and money. Furthermore, while the demonstration that selection has occurred is not excessively difficult, the nature of action and the causes of the selective process are much harder to discover or to prove.

Experiments which can provide valid evidence on the existence and dynamics of natural selection do not have to be performed under completely undisturbed natural conditions. After all, many of the evolutionary changes which are taking place at the present time are doing so under the influence of new environmental conditions brought about by the activity of man. The essential differences between natural selection as it occurs in evolution and artificial selection as practised by animal and plant breeders is that natural selection is the result of the elimination or non-reproduction of individuals through the agency of the environment, whether completely natural or man-made, while in artificial selection certain individuals are selected and segregated consciously from the population by the hand of man, and only the offspring from these individuals are allowed to form the next generation. The experiments to be discussed, therefore, all involve subjecting populations containing different genotypes to various more or less artificial environments of which certain features are known and recorded, and observing the effect of these environments on the genotypic composition of the populations concerned.

Since the normal type of population in nature consists of heterozygous, cross-breeding individuals, experiments with this type of material will in the long run tell more about the role of selection in evolution than will those with self-fertilized species, or with asexually reproducing clones. Unfortunately, however, only two experiments dealing with such cross-fertilized plants are known to the writer. Sylvén (1937) showed that when strains of white clover (*Trifolium repens*) originating from Denmark and Germany were planted in the more severe climate of southern Sweden, they became adjusted to this climate over a period of two years through selective elimination of the less hardy individuals. This adjustment became evident through a marked increase in yield of green matter. The increase, however, took place only in two varieties which had not previously been subjected to artificial selection, and which were therefore highly heterozygous at the start of the experiment. A more highly selected, relatively homozygous strain did not respond to this change in climate.

The second experiment is that of Clausen, Keck, and Hiesey (1947) on *Potentilla glandulosa.* They planted under cultivation in their three standard environments — Stanford, Mather, and Timberline — clonal divisions of each of 575 individuals belonging to an F_2 progeny derived from a hybrid between subsp. *nevadensis* and subsp. *reflexa.* Over a period of 5 to 8 years some of the individuals were eliminated at each station, and as expected these were different in most cases. On the other hand, some of these F_2 genotypes showed the unexpected ability to survive and produce seed at all three stations, which was not true of either of their original parents. In this experiment only one generation was tested in this fashion. The continuation of such an experiment over several generations of natural reproduction would be very desirable.

A far larger number of experiments is available which deal with competition between constant biotypes such as are found in apomictic species, like *Taraxacum,* and in essentially homozygous, self-fertilized species like the cereal grains. This is partly because of the economic importance of such plants, and partly because the technique of experimentation and the recording of results is far easier in experiments with them than it is in experiments with cross-fertilizing species. The results of experiments on intervarietal competition are therefore presented here in detail not because they represent the best possible evidence on the action of natural selection in plants, but because they are the best which is available at present.

The experiments of Sukatschew (1928), showing differential survival of different apomictic clones of *Taraxacum* under different cultural conditions, have already been discussed in detail by Dobzhansky (1941). The most significant features of Sukatschew's results are, first, that the survival of any race depended on a complex combination of factors, including type of soil, density of planting, and the amount and type of competition, and, second, that the ability of a race to produce abundant seeds and so perpetuate itself was not directly related to the ability of the individual plants to survive.

These results immediately emphasize the point that two equally important but somewhat different factors are involved in the success of a race or species under natural conditions; first, the ability of the individual plants to survive and grow to maturity, and, second, their capability of reproducing and disseminating their offspring. This point will be discussed in more detail later in this chapter.

The most extensive experiment yet performed on natural selection in the higher plants is that of Harlan and Martini (1938) on barley. In this experiment, a mixture containing equal quantities of seed of eleven different commercial varieties of cultivated barley (*Hordeum vulgare*) was planted in each of ten different experiment stations located in various parts of the United States. The seeds were sown on field plots prepared under as nearly uniform cultural conditions as possible. At the end of each growing season, the mixed crop was harvested in bulk, the seeds were thoroughly mixed, random samples of 500 seeds each were extracted, sorted into varieties, and the number of seeds of each variety was counted. The remainder of the mixture was saved for planting in the following spring at the same rate as that of the preceding season. The length of the experiment varied from four to twelve years, depending on the locality.

A part of the results is shown in Table 2. The most obvious

TABLE 2

SURVIVAL OF BARLEY VARIETIES FROM THE SAME MIXTURE AFTER REPEATED ANNUAL SOWINGS AND COMPETITION OF VARIOUS DURATION [a]

	Arlington, Va. (4 yrs.)	Ithaca, N.Y. (12 yrs.)	St. Paul, Minn. (10 yrs.)	Moccasin, Mont. (12 yrs.)	Moro, Ore. (12 yrs.)	Davis, Calif. (4 yrs.)
Coast and Trebi [b]	446 T	57 T	83 T	87	6	362 C
Hannchen	4	34	305	19	4	34
White Smyrna	4	0	4	241	489	65
Manchuria	1	343	2	21	0	0
Gatami	13	9	15	58	0	1
Meloy	4	0	0	4	0	27

[a] Condensed from Harlan and Martini 1938.

[b] The records on these two varieties were combined because of the difficulty of distinguishing between their seeds. The figures marked T indicate a predominance of Trebi, those marked C, predominance of Coast.

feature of this table is that in nearly every station the majority of the seeds at the end of the experiment consisted of one variety and that the surviving variety differed according to the locality. Equally significant, however, are two other facts. In the first place, the number of varieties which survived varied considerably from one locality to another, even when these localities differed relatively little in their climate. For instance, the locality at which the largest number of varieties survived in reasonably large percentages was Moccasin, Montana. Yet, in Moro, Oregon, with a similar dry, cold climate, and having the same dominant variety, namely, White Smyrna, the number of surviving varieties is the smallest of any of the localities. Secondly, the individual varieties, which approximate pure lines consisting of one or a few very closely related biotypes, differ considerably not only in their tolerance to a particular climate but also in their ability to tolerate a variety of climatic conditions. For instance, the variety Hannchen, though dominant only at St. Paul, Minnesota, survived at least in small percentages at each locality. On the other hand, Manchuria, which was highly successful at Ithaca, New York, apparently has a much narrower range of tolerance and was a failure at nearly every other locality.

This experiment provides suggestive analogies to competition between the different species in any community, as well as that between different biotypes of one cross-breeding species. The first analogy is obvious. Examples of localities rich as well as poor in species are well known to students of plant distribution, as are also species with wide and with narrow ranges of climatic tolerance (Mason 1936, 1946, Cain 1944). The analogy to competition between interfertile biotypes of the same species is indirect, but possibly of greater significance. Some of these barley varieties differ from each other as much as do the different subspecies of a wild species, but in the degree of their morphological and physiological differences from each other some varieties are more comparable to the different individuals of a cross-breeding population. The results of this experiment might lead one to predict, therefore, that if a similar experiment were performed on a heterozygous, cross-breeding population, some localities would retain a high degree of genetic variability, while others, though containing at the end of the experiment a large number of vigorous, productive plants, would nevertheless select out a population containing many fewer genetic variants. Furthermore, natural selection should permit the survival of some genes and gene combinations over a wide range of environments, while restricting others to particular localities, where they might nevertheless be highly successful. Actually, such conditions do exist in nature, so that the experiment of Harlan and Martini suggests a way of showing how patterns of this type of intraspecific variation can be produced by natural selection.

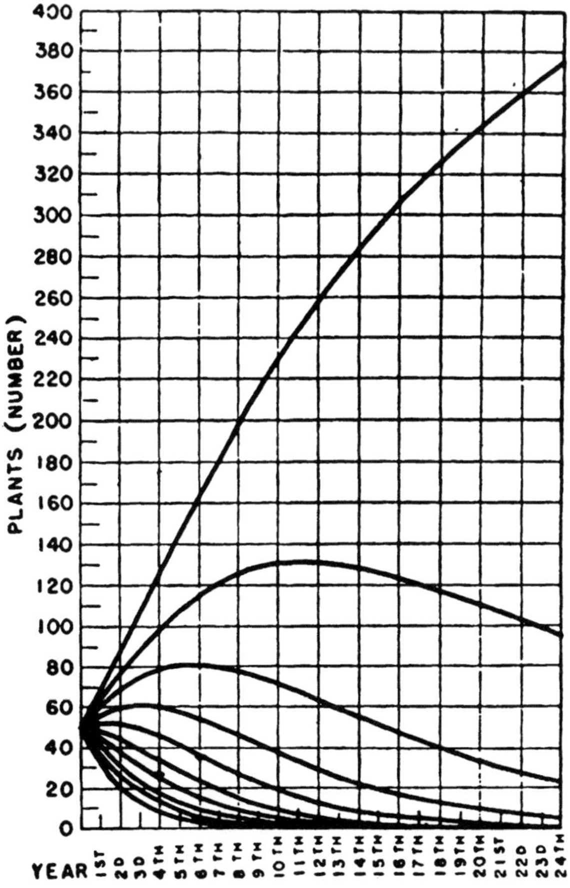

FIG. 13. Theoretical curves of natural selection based on an equal mixture of ten varieties of cereals differing by five kernels each in their productivity per plant, the poorest plant producing 45 seeds. From Harlan and Martini 1938.

Another set of facts brought out by Harlan and Martini concerns the rate at which the unsuccessful and the partly successful varieties become eliminated. Three different types of curves were found for the yearly change in percentage of seeds of a variety (Figs. 13, 14). The most successful ones rose rapidly and rather evenly throughout the experiment. Those of intermediate adaptive value at any locality rose rapidly at first, reaching a rather low peak, then declined equally rapidly, but the rate of this decline always slackened toward the end of the experiment.

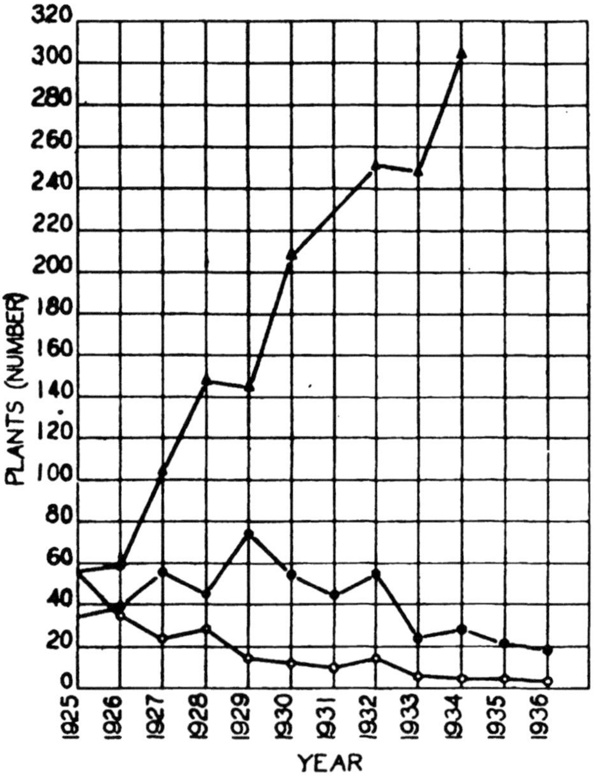

FIG. 14. Actual curves showing the change in number of plants of representative varieties of barley initially sown in equal quantities in a mixture. Triangles, a dominant variety, Hannchen at St. Paul, Minn.; solid circles, a better than average variety, Hannchen at Moccasin, Mont.; open circles, a poor variety, Meloy at Moccasin, Mont. Redrawn after Harlan and Martini 1938.

Finally, the unsuccessful types declined rapidly at the beginning of the experiment and then more slowly, so that toward its end some of them retained a very small number of seeds in the mixture for several generations. These results, which agree in general with theoretical expectation, form a practical demonstration of a principle similar to that brought out by Fisher (1930) and Haldane (1932a) for the survival of alleles in a heterozygous population. Selection acts very rapidly at intermediate gene frequencies, but much more slowly at very high and very low ones. The fact that a large proportion of rare and localized wild species are actually ancient relicts (Fernald 1931, Cain 1944, Chap. XV) is due at least in part to this principle. Many of these rare types, by avoiding the keenest regions of competition and the most rigid types of selection, have lengthened greatly their existence.

A final fact of practical as well as theoretical importance is that the variety surviving at any locality is not necessarily the one which gives the highest yield at that locality when grown in a pure stand by the farmer. For instance, the most successful commercial variety near Ithaca is similar to Hannchen, which in the mixture survived best at St. Paul, Minnesota, while at the latter locality the best agricultural variety is Manchuria, which in Harlan and Martini's experiment survived best at Ithaca. Suneson and Wiebe (1942) tested four different varieties of barley and five of wheat over periods of four and eight years, respectively, in pure stands and in a mixture grown under identical conditions at Davis, California. They found that of the four barley varieties Vaughn, which survived in the smallest percentage in the mixture, had in a pure stand the highest mean yield over the eight-year period. On the other hand, Atlas and Club Mariout, which ranked first and second in the mixture, were second and fourth in yield in pure stands. The differences among the wheat varieties were comparable. These results are comparable to those of Sukatschew, when he found that an apomictic strain of *Taraxacum* which gives a large amount of seed when competing with itself in a pure stand may grow very poorly when raised at the same density in competition with other strains. Together these experiments emphasize the complexity of the factors entering into natural selection and the difficulties involved in finding out the specific basis of selection of any particular characteristic.

Suneson and Wiebe, although well acquainted with the morphological characteristics and the manner of growth of the varieties they used, were unable to explain satisfactorily the reason for the differences between the behavior of these varieties in mixtures and in pure stands.

Laude and Swanson (1942), in similar experiments involving competition between two varieties of wheat (Kanred vs. Harvest Queen and Kanred vs. Currell), found that during the first two years the change in percentage of the two varieties was relatively slight, that for the next five years the decline of the less competitive variety was rapid, but that for the final two or three years of the experiment the rate of change was again slow in the event that one variety had not yet been eliminated. They also showed that elimination or reduction in frequency could occur at the vegetative stage, at the reproductive stage, or both, and that the various causes of elimination were not necessarily correlated with each other. Competitive ability was therefore found to be the result of the interaction of several different, independently inherited characteristics, and as in the other experiments was not necessarily related to the yield of the variety in a pure stand.

In contrast to the higher plants, some of the fungi and bacteria are particularly favorable material in which to study natural selection, because of the rapidity of their cell generations and the relative simplicity of their growth and development. Winge (1944) and Pontecorvo and Gemmell (1944) have independently pointed out that in colonies of yeasts and molds which have originated from a single spore and are growing at an even rate on a uniform artificial medium, the selective value of mutations in relation to the wild type can be determined by the shape of the sector they form in the colony. Unsuccessful mutations form a lens-shaped inclusion within the circular mass of the colony; slightly inferior ones form a sector with convex margins, due to the slackening of their growth as the colony gets older, and their position is marked by a slight indentation of the margin of the colony. Mutants with a growth rate equal to that of nonmutated cells form straight-edged sectors, while the superior ones form sectors with concave margins which become expanded near the periphery of the colony and produce a bulge in its margin (Fig. 15). Actual examples of all these types of mutations are illustrated by Winge in yeast and by Pontecorvo and Gemmell in *Neurospora,* while other fine examples are shown by Skovsted (1943) in the yeast relative *Nadsonia richteri.* These organisms are excellent material for experimental studies of the selective basis of individual muta-

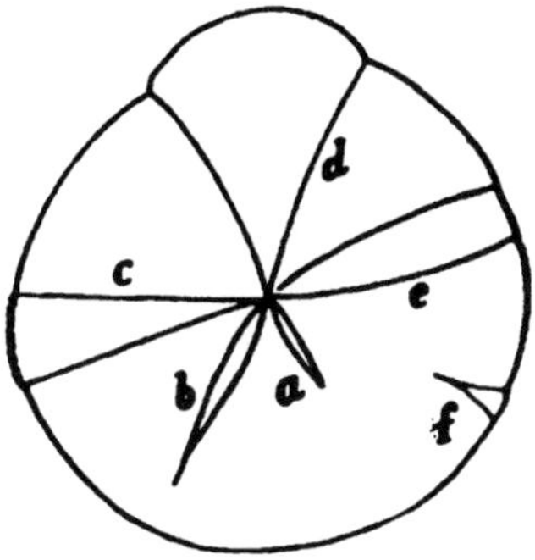

FIG. 15. Different types of mutant sectors in yeast colonies: a, b, and e, with selective value lower than the nonmutant strain; c, with the same selective value; d and f, with selective value higher than the nonmutant strain. From Winge 1944.

tions, as well as of the dynamics of selection, but as yet they have been little used for this purpose.

More accurate quantitative data on selection rates can be obtained in the case of mutations which affect the ability of the organism to grow in the presence of some known chemical component of the substrate. For instance, Spiegelman and Lindegren (1944) showed that in a genetically heterogeneous haploid culture of yeast, adaptation to the fermentation of the sugar galactose, with consequent ability to grow on this medium, increases over a 48-hour period at a regular rate. This rate can be expressed in linear fashion by a logarithmic equation, which was derived on the basis of the assumption that certain cells of the original culture were growing faster than others because of their greater ability to ferment galactose. In the bacterial species *Staphylococcus aureus* Demerec (1945a,b) showed that resistance to low or moderate concentrations of penicillin increased at a rate which would be expected on the assumption that mutations to resistance were occurring and being selected. He further proved experimentally the random occurrence of these mutations and the greater resistance of the strains which had been selected. In this organism, the mutation and selection is progressive, each selected mutant having a greater resistance to penicillin than its progenitor. Here, therefore, we have an example of directed evolution by means of random mutation and selection through a particular environmental factor, all taking place in a single test tube in the course of a few days.

In a similar manner Braun (1946a,b) demonstrated that the phenomenon of dissociation in *Brucella abortus,* involving the appearance and spread of rough cells in bacterial cultures of the smooth type, is due to mutation and selection. Different clones of this organism have different and characteristic rates of dissociation. These appear to be determined only to a minor extent by the rate of mutation. Genetically determined growth rates, which can be modified by various types of changes in the environment, are much more significant. In particular, the rate of dissociation is affected by the period at which the number of viable cells the medium can support reaches its maximum, and after which there is a steadily increasing ratio between the total number of cells and the number of viable ones. Because of their greater viability in relation to smooth types, the rough cells increase relatively rapidly under the latter set of conditions.

Even more striking from the evolutionary point of view are the experiments of Emerson and Cushing (1946, Emerson 1947) on the resistance of *Neurospora* to drugs of the sulfonamide type. Cultures of this mold grown on a medium containing these drugs became adapted to growth on this medium by means of the occurrence and establishment by selection of mutations resistant to its toxic effects. One of these mutations, moreover, mutated still further into a strain with such highly specialized properties that its optimal growth occurred only on a medium containing sulfanilamide. Here we have an example of extreme specialization occurring within the course of a few days.

Finally, Ryan and Lederberg (1946) have demonstrated competition and selection between nuclei of different genetic constitutions within the same organism. In the mold *Neurospora,* as in other Ascomycetes, hyphae from mycelia of different genetic constitutions may fuse to form a heterokaryon, that is, a mycelium containing within its coenocytic hyphae nuclei derived from the two mycelia which have fused and therefore are different in their genetical constitutions. Similar heterokaryons will be produced by mutation in single nuclei of such coenocytic mycelia. Ryan and Lederberg found that the "leucineless" mutant — that is, one unable to grow in a medium lacking the amino acid leucine — could be combined in a heterokaryon with the normal wild type, and that this heterokaryon would grow normally on a minimal medium lacking leucine. In such a medium, however, the nuclei derived from the wild type mycelium divide so much more rapidly than those containing the leucineless mutant that the latter become eliminated, and after reaching a size of about 4 cm in diameter the mycelium was no longer a heterokaryon but a typical "wild" mycelium. Physiological adaptation in microorganisms, therefore, in colonies and even within single organisms, is very often produced by mutation and selection, and sometimes in a series of successive steps forming a miniature evolutionary sequence.

HISTORICAL EVIDENCE FOR NATURAL SELECTION

Because of the difficulty of carrying out adequate experiments on natural selection in the higher animals and plants, much of our knowledge about the action of selection in these organisms will have to be derived from studies of changes in the genetic composition of populations over periods of many years which accompany recorded changes in the character of the external environment. Examples of such historic changes in the composition of populations are well known in animals and have been fully discussed by Dobzhansky (1941, pp. 190–196). Even more striking examples have been described recently. In the hamster, a gopherlike rodent of eastern Europe, Gershenson (1945a,b) described marked fluctuations in the frequency of the melanic form, which differs from the normal brown type by a single gene. The frequency of melanics in parts of southern Russia varies both from year to year and from one season to another of the same year. These variations are regular and are very similar in different districts; they cannot be ascribed to random fluctuation or to migration. Some of the changes, particularly the seasonal ones, are so rapid that very high selective pressures must be acting to produce them. Similar high selection pressures must be responsible for seasonal changes in frequency of different inversion types in populations of *Drosophila.* This has been demonstrated experimentally by Dobzhansky (1947a,b, Wright and Dobzhansky 1946) for western American populations of *D. pseudoobscura.* Dubinin and Tiniakov (1945, 1946) have data on *D. funebris* in Russia which can be explained in a similar fashion.

The only comparable evidence in higher plants known to the writer is that given by Kemp (1937) as a result of observations on a pasture in southern Maryland. This pasture, after having been seeded to a mixture of grasses and legumes, was divided by its owner into two parts, one of which was protected from livestock and used for hay, while the other was heavily grazed. Three years after this division had been made, Kemp dug up plants of blue grass (*Poa pratensis*), orchard grass (*Dactylis glomerata*), and white clover (*Trifolium repens*) from each half of the pasture and transplanted them to uniform conditions in his experimental garden. He found that the grazed half of the pasture contained a high proportion of biotypes of each species which produced a dwarf, procumbent growth under experimental conditions, while those taken from the ungrazed half were vigorous and upright. This, like the experiments on animal material, indicates that very high selective pressures can be operating in these populations.

THE ADAPTIVE VALUE OF DIAGNOSTIC AND DISTINGUISHING CHARACTERISTICS

The evidence presented above, as well as many other examples from the animal kingdom, leaves no doubt that natural selection is an active, directive force in evolution. Those differences between varieties, species, and higher categories which represent either adaptations to different environments or different ways of becoming adapted to the same environment must all have originated under the direction of natural selection. Obviously, therefore, a final estimate of the importance of selection in evolution must depend largely on determining what proportion of intersubspecific and interspecific differences are of such an adaptive nature. Unfortunately, however, the determination of the adaptive character of many types of differences between organisms is one of the most difficult problems in biology. Failure to realize this fact has been responsible in large measure for differences in opinion regarding the importance of selection in evolution. The early followers of Darwin, believing that all differences between organisms have an adaptive value which can easily be seen and interpreted, made many unwarranted speculations based on superficial observations of differences between plants and animals in color and form, and many of their examples of protective and warning devices have been justly criticized. But on the other hand a greater number of recent biologists, among whom may be mentioned Schaffner (1929, 1930, 1933), McAtee (1932), and Robson and Richards (1936), with the same oversimplified concepts of the action of selection, have argued that because the adaptive nature of certain characteristics cannot be easily seen or proved, it therefore does not exist.

Conclusive evidence against both of these points of view is provided by the experimental and observational studies of Gershenson, Dubinin, Dobzhansky, and many others, some of which have been discussed in the preceding section. These results show that carefully conducted experiments and observations of changes in the composition of populations can prove convincingly the existence of natural selection as an active force, but that the demonstration of how selection acts, and of the reason for the selective value of a particular character, is a much more difficult task. We are therefore no longer justified in assuming either that all characters are adaptive and can be demonstrated as such or that character differences must be considered nonadaptive and not influenced by natural selection until the basis of selection has been discovered and proved. If we are ignorant of the life history, development, and ecological relationships of a species we must maintain a completely open mind and an agnostic position concerning the adaptiveness or nonadaptiveness of its distinguishing characteristics. Even in the case of better-known species, neither the adaptive nor the nonadaptive quality of a particular character should be assumed unless definite evidence is available concerning that character.

In the case of species in which a large number of biotypes has been compared as to their morphological and physiological differences as well as their reaction to different climatic conditions, fairly safe conclusions may be reached as to the adaptive or nonadaptive quality of most of their distinguishing characteristics.

For instance, in the genus *Achillea,* Clausen, Keck, and Hiesey (1948) were able to conclude that height of plant, leaf texture, size of heads, and number of florets per head are all correlated directly with the environment and therefore must have survival value, while differences in length and width of the ligules of the ray florets, flower color, leaf cut, herbage color, and branching pattern show parallel variations in different habitats and therefore are probably not adaptive in character. Nevertheless, even those characters which are adaptive vary so much in any one locality that their adaptiveness cannot be considered as a fixed quantity. In the words of Clausen, Keck, and Hiesey (1948, p. 108):

> The analysis of *Achillea* populations suggests that natural selection is far from being an absolutely rigid process. Many compromises are tolerated, and the fitness of a particular plant depends not so much upon a single character as upon a combination of several. Such a compensatory system of adaptation is flexible, for a relative lack of fitness in one character may be compensated by special suitability in another.

This principle of compensatory systems can explain many apparently irrational features of adaptation. For instance, the greater sensitivity to frost of the vegetative parts of the alpine ecotype of *Achillea* may be of positive selective value, as a means of causing the plant to go into dormancy sufficiently early in the fall to escape severe frost injury. On the other hand, frost resistance of the leaves probably has a positive selective value in alpine races of slow-growing species with thick or leathery leaves, like many of the Ericaceae and some species of *Eriogonum* and *Pentstemon,* which survive the winter as rosettes. There is every likelihood, therefore, that the same morphological or physiological characteristic may under the same environmental conditions have either a positive or a negative selective value, depending upon the other characteristics with which it is associated.

The above data and many more of a similar nature make highly probable the concept that many and perhaps most of the differences between related subspecies and species in their vegetative characteristics have arisen under the guiding influence of natural selection. The systematic botanist and the student of comparative morphology, however, rightly place more emphasis in classification on characteristics of the reproductive structures than on the vegetative structures. For this reason the most important questions concerning the role of selection in plant evolution center about the degree and the manner in which selection has entered into the origin of the differences between species, genera, and families of plants in their reproductive structures.

The final answer to this vastly more complex question of the selective value of reproductive characteristics cannot be given until a much greater array of precise observational evidence, and particularly experimental evidence, is available on the ecology and the genetics of these differences in reproductive structures. Nevertheless, a number of facts are available which point the way to a solution of this problem.

A very significant observation in this connection has been made by Rick (1947) on a mutant of the tomato, *Lycopersicon esculentum.* This species, along with others of its genus, has the anthers held together into a tube by a series of lateral hairs. The stigma, protruding from the center of this tube, receives pollen directly from the anthers, which normally results in self-fertilization. In a commercial planting of tomatoes Rick found a recessive mutation which was conspicuous on account of its unfruitfulness. This was caused by the failure of the mutant to develop the lateral hairs on the anthers, which consequently were spread wide apart at anthesis and failed to deposit pollen on the stigmas. When cross-pollinated artificially or by insects, the mutant set an abundance of good fruit. Obviously, the presence or absence of lateral hairs on the anthers, one of the reproductive characteristics diagnostic of the genus *Lycopersicon,* is of great selective importance to its species, although this importance would not be realized by anyone studying only normal plants.

THE INDIRECT ACTION OF NATURAL SELECTION

The observations reviewed in the last section lead to the hypothesis that the action of natural selection in directing the origin of differences between species, genera, and families in reproductive characteristics, has been more often indirect than direct. An understanding of the indirect action of natural selection may be obtained by studying three processes through which it works. These are *developmental correlation, adaptive compensation,* and *selective correlation.*

The principle of *developmental correlation* was well recognized by Darwin, and was mentioned several times in the *Origin of Species.* Modern evidence on the nature of gene action, as discussed in Chapter III, tends to emphasize more than ever the importance of this principle. In terms of modern genetics, it can be stated about as follows. The direct action of genes is on the processes of development and metabolism; hereditary differences between adult individuals in visible characters are produced indirectly through the effects of genes on developmental and metabolic processes. Because of this fact, character differences which are affected by the same developmental or metabolic process are necessarily correlated with each other, since they are influenced by the same genes. This is the basis of most if not all of the phenomena which have been termed by geneticists pleiotropy, or the production of manifold effects by a single gene.

The effects of such correlations on the alteration of characters by natural selection are undoubtedly very great, but are practically unexplored. If, for instance, long roots should, because of a change in the environment, acquire a high selective value in one race of a species, mutations producing cell elongation in the roots would become established in that race. Some of these mutations would almost certainly affect cell elongation in leaves and flowers, and would therefore produce interracial differences in respect to these parts. In plants, which produce serially large numbers of similar organs, developmental correlation is to be expected with particularly high frequency, and it is very likely responsible for the origin by selection of many character differences which by themselves have no selective value.

The principle of *adaptive compensation,* the importance of which was recently suggested by Clausen, Keck, and Hiesey (1948), is a direct consequence of the effects on selection of developmental correlation. A gene or series of genes may be favored by selection because of one of their effects, but other effects of the same genes may be disadvantageous to the organism. Under such conditions, still different genes which tend to compensate for the harmful effects of the original ones will have a high selective value. An example of this type is provided by Tedin's (1925) analysis of genetic differences between races of the weed, *Camelina sativa,* as will be discussed in the next section of this chapter.

Once a group of genes have been established in a race because of their role in adaptive compensation they and the initially valuable genes form an adaptive system which must be maintained as a unit if the race or species is to retain its adaptiveness. As Tedin has pointed out, groups of characters will be kept together by *selective correlation.* The relation between these three secondary effects of selection may therefore be stated as follows. Developmental correlation brings about adaptive compensation which results in selective correlation.

The indirect action of selection can be explored both in relation to the differentiation of flowers as efficient mechanisms for securing the transfer of pollen from one plant of a species to another one of the same species and in relation to the differentiation of fruits and seeds as efficient mechanisms for securing the maturation, dispersal, and germination of seeds, as well as the early development of the seedlings derived from them. The discussion of different types of flowers will be deferred to Chapter VI, where they will be taken up in connection with the problem of the origin of species; and to Chapter XIII, where the role of selection in the origin of genera, families, and orders of flowering plants will be discussed. The three following sections will be devoted chiefly to a discussion of the direct and indirect role of natural selection in the origin of differences in fruits and seeds.

THE GENUS *Camelina* AS AN EXAMPLE OF THE ACTION OF SELECTION

The example which provides the most evidence on the indirect action of natural selection is that of adaptation in the genus *Camelina.* This example is chosen because it has been carefully studied by three workers, because the adaptive differences within the group involve reproductive as well as vegetative structures, and because the particular selective agents which have been at work in the group are comparatively well known. The present account has been compiled from the careful and discerning study of the different forms of *Camelina* growing in Russia by Zinger (1909), the later more detailed account of the distribution of the same forms by Sinskaia and Beztuzheva (1931), and from the genetic data obtained on the Swedish forms of the same species by Tedin (1925).

The genus *Camelina,* of the family Cruciferae (mustard family), consists of seven or eight species, the majority of which are confined to southeastern Europe, Asia Minor, and southwestern Asia. All are annuals, and all but one are winter annuals, the seeds of which germinate in the fall, form a rosette of leaves existing through the winter, and flower in the following spring and early summer. They are all self-pollinated. The group of forms which has been studied has been classified differently by nearly every systematist who has treated it; of the two most important recent treatments in floras, that of Vassilchenko (1939) recognizes six distinct species, while Hegi (1906) recognizes only one species, with four subspecies. Four entities seem to be sufficiently distinct so that they are generally recognized, as follows (see Figs. 16 and 17):

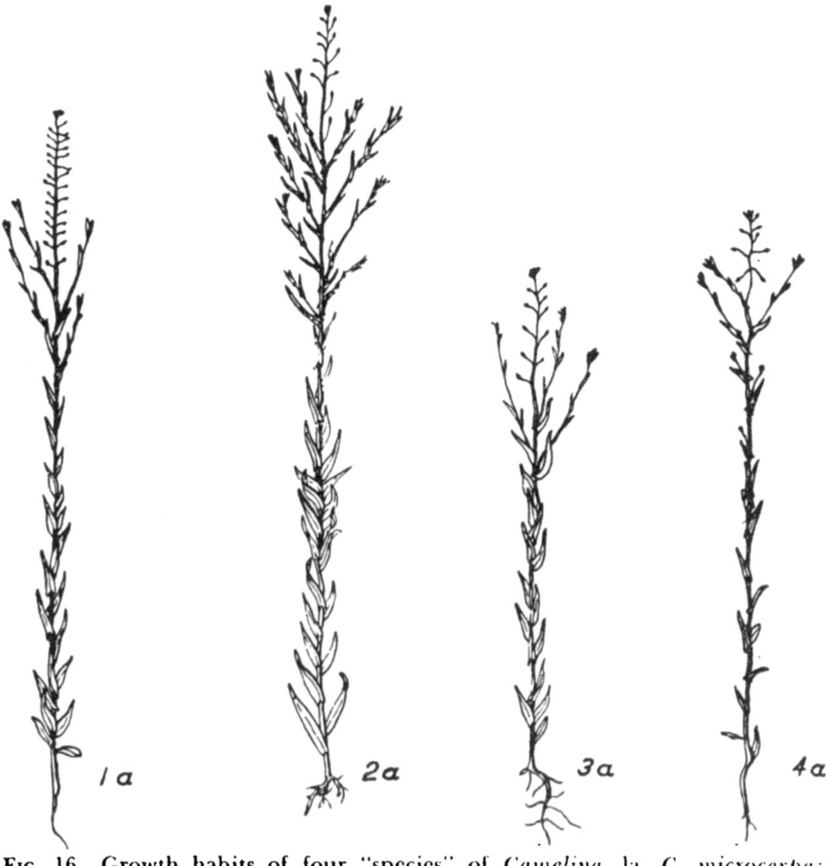

FIG. 16. Growth habits of four "species" of *Camelina*. 1a, *C. microcarpa*; 2a, *C. pilosa*; 3a, *C. sativa*; 4a, *C. sativa* subsp. *linicola*. Redrawn from Zinger 1909.

C. microcarpa Andrz. — A winter annual with freely branching stems, dense pubescence, inflorescences with numerous, rather crowded pods on ascending peduncles, and rather small pods (mostly 4–6 mm long) with relatively numerous small seeds. This

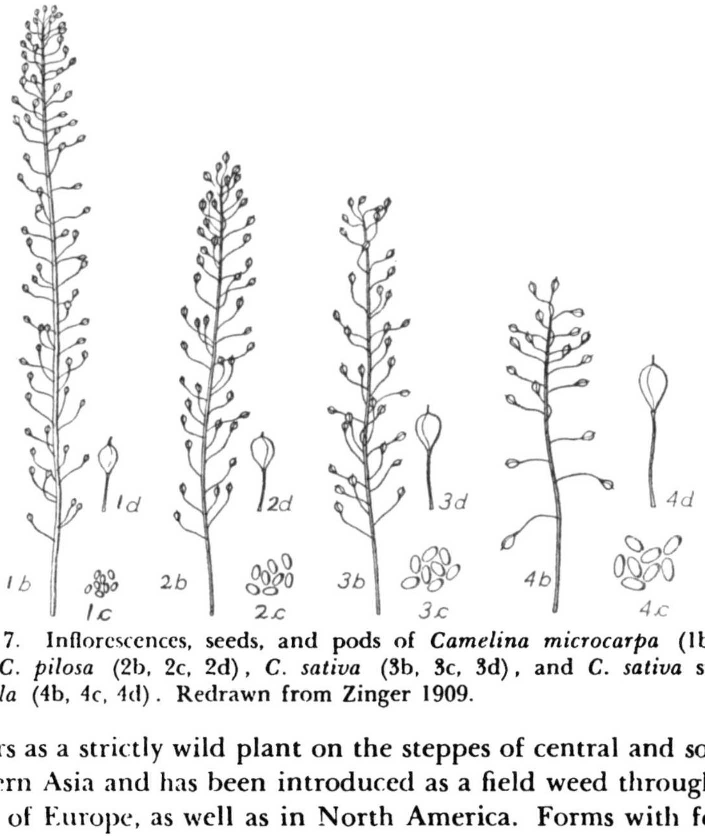

FIG. 17. Inflorescences, seeds, and pods of *Camelina microcarpa* (1b, 1c, 1d), *C. pilosa* (2b, 2c, 2d), *C. sativa* (3b, 3c, 3d), and *C. sativa* subsp. *linicola* (4b, 4c, 4d). Redrawn from Zinger 1909.

occurs as a strictly wild plant on the steppes of central and southwestern Asia and has been introduced as a field weed throughout most of Europe, as well as in North America. Forms with fewer somewhat larger pods less easily dehiscing have been separated as *C. sylvestris* Wallr.

C. pilosa (DC) Zinger. — A winter annual with spreading pubescence and large pods approaching in size those of *C. sativa.* Its distribution in eastern Europe is the same as that of *C. microcarpa,* but apparently it is not found in Asia. This form is intermediate between *C. microcarpa* and *C. sativa* and apparently forms exist which connect it with both of the latter species.

C. sativa (L) Crantz. (*C. glabrata* DC and of Russian authors). — A summer annual with less pubescence than the two previously mentioned types and rather large pods (mostly 7–8 mm long). It is found in northern Eurasia and North America, but everywhere as a weed in fields. Tedin found that F_1 hybrids between this as well as the following subspecies and *C. microcarpa* from Sweden were highly sterile, suggesting that the two are genetically distinct ecospecies.

C. sativa subsp. *linicola* (Sch. et Sp.) (*C. foetida* Fr., *C. alyssum* Thell.). — A summer annual differing from typical *C. sativa* in its slenderer, less branched stems with longer internodes; narrower, more glabrous leaves; inflorescences with fewer flowers on spreading peduncles; even larger pods (mostly 9–11 mm long); and large seeds. It is found exclusively as a weed in flax fields. Along with *Spergula maxima, Silene linicola,* and several others, this subspecies forms a group of flax "mimics," which closely resemble flax in their vegetative characteristics, have large seeds which simulate those of flax, and fruits which dehisce with difficulty, so that the seeds are harvested with the flax crop and threshed out along with the flax seeds. All these characteristics make these types unfitted to grow anywhere except as weeds infesting flax fields. The explanation of their origin has been the principal aim of the students of this group. Tedin has shown that hybrids between *Camelina sativa* subsp. *linicola* and typical *C. sativa* are fully fertile and segregate in the second generation to produce a great array of intermediate types. Sinskaia and Beztuzheva have given abundant evidence that these intermediates exist in nature.

Two additional subspecies of *C. sativa,* subsp. *caucasica* Sinsk. and subsp. *crepitans* Sinsk., infest flax fields in certain parts of Russia. They will be discussed in more detail later.

About the origin of the first three types only a few speculations can be made. Cytologically, *C. microcarpa* and *C. sativa* have been found to possess $2n = 40$ chromosomes (Manton 1932), and *C. pilosa* probably has the same number. They therefore are probably polyploids and may very well be allopolyploids of ancient hybrid origin. Before the advent of man, *C. microcarpa* and its relatives probably existed as wild plants in the steppes of southeastern Europe and southwestern Asia. Then, with the dawn of agriculture in these regions, *C. microcarpa,* along with many other species of plants, migrated into cultivated fields as a weed. Soon, however, it must have been noticed by man and itself brought into cultivation for the sake of its oil-bearing seeds.

Zinger cites the discovery of seeds of *Camelina* in cave deposits of neolithic age in Austria. In the opinion of Zinger, the absence in a wild state of *C. sativa* or any other summer annual of this genus suggests that the summer annual habit arose as a result of selection by early man. In support of his hypothesis, he showed that when *C. microcarpa* is seeded in early spring, occasional individuals flower and produce seed during the first season. Seed from these individuals produces an even higher percentage of these early-maturing types, while in a similar experiment with *C. pilosa,* Zinger selected in two generations from the original winter annual type a line which bloomed only a little later than typical *C. sativa.* At the time of the beginning of flax cultivation, therefore, *C. sativa* probably existed as a summer annual, grown under cultivation and spontaneous as a weed among other crops.

The opinion of Zinger, that the first type to enter flax fields as a weed was *C. sativa,* and that subsp. *linicola* originated in this habitat as a result of the selective forces operating in it, is well supported by the data of Sinskaia and Beztuzheva. They found that in the parts of Russia in which flax growing is intensive and agricultural conditions are good, extreme *linicola* types are found, while in such regions as southwestern Russia, where flax is grown only occasionally, it may be infested by forms which differ little from typical *C. sativa.* We may safely assume that the earliest cultivation of flax for fiber was performed much less efficiently than is this highly specialized culture at present, so that under these conditions *C. sativa* probably was well enough adapted to living in flax fields. In fact, Sinskaia and Beztuzheva report that in parts of the Caucasus and in Armenia, flax and *Camelina* are at present seeded and cultivated together in fields, and their seeds are used for oil. The *Camelina* type sown in the former region is the variety *caucasica,* which deviates in some respects from both typical *C. sativa* and subsp. *linicola,* but has fewer specialized features than typical representatives of the latter subspecies.

Once *Camelina* had entered into flax fields as a weed, the selective forces operating on it, as pointed out by Sinskaia and Beztuzheva, were of three types: climatic factors, phytosociological factors, and the effects of threshing and winnowing the seeds. Of these three, the last two were most important in producing the difference between *C. sativa* and subsp. *linicola,* while climatic factors have been active chiefly in adapting subsp. *linicola* to the different regions in which flax is grown. The rhythm of growth has been regulated by the selection of combinations of genetic factors which cause the *Camelina* to flower and ripen its seeds at the same time as the flax growing in the same region. Strains of subsp. *linicola* from northern Russia are relatively early, and those from farther south progressively later. Those of the north, growing under relatively moist conditions, have relatively larger leaves and less pubescence than southern strains of subsp. *linicola*. This subspecies, therefore, contains the same type of ecotypic variation in relation to climate that is found in wild species.

The phytosociological factors, resulting from competition in growth with the flax plant, have been the chief selective agents responsible for the differences in vegetative characters between subsp. *linicola* and typical *C. sativa*. The stems of flax, particularly that cultivated for fiber, grow very straight and dense, so that they shade strongly the ground on which they grow. This reduces both the intensity of light for photosynthesis and the amount of water lost from the soil. The only plants which can compete successfully under these conditions are those which grow rapidly, have straight, unbranched stems, and a sufficiently large leaf surface. On the other hand, pubescence and a large amount of supporting fibrovascular tissue are of relatively little value to a plant growing in this habitat.

Zinger raised the progeny of a single plant of *C. sativa* in a garden bed, half of which was planted also to a dense stand of flax. The *C. sativa* individuals growing in competition with flax had slenderer stems, longer internodes, narrower leaves, and less pubescence than their sisters which were not subjected to this competition. Furthermore, the degree of difference in these characteristics, even down to such details as the size of the stellate hairs and the amount of vascular tissue in the stem, was nearly the same in these two lots of sister plants of *C. sativa* as the difference which prevails between typical *C. sativa* and normal subsp. *linicola* when grown under uniform conditions. In other words, if *C. sativa* is subjected to competition with flax, it adapts itself by means of environmental modifications in vegetative characteristics which in character and degree closely parallel the condition determined genetically in subsp. *linicola*. In the latter subspecies this specialized habit is retained indefinitely when this subspecies is grown apart from flax, as was shown by both Zinger and Tedin.

This parallelism between environmental modification and adaptive hereditary differences is so close that Zinger was led to consider the possibility of the inheritance of the acquired modification. Nevertheless, he believed that the hypothesis of selection of small variations in the direction of these characteristics was more plausible, and this hypothesis is strongly supported by the genetic work of Tedin. The latter author found that a typical pubescent *C. sativa* differs from the most glabrous type of subsp. *linicola* by three Mendelian factors for pubescence, but that other lines of subsp. *linicola* may possess only one or two of the factors for lack of pubescence. Length of internode and width of leaves were not studied genetically by Tedin, but in respect to absolute height, which must bear a direct relation to internode length, the two subspecies differ by a large number of factors, forming a typical multiple factor or polygenic series. And there appears to be no correlation between the factors for height and those for pubescence. We can therefore postulate that the first plants of *C. sativa* which infested flax fields adapted themselves by means of environmental modification of the phenotype, but that very soon genotypes became established which forced the plant into the habit of growth best adapted to these conditions. These mutations gave the plants enough extra vigor so that they became selected. This is another example, similar to those given by Turesson (1922b) of genetic variations in an ecotype which are paralleled by environmental modifications.

The explanation of the origin of these differences as a result of the creation and maintenance by natural selection of a favorable combination of independent genetic factors which arose by mutation is further supported by the evidence of Sinskaia and Beztuzheva on subspp. *caucasica* and *crepitans*. The first subspecies has the slender stems, long internodes, and narrow leaves of subsp. *linicola*, but it is strongly pubescent and much branched. This is associated with and probably results from the fact that in the Caucasus the flax is grown only for seed and oil, and therefore is itself a relatively low, branched type which is grown in more open stands and is often very strongly mixed with *Camelina*.

Subspecies *crepitans* is a type which is now very local, infesting only the fields of *Linum crepitans,* an ancient relict crop now confined to a few districts of southern Russia (Vavilov 1926). This rare subspecies of *Camelina* is well branched when grown by itself, but becomes phenotypically modified to an unbranched type when grown with flax, just as does typical *C. sativa.*

The third set of factors, those connected with threshing and winnowing, are of the greatest interest, since they have played the largest role in causing the differences between the two subspecies in reproductive characteristics. The failure of dehiscence of the pod would have the highest selective value in this connection, since without this character the seeds would drop from the *Camelina* plants before threshing, and therefore would never become mixed with flax seeds. This is verified by the distribution of subsp. *crepitans,* which is restricted in its distribution to the few small areas in Russia in which persists the culture of *Linum crepitans,* a flax with strongly shattering seed capsules. In *Camelina* subsp. *crepitans* the pods dehisce and shatter even more easily than in the wild *C. microcarpa.* This is an adaptation to the procedure of harvesting *Linum crepitans* slightly before maturity so as not to lose its seeds, which are used for oil. If the pods of *Camelina* did not shed their seeds after this treatment as readily as those of the flax, they would not become mixed with them.

More fundamental are the results of winnowing the flax seeds after threshing. By this process most foreign materials, including weed seeds that are very different from those of flax, are eliminated. Winnowing, therefore, exerts a selective pressure in favor of a certain type of seed size and shape, and this pressure becomes progressively stronger as the winnowing is more thorough. Zinger believed that the indirect effects of selection for seed size were directly responsible for the changes in size and shape of the pods which accompanied the origin of subsp. *linicola.* Tedin, however, showed that the size of seeds is governed by a multiple factor series which is independent of those controlling the size and shape of the pods, and furthermore that the factor series controlling length, width, and thickness of pods are likewise independent of each other and of those controlling the position of the peduncles. For this and other reasons Sinskaia and Beztuzheva rightly question the hypothesis of Zinger. Nevertheless, Tedin found upon examining a large series of herbarium specimens that certain correlations between these genetically independent characters do exist in nature, and that a large proportion of the recombination types which appeared abundantly in the segregating progeny from his hybridizations are actually absent or very rare in natural populations. These correlations he designated *selective correlations,* and they are doubtless what Zinger observed. An understanding of their nature should provide a clue to the way in which selection modifies reproductive characteristics in this and other plant groups.

The primary selective factor which affects the characteristics of the pods and seeds is the size and shape of the latter, as it determines their reaction to the winnowing process. Typical subsp. *linicola* has seeds considerably larger than typical *C. sativa,* and for this reason Zinger assumed that the primary basis for selection was seed size. Sinskaia and Beztuzheva, however, found that the geographic distribution of differences in seed size is exactly the opposite in flax from that in the races of *Camelina* infesting flax fields. In flax, the northern races have the smallest seeds and those from more southerly regions usually have larger seeds. In *Camelina,* on the other hand, the largest seeds are found in the *linicola* types of the north, and the smallest in those of the south, particularly in subsp. *caucasica.* Thus, while seeds of the northern forms of subsp. *linicola* approach in size those of the flax with which they grow, those of subsp. *caucasica* are many times smaller than the flax seeds with which they are associated.

Sinskaia and Beztuzheva found that the explanation for this apparent anomaly lies in the fact that the ability of *Camelina* seeds to become mixed with those of flax depends not on their size per se, but on whether or not they are blown to the same distance by the winnowing machine. This is determined by the relation between size of surface and total weight. Obviously a flat, thin seed will be blown a long distance and a thick, round, or angular seed a relatively short distance, regardless of the total seed size. Now, the relatively smaller seeds of fiber flax grown in northwestern Russia are of the flat, thin type, while the flax seeds grown for their oil content in more southern regions are relatively thick. In the case of *Camelina,* the large seeds of subsp. *linicola* of the north have a relatively large surface area, and so approach those of fiber flax in size, shape, and reaction to winnowing. But the seeds of the southern and particularly the Caucasian races of *Camelina* are not only smaller but also have a reduced surface area in relation to their weight. They therefore are blown only a short distance by the winnowing machine, as are the flax seeds with which they are associated. In some regions, however, particularly in southwestern Russia and the Urals, small, light seeds of flax are associated with small, relatively thick seeds of *Camelina*. In these regions flax culture is relatively little developed, so that selection has been less intense and has acted over a shorter period of time.

The variations in size and shape of pods which characterize the different races of *Camelina* were in all likelihood produced by the same selective factors which caused the changes in seed size. Tedin has shown that a developmental correlation exists between seed size and number of seeds per pod; the genes for increased seed size tend to reduce the seed number, provided that the genic complex for pod size is unaltered. There is, however, an independent series of genes for pod size, so that selection simultaneously for mutants producing larger seeds as well as those increasing pod size would enable the race to increase its seed size without suffering too great a loss in seed number and therefore in reproductive capacity. Since, however, the genetic factors affecting the different dimensions of the pod are independent of each other, pod size can be increased in a number of different ways. And the evidence from the different selective correlations found by Tedin suggests that this has actually taken place. The two Scandinavian lines of subsp. *linicola* which he found to have the largest seeds and pods were those he designated Nos. 1 and 4. In line No. 1, the pods were relatively long, narrow, and thin, the approximate modal dimensions being 9.7, 5.7, and 3.7 mm, while in line No. 4 the pods were relatively short, broad, and thick, the corresponding dimensions being 8.5, 6.6, and 5.5 mm (Fig. 18). It can be seen that the volume of the pods in line No. 4 would be about 1½ times that in No. 1, and since the size of their seeds is about the same, the number of seeds per pod is correspondingly greater in line 4. On the other hand, line 1 may actually produce as many or more seeds per plant as line 4, since according to Tedin's illustration the inflorescence of line 1 is considerably longer than that of line 4 and contains about twice as many pods. This may be associated with the fact that line 1 has a modal height of 95 to 100 cm, while that of line 4 is only 55 to 60 cm. Most of this difference in height is due to the fact that

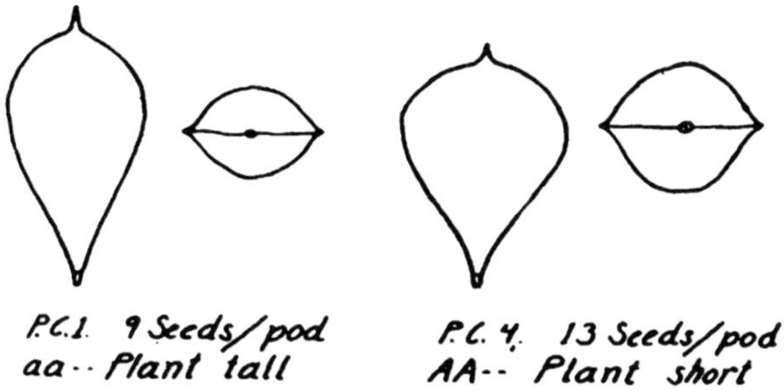

FIG. 18. Pods of the two most common Scandinavian forms of *Camelina sativa* subsp. *linicola*. Reconstructed from the data of Tedin 1925.

line 1 possesses a factor aa, which both increases height and, either through developmental correlation or genetic linkage, produces entire leaves; while line 4 possesses its opposite allele, AA, which produces lower growth and lobed leaves. The pod characteristics of line 1 are those which Scandinavian botanists have designated as var. *macrocarpa,* while line 4 is the type called var. *foetida.*

Tedin grew and classified 215 different lines and studied 290 different herbarium specimens of subsp. *linicola* from Scandinavia. Of these, all but 38 belonged to either the *macrocarpa* or the *foetida* type. This is in striking contrast with the fact that the F_2 generation of the artificial hybrid between lines 1 and 4 produced an enormous array of segregates of all sorts, and the original types were reconstituted in only a small percentage of the progeny. Furthermore, of the 65 lines with pods of the *macrocarpa* type, 41 were more than 70 cm and only 24 less than 70 cm tall, while of the 130 *foetida* lines only 13 were more than 70 cm tall. As expected, there was a larger proportion of lines with lobed leaves among those of the *foetida* type, but this difference was less marked.

We can therefore conclude that in the Scandinavian population of subsp. *linicola* the problem of increasing seed size and maintaining at the same time an adequate number of seeds per plant has been solved in two somewhat different ways, which may have originated independently. Both of them illustrate very well the principle of selective correlation as well as that of adaptive compensation. On a small scale, these types of *Camelina* illustrate the probable way in which the same selective agent, acting on somewhat different genetic material, can produce different adaptive systems affecting a number of characteristics besides the one which is the immediate object of selection. The origin of the larger differences in reproductive structures between species and genera is probably of a similar nature.

SELECTIVE CORRELATION AND SEED CHARACTERISTICS

Another good example of selective correlation is provided by the work of Gregor (1946) on *Plantago maritima.* He found that plants of this species which are selected under the good growing conditions of a garden or a natural habitat at a high edaphic level have hereditary tendencies toward erectness, numerous scapes and flowers, and relatively large seeds. On the other hand, plants adapted to poorer conditions of growth tend to be decumbent and somewhat dwarfed in their hereditary constitution. The erect, large-seeded plants have a much greater reproductive capacity and therefore a pronounced selective advantage when growing under good conditions. This advantage is increased when they are competing with dwarf plants. On the other hand, the decumbent plants are at a relative advantage under poor conditions, because their seed setting is the least affected by deficient nutrition. In *Plantago,* as in *Camelina,* the habitats available to the plant select out not individual characters, but character combinations, and certain characteristics which appear on the surface to have no imaginable selective value are nevertheless strongly affected by the selection process.

Salisbury (1942) has provided valuable evidence on the action of natural selection in directing the origin of differences in fruits and seeds by showing that among the wild plant species of Great Britain there is a marked correlation between seed size and the habitat of the species. In fact, the contents of his book may be regarded as a model of the type of information which must be gathered and tabulated on an even larger scale before any definite hypotheses may safely be made about the nature of the selective forces responsible for the differentiation of the various types of reproductive structures found in the plant kingdom. By measuring seed size in several hundred species of the British flora, Salisbury demonstrated that the plants of open habitats, such as fields and areas of disturbed soil, have on the average smaller seeds than those of semiclosed or closed communities, like those of turf and pasture, while larger and larger seeds are possessed in turn by species of scrub and woodland margin, by shade species of the herbaceous flora of woodlands, and, finally, by woodland shrubs and trees. The size of the seed, and consequently the amount of food it contains, is inversely correlated with the amount of light available to the young seedling for the manufacture of its own food by photosynthesis. Exceptions are found where expected, namely, among shade-loving saprophytes and parasites and among Orchidaceae, Ericaceae, and other groups in which the young seedlings are nourished by mycorhizal fungi.

Examination of the species lists provided by Salisbury shows that the distribution of certain fruit types, and through them of certain families, in the various communities which he describes is according to expectation if it is assumed that the controlling selective factor is seed size. Salisbury (p. 174) points out that in the larger seeds of the closed communities special devices for seed dispersal, such as the hooks on the fruits of *Galium* species, are particularly frequent. Even more striking is the fact that baccate fruits, ideally adapted for dispersal by animals, are absent or rare in the more open habitats, while they are possessed by 25 percent of the herbs of the ground flora of woodlands and by 76 percent of the woodland shrubs.

Turning to the distribution of families in these lists, we find an interesting comparison between the members of the Scrophulariaceae and of the Labiatae. These two families are similar in many respects, the most striking difference between them being that the former family possesses capsular fruits with very numerous small seeds, while the latter has fruits bearing four relatively large nutlets. As expected, the Scrophulariaceae are more common than the Labiatae among the species of open habitats. In his table Salisbury lists 18 species of Scrophulariaceae and two of Labiatae. Among the species of semiclosed or closed communities are listed one of the Scrophulariaceae and five of

Labiatae; among the species of scrub and woodland margin the numbers are three and five; while among the shade species of woodlands there are no Scrophulariaceae and one species of Labiatae. The Caryophyllaceae, with their central placentation, have an ovary structure adapted to the production of numerous small seeds. It is significant, therefore, that they represent 14 percent of the species of open habitats, 5 percent of those of closed or semiclosed habitats, 6 percent of those of scrub and woodland margin, and are not listed among the shade species of woodlands. The Leguminosae, with their single carpel and placentation confined to a suture, are adapted to the production of few, large seeds, often without well-developed means of dispersal. According to expectation, they comprise 4 percent of the species of open communities, 20 percent of the species of closed or semiclosed communities, 10 percent of those of scrub and woodland margin, and are absent from the forest community. Data such as these are obviously of great significance in relation to the selective value of differences between families and genera, and will be referred to in a later chapter.

It is obvious that the conclusions of Salisbury cannot be applied uncritically to plant communities in regions possessing very different climates from that of Great Britain. For instance, the species found in the open communities of the drier parts of California often have exceptionally large seeds. This may be due to the need for the young seedling to establish an extensive root system rapidly in order to obtain water, as was postulated by Salisbury for the seeds of dune plants. Similar differences may exist in relation to reproductive capacity. Salisbury showed that under the relatively favorable conditions for plant growth which prevail in Britain, the reproductive capacity of related species of the same genus, expressed in terms of the number of seeds produced per plant, bears a direct relation to the frequency of their occurrence and the extent of their distribution. But many of the commonest species of the drier parts of California produce relatively few seeds per plant, particularly in dry years. In climates where moisture rather than competition is the most important factor limiting plant distribution, the most significant correlation may well be that between reproductive capacity and available moisture. For this and many other reasons, extension of the data provided by Salisbury is a prime desideratum.

SELECTION AND DIFFERENTIATION IN THE COMPOSITAE AND THE GRAMINEAE

In two widely different families of flowering plants, the Compositae and the Gramineae, the writer has noticed correlations of character differences which have led him to the hypothesis that natural selection, operating through the medium of both developmental correlation and selective correlation, has been responsible for the origin of a large proportion of the differences in reproductive characteristics which separate species and genera. Examples will be given from both families.

In the tribe Cichorieae of the family Compositae, to which belong the lettuce (*Lactuca*), the dandelion (*Taraxacum*), the hawkweed (*Hieracium*), as well as *Crepis* and several other genera, the structure of the individual florets at anthesis varies little throughout the tribe. The great diversity of variation, on which genera and species are largely based, is in the involucral bracts or phyllaries, which surround the head of florets, and in the mature fruits, particularly in the system of scales or bristles known as the pappus, which crowns their apex. Many of these structures are obviously adapted to the protection of the seed while it is growing, and especially to its efficient dissemination when ripe.

The first selective correlation evident in this tribe is between the habit of growth or length of life of the plant and the degree to which these protection and dispersal mechanisms are developed. In the genera *Dubyaea* and *Prenanthes,* as well as in the more primitive sections of *Crepis,* all the species are long-lived perennials, living for the most part in the great forest belts which have had a stable climate for long periods of time (Stebbins 1940b, Babcock 1947a). These species are little specialized in their involucres and achenes and for the most part have relatively inefficient methods of seed dispersal. Their seeds are large and heavy and the pappus bristles are relatively coarse and few, so that they are not easily borne by the wind. On the other hand, all the groups of rapidly growing, short-lived annual or biennial species have developed some types of specialization in their involucres or achenes or both. The same high specialization has been developed in groups like *Taraxacum* which, although they have remained perennial, have taken to colonizing disturbed habitats, many of which are more or less temporary. This suggests that the coexistence in modern times of unspecialized as well as specialized types is a result of adaptive compensation between vegetative and reproductive efficiency. In those types which on the basis of their vegetative structures are fitted to live for a long time in a stable habitat, selection pressures in favor of efficient seed production and dispersal are relatively low, and mutations which might start evolution in this direction do not have a chance to become established. But once the species migrates into or becomes exposed by climatic change to a more unstable habitat, structures which enable it to move about more easily in response to climatic and edaphic changes immediately acquire a relatively high selective value.

The particular mechanisms which are developed must depend partly on the nature of the mutations which happen to occur first in a particular line and partly on the conditions of the environment surrounding the species. In the Cichorieae, nearly all the shorter-lived species of northern and mesophytic regions have highly developed mechanisms for seed dispersal by the wind. The most familiar of these is the achene of the dandelion, *Taraxacum,* with its long beak and spreading pappus bristles, forming a parachute. Nearly as efficient are the achenes of various species of *Crepis, Lactuca,* and *Agoseris,* all of which have developed similar adaptations independently of each other. On the other hand, many of the species of this tribe living in warmer, drier regions have developed mechanisms which serve for dispersal either by animals or by both agencies. This has been true in the genus *Sonchus,* of which the largest number of species are found in Africa. One of the best diagnostic features of this genus is the possession of pappus bristles of two types, one straight and coarse, the other fine and crisp. In many species of the genus, like the common sow thistle, *Sonchus oleraceus,* the latter type of hairs, like the similar lint on cotton seeds, adhere to the fur of animals and the clothing of man and so disperse the seeds very efficiently without the aid of the wind. Obviously, the elevation of these adherent hairs on a slender beak, like that of the dandelion and the lettuce, would be a detriment rather than an aid to seed dispersal in the sow thistle, since contact with animals would tend to break this beak and so separate the seed from its dispersal mechanism. There is good reason to believe, therefore, that the diagnostic characters generally used to separate *Sonchus* from *Lactuca* and *Taraxacum,* namely, the presence or absence of a beak and the character of the pappus, are in each case two parts of an integrated mechanism of seed dispersal and therefore originated through the guidance of natural selection acting on different initial mutations. Furthermore, the direction of evolution taken by ancestors of each of these groups toward more efficient seed dispersal may have been determined by the environment under which they existed.

In other annual species of this tribe, such as *Hypochaeris glabra,* the achenes are of two types. The ones in the middle of the head have the beak and parachute mechanism for wind dispersal common in the tribe, while those on the periphery have no beak and a pappus that is modified into a cobwebby structure. On the other hand, the latter have a sharp pointed base and their surface is covered with tiny upward pointing barbs. All of these are efficient means for securing dispersal by animals, as anybody can testify who has walked through or lain in a patch of the common cat's ear, *Hypochaeris glabra.*

In the genus *Crepis,* the dimorphic achenes present in several of the annual species appear to have a different value in connection with seed dispersal. In such species as *C. foetida, C. sancta, C. dioscoridis,* and *C. vesicaria* the marginal achenes differ from the inner ones in being nearly or entirely beakless, in having sometimes a reduced pappus, and in being more or less firmly enclosed in the inner bracts or phyllaries. They are not usually dispersed by the wind, but tend to remain in the involucre. There they are protected by the hard enclosing phyllaries and ensure the germination of seedlings in the site occupied by the parent plant. Occasionally the phyllaries may break away from the involucre and may be transported with their enclosed achenes by means of the adherence of their rough hairy outer surface to animals or man. The difference in this tribe between species having monomorphic achenes and those with dimorphic achenes is not merely for the convenience of the classifier, nor is it a random meaningless one. It represents differences in methods of seed dispersal. This, of course, does not determine the life or death of an individual or even of a species, but it does have a great effect on the ability of the species to spread and consequently to develop the geographic isolation which is the usual prelude to the formation of new species.

In the Gramineae or grass family, as in the Compositae, the parts of the individual flower vary little from one species or genus to another, and the great diversity is in the inflorescence, including the scales or bracts, known as glumes and lemmas, which envelop the flower and seed. And as in the Cichorieae, most of the differences in these structures are connected with different methods of seed dispersal.

A particularly significant fact, therefore, is that in many tribes or genera of this family there also exists an inverse correlation between the persistence and ability for vegetative reproduction of the plant and the degree of development of its mechanism for

TABLE 3

RELATION BETWEEN GROWTH HABIT AND SEED DISPERSAL MECHANISMS IN 215 SPECIES OF THE GRAMINEAE, TRIBE HORDEAE

	Rhizomatous perennials	Caespitose perennials	Annuals	Total
Awn shorter than lemma, glumes not elongate	48	59	6	113
Awn longer than lemma, glumes not elongate	4	56	7	67
Awn usually longer than lemma, glumes aristate, elongate	0	18	17	35
Total	52	133	30	215
Rachis continuous	52	115	3	170
Rachis fragile	0	18	27	45
Total	52	133	30	215

seed dispersal. In the tribe Hordeae, for instance, the species fall into three types on the basis of their habit of growth: rhizomatous, or sod-forming, perennials; caespitose perennials, or bunch grasses; and annuals. The two principal methods of efficient seed dispersal in the tribe are, first, the roughened awns or "beards" on the ends of the glumes and lemmas, which cling to various parts of animals, as well as to human clothing, and, second, the tendency of the entire inflorescence or head to break into pieces or to shatter at the nodes of its rachis. This either facilitates the transport of seeds by animals or, as in the squirreltail grasses (*Sitanion*) of the western United States, causes the joints with their clusters of elongate awns and fertile florets to be blown by the wind, scattering their seeds as they go, as in the well-known tumbleweeds.

The distribution of the different types of seed dispersal in relation to habit of growth in this tribe is shown in Table 3. This was compiled from the grass flora of the United States by Hitchcock, the flora of the U.S.S.R. by Komarov, and Boissier's *Flora Orientalis,* and therefore includes the two regions of the greatest concentration of species of this tribe, namely, the western United States and central to southwestern Asia. This table shows that the efficiency of seed dispersal increases in relation to the decrease in vegetative vigor. That this increase through the development of specialized structures has occurred independently in a number of different evolutionary lines in this tribe is shown both by morphological evidence and by genetic evidence. In the most highly developed of these lines, which culminates in the genus *Hordeum,* or barley, accessory awns for the more efficient dispersal of seed have developed through the sterilization of two of the three spikelets at each node of the spike or head and the reduction of their glumes and lemmas to prolonged awns (Fig. 19, 1A–1C). In another line, represented by the North American genus *Sitanion,* the awns are also much prolonged, so that the head has a "bearded" appearance like that of many species of *Hordeum,* but all the spikelets have remained fertile, and accessory awns have developed through the division of each of the sterile scales or glumes into several parts (Fig. 19, 2A–2C). The evidence from chromosome pairing in interspecific hybrids shows that *Sitanion* is much more closely related to species of *Elymus* and *Agropyron,* which lack these specializations, than it is to *Hordeum* (Stebbins, Valencia, and Valencia 1946a,b, Stebbins unpublished). A third line which has developed a similar type of specialization is represented by the Old World genus *Aegilops.* In this genus, however, the accessory awns are developed through the

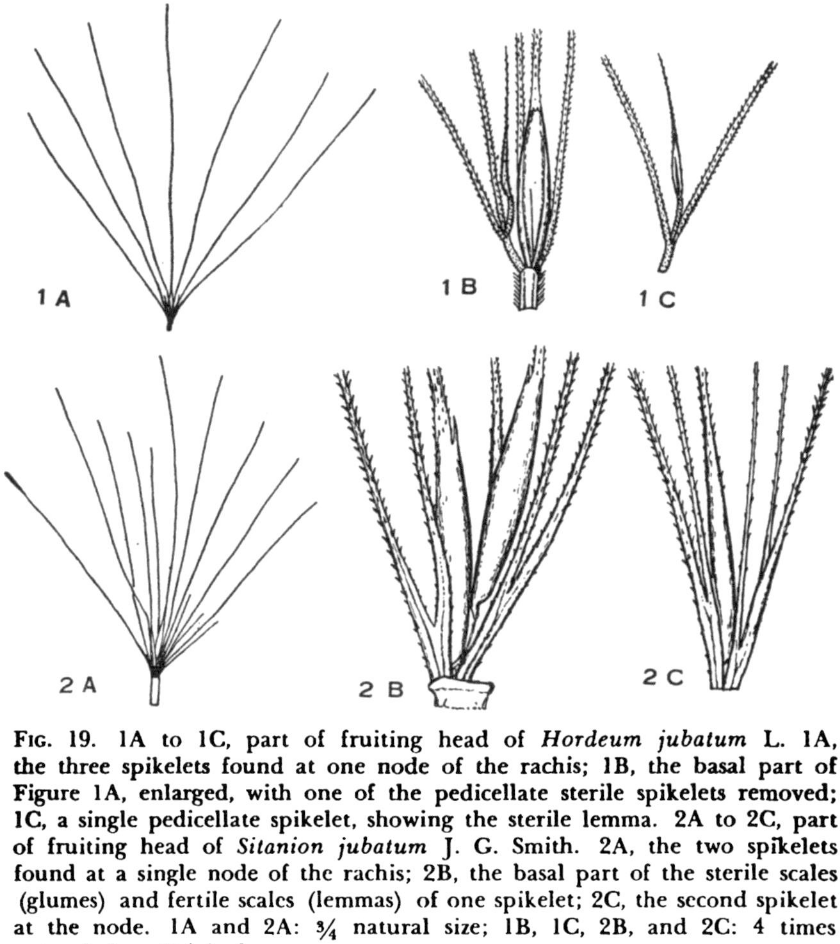

FIG. 19. 1A to 1C, part of fruiting head of *Hordeum jubatum* L. 1A, the three spikelets found at one node of the rachis; 1B, the basal part of Figure 1A, enlarged, with one of the pedicellate sterile spikelets removed; 1C, a single pedicellate spikelet, showing the sterile lemma. 2A to 2C, part of fruiting head of *Sitanion jubatum* J. G. Smith. 2A, the two spikelets found at a single node of the rachis; 2B, the basal part of the sterile scales (glumes) and fertile scales (lemmas) of one spikelet; 2C, the second spikelet at the node. 1A and 2A: ¾ natural size; 1B, 1C, 2B, and 2C: 4 times natural size. Original.

prolongation of the nerves of the sterile scales, or glumes, and also of the fertile scale, or lemma. Evidence from many interspecific hybrids (Kihara 1940, Aase 1946, Sears 1941a,b) has shown that *Aegilops* is very closely related to *Triticum,* and through it to various Old World species of *Agropyron* which completely lack awns of any kind.

In this group of grasses, therefore, as in the tribe Cichorieae, some of the most conspicuous of the reproductive characteristics used for the separation of species and genera consist of a series of different devices for the more efficient dispersal of seed, which have probably arisen in response to increased selection pressures placed upon the ancestral species by changes in its growth habit, its environment, or both. Some acquaintance with other species groups in the Compositae and in the Gramineae has convinced the writer that these are not isolated instances, but represent widespread trends. To be sure, there are undoubtedly both apparent and real exceptions to them in certain species and genera, but nevertheless the correlations hold strongly enough so that natural selection must have played a large part in the development of these groups.

NATURAL SELECTION AND MORPHOLOGICAL DIFFERENCES: SUMMARY

Based partly on the examples given above, we may now summarize the various ways in which the origin of differences in external morphology is related to the influence of natural selection.

First, the direct influence of selection, through the immediate adaptive value of the visible changes. This is true of most of the differences between climatic and edaphic ecotypes, as exemplified by *Potentilla* and *Achillea.*

Second, the origin of differences not adaptive of themselves, but developmentally correlated with adaptive characters. A glance at the examples of developmental correlation discussed on pages 87 to 89 will serve to show how far the effects of such correlation can extend. Furthermore, the particular effect of a gene which gives it an adaptive value is probably in many cases very different from the one which is easily seen and measured. An example of developmental correlation which by itself produces a disadvantageous effect is that between increasing seed size and decreasing number of seeds per pod in *Camelina.* Since this negative correlation is probably widespread throughout the higher plants, reduction in number of seeds per ovary, a very common tendency in the seed plants, may often have been caused by selection for increased seed size.

Third, characters not directly adaptive may acquire a selective value as part of a compensatory system of adaptation. Thus, in the origin of *Camelina linicola,* genes for increasing pod size acquired a selective value as a means of offsetting the effect of developmental correlation in reducing seed number along with selection for increased seed size. The development of more efficient means of seed dispersal in the tribe Hordeae may be conceived as a means of offsetting the disadvantages of reduced vegetative vigor and shortened life in the annuals.

Fourth, strictly nonadaptive characters may become established in small populations through the effects of chance. This possibility will be discussed in greater detail in the following section.

THE DYNAMICS OF SELECTION AND RANDOM VARIATION

The action of natural selection and its relation to other evolutionary processes will not be fully understood without some comprehension of the rate at which selection acts to change the frequency of genes and gene combinations in populations. The most important factor controlling this rate is obviously the particular selective advantage of individual genes and gene combinations, which has been expressed as the selection coefficient (Haldane 1932a, Wright 1940a, Dobzhansky 1941, p. 219). The relation between selection, mutation frequency, and gene frequency in populations of different sizes has been calculated by the use of mathematical formulae, which estimate the rate of evolution under various assumed values for mutation rate and the intensity of selection.

Some very important conclusions have been reached on the basis of these calculations. In the first place, the same intensity of selection will change the frequency of a gene in a population many times more rapidly when this frequency is in an intermediate range than when it is very low or very high (Haldane 1932a, Dobzhansky 1941, p. 220). This has placed on a quantitative basis the long-recognized fact that individual genes or gene combinations, if present in a very low concentration in large populations, will have difficulty becoming established even in the presence of favorable selection. Another conclusion is that if relatively low rates of adverse selection are operating, these may be partly counteracted by the effects of recurrent mutation, so that unfavorable genes may be present in small concentrations in populations. Finally, Wright has reached the conclusion that in small populations the effects of random fluctuation in gene frequency reduce considerably those of selection (Dobzhansky 1941, pp. 332–336).

The extent to which these conclusions can be applied to actual situations in natural populations depends primarily on three factors; first, the intensity of selection pressures which prevail at various times, second, the heritability, and, third, the sizes of interbreeding populations. At present, far too little is known about any of these factors. Recent experimental data, however, particularly those of Gershenson (1945) on the hamster and of Dobzhansky (Wright and Dobzhansky 1946, Dobzhansky 1947a,b) on inversion types in *Drosophila pseudoobscura* have demonstrated the existence as a normal phenomenon of selection coefficients many times higher than those postulated by any of the mathematical students of evolution. Selection coefficients under the normal seasonal fluctuations in the environment may range from low figures like 0.001 to very high ones like 0.8 (Gershenson 1945b). The latter figure indicates that under certain conditions the genotype selected against may be considered semilethal (in Gershenson's example this was the normal color form of the hamster as opposed to the melanic form), while the same data have shown that under other conditions the opposite genotype (the melanic hamster) may be equally semilethal. Obviously, therefore, our knowledge about the dynamics of selection will at present be most increased if we obtain actual data on the intensity of selection for various genotypes under different natural conditions.

The third factor, population size, is important chiefly in connection with the effects of random fluctuations in gene frequency. Numerous workers (cf. Dobzhansky 1941, pp. 161–165) have shown that while in infinitely large populations gene frequencies tend to remain constant except for the effects of mutation and selection pressure, in populations of finite size there is a gradual reduction in variability owing to chance fluctuations in gene frequency and random fixation of individual alleles. This phenomenon of "random fixation," "drift," or the "Sewall Wright effect" (Huxley 1942, p. 155) is undoubtedly the chief source of differences between populations, races, and species in nonadaptive characteristics. The extent to which it operates depends, of course, on the sizes of interbreeding populations and the length of time over which these sizes are maintained. As indicated in Chapter I (page 38), our knowledge of these sizes is at present very limited. Nevertheless, such information as we have indicates that in cross-breeding plants natural populations are rarely maintained for a sufficient number of generations at a size small enough to enable many of their distinctive characteristics to be due to random fixation. The formula developed by Wright (1931) states that in a population of N breeding individuals, the proportion of heterozygosity would decrease at the rate of $1/2n$ per generation. This would mean, for instance, that in a population with an effectively breeding size of 500 individuals and containing 200 heterozygous loci, one gene would be fixed or lost about every five generations, although this rate would be slower if the population occasionally received immigrants from outside. Completely isolated populations of this small size are rarely found in the higher plants and are still less often kept at such sizes for many generations. In woody plants and long-lived perennial herbs it is well-nigh impossible. Completely isolated populations of such plants, at least in temperate regions, nearly always contain at least several hundred individuals, as do the groves of the Monterey cypress, *Cupressus macrocarpa*, on the coast of California, and of the giant *Sequoia* in the Sierra Nevada. Since the usual length of a generation in such plants is measured in scores or hundreds of years, the time required for the fixation of even two or three genetic factors would often be measured in tens of thousands of years. In periods of this length the environment would undoubtedly undergo considerable variations, subjecting the population to different selection pressures, which would tend to outweigh the effects of random fixation. In the tropics, where the number of species is much larger and the individuals of a given species in a particular area correspondingly fewer, there would be more likelihood of the establishment of nonadaptive differences between races by means of random fixation. This would be particularly true of tropical oceanic islands, with their small size and diversity of ecological niches. As yet, no plant group has been studied on such islands with Wright's concepts in mind. But such systematic studies as those made in Hawaii, for instance, make very likely the supposition that in some groups of the higher plants, such as *Gouldia, Cyrtandra,* and *Bidens,* many nonadaptive differences have been established by random fixation, as seems almost certainly true in land snails, based on the data of Crampton, Gulick, and Welch.

The only plant example known to the writer in which the action of random fixation or drift seems to have taken a prominent part in the differentiation of isolated small populations is that of the complex of *Papaver alpinum* in the Swiss Alps, as described by Fabergé (1943). This species is confined to a series of highly specialized habitats, namely, talus or "scree" slopes of limestone or shale, and for this reason nearly all of its populations are very small and strongly isolated from each other. There is a remarkable amount of geographic variation, much of it in respect to flower color and the character of the pubescence of the scapes, in which alternate characters are distributed more or less at random and seem to be nonadaptive. Furthermore, in contrast to most interracial differences in other species, the character differences in *P. alpinum* are determined to a large extent by one or two pairs of genes. All these characteristics are what would be expected under the action of random fixation of genes.

Although the differentiation of continuously small, completely isolated populations through the action of random fixation of genes is probably a relatively uncommon phenomenon, several other situations occur which would strongly favor the establishment of nonadaptive differences by means of this process. The first of these is the occurrence of great fluctuations in population size. As Wright (1940a,b) has pointed out, the effective breeding size of such a fluctuating population is near the lower limit of its variation in numbers. Hence, if a population is periodically reduced to a few score of breeding individuals, random fixation would take place in it even if the normal size of the population were very much larger. Elton (1930, pp. 77–83) has given a graphic picture of how the completely nonadaptive "blue" mutant could become established in populations of arctic foxes by this method through periodic decimations of the population by famine, and Spencer (1947) has described an example in *Drosophila immigrans* in which the high frequency of a nonadaptive mutant gene is almost certainly to be explained on this basis. In the higher plants, however, this situation would arise relatively infrequently. Their populations are often greatly reduced by extreme cold, drought, or other types of catastrophe, but the plants which perish usually leave behind them numerous resistant, long-lived seeds. Therefore, the new population which is established upon the return of favorable conditions usually is derived from seeds of a considerable proportion of the original population, rather than seeds from the few last survivors.

An example which might be explained on the basis of long-continued reduction in population size producing *random association* of adaptive characteristics and selectively neutral characteristics is that described by Melchers (1939) in the genus *Hutchinsia* of the European Alps. Systematists have usually recognized in this small genus of the Cruciferae two subspecies in the Alps, *H. alpina* and *H. alpina* subsp. *brevicaulis.* These are distinguished by the shape of their petals, which are clawed in *H. alpina* and spatulate in subsp. *brevicaulis.* The two "species" (as he designated them) were demonstrated by Melchers to be interfertile, and he noted many intermediate forms where they occur together, thus supporting the opinion of most systematists that they are subspecies rather than species. There is no apparent adaptive significance, either direct or indirect, in the difference in petal shape. But in the Swiss Alps *H. alpina* and subsp. *brevicaulis* are separated by a difference which, though invisible, is vastly more important from the point of view of evolution and selection. The distribution of *H. alpina* is entirely on limestone, while subsp. *brevicaulis* grows on granite and other igneous rocks. Melchers showed that this distribution is due to genetically controlled differences in the physiological requirements of the two subspecies, since subsp. *brevicaulis,* when grown in artificial water cultures, will tolerate a much lower concentration of calcium than will *H. alpina.* In the Dolomite Mountains of the southeastern part of the Alpine region, on the other hand, there exists a form of *Hutchinsia* which agrees in petal shape with subsp. *brevicaulis,* but grows on limestone and has the relatively high calcium requirement in artificial culture that is characteristic of *H. alpina* (Fig. 20). Melchers found that the differences both in petal shape and in the requirement for calcium are governed by a relatively small number of genes, perhaps two or three. Furthermore, these genes segregate independently of each other, and there is no evidence for either developmental correlation or genetic linkage.

These facts could be explained on the assumption that at some time in the past, probably during the Pleistocene ice age, the Alpine populations were much reduced and consisted of three

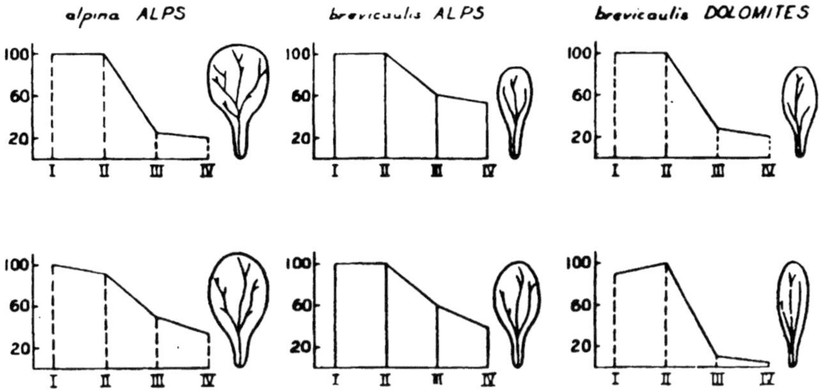

FIG. 20. Diagrams showing reaction to calcium as well as petal shape in six strains of *Hutchinsia*. Roman numerals represent culture solutions containing progressively decreasing concentrations of calcium, as follows: I: 0.000505 molar; II: 0.000253 molar; III: 0.000126 molar; IV: 0.000063 molar. Vertical lines represent percentage of normal growth achieved by the strain represented in the solution concerned. At the right of each diagram is drawn a petal of the strain represented by the diagram. At left, *H. alpina,* from Gschnitztal, northern Tirol. In the center, *H. alpina* subsp. *brevicaulis;* above, from Matreier Tauerntal, Austrian Alps, below, from Vikartal, south of Innsbruck, Austria. At right, subsp. *brevicaulis;* above, from Monte Cristallo, Dolomites, below, from Grödener Joch, Dolomites. After Melchers 1939.

relatively small and strongly isolated fragments, a central one on igneous rock and a northern as well as a southeastern one on limestone. The northern one on limestone acquired by random fixation a high concentration of the gene for clawed petals, while the central and the southeastern one, the former on igneous rock and the latter on dolomitic limestone, both acquired the gene for spatulate petals either by random fixation or through the survival of the original condition of the species. Then, when more favorable conditions allowed the spread of *Hutchinsia* into the glaciated territory, the northern population spread quickly into all of the limestone areas adjacent to it, carrying along the genotype for clawed petals, while the other two populations similarly carried with them into all of the regions available the genotype for spatulate petals.

Another situation in which random fixation and random association of unrelated characters may be expected is during the advance of a species into a territory newly opened to colonization.

Timofeeff-Ressovsky (1940) has pointed out that such an advance will not necessarily be a regular progression of a large population, but that isolated "advance guards" of the incoming population may at first establish small populations, which will then spread and merge with each other. If, for instance, we consider the colonization by plants of the regions left bare by the retreating Pleistocene ice sheets, we realize that this terrain was far from even. There must have been many small pockets with particularly good soil conditions or with protection from winter storms. In such places, some of them perhaps scores of miles north of the southern limit of glaciation, the seeds of a few plants would be transported by wind or animals, and the hardiest of the seedlings coming from these seeds would form the first small plant populations. The sites most favorable for a particular species might be many miles away from each other, so that the species would exist at first in the form of small, isolated populations. These would be subject to rigid selection for certain characteristics, but others, like flower color and petal shape, might be unaffected by selection. Their nature in each newly established population would be determined by the genetic constitution which the ancestors of the population happened to possess. As conditions became more favorable, the small populations would grow and merge, always maintaining their original characteristics. The final result would be a new population which in certain characteristics showed strict adaptation to various environmental conditions, but in others was characterized by regional variation of a purely nonadaptive character. This nonadaptive pattern of variation might be retained in the population for a long time or even indefinitely by means of "isolation by distance."

In *Linanthus parryae,* a small annual species of the Mohave Desert in Southern California, Epling and Dobzhansky (1942) have given an example which may be explained by the past history of the population, as well as by the action of random fixation through isolation by distance. In favorable years the species forms within the area studied a continuous population over an area about seventy miles long and five to fifteen miles broad. Most of the plants in this population have white flowers, but in three distinct areas occur varying concentrations of plants with blue flowers. Within these areas, however, the frequency distribution varies rather irregularly. Samples containing all blue plants were sometimes found only a mile or less from those with no blues (Figure 21). Summation of all the samples showed a preponderance of those with 0 and with 100 percent of blues and fewer with inter-

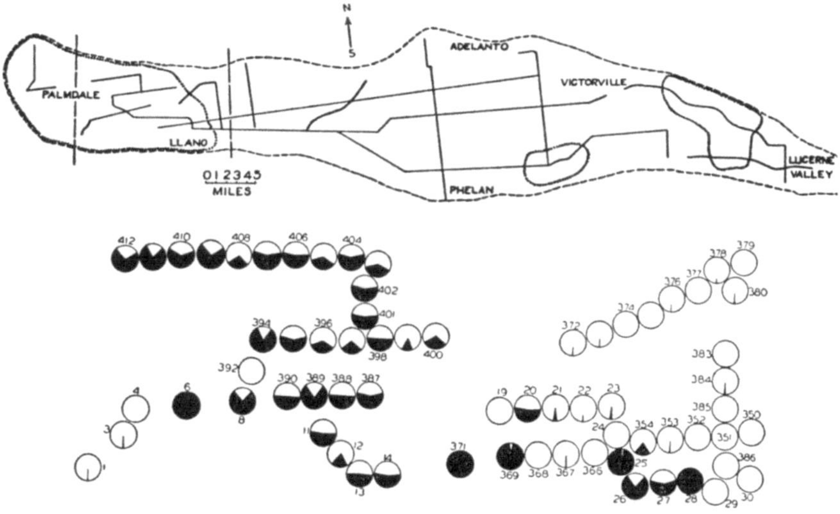

FIG. 21. Above, map showing the distribution of *Linanthus parryae* in Southern California (broken line), the location of the three areas in which blue-flowered plants are found (dotted lines), and the routes taken for sampling the population (continuous lines). Below, enlarged diagram of a part of the western end of the area shown in the map above (that bounded on the map by two vertical broken lines), showing the relative frequency of white- and blue-flowered plants at each collection station in that area. Numbers are those arbitrarily assigned to the stations. Black sectors, blue flowers; white sectors, white flowers. From Epling and Dobzhansky 1942.

mediate percentages, a result which would be expected on the basis of random fixation. A likely explanation is that this population has in past successions of unfavorable years been broken up for many generations into isolated smaller populations, and that the survivors happened to have white flowers in most of these, but blue ones in a few of them. Isolation by distance, resulting from the fact that most plants receive pollen from their immediate neighbors, has kept the random pattern of variation in the population intact through the years when it has been large and continuous.

A final way in which random variations could become established in populations is through restriction of cross-breeding by the plants themselves. Many plant species are largely self-pollinated, but occasionally undergo crossing between selfed lines. The effects of such a genetic system, as Wright (1940b) has pointed out, will be the same as those of great fluctuations in size in a cross-breeding population. The further implications of this and other such systems are discussed in the following chapter.

The material presented in this chapter is intended to show that individual variation, in the form of mutation (in the broadest sense) and gene recombination, exists in all populations; and that the molding of this raw material into variation on the level of populations, by means of natural selection, fluctuation in population size, random fixation, and isolation is sufficient to account for all the differences, both adaptive and nonadaptive, which exist between related races and species. In other words, we do not need to seek unknown causes or motivating agencies for the evolution going on at present. And the differences between genera, families, orders, and higher groups of organisms, as discussed in Chapter XIII, are similar enough to interspecific differences so that we need only to project the action of these same known processes into long periods of time to account for all of evolution. The problem of the evolutionist is no longer that of finding unknown causes for evolutionary progress or direction, but of evaluating on the basis of all available evidence the role which each of the known forces has played in any particular evolutionary line. In the future, new generalizations about evolution will come chiefly from comparisons made between specific examples throughout the plant and animal kingdoms.

CHAPTER V

Genetic Systems as Factors in Evolution

In the preceding chapter the fact was emphasized that in the higher organisms under discussion selection acts primarily on gene combinations rather than on single mutations. This fact inevitably carries with it the corollary that as a force in determining the rate and direction of evolution, recombination is of equal or greater importance than mutation. For this reason, the various factors which influence the rate and nature of recombination must be given separate consideration. These factors are of two sorts, external and internal. The external factors affecting recombination, which consist mainly of the size and structure of the species population, were discussed briefly in the last section of the preceding chapter.

The internal factors together constitute the *genetic system*, as recognized by Darlington (1939, 1940) and Huxley (1942, p. 126). That the chromosomal machinery is only a part of the genetic system, while the type of reproduction — whether sexual or parthenogenetic, with self- or cross-fertilization, and so forth — is equally or more important, was pointed out by Huxley. Both Darlington and Huxley emphasize the fact that the genetic systems of different organisms differ widely from each other and that selection acts to maintain or alter this system, just as it acts on the characteristics of external morphology. We may therefore think of the evolution of genetic systems as a course of evolution which, although running parallel to and closely integrated with the evolution of form and function, is nevertheless separate enough to be studied by itself. In plants, a study of this thread of evolution is particularly important since, as Anderson (1937a) and Turrill (1942a) have pointed out, the diversity of the genetic system in the plant kingdom is much greater than that found among animals.

As Huxley (1942) has indicated, comparative genetics and comparative evolution are for the most part sciences for the future, and most statements concerning them must at present be largely hypothetical. Nevertheless, the gathering together of facts and opinions on these subjects is desirable at this early date, even if only to form a basis for future discussion and for the design of critical experiments. Viewed in this light, the last chapter of Darlington's book *The Evolution of Genetic Systems* stands out as a pioneering effort in which the most important principles of these sciences have already been stated.

The first of these principles, which brings out the importance of recombination in relation to the genetic system, is given on page 127 of this book, as follows: (The italics are those of the present author.)

> In a word, the unit of variation is not the unit of selection. Changes in the chromosomes are determined by conditions of molecular stability. They are biologically at random. *The combinations of these changes together with the selection of environments is what takes us from the chemical level of mutation to the biological level of adaptation.*

Next in importance is the principle of compromise, which, as Darlington points out, is a "corollary of integration in the genetic system." The most important compromise is that between the need for constancy as opposed to that for regulated change. The latter, as Cuénot (1911) and Goldschmidt (1933, 1940), as well as Darlington, have pointed out, involves the need for the formation of preadapted gene combinations. This need for preadaptation requires that all species populations of the more complex organisms should include individuals which are not perfectly adapted to the immediate contemporary environment of the species, but may be adapted to new environments to which the species will become exposed. As shown by Mather (1943), immediate selective advantage is gained by the *fitness* of the particular individuals exposed to the selective forces, but the survival of the evolutionary line in a changing environment depends on its *flexibility*, expressed in terms of its ability to produce new gene combinations at a definite rate per unit of time. In all evolutionary lines, a compromise is necessary between the conflicting requirements of fitness and flexibility. Furthermore, no one level of compromise is equally advantageous for all types of organisms. Depending on the length of their life cycle, the complexity of their development, and other factors, different groups have established compromises at very different levels, some in the direction of maximum stability and fitness at the expense of flexibility and others with immediate fitness of all offspring sacrificed to the maintenance of flexibility.

The third principle established by Darlington is that of anticipation. This is similar to preadaptation as defined by Cuénot and Goldschmidt, but emphasis must be placed on the fact that the selective advantage of a particular genetic system benefits, not the individual in which it arises, but its posterity. Whether an individual produces offspring of many different types or all similar makes no difference to the organism itself, but may have a profound effect on the survival of the race. Hence, the evolution of genetic systems involves competition, not between individuals, but between evolutionary lines. The time factor is therefore of vastly greater importance than it is in the evolution of morphological and physiological adaptations. The various mechanisms composing the genetic system must have arisen before they possessed selective value as members of this system. Either they arose by chance, or else they possessed at the beginning a different use from their present one.

Another important point about the genetic system is that it affects not only the survival of a group of organisms but also their capacity for evolutionary change. In many cases organisms may survive over long periods of time even if their genetic system is ill-adapted to their evolutionary progress. Such species are doomed to conservatism and are tied to the habitats in which they originally existed, but even with such restrictions they may survive for millions of years. Hence, in discussing the selective value of genetic systems we must consider primarily the advantages a particular system gives to the progeny of those who have it and the evolutionary possibilities which it holds out to the race. The immediate advantages or disadvantages of the system are of secondary importance.

Before a survey can be made of the genetic systems adopted by organisms of different types, the concept of flexibility must be defined more explicitly. It may be thought of as evolutionary opportunism and expressed in terms of the number of gene combinations which a population occupying a given area can form in a given unit of time. The latter must be expressed chronologically rather than in terms of generations of the organism concerned, since alterations of the external environment are largely independent of the character of the life cycle of the organism undergoing selection. Furthermore, the area which a population occupies is more important than its numbers, since competition restricts environments spatially and with reference to the available food supply. If flexibility is defined in these terms, then the size of the organism and the length of its generations are among the major factors affecting this quality. Small organisms with short life cycles will possess a considerable degree of flexibility simply on account of these characteristics, while in large, slowly developing ones other factors favoring flexibility will have a correspondingly higher selective value. These factors consist chiefly of the methods of reproduction, in so far as they affect cross-fertilization and panmixy. Mutation rate, however, is by no means a negligible factor, particularly in small, rapidly breeding organisms.

MUTATION AND EVOLUTION IN ASEXUAL ORGANISMS

The genetic system most strongly promoting immediate fitness at the expense of flexibility is that in which sex is absent. This is found as an occasional aberration in higher plants, which will be discussed in Chapter X, on apomixis, and occurs similarly among animals in some phyla of Metazoa. It is, however, much more widespread among the thallophytes, both algae and fungi, and in the Protozoa.

Biologists have commonly believed that two classes of organisms, the bacteria and the Cyanophyceae or blue-green algae, are universally and primitively asexual (Fritsch 1935, Copeland 1938). This belief has been based on the facts that no complete sexual cycle has as yet been observed in any of these organisms and that their nuclear substance seems to be relatively poorly organized and incapable of undergoing the precise type of mitotic division found in the higher organisms. Recently, however, both genetic and morphological evidence for the existence of sex in some bacteria has been brought forward (Lederburg and Tatum

1946, 1947, Lederburg 1947, Dienes 1947). How far our concepts concerning these organisms will have to be revised is not yet clear, but the assumption is still fairly safe that sexual fusion is very rare or absent in many, if not the great majority, of bacteria.

In other lower organisms, such as the flagellate Protozoa, many of the unicellular green algae, some multicellular algae (such as the golden greens or Chrysophyceae), and most of those fungi classed as *fungi imperfecti,* sexual reproduction is unknown or extremely rare. And as we ascend to more complex plants, such as lichens and bryophytes, we find many examples of species in which the sexual cycle is well known, but asexual reproduction by means of various types of vegetative propagules is far commoner than the sexual method. There is no doubt that asexual reproduction is far more widespread in plants than it is in animals and that in the nonvascular plants its frequency is inversely correlated with the complexity of the vegetative plant body.

HAPLOIDY AND THE EVOLUTION OF DIPLOIDY

The most striking difference between the sexual cycle of animals and those found in plants is that, with the exception of a few Protozoa, animals are diploid at all stages and their only haploid cells are the gametes themselves; while nearly all plants possess a haploid stage of greater or less duration. Furthermore, the sequence of types of alternation of generations, involving the gradual increase in emphasis on the diploid generation, is one of the best-known features of plant evolution. The striking fact is that in each major phylogenetic line of plants this increase in diploidy has occurred independently, in every case correlated with the increase in complexity of the plant body.

The simplest nucleate organisms, the flagellates, at least those with sexual reproduction, are strictly haploid except for the zygote. The same is true of most of the filamentous algae, green and brown, and most of the fungi with the least differentiated plant bodies, namely, the Phycomycetes.

The diploid generation has undoubtedly evolved independently many different times, apparently through some physiological adjustment changing the first division of the zygote from meiosis into a diploid mitosis (Svedelius 1929, Kylin 1938). The opposing theories of plant morphologists as to whether this generation first appeared as an entirely new set of vegetative structures interpolated into the life cycle of the plant (Bower 1935, Chap. XXIV) or whether a series of existing morphological structures merely became transferred from the haploid to the diploid condition by a delay in the timing of meiosis (Svedelius 1929) will be discussed later. It may be remarked here, however, that much recent experimental evidence, which is discussed in Chapter VIII, has shown that the transformation of tissues or organs from the haploid to the diploid or polyploid condition never produces a radical change in their external morphology, so that the mere formation of a diploid generation could not give that generation a distinctive external appearance unless mutations should occur which affected it alone.

In most unicellular organisms, both plant and animal, the life cycle is either entirely haploid except for the zygote or strictly diploid except for the gametes. This indicates that the change from haploidy to diploidy in these organisms took place in a single step, without the intermediate condition of an alternation between haploid and diploid generations, such as that existing in many higher plants. There is no doubt that the diploid Protozoa and the unicellular algae known to be diploid are in general more complex in structure than haploid unicellular organisms. The ciliate Protozoa or Infusoria have particularly large and structurally complex cells, while the diatoms, which are diploid, are probably the most complex and structurally diversified of the unicellular algae. In both groups, the independent origin of diploidy accompanying an increase in structural complexity seems to be a likely hypothesis.

Among the groups with an alternation of generations, three different types can be recognized: those with similar haploid and diploid generations (*Ulva, Dictyota, Polysiphonia*), those with a relatively small diploid or sporophytic stage (*Cutleria, Nemalion* and other red algae, Bryophyta, Ascomycetes), and those with the sporophyte large and conspicuous (Laminariales and many other brown algae, Basidiomycetes, vascular plants). The first type is found only among the algae. Among the algae showing this type are filamentous forms, like *Ectocarpus,* as well as forms with more complex thalli, like *Polysiphonia,* but most of these algae have a more complex vegetative structure than the algae which are strictly or predominantly haploid. The existence of these forms with morphologically identical haploid and diploid generations is perhaps the best evidence in favor of an explanation of the evolution of diploidy based primarily on the genetic properties of the diploid condition. In mosses, ferns, seed plants, and even in the predominantly diploid brown algae, the origin of the diploid sporophyte might be interpreted on the assumption that this generation, because of its new structural characteristics, performed functions of value to the plant which could not be carried out by the haploid gametophyte. But in the case of such algae as *Cladophora* and *Polysiphonia* such an interpretation is impossible, since the sporophyte, except for its reproductive characteristics, is structurally and functionally equivalent to the gametophyte. The establishment of a mutation or mutations producing this identical diploid generation must be based on the existence of a selective advantage for this generation, simply and solely because of its diploid condition. And if this advantage does not result from its structural or physiological properties, it must reside in the genetical properties inherent in the diploid sporophyte. That this sporophyte could have evolved and become established independently in a number of different groups purely by the chance accumulation and establishment of mutations is highly improbable.

In those groups with a prominent gametophyte and a reduced sporophyte, not only is the plant as a whole more complex than that in strictly haploid groups, but in addition the sporophyte usually includes the most complex and diversified structures of the plant body. The gametophyte of the mosses and the liverworts is undoubtedly more complex than the plant body of the haploid green algae, from which they are probably descended (Fritsch 1916). But at least in mosses, the elaborate structures of the diploid spore capsule are more complex than any part of the gametophyte, and it is these diploid structures that are sufficiently diverse to provide diagnostic characteristics for families, genera, and in some cases species.

A life cycle with the diploid generation predominant or exclusively present has appeared at least three times in the evolution of multicellular plants: in the order Siphonales of the green algae, in the brown algae, and in the vascular plants. Although these three groups are as different from each other as plants can be in their structure and physiology, they have one characteristic in common: each group contains on the average larger and more complex plants than do the nearest relatives of that group.

This fact is least obvious in the Siphonales, but such genera as *Codium* and *Caulerpa* certainly have among the most complex plant bodies of any green algae. The Laminariales and the Fucales, on the other hand, contain unquestionably the largest and most complex plant bodies of all the algae. In fact, such genera as *Laminaria, Nereocystis,* and *Postelsia* are the only plants outside of the Tracheophyta or "vascular plants" which have a marked differentiation of tissues, particularly within their stalks.

In the vascular plants, the increasing dominance of the diploid sporophyte in groups successively better adapted to life on land formerly led many botanists to think of adaptation to land life as the direct cause of the rise of this sporophyte (Bower 1935, Pincher 1937). Now, however, the well-known diploid condition in the strictly aquatic algae mentioned above, as well as the haploid state of the vegetative body of such strictly terrestrial forms as most Ascomycetes and in particular lichens, shows us clearly that the causal relationship between diploidy and life on land in the vascular plants is an indirect, not a direct, one. This connection is through the fact that a plant cannot attain any appreciable size on land unless it has a relatively complex vegetative body, equipped with conducting, supporting, and storage tissues.

In the fungi, comparison between the ascomycetes and the basidiomycetes gives us just as good an example of increase in size and complexity following the appearance of a condition similar to diploidy as does comparison of the bryophytes and vascular plants among the archegoniates. In most fungi, however, "diploidy" occurs in an anomalous form, the dikaryotic condition, in which the parental haploid nuclei do not fuse, but remain side by side in each cell and divide simultaneously as new cells are formed. As Buller (1941) has pointed out, the genetical properties, and consequently the evolutionary possibilities, of the dikaryotic diploid cell are very similar to those of the synkaryotic cell. The dikaryotic condition is advantageous in the fungi chiefly because, as Buller (1941) has shown, it is the most efficient way of fertilizing and converting to diploidy fully developed mycelia.

The mycelium, or vegetative portion of the plant body, in the Basidiomycetes differs from that in most Ascomycetes only in its prolonged life in the dikaryotic condition. The increase in prominence of the dikaryotic diploid generation in this group is associated with the increase in size and complexity of the basidiomycete reproductive body. As a matter of fact, the elaboration of these large, relatively complex structures, such as the mushroom of the Agaricaceae and the perennial "bracket fungus" of the Polyporaceae, was probably made possible by a gradual increase in size and longevity of the mycelia from which they arise.

HETEROKARYOSIS AS A GENETIC SYSTEM

In certain fungi, particularly the Ascomycetes and their asexual derivatives classified as *fungi imperfecti,* the plant body may contain nuclei of two different types and may therefore exhibit segregation by means of a series of processes not involving sexual fusion at all. This is through the fusion of hyphae belonging to different mycelia and the passage of nuclei from one mycelium to another, regardless of the sex or mating type of the mycelium, a phenomenon known as *heterokaryosis.* The implications of its existence have been realized by mycologists for some time (Brierley 1929), but it was first clearly demonstrated by Hansen and Smith (1932) in the imperfect fungus *Botrytis cinerea.* When single-spore cultures were isolated repeatedly, three different types of mycelium were obtained. One of these produced normal conidia, or asexual spores, in abundance, and mycelia from these conidia were uniformly conidial. A second type produced extensive mycelial growth, but very few conidia, and remained constant like the first type. The third type was intermediate in its amount of conidia formation and continued to segregate conidial, mycelial, and intermediate types. When conidial and mycelial strains were grown in mixed cultures, isolates from such cultures produced all three types. The authors observed frequent anastomosis of hyphae in their cultures and concluded that the intermediate type of mycelium owed its peculiar properties to the presence in the coenocytic hyphae of two genetically different types of nuclei, one capable of producing the conidial strain and the other the mycelial type of strain. The coenocytic character of these fungi makes it possible for many nuclei to exist side by side in the same cytoplasm and to pass from one end of the mycelium to the other with the streaming of the protoplasm. Although cross walls are present, they are perforated and serve only to strengthen the hyphae.

These observations were extended by Hansen and Smith (1935) and Hansen (1938) to a large number of different species of imperfect fungi, including thirty genera. In some species, the phenomenon was found in only a small proportion of the strains analyzed, but in others, such as *Phoma terrestris,* it is apparently always present. Hansen concludes that in these coenocytic organisms the nucleus, rather than the cell, must be considered the basic unit of the individual. A multinucleate spore is analogous to a group of individuals in other organisms, and the "segregation" observed in cultures derived from conidia of the intermediate type strains is actually the separation of individual genotypes from a mixed colony enclosed in a single cell wall. But, as will be pointed out below, this concept cannot be strictly held to, since in respect to its physiological properties, a heterokaryon containing two different types of nuclei behaves like a genetic heterozygote.

The nature of the heterokaryotic condition has been greatly clarified by observations on the genus *Neurospora,* in which the true genetic behavior connected with the sexual cycle is well known, and in which many mutants, both morphological and biochemical, have been isolated, as well as produced artificially. Lindegren (1934) found that certain mycelia of *N. crassa* contained both (+) and (−) nuclei, but were unable to produce perithecia and ascospores because of the presence of self-incompatibility ("self-sterility") factors. Dodge (1942) produced in *N. tetrasperma* heterokaryons between an X-ray induced dwarf mutant and a number of normal, though slow growing, strains, both of the same sex and of the opposite sex. These showed conspicuously greater growth than either of the homokaryotic strains from which they were obtained, thus demonstrating the condition of heterokaryotic vigor, which is entirely analogous to hybrid vigor or heterosis in genetic heterozygotes. Beadle and Coonradt (1944) showed that heterokaryons containing nuclei which bear mutations for the inability to synthesize specific growth factors plus other nuclei bearing their wild-type alleles grow normally in the absence of these growth factors, and that the heterokaryon is maintained by natural selection in a medium not containing these factors. They concluded, therefore, that dominance is exhibited in the relationships of two allelomorphs contained in the different haploid nuclei of a heterokaryon, just as it is in typical nuclei of other organisms. Pontecorvo (1947) has described heterokaryons in *Penicillium* and *Aspergillus,* and has shown that when the uninucleate conidiospores are produced in the latter genus, they may in the case of some color mutants assume the color determined by the joint action of the two genes present in the previous heterokaryon, and in the case of other such mutants the color may be determined by the gene present in the single nucleus which has become segregated into the conidial hypha ("autonomous" gene action).

Data of considerable evolutionary significance have been obtained in respect to the relative frequency of the two types of nuclei in a heterokaryotic mycelium. Pontecorvo (1947) found that in a heterokaryon between two X-ray mutant strains of *Aspergillus oryzae* the ratio between the two types of nuclei was relatively constant, fluctuating between 1:2.7 and 1:3.1. Beadle and Coonradt (1944), on the other hand, found that the ratio of two types of mutant nuclei in different isolates of the same heterokaryotic combination of *Neurospora crassa* varied in one example between 1:1.6 and 1:17.6, and in another between 1:2.2 and 5:1, but that these variations had no effect on the growth rate of the heterokaryon. Finally, Ryan and Lederburg (1946) and Ryan (1947) found that in *N. crassa* heterokaryons between a normal strain and one containing a mutant unable to synthesize the amino acid leucine, although able to grow normally on a minimal medium without leucine, showed depressed growth in a medium containing a limiting concentration of leucine. They demonstrated that this is due to the fact that in the presence of leucine, selection occurs in favor of the leucineless nuclei, and those parts of the mycelia which have grown the length of a test tube on such a medium contain only nuclei of the latter type. This shows that heterokaryotic mycelia have one important property of populations of individuals: their genetic composition can be altered by natural selection.

The evolutionary possibilities of the heterokaryon have been discussed by Lindegren (1942), Beadle and Coonradt (1944), Sansome (1946), and Pontecorvo (1947). Since dominance relations hold here as in diploid plants, recessive genes can persist indefinitely in populations, and the possibilities for recombination are on genetic grounds as great as they are in diploid organisms. Heterokaryons thus have the flexibility of diploids. But they have also the immediate fitness of haploids, since the heterokaryon can at any time be resolved into its haploid components, provided these are adapted to the surrounding medium, and furthermore the mycelium can become homokaryotic by natural selection. From the genetic and physiological point of view, therefore, heterokaryons have all the advantages of both haploid and diploid organisms, a condition amply shown by the number, diversity, and ubiquitousness of saprophytic and pathogenic species of Ascomycetes and *fungi imperfecti.* But Lindegren has pointed out that heterokaryons are at a disadvantage when it comes to the formation of complex morphological structures, since the presence of mixtures of nuclei militates against integration of a definite series of genetically controlled growth processes, which is necessary for the development of such structures. It is perhaps significant in this connection that in 30 isolates representing four genera of Ascomycetes, all of which are regularly sexual and most of which form relatively complex fruiting bodies, Hansen (1938) found all to be homokaryotic.

Beadle and Coonradt (1944) have suggested that heterokaryosis might represent a stage in the evolution of sexual reproduction, although they state clearly that the fungi could hardly represent the group intermediate between asexual and sexual organisms. But Sansome (1946), who found that in *Neurospora crassa* heterokaryons between mycelia of opposite sex are difficult or impossible to obtain, concluded that the heterokaryotic condition could not have been a step in the evolution of the heterothallic sexual system, since the sexual, or "mating type," factors tend to suppress heterokaryosis.

Lindegren (1942) has pointed out that the dikaryotic condition existing in the Basidiomycetes, in which each cell contains just two nuclei of opposite sex, approaches the diploid condition of the higher plants and animals in that cells of this type can cooperate to produce the integrated development of complex morphological structures. But a conspicuous difference is in the possibilities for competition between nuclei of the same sex. In diploid organisms, the zygote and the resulting individual must be entirely the product of a single fusion between two particular haploid nuclei. But in a dikaryotic mycelium, fusion with other mycelia of different genetic constitution can occur at any time, and one of the nuclei of the original dikaryon may be replaced by "invading" nuclei of the same sex but genetically different. Thus, the vegetative mycelium of a dikaryon possesses some of the flexibility afforded to the heterokaryon by natural selection between nuclei of the same individual mycelium. Fungi, therefore, possess two genetic systems which are *sui generis* and are adapted to their particular method of vegetative growth. These systems form most interesting intermediate conditions between haploidy and diploidy.

CROSS-FERTILIZATION AND SELF-FERTILIZATION IN THE HIGHER PLANTS

In nearly all animals, cross-fertilization is necessary. Although some of them are hermaphroditic, the great majority of these are incapable of fertilizing themselves. In plants, on the other hand, the hermaphroditic condition is the usual one, and species which regularly fertilize themselves are by no means uncommon. But in nearly every major subdivision of the plant kingdom some species require cross-fertilization, and if self-fertilized produce either no offspring or else weak and degenerate ones, while other species, often closely related, are regularly self-fertilized and seem to suffer no ill effects from this continued close inbreeding. This fact is particularly obvious among higher plants in the grass family, in which a relatively large number of species has been studied genetically (Beddows 1931, Jenkin 1931, 1933, 1936, Nilsson 1933, Smith 1944). The fact is well known that cereal rye (*Secale cereale*) is normally cross-pollinated, is highly self-incompatible, and produces weak and abnormal offspring when forcibly inbred, while wheat (*Triticum* spp.), which is so closely related to rye that the two can be intercrossed, is normally self-pollinated, so that the older commercial varieties of this crop are almost completely homozygous when kept pure. Similar examples may be found among many wild species of this and other families of angiosperms. They might be explained purely on the basis of chance, that is, random fixation of genes for self-incompatibility in some species and for self-fertilization in others. In that case, however, the two types of reproduction should be distributed at random with respect to other characteristics of the species concerned. This is by no means the case. To illustrate this point, Table 4 has been prepared, showing the degree of cross- or self-pollination in the species found in the United States of the genera *Bromus, Festuca, Lolium,* and the tribe Hordeae as compared with their habit of growth and the length of their life cycle.

TABLE 4

TYPE OF FERTILIZATION AND GROWTH HABIT OF CERTAIN GRAMINEAE IN THE GENERA *Bromus*, *Festuca*, *Lolium*, *Elymus*, *Agropyron*, *Sitanion*, AND *Hordeum*

	NUMBER OF SPECIES			
Type of fertilization	Rhizomatous perennials	Caespitose perennials	Annuals	Total
I. Plants more or less self-incompatible, predominantly cross-fertilized	13	13	0	26
II. Plants self-compatible, but flowers chasmogamous and cross-fertilization frequent	1	39	3	43
III. Plants self-compatible, flowers often or always cleistogamous, normally self-pollinated	0	5	27	32
Total	14	57	30	101

These groups have been selected merely because they are better known to the writer than other grasses. Some familiarity with other genera throughout the family indicates that many of them would have served equally well.

The following generalizations may be made from this table and a study of the distribution of the species included. In the first place, as is well known for angiosperms as a whole (Raunkiaer 1934), annual species are relatively uncommon in cool temperate regions and predominant in warm, dry ones, with seasonal rainfall. Second, while the perennial species may be either cross- or self-fertilized, depending on the species, the annuals are almost exclusively self-pollinated. Third, among the perennial species those with rhizomes are almost exclusively self-incompatible and cross-fertilized, while the caespitose, "bunch-grass" types show various degrees of self- or cross-fertilization. These last two generalizations hold with few exceptions throughout the grass family. Finally, among the caespitose perennials, the 44 species inhabiting cool temperate regions include 12 (27 percent) self-incompatible cross-fertilizers, while among the 12 species of hot, dry climates with winter rainfall only one (8 percent) is of this type. This relationship becomes more evident when other groups of grasses are considered. For instance, such familiar caespitose grasses of cool temperate regions as *Dactylis glomerata, Phleum pratense,* and many perennial species of *Poa* are self-incompatible, while the xerophytic "bunch grasses" in such genera as *Stipa, Aristida,* and *Triodia* are largely self-fertilized.

THE CHROMOSOMAL MECHANISM AND THE GENETIC SYSTEM

The three primary elements of the genetic system which have been discussed already are the presence or absence of sex, haploidy vs. diploidy, and cross-fertilization vs. self-fertilization. The fourth and final one of these elements is the chromosomal apparatus, particularly the number of chromosomes and their behavior at meiosis. Darlington (1939, p. 77) has pointed out that the amount of genetic recombination in any particular intermating group is determined by the chromosome number of the species and by the amount of crossing over in each chromosome, which is determined in turn by the chiasma frequency. He has therefore suggested that the "sum of the haploid number of chromosomes and of the average chiasma frequency of all the chromosomes in a meiotic cell" could be considered as a *recombination index.*

It is obvious that the larger the recombination index, the greater is the number of new gene combinations which can be formed by segregation and recombination *in a limited number of generations.* The smaller this index, the greater is the number of generations over which a given favorable gene combination could be expected to persist. A high recombination index, therefore, promotes flexibility, and a low one fitness over a short time.

The factors making up the recombination index have not been studied carefully or extensively in any of the cryptogams except for a few isolated species of fungi. The pattern of distribution of these factors, therefore, is known only in certain groups of animals and seed plants. Here, however, evidence is available that this distribution is not a random one, but, like the distribution of sex, diploidy, and self- or cross-fertilization, is correlated with various characteristics of the growth and longevity of the plants themselves.

The writer has noted previously (Stebbins 1938b) that among the flowering plants or angiosperms the basic haploid number of woody genera, regardless of their phylogenetic position, is significantly higher than that of herbaceous genera. This suggests that among long-lived plants, a high recombination index is favored. On the other hand, no significant difference was found between the basic chromosome number of herbaceous perennials and that of annuals. This apparently contradictory evidence can be largely explained by the fact mentioned above that a large proportion of annual species are predominantly self-pollinated. The relation between self-pollination and the recombination index will be discussed below. Evidence that within strictly cross-pollinated groups annual species tend to have lower recombination indices than perennial ones is difficult to obtain at present, but certain data from the tribe Cichorieae of the family Compositae are suggestive. The fact that the annual species of *Crepis* have on the average lower chromosome numbers ($x = 4$ and 3) than the perennial ones is now well known (Babcock and Cameron 1934, Babcock and Jenkins 1943). Furthermore, cytological studies of annual species, as well as closely related perennial species, of *Crepis* have shown that they all have relatively low chiasma frequencies (Koller 1935, Richardson 1935, Jenkins 1939, Tobgy 1943). Such annual species as the three-paired *C. capillaris* and *C. fuliginosa,* therefore, have the lowest chiasma frequencies which the genetic mechanism is capable of producing. In such strictly perennial genera as *Prenanthes, Hieracium,* and *Taraxacum,* on the other hand, the basic chromosome numbers are consistently $x = 9$ or 8, and such evidence as is available (Gustafsson 1932, 1935a) indicates that the chiasma frequencies of the sexual species are also relatively high. Furthermore, in *Lactuca* and

Sonchus there are many annual species with haploid numbers of 8 or 9 or even (as in *S. oleraceus*) with polyploid multiples of these. Many of these species, however, have been found to be self-compatible, so that in them constancy can be secured by self-pollination and homozygosity. The possibility exists, therefore, that most annual species possess one of two types of genetic systems: either they are largely self-fertilized and have a high recombination index or they are cross-fertilized and have a low one.

THE SELECTIVE VALUE OF GENETIC SYSTEMS

The material summarized in the preceding pages illustrates two points. First, we can find among plants many different types of genetic systems, in which the elements of sexuality, diploidy, cross-fertilization, and free recombination vs. genetic linkage are developed to various degrees. Furthermore, although many evolutionary lines have progressed from one genetic system to another, few if any of the systems now found among plants can be considered obsolete or inadaptive. Nearly all are found among at least some groups of common and widespread organisms. Second, the distribution of types of genetic systems is by no means at random throughout the plant kingdom, but is correlated with various features of the adult morphology, life history, and habitat of the plants themselves. These points lead to the hypothesis that in each group of organisms a particular type of genetic system is established as a result of the genetic potentialities of that group and of the action of natural selection. The most important feature of each genetic system is the level of compromise it establishes between the need for fitness, in order to secure immediate survival, and that for flexibility, as a means of potential adaptive response to future changes in the environment. The length of time during which any evolutionary line can exist on the earth and the amount of diversification and evolutionary progress of which it is capable are influenced to a great extent by the degree of harmony achieved between the level of compromise in the genetic system and the morphological and developmental characteristics of the organism.

The discussion already presented of the five primary elements of the genetic system has brought out two important correlations, which hold more or less regularly throughout the plant kingdom.

The first is that between the level of compromise and the length of life of the individual organism. In short-lived, rapidly reproducing organisms the genetic system is usually one which favors fitness at the expense of flexibility, while an increasing length of life, and particularly of the sexual cycle of generations, raises the selective value of genetic systems which favor flexibility at the expense of the immediate fitness of all of the sexual progeny of an individual.

The second correlation is that between the level of compromise and the developmental simplicity or complexity of the organism. In organisms which are structurally or developmentally simple, particularly in terms of the number of different kinds of cells and tissues which are elaborated during development, the level of compromise tends to favor fitness. But with increasing complexity of structure and development, the selective advantage of flexibility, considered in terms of evolution over long periods of time, correspondingly increases.

THE ORIGIN AND DEGENERATION OF SEX AS AFFECTED BY NATURAL SELECTION

Our present knowledge concerning the genetic consequences of sexual reproduction and its distribution within the plant and animal kingdoms makes possible only one type of conclusion concerning its biological and evolutionary significance. There is no longer any reason for believing, as did some biologists of the last generation, that sex exists as a means of rejuvenating the protoplasm, or for any other reason than its function in securing a great variety of genetic recombinations, by which the evolutionary line may adapt itself to new and varied environments. The prime importance of this function of sex, which has been realized by many biologists beginning with Weissmann, has been very aptly stated by Muller (1932) as follows:

> It is not generally realized that genetics has finally solved the age-old problem of the reason for the existence (i.e., the function) of sexuality and sex, and that only geneticists can properly answer the question, "is sex necessary?" There is no basic biological reason why reproduction, variation, and evolution can not go on indefinitely without sexuality or sex; therefore sex is not, in an absolute sense, a necessity, it is a "luxury." It is, however, highly desirable and useful, and so it becomes necessary in a relativistic sense, when our competitor-species

are also endowed with sex, for sexless beings, although often at a temporary advantage, can not keep up the pace set by sexual beings in the evolutionary race, and when readjustments are called for, they must eventually lose out.

If we view sexual reproduction in this light, then we must conclude that those evolutionary lines which have been highly successful over long periods of time in spite of the absence of sex have achieved this success either by securing the necessary flexibility in some other way or because flexibility in terms of the formation of new genetic combinations is less important for their existence or evolutionary progress than it is in sexual organisms. Both of these factors have been important.

Organisms with a short life cycle and rapid reproduction can achieve a certain degree of flexibility simply because of these characteristics. Changes in the environment occur at a definite rate in terms of chronological time and affect all organisms equally at this rate. But the degree of development of sexual reproduction and cross-fertilization determines the rate at which new recombinations can be produced *per generation*. Hence, if the sexual cycle is shortened, the number of gene recombinations produced per generation can be correspondingly reduced without affecting flexibility in terms of the number of gene combinations available in a given unit of chronological time.

The importance of this time-developmental factor may be estimated by comparing directly the length of life cycle of different organisms. In bacteria, the generation may be as short as thirty minutes, making 48 generations possible per day. This rapid rate of reproduction is never kept up, but a constant rate of 100 generations per week would not be excessive for a bacterial population living in a changing environment that would promote evolution. Given equal rates of mutation and recombination, therefore, such a bacterial population could evolve 5,000 times as rapidly as an annual seed plant and 50,000 to 100,000 times as rapidly as the average forest tree. Therefore, except in the highly improbable event of a rapid increase in the rate of useful mutations in these higher organisms, they would have to produce new gene combinations at a rate many thousands of times as high as that of the bacteria attacking them.

Furthermore, in organisms with a short life cycle fitness itself carries a relatively high selective advantage in relation to flexibility. Such organisms, like bacteria, fungi, and the smaller algae, are usually small in size, are destroyed in huge numbers by their enemies, and depend for their survival chiefly on their ability to build up large populations rapidly in a favorable medium. During these periods of increase the production of any organism not adapted to this medium is a wasted effort on the part of a population whose very life depends on reproductive efficiency. Such a growing population cannot experiment with new gene combinations; it must sacrifice flexibility to immediate fitness.

On the other hand, many of the larger plants maintain populations whose size varies relatively little over periods of thousands or even millions of years. In any population of forest trees, for instance, a seedling can grow to maturity only when an old tree dies, leaving room for it in the forest, or in the rare (previous to man) event of a landslide, fire, or other catastrophe. During normal times, therefore, thousands of competitors are available for every niche which may be occupied by an adult individual, and a huge wastage of young in terms of ill-adapted gene combinations is little or no disadvantage to the population. We can say, therefore, that in organisms with a long life cycle a genetic system promoting flexibility is of high selective value, while simple, short-lived, and rapidly reproducing organisms need a genetic system favoring fitness.

The lower selective value of new and complex gene recombinations in many of the asexual organisms is very likely due also to their ability to adapt themselves to new environments in relatively simple ways. Although it is true that many bacteria and protozoa are highly complex and specialized in respect to the physiological activity of their cells, all of them are developmentally simple, in that their ontogeny involves the differentiation of only a very small number of different kinds of cells. If, as was emphasized in Chapter III, we consider the action of genes in terms of their effect on processes rather than on adult characteristics, this fact has important implications. If all of the cells of the organism are alike physiologically, then a particular change in any developmental process will have a uniform effect on the organism as a whole and is likely to have a particular selective advantage or disadvantage as a unit, independently of its relation to other changes in developmental processes. But if the organism possesses a large number of different types of cells, which are produced by a great variety of physicochemical processes acting at different intensities during the various stages of development, then a single change in one process may affect different parts of the same organism in very different ways, some of them advantageous, and some deleterious. Under such conditions, a change which may benefit the entire organism by adapting it to new environmental conditions is most easily achieved through the combined action of several different changes in various ones of the numerous developmental processes. The disadvantageous effects of one change are balanced by the favorable effects of another. The principles of developmental correlation and adaptive compensation, discussed in Chapter IV, are obviously special cases of this more general principle.

For these reasons, we should expect that genetic systems favoring recombination would arise in organisms with structural and developmental complexity even more rapidly than in those with increasing lengths of the sexual generation. This would explain the greater development of sex in animals than in plants, since animals are in general much more complex than plants in both the number and the degree of differentiation of their tissues and in their developmental processes.

A GENETIC THEORY FOR THE ORIGIN OF THE DIPLOID STATE

The idea that the diploid condition became established in the higher plants because of its greater selective value in respect to the production of gene combinations was first advanced by Svedelius (1929). He pointed out that in a strictly haploid organism, in which the zygote undergoes meiosis without any intervening mitotic divisions, each act of fertilization can yield in the following generation only four different gene combinations, since only one meiotic division is possible. On the other hand, the production of a large diploid organism between the time of fertilization and meiosis makes possible the occurrence of hundreds of meiotic divisions in the tissues derived originally from a single zygote. Since the segregation of the genes in each of these divisions will be somewhat different from that in any of the others, each act of fertilization ultimately produces a great variety of genetically different gametes and so increases the potentialities for gene recombination.

Another important and perhaps more significant factor in the evolution of diploidy in the more complex organisms is the greater selective value in them of genetic and evolutionary flexibility. The diploid condition brings about an increase in flexibility because it makes possible the condition of genetic dominance and recessiveness. In a haploid organism every new mutation is immediately exposed to the action of selection, and if this action is deleterious or even only slightly disadvantageous at the time it originates, the organism is likely to be lost, even though selection might be advantageous in some environment to which the descendants of the organism might be exposed or in combination with some other gene present in the population. In a diploid organism, on the other hand, each new mutant arises as a heterozygote and, if recessive, is sheltered from selection. If the mutation occurs repeatedly, it will eventually become established in a small percentage of the population, even if it is deleterious or actually lethal. The high frequency of heterozygosity for such lethal or semilethal genes in all populations of cross-fertilized organisms which have been examined for them is well known to geneticists (Dobzhansky 1941). Among the neutral and the semilethal genes present in such populations, there are likely to be some with potential selective advantages under environmental conditions the population may be called on to face in the future, and others capable of forming favorable combinations with other mutations which may arise in different individuals of the population.

A simple example will illustrate the greater effectiveness of diploidy in securing new adaptations by means of gene combinations. Suppose that the gene combination AB represents an adaptive "peak" for a particular population or race and that the allelic condition ab is another such "peak," toward which under certain environmental conditions the race can or must evolve. If the organism is haploid, this evolution can take place only if the intermediate conditions, Ab and aB, are equal or superior in selective value to the original condition AB. But if the organism is diploid, and the genes a and b are recessive, then the mutant forms AaBB and AABb can become established in the population regardless of the selective value or disadvantage of a and b. Crossing between two such mutants will produce the double heterozygote AaBb, which can cross with similar double heterozygotes to produce the desired new adaptive combination. The selective value of diploidy, therefore, depends directly on the degree to which gene combinations rather than single genes produce the genotypic changes necessary for evolution. And this is, of course, directly connected with the degree of developmental complexity of the organism.

Although the genetical explanation suggested here to account for the origin and rise of the diploid sporophyte in the higher plants cannot provide decisive evidence on the original nature of this sporophyte, its acceptance does increase the probability of one of the two current hypotheses. The "antithetic" or "interpolation" theory of the origin of the vascular sporophyte (Bower 1935) maintains that this generation first appeared as a relatively small, simple structure, similar to that found in the modern genus *Anthoceros* and its relatives, and gradually increased in size, presumably through the establishment of mutations delaying the onset of meiosis. The "homologous" theory, on the other hand, (Eames 1936) conceives of the sporophyte and gametophyte of the ancestral vascular plants as essentially alike. The sporophyte is viewed, not as an entirely new structure, but as a modification of some preexisting haploid form. The differentiation between sporophyte and gametophyte is believed to have taken place before the origin of vascular plants as such. Further evolution has consisted both of increase in size and complexity of the former and reduction of the latter generation. The morphological evidence in favor of this theory has been well summed up by Eames (1936, pp. 392–397). It accords better with the genetical theory for the following reasons. As Fritsch (1916) has pointed out, the modern haploid green algae include forms with many diverse types of life history, and their remote ancestors were probably similar in this respect. Different types of alternation of generations could have arisen by the conversion to diploidy of different parts of the haploid life cycle. Since, as has already been pointed out, mutations producing a diploid sporophyte undoubtedly became established many different times in the evolution of the plant kingdom, there is no reason to assume that such mutations arose only once in the haploid archegoniates, which were the presumable common ancestors of the Bryophytes and the vascular plants. If the diploid state is of selective value because of the evolutionary potentialities it gives to the plant in terms of greater diversity and complexity, then a great initial advantage would have been granted those archegoniates in which a relatively large proportion of the life cycle was converted at one step to diploidy. Hence, it is easiest to conceive of the vascular plants as having evolved from haploid archegoniates in which mutations at once produced an alternation between similar generations like that found in the modern algae *Ulva* and *Polysiphonia.* Bryophytes, on the other hand, could owe their relatively slight success as a group to the fact that in their ancestors, diploidy first appeared in only a small portion of their life history.

CAUSES FOR THE DEGENERATION TOWARD SELF-FERTILIZATION

The occurrence of self-fertilization as well as cross-fertilization in various groups of plants must be explained by a hypothesis different from those used for the origin of sex and of diploidy. There is little doubt that in this case the genetic system promoting greater flexibility, namely, cross-fertilization, is the ancestral one, and that self-fertilized types have evolved repeatedly in various phyla and classes of the plant kingdom, presumably in response to selection pressure in favor of immediate fitness. There are probably two principal reasons for the selective advantage of self-pollination in certain types of organisms. In the first place, short-lived plants with no effective means of vegetative reproduction must produce a good crop of seeds or perish. If an annual plant cannot be self-fertilized, and if in some abnormal season conditions are unfavorable for cross-fertilization at the time when the plant is sexually mature, it may perish without leaving any offspring. We can expect, therefore, that selection in favor of self-fertilization is very likely to occur in all annuals and other short-lived plants living in deserts, alpine or arctic regions, or other extreme environments. These must sacrifice their evolutionary future for the sake of their immediate survival. That this is not the only reason for the evolution of self-fertilization is suggested by two facts. In the first place a number of groups of short-lived annuals, such as many species of the genus *Crepis* and of various genera belonging to the tribe Madinae in the family Compositae, are self-incompatible and must be cross-fertilized, yet set a high percentage of seed even in dry years when their plants are greatly stunted. Second, some long-lived perennials, such as various species of grasses in the genera *Stipa, Danthonia,* and many others, have evolved mechanisms promoting a high degree of self-fertilization, in spite of the fact that the production of a heavy crop of seeds is by no means essential to their survival. An additional hypothesis is therefore necessary to account for the evolution of self-fertilization in such perennials, as well as in annuals living in a climate that is rarely or never unfavorable for cross-fertilization.

Such a hypothesis may be based on the fact that most of the species concerned live in habitats characterized by a great fluctuation in climatic conditions from season to season or in pioneer associations which are constantly changing their extent and position. The populations of such species are therefore greatly reduced in size periodically, and in certain seasons they must be built up very rapidly from the seed available in the ground. Such drastic changes in the populations of annual species in the San Joaquin Valley of California have been recorded by Talbot, Biswell, and Hormay (1939) and are well known to botanists in this and similar regions.

When plants are living under such conditions, they can build up their populations successfully in favorable years only if a large supply of seed is available and if the seedlings produced from these seeds are all or nearly all very well adapted to the conditions which prevail at such times. Such species cannot afford to experiment with a large number of new and possibly ill-adapted gene combinations. In them, a high premium is placed on a genetic system which favors opportunism and enables a favorable gene combination, once it has been achieved, to spread over a large number of genetically similar individuals. Genetic similarity is further favored by the fact that annual species, at least those which have been grown by the writer, have a relatively high degree of phenotypic plasticity as compared with perennials, so that the individual genotype can be enormously modified in respect to size, seed production, and other characteristics, depending on whether the environment is favorable or unfavorable. The great predominance of self-pollination in annual species of grasses and many other flowering plants can therefore be best explained on the basis of the selective advantage this genetic system gives to the species populations of such plants, as well as to the need for security in seed production.

Nevertheless, these predominantly self-pollinated annuals have not lost their capacity for forming new gene combinations. It is a well-known fact that in cultivated wheat from 1 to 3 percent of outcrossing may occur in some varieties under certain climatic conditions (Hayes and Garber 1927, p. 97). And in wild populations of predominantly self-pollinated grasses, the same is apparently true. Harlan (1945a,b) has demonstrated this for the perennial *Bromus carinatus,* while Knowles (1943) has shown that in the annual *B. mollis* the populations found in California, although they have been introduced into the state only about sixty years ago (Robbins 1940), are already differentiated geographically into ecotypes which behave differently when grown in a uniform environment. This would not be possible unless a certain amount of cross-pollination and heterozygosity was present in the species.

The genetic mechanism present in these annual plants apparently works about as follows. A particular family or racial line can maintain itself by self-pollination for scores or hundreds of generations. During this time it becomes completely homozygous and exists as a single biotype consisting of scores, hundreds, or even thousands of genetically similar individuals. Occasionally, however, an individual of such a biotype will produce a few seeds which are the result of cross-pollination with another biotype. If these seeds grow to maturity, they will form hybrid plants of a new morphological type, which will be highly heterozygous. Progeny of these F_1 interracial hybrids segregate extensively, producing a great variety of new genotypes. Some of these may be better adapted to the environment occupied by the parental biotype, while others may be capable of occupying new environmental niches. Following each chance interracial hybridization, therefore, there will ensue a burst of evolution lasting for six to eight generations, during which variability through segregation and consequent natural selection will be at a maximum. The surviving racial lines will gradually lose their remaining variability through inbreeding and will remain static until a new chance hybridization initiates another short burst of evolution.

That this type of evolution has in many instances been highly successful is evident from the fact that many groups of grasses with predominant self-pollination have become differentiated into a large number of species, many of which possess highly distinctive morphological characteristics. Good examples are the subgenera *Eubromus* and *Bromium* of *Bromus,* the subgenus *Vulpia* of Festuca, and the genus *Aegilops.* There is little doubt that a careful study of some of the annual genera in such families as Cruciferae, Leguminosae, Boraginaceae, Scrophulariaceae, and Compositae would reveal genetic and evolutionary patterns similar to the one just outlined for the grasses.

There is, furthermore, evidence that the evolution of a genetic system favoring constancy and fitness has taken place independently in a group of plants entirely unrelated to the angiosperms. Buller (1941) has produced evidence that in the Basidiomycetes, the primitive sexual condition is the heterothallic one, in which cross-fertilization and heterozygosity are obligatory. The heterothallic condition is general throughout the class of fungi, and the evolution of many genera has involved only heterothallic species. Homothallic species, which carry out self-fertilization through fusion of hyphae or gametes derived from the same mycelium, are scattered through various genera and usually appear like sexually degenerate types.

In many groups of fungi, such as the genus *Coprinus* among the mushrooms, no special feature of the homothallic species has been observed which might be looked upon as a causal agent for their evolution in the direction of sexual degeneracy. But in the genus *Puccinia* among the rusts there is a definite correlation between the length of life cycle and the presence of homo- or heterothallism (Jackson 1931). All the rusts of the long-cycle type, like the well-known wheat rust, *P. graminis tritici,* which has alternate hosts and the full complement of five types of spores, are heterothallic. But there are other species of *Puccinia* which are known as short-cycle rusts because they have only one host and three types of spores, the uredospores, the teliospores, and the basidiospores. These are homothallic. In both their life cycle and their genetic system they may be compared with the annual, self-fertilized species of seed plants.

THE SELECTIVE VALUE OF THE RECOMBINATION INDEX

Earlier in this chapter a possible correlation was suggested between longevity and a high recombination index on the one hand, and between the annual habit and a low one on the other. This correlation is an expected one, on the basis of the general principle developed in this chapter. As mentioned previously, genetic homogeneity is likely to have a high selective value in all organisms which are subject to great fluctuations in the size of their populations. But if this homogeneity is achieved by permanent homozygosity through enforced self-fertilization, the evolutionary future of the race is greatly jeopardized. On the other hand, if self-incompatibility and cross-fertilization are retained, an increase in the amount of genetic linkage, through reduction in chromosome number, in crossover frequency per chromosome, or both, can produce an increased homogeneity over periods of a few generations. This is a long enough period to build up the population in response to favorable environmental conditions. But with such a genetic system, new gene combinations can be continuously produced by crossing over, even if at a reduced rate, and the evolutionary possibilities of the race are relatively little affected. We should expect, therefore, that the greatest amount of morphological and physiological diversity among annual species would be found among those with cross-pollination and with relatively low chromosome numbers and chiasma frequencies.

Data recently obtained on chiasma frequencies in certain grass species (Stebbins, Valencia, and Valencia 1946a) suggest that the recombination index most advantageous for self-fertilized species may be much higher than that suited to cross-fertilized species with the same growth habit. The species *Agropyron parishii*, *A. trachycaulum*, *Elymus glaucus*, and *Sitanion jubatum*, although they all agree in the number and size of their chromosomes, differ widely in their metaphase chiasma frequencies. These differences are correlated with the degree of self-pollination, since *Agropyron parishii*, which is to a large degree cross-pollinated, has a mean frequency of 2.01 chiasmata per bivalent, while the predominantly self-pollinated *Sitanion jubatum* has 2.79 chiasmata per bivalent, and the two other species are intermediate in both respects. A probable explanation is that in the predominantly self-fertilized species the ability of the rare heterozygous interracial hybrids to produce many new gene recombinations is of particularly high selective value as a means of adapting the species to new conditions. In these species, therefore, fitness and constancy are secured entirely through self-pollination and homozygosity, while the mechanism for flexibility and adaptation to new situations consists of short periods of great heterozygosity, following interracial hybridization. During these periods natural selection is acting with great intensity, and a high recombination index is of great value.

The genetic systems discussed in this chapter are those found in organisms with normal sexual reproduction and without pronounced abnormalities in the pairing and segregation of the chromosomes. In plants, however, the genetic system is often modified by duplication of the chromosome sets, or polyploidy, by substitution of asexual for sexual reproduction, or apomixis, and by permanent heterozygosity for extensive differences in the arrangement of the chromosome segments, or structural hybridity. These phenomena are all of such great importance that they must be treated in separate chapters, and since they greatly affect the nature of species in the groups in which they occur, discussion of them is postponed until Chapters VIII–XI, following the chapters dealing with the species problem in normal diploid organisms.

GENETIC SYSTEMS IN PLANTS AND ANIMALS

One of the most remarkable features of the succession of discoveries which have established and advanced the science of genetics during the past half-century has been the great similarity between the phenomena in animals and those in plants. Segregation, recombination, linkage, crossing over, nearly all the complex cytological processes involved in meiosis, and all the types of progressive change, such as mutations and chromosomal aberrations, have been found in plants as well as in animals, often in the same form. Furthermore, in many respects the difference in genetic behavior between different types of plants is as great or greater than that between certain plants and certain animals. It is for this reason that genetics must be regarded as a branch of biological science separate from either botany or zoology, but forming a powerful connecting link between them.

Nevertheless, there are important differences between the genetic systems found in plants and those characteristic of animals. These differences are correlated with and probably caused by the very different modes of life of members of the two kingdoms. The three differences between plants and animals which have most strongly affected their genetic systems are in complexity of development, in longevity and ability for vegetative reproduction, and in mobility (Anderson 1937a, Turrill 1942a).

The differences in complexity of development and in the diversity of the cells and tissues of the mature organism are very great and extend from the highest to the lowest members of each kingdom. Not only are the systems of organs and tissues vastly more complex in a fly or a man than they are in a dandelion or an oak tree; the complexity of a sponge or a jellyfish is equally great when compared with that of *Spirogyra* or *Neurospora.* The complexity of development is further increased by the different methods of growth and development which prevail in the two kingdoms. The "closed" system of growth found in animals involves the precise, orderly succession of a long, complex series of physicochemical developmental processes, all occurring during the short period of early embryonic growth. Differentiation, youth, middle age, and old age occur in nearly all of the tissues simultaneously. If the genes of an animal are not able to act harmoniously to produce the integration of this complex developmental system, death results. Single mutations with a radical effect on any one process will almost certainly throw the system so far out of harmony that it will not be able to function properly and will produce either death or an ill-adapted monster. New combinations of genes, each having a relatively slight effect, are almost essential to evolutionary progress even in the simpler animals.

On the other hand, the open system of growth characteristic of plants necessitates a much less complex and precise integration of developmental processes. Differentiating and youthful tissues are confined to localized areas like the tips of the shoots, the roots, and the cambial zones. A relatively small proportion of the cells of the plant is undergoing differentiation at any one time. In many algae and fungi, there are never more than one or two types of cells being differentiated at any moment. Even in the vascular land plants the number of physicochemical processes involved in these developmental successions is probably less than that needed for most animals, and the need for the precise integration of the action of many different genes must be correspondingly less. Single mutations could under such circumstances play a relatively important role in evolutionary progress, and the number of adaptive combinations which could be obtained from a relatively small number of genes is probably larger.

All these factors contribute to make gene recombination considerably more important in animals than it is in plants. This is probably responsible for the fact that all animals are diploid (except for the males of some insects) and that the great majority of them, even hermaphroditic types like snails and worms, are normally cross-fertilized. In most animals, indulgence in continued self-fertilization appears to represent evolutionary suicide; in plants it may slow down evolutionary progress, but not necessarily stop it. In animals, furthermore, we should expect basic chromosome numbers and chiasma frequencies to run higher than in plants. This is certainly true of vertebrates as compared with flowering plants; the only numbers in the latter class which equal the ones usually found in vertebrates are obviously polyploid derivatives of the basic numbers, and therefore in a special class. The insects present a different and more complex picture. There is little doubt that a careful comparative cytogenetic study of the orders of insects in relation to their general biology would reveal much concerning the evolutionary significance of genetic systems determined by the recombination index.

The difference between plants and animals in longevity and the development of asexual, vegetative reproduction affects their genetic systems in an entirely different way. In animals, sexual reproduction is as important for maintaining the population of the species as it is for producing new variant gene combinations. This is also true in many plants, such as the annual seed plants, but in other types of plants, particularly many thallophytes, such as various saprophytic fungi and lichens, it is not true at all. A few facts will give an idea of the difference in this respect between animals and many types of plants.

In by far the majority of animal species the length of life of any individual is less than five years, and species of which the individuals live more than a hundred years are truly exceptional. But if we regard the individual from the point of view of its heredity and consider the life span of an individual genotype which has been derived from a single fertilized egg or zygote, then we must recognize that there are many plant species in which a life span of a hundred years is commonplace and a longevity of one or even many thousand years may not be unusual. The most spectacular examples of longevity, and those to which one's attention is most often called, are the large trees like the species of *Sequoia* and *Eucalyptus* (Molisch 1938). But these examples are unusual only if we consider an individual as a single growing shoot or shoot system. There are many species of shrubs and trees which sprout repeatedly from the root crown, so that the age of the individual is far greater than that of any one of its trunks. The species of willows (*Salix*), poplars (*Populus*), and many oaks (*Quercus*) are familiar examples, and even more spectacular ones are found in the crown-sprouting shrubs of dry areas, like the manzanitas (*Arctostaphylos* spp.) of California (Jepson 1939, pp. 29–34), which can be burned repeatedly without being killed. The life span of such individuals cannot be estimated, but barring the interference of man they can be destroyed only by disease, competition with other plant species, or a radical change in their habitat. Molisch (1938) has noted evidence of aging in many species of plants, but excepts those which have natural means of vegetative reproduction or regeneration. In stable plant communities seriously diseased plants are rare, so that the life span of many individuals which are able to sprout from the roots or crowns must approach that of the community itself. There is little evidence that the undifferentiated meristematic tissues of such plants are subject to senescence, as are the completely differentiated tissues of adult animals.

The same arguments apply with equal or greater force to the clones which represent the individual genotypes of perennial plants with effective means of vegetative reproduction, such as stolons, rhizomes, corms, and bulbs. As genetic individuals, these plants are not only immune to such agencies as age, frost, and fire, which destroy their individual shoots. Their separate vegetative parts or propagules can be carried for considerable distances by water, animals, or other agencies. In this way the individual as a whole is immune to the diseases, competition, or environmental change which may destroy some of its parts; it is practically immortal. Hitchcock and Chase (1931) state that many individual plants of buffalo grass on the western plains are probably the same ones which colonized these plains after the retreat of the glaciers.

The age of the clones of such vegetatively reproducing plants is impossible to estimate unless historical records are available. One such record is that of the saffron crocus (*Crocus sativus*). This "species" is highly sterile as to pollen and seed and is known chiefly in cultivation. Under these conditions, it is normally propagated by its corms. Its minimum age may be estimated from the fact that records of its cultivation for the yellow dye derived from its flowers date back to the ancient Greeks. It must therefore be at least 2,500 to 3,000 years old, and some strains may represent genetic clones of that age. There is every reason to believe that numerous clones of wild plants approach this age. By measuring the size of the clones and the rate of yearly spread, Darrow and Camp (1945) have estimated the age of a clone of *Vaccinium* as at least 1,000 years.

In such perennial plants with efficient vegetative reproduction a high degree of seed sterility is no detriment to the immediate existence of either the individual or the population. Examples are not uncommon of widespread and abundant wild species which over large areas reproduce entirely by vegetative means (Ernst 1918, pp. 569–588). A familiar example is the sweet flag, *Acorus calamus,* throughout Europe and western Asia. Others are such aquatics as *Eichhornia crassipes, Anacharis* (*Elodea*) *canadensis,* and *Lemna* spp.; bulbous plants like the sterile triploid form of *Fritillaria camschatcensis* in Japan (Matsuura 1935); and numerous species of lichens in the genera *Usnea, Ramalina, Cladonia,* and many others. Examples of vegetative as well as agamospermous apomixis, in which asexual reproduction has partly or completely replaced the sexual method, will be cited in Chapter X. In all of these instances, the reproductive potential and the spreading capacity of the species are secured largely through asexual means. Sexual reproduction serves not so much to maintain and increase the species as to produce the necessary genetic variability.

The most important result of this condition is the degree to which it lessens the selective value of complete fertility in sexual reproduction. Highly sterile genotypes may live for such a long time and spread so far by asexual means that they can under some conditions compete successfully with their fully fertile relatives. There is a striking example of this situation in the grass genus *Elymus* in California. *E. triticoides* is typically a rhizomatous grass of bottom lands in alkali soil, and although often not fully fertile, nevertheless as a rule it sets a fair amount of seed when different clones of this self-incompatible species are growing close together. *E. condensatus* is a giant caespitose grass found chiefly along the coast of central and southern California and preferring well-drained hillsides. Although its seed production is frequently very low, this phenomenon appears to be largely environmental rather than genetic, since most plants of this species grown under cultivation set seed rather freely. In addition there are found on hillsides and particularly along railroad embankments and highway margins plants which resemble robust forms of *E. triticoides,* but in their broader, greener leaves and particularly in characteristics of their inflorescences and spikelets approach *E. condensatus.* Since some of these closely resemble the artificial hybrid between *E. condensatus* and *E. triticoides* produced by the writer, they are most likely of hybrid origin. They are very highly sterile under all conditions, and their anthers contain almost entirely aborted and shriveled pollen, in this respect often exceeding the sterility of the artificial F_1 hybrid. These sterile hybrids and hybrid derivatives, because of their great vegetative vigor and their ability to spread by rhizomes, have proved better competitors under the environments produced by man's activity than have either of their parental species and they have spread into areas where neither *E. condensatus* nor *E. triticoides* is found. Some of them have retained the chromosome number 28 found in both parental species, while others have 42 chromosomes in their somatic cells (Gould 1945). Such a situation, which is fraught with considerable evolutionary possibilities, could arise in only a few types of animals. A further discussion of this and of similar situations is presented in Chapter VII.

The mobility of animals, as compared to the sedentary nature of plants, affects chiefly the structure of their populations. The difference between the two kingdoms, however, may not be so great as would seem at first sight. Animals move about mainly in search of food, and in a large proportion of the species of various groups the individuals tend to breed in nearly the same place as that occupied by their parents. This is well known for birds and many species of mammals. On the other hand, plants possess a considerable amount of passive mobility, which is usually most effective in connection with the sexual cycle. There is little doubt that the pollen of many species of trees is borne for many miles by the wind and may function perfectly after such journeys. Seeds are also carried for considerable distances, so that the offspring may grow up and find a mate in some region more or less distant from that occupied by its parents. Furthermore, in plants with effective vegetative reproduction, particularly with such structures as tubers, corms, and small bulbs, these parts may be carried considerable distances, so that a single genotype may simultaneously be fertilizing several other genotypes separated from each other by many miles. In plants as in animals, fertilization is usually between individuals which grew up near each other, but in species of both kingdoms mating between individuals originally separated from each other by long distances occurs fairly regularly, although with a relatively low frequency. There are, however, some species of animals such as ducks (Mayr 1942, p. 241) which because of their mobility and promiscuity have interbreeding populations derived from a much larger area than is ever the case in plants, and it is probably true in general that on the average effectively interbreeding populations of vertebrate animals are spread over a larger area and may reach a larger size than in plants. Accurate comparative data on this point are badly needed.

If this last generalization is true, then it follows that direct adaptation to the environment should be more regularly present in the larger animals than it is in plants, and the effects of isolation and chance fixation of genes should be less widespread. The discussion in Chapter II brought out the fact that in many groups of seed plants the species have a narrower geographical range than is typical for the higher animals and show less tendency to consist of well-defined subspecies. In the following chapter the importance will be brought out of small populations or of "bottlenecks" involving temporary reductions in population size in the origin of the isolating barriers which separate species. These might be expected to occur more frequently in plants than in animals, since animals could migrate more easily to escape temporary vicissitudes of the climate. It is perhaps noteworthy that the nearest approach among plants to the variation patterns found in most of the higher animals is found among forest trees, which have large populations and a maximum mobility of the gametes through wind transportation. Here again, a comparative study of habits of ecology, population size, and the nature of species and the genetic system in various orders of insects would doubtless be very enlightening. It may be that differences in these respects as great as those found between the higher plants and the vertebrate animals exist between different orders of insects. Nevertheless, comparisons between the genetic systems of animals and plants and their relation to the types of evolution found in the two kingdoms will undoubtedly be very fruitful subjects for future research.

CHAPTER VI

Isolation and the Origin of Species

ONE OF THE MOST important principles which has emerged from the recent studies of evolution is that the terms "evolution" and "origin of species" are not synonymous, as implied by the title of Darwin's classic. The foundation of this principle is the more precise, objective concept of the nature of species which has been obtained through more careful systematic and particularly cytogenetic studies of interspecific differences and of the barriers between species (Clausen, Keck, and Hiesey 1939, Dobzhansky 1941, Chap. XI, Mayr 1942, Chap. VII). The new definitions of the species which have been based on these studies are numerous; a different one can be found in each of the three publications cited in the previous sentence, while the group of essays compiled by Huxley (1940) contains nine more. Comparing these definitions, however, one is struck, not by their diversity, but by the large common ground of agreement between them. All of them stress the importance of genetic and morphological continuity within species, and recognize at the same time that a species may include within its limits an array of morphologically and physiologically diverse genetic types. They also agree that the boundaries between the species of sexually reproducing organisms are real, objective phenomena, and that they are produced by isolating mechanisms which prevent or greatly restrict the exchange of genes between the members of different species. The common ground of agreement between these definitions may be expressed as follows. In sexually reproducing organisms, a species is a system consisting of one or more genetically, morphologically, and physiologically different kinds of organisms which possess an essential continuity maintained by the similarity of genes or the more or less free interchange of genes between its members. Species are separated from each other by gaps of genetic discontinuity in morphological and physiological characteristics which are maintained by the absence or rarity of gene interchange between members of different species. The above sentences are not to be construed as this author's definition of a species, since several different species definitions are possible within the framework of their meaning.

In the light of this concept, the processes involved in "descent with modification," to use Darwin's classic phrase, can be shown clearly to apply to differentiation within species, as well as to the further divergence of species and higher categories once they have become separated from each other. The processes peculiar to the origin of species are those involved in the building up of isolating mechanisms which restrict the interchange of genes between different species. One purpose of this chapter is to show that the processes responsible for evolutionary divergence may be entirely different in character and genetically independent of those which produce isolating mechanisms and, consequently, distinct species.

A COMPARISON OF DIFFERENCES WITHIN SPECIES AND BETWEEN SPECIES

The first question that arises is whether the characteristics of external morphology and physiology which distinguish species are different in kind from those between different races or subspecies of the same species, or whether they differ only in degree. The answer to this question is unequivocal, and the data from cytogenetic studies support the opinion held by most systematists. The differences between closely related species are nearly all duplicated by or paralleled by differences between races or subspecies of a single species. In many instances the traits which characterize genera or even higher categories can be found to vary within a species.

Some examples to illustrate this fact have already been presented in the discussion of ecotype differentiation in Chapters I and II. In *Potentilla glandulosa,* for instance, the differences between ecotypes include characteristics of the size and shape of sepals and petals, and the size, shape, and color of the achenes, which are quite similar to characters used in other subgenera for the separation of valid species. In *Crepis*, Babcock and Cave (1938) found that the species previously classified by systematists as *Rodigia commutata* is able to exchange genes freely with the widespread species *Crepis foetida*. Thus, the character difference which had previously been assumed to be of generic significance, namely, the presence or absence of paleae on the receptacle, was found to be an actual interracial difference within the limits of a single species. A similar example is that of *Layia glandulosa* var. *discoidea*, described by Clausen, Keck, and Hiesey (1947). This form is so strikingly different from its relatives in characteristics generally assumed to be of generic or even tribal significance that its original collectors were hesitant to place it in the tribe Madiinae. But when crossed with the well-known *Layia glandulosa* it formed fertile hybrids, which in the second generation yielded a great array of vigorous and fertile individuals, segregating for the characteristics of involucral bracts and ray florets, which are commonly used as diagnostic of genera and tribes in the family Compositae. Further examples of differences which have commonly been assumed to separate species, but may be only of racial or subspecific significance may be found in the discussion in Chapter II concerning the variation patterns within the genera *Aquilegia* and *Quercus*. From his hybridizations in the genus *Salix*, Heribert-Nilsson (1918) found that in their genetic basis, interspecific differences in that genus differ from intervarietal ones only in magnitude and degree of complexity.

The objection might be raised here that the examples given in the above paragraph apply only to the diagnostic characters which have provided convenient "handles" for the systematist, not to the perhaps more "fundamental" physiological and biochemical differences between species. In answer to this objection, however, we can refer to the experimental studies which show that different races or ecotypes of the same species may have entirely different systems of reaction to their environment. *Potentilla glandulosa* subsp. *typica*, which grows actively throughout the year, differs radically in its physiological reactions from the alpine subsp. *nevadensis*, which even in the environment of *typica* remains dormant for the winter months. But these two subspecies are completely interfertile, are connected by a whole series of naturally occurring intermediate types, and the physiological differences between them show genetic segregation of exactly the same type as that which characterizes the recognizable differences in external morphology. Another important physiological difference, namely, the reaction to photoperiodism, is most often of specific value, but in *Solidago sempervirens* (Goodwin 1944), *Bouteloua curtipendula* (Olmsted 1944), and probably in other species it is a characteristic difference between races of the same species. Self-incompatibility ("self-sterility") and self-compatibility ("self-fertility") usually separate species, but in *Potentilla glandulosa, Antirrhinum majus* (Baur 1932), and several other species self-incompatible and self-compatible races exist within the same interfertile population system.

The same may be said of biochemical and serological differences. In general, these are more profound when species are compared than when comparison is made between races, but the difference is only in degree. Differences exist within species of *Dahlia, Primula, Papaver, Streptocarpus,* and other genera in the biochemistry of their flower pigments (Scott-Moncrieff 1936, Lawrence and Scott-Moncrieff 1935, Lawrence, Scott-Moncrieff, and Sturgess 1939), of tomato in vitamin content, and of maize in aleurone and other chemical substances in the grain. Differences in antigenic properties have been used to determine the relationships between species, genera, and families both of animals and of plants (Irwin 1938, Mez and Siegenspeck 1926), but the same type of difference can be found on a smaller scale between races of the same species, as is evident from the work of Arzt (1926) on barley (*Hordeum vulgare* and *H. spontaneum*), as well as the studies of blood groups in man and of the still more complex serological differences between races of cattle (Owen, Stormont, and Irwin 1947). That morphological and biochemical differences should run parallel to each other is of course what one would expect, since differences in external morphology are simply the end products of different biochemical processes occurring in the development of the individual.

Further evidence for the essential similarity between interracial and interspecific differences is found in the example of "cryptic species," that is, of population systems which were believed to belong to the same species until genetic evidence showed the existence of isolating mechanisms separating them. A typical example in genetic literature is that of *Drosophila pseudoobscura* and *D. persimilis* (Dobzhansky and Epling 1944). In plants, two of the best examples are the complex of *Hemizonia* (or *Holocarpha*) *virgata-heermannii–obconica* (Keck 1935) and *Crepis neglecta–fuliginosa* (Tobgy 1943). Both of these are in the family Compositae. In regard to the former, Keck writes as follows in his description of a new species, *H. obconica,* " . . . its distinctive characteristics were not observed clearly until cytological studies had shown that there were two species in the *H. virgata* complex. *Hemizonia obconica* is a species with a haploid chromosome number of 6 (like *H. heermannii*), while in *H. virgata* the number is 4. The species do not form fertile hybrids whether growing side by side in nature or as the result of artificial garden crossings." The differences in external morphology, as well as in physiological characteristics of growth and development, between *Crepis neglecta* and *C. fuliginosa* are in every way comparable to those between different subspecies of certain other species, such as *C. foetida* (Babcock 1947). In both of these examples, the specific status of the entities concerned is based largely on discontinuities rather than large differences in morphological characteristics, and particularly on numerical and structural differences in the chromosome complements which cause the sterility of the F_1 hybrids. Such obvious chromosomal differences are not, however, essential accompaniments of the differentiation between species, as will be evident from examples presented later in this chapter.

The concept that species differences are of a different order than interracial differences within the species has been advocated most strongly in recent times by Goldschmidt (1940). The examples which he gives are drawn mostly from zoological literature, with particular emphasis on his own experience with the moth genus *Lymantria.* Mayr, however, has pointed out (1942, pp. 137–38) that in this example Goldschmidt dealt with only three of the numerous species of this genus, and that these three species are related most closely, not to each other, but to different ones of the species not studied by Goldschmidt. *Lymantria,* therefore, is not well enough known as a genus so that any conclusions can be drawn about the nature of the species forming processes in it. The only plant example which Goldschmidt cites in this particular discussion is that of *Iris virginica* and *I. versicolor,* quoted from Anderson (1936b). This was certainly an unfortunate choice. As is pointed out below, in Chapter IX, *I. versicolor* is an allopolyploid which contains the chromosomes of *I. virginica* combined with those of *I. setosa* var. *interior* or a related form. These species, therefore, represent a type of evolution which has nothing to do with the origin of species on the diploid level. Elsewhere in his book Goldschmidt has stated that species represent different "reaction systems," but this term is nowhere clearly defined. When applied to species, it can have only one meaning which agrees with the factual evidence. This is that members of the same species are genetically compatible, in that they can intercross freely and produce abundant fertile, vigorous, segregating offspring; while members of different species often (but not always) react with each other in such a way that hybrids between them either cannot be obtained or are incapable of producing vigorous, fertile progeny.

This similarity between interracial and interspecific differences carries with it the implication that species may be derived from previously existing subspecies. The converse statement, that all subspecies are destined eventually to become distinct species, is, however, very far from the truth. The isolating barriers which separate species arise only occasionally, and until they appear the different subspecies of a species will be firmly bound to each other by ties of partial morphological and genetic continuity.

A further statement which may safely be made on the basis of existing knowledge is that many subspecies may become species without any further divergence in morphological characteristics. If, for instance, environmental changes should wipe out all of the populations of *Potentilla glandulosa* except those inhabiting the coast ranges and the high Sierra Nevada of California, then the surviving populations would have all the external characteristics of two different species. Furthermore, if the same changes should bring about the establishment in one of these population systems of chromosomal differences which would cause it to produce sterile hybrids with its relatives, then two species would have been differentiated out of one preexisting specific entity. At least some differentiation and divergence in morphological and physiological characteristics is the usual and perhaps the universal accompaniment of the process of species formation. But this divergence need not be different in kind or greater in degree than that responsible for the differentiation of races and subspecies within the species. The critical event in the origin of species is the breaking up of a previously continuous population system into two or more such systems that are morphologically discontinuous and reproductively isolated from each other.

THE EVOLUTIONARY SIGNIFICANCE OF SPECIES FORMATION

The characterization given above of the species-forming processes might seem at first glance to relegate speciation to a comparatively minor role in the drama of evolution. But this is by no means the case. As Muller (1940) has pointed out, the segregation of a previously interbreeding population system into two reproductively isolated segments tends to restrict the supply of genes available to each of these segments and tends to canalize them into certain paths of adaptation. Evolutionary specialization is therefore greatly furthered by the process of speciation.

This concept gains further importance when we realize that by far the greatest proportion of the diversity among living organisms reflects, not their adaptation to different habitats, but different ways of becoming adapted to the same habitat. Biological communities consisting of scores or hundreds of different species of animals and plants can exist in the same habitat because each species exploits the environment in a different way than its associates. The specificity of each different organism-environment relationship is maintained by the failure of the different species in the same habitat to interbreed successfully (Muller 1942). Speciation, therefore, may be looked upon as the initial stage in the divergence of evolutionary lines which can enrich the earth's biota by coexisting in the same habitat.

TYPES OF ISOLATING MECHANISMS

A classification of the different isolating mechanisms which form the barriers between species is given by Dobzhansky (1941,

p. 257). This has been somewhat simplified and adapted to the phenomena found in plants by Brieger (1944a), but his system has some undesirable characteristics, such as the failure to recognize the fundamental distinction between purely geographical or spatial isolation and the various types of reproductive isolation. The following classification is intended to combine the simplicity introduced by Brieger with the more advantageous arrangement established by Dobzhansky.

External Barriers:

I. Spatial isolation

II. Physiological isolation

A. Barriers between the parental species
1. Ecogeographical isolation
2. Ecological separation of sympatric types
3. Temporal and seasonal isolation
4. Mechanical isolation

Internal Barriers:

5. Prevention of fertilization

B. Barriers in the hybrids
1. Hybrid inviability or weakness
2. Failure of flowering in the hybrids
3. Hybrid sterility (genic and chromosomal)
4. Inviability and weakness of F_2 and later segregates

Grouping these isolating mechanisms in a somewhat different way, Darlington (1940), Muller (1942), and Stebbins (1942) have recognized two major subdivisions, external and internal isolating mechanisms. The extreme examples of these two categories, such as spatial isolation on the one hand and hybrid sterility on the other, are readily classified on this basis, but intermediate types, particularly the barriers in the reproductive phase of the parental species, are more difficult to place. Nevertheless, there will be reason in the following pages to discuss collectively all of these barriers which after artificial cross-pollination prevent the formation of hybrids or reduce their fertility. These will be termed internal barriers, since their principal action is within the tissues of the plant, as contrasted with external barriers, which prevent or reduce the frequency of cross-pollination between different species populations in nature. In the tabulation above, the external barriers are those under the headings I and IIA, 1–4, while the internal ones are listed under IIA, 5, and IIB, 1–4.

SPATIAL AND ECOLOGICAL ISOLATION IN RELATION TO SPECIES FORMATION

As a generalization, purely spatial or geographical isolation is, as Dobzhansky points out, on an entirely different plane from all the different reproductive isolating mechanisms. When applied to specific examples, however, the reality of this distinction is vitiated by two facts. In the first place, this type of isolation may persist over long periods of time without causing the isolated populations to diverge from each other enough to become recognizably distinct or even different. Species with disjunct distributions are well known to phytogeographers, and numerous examples are given by Fernald (1929, 1931), Cain (1944), and many other authors. Furthermore, as is pointed out in Chapter XII, the nature of many distributions strongly suggests that some of these disjunct segments of the same species have been isolated from each other for millions of years. There is some reason for believing, therefore, that geographic isolation alone often does not result even in the formation of subspecies. In continental areas, this type of isolation is important chiefly in conjunction with the divergent trends of selection produced by different ecological factors. In small populations, which are particularly frequent on oceanic islands, spatial isolation is the usual precursor to divergence in nonadaptive characters by means of genetic drift or random fixation.

Furthermore, geographically isolated races, as well as allopatric species, are usually separated from their relatives in adjacent regions by ecological barriers as well as by geographic ones. This is due to the fact that different geographic areas usually have different climatic and ecological conditions. In the great majority of examples of races, subspecies, or species which are kept apart because they live in different regions, the question may be justly asked if distance is the primary isolating factor, or if the initial separation and divergence has not been due to the response of the ancestral population to the selective effect of ecological differentiation in various parts of its originally more uniform environment. Most geographical races, therefore, may have arisen primarily as ecotypes, and their separate distributions may be an incidental secondary result of this ecotypic differentiation.

In many groups of plants, particularly those with very effective methods of seed dispersal, this distinction between strictly geographical isolation and ecogeographical isolation is of more than academic interest. For instance, a species of grass, pine, or goldenrod may be divided into two subspecies, one occurring in a more northerly area in a relatively cool, moist climate, and the other farther south in a warm, dry region. If one plots the distribution of the two subspecies on a map, one gains the impression that the isolation between them is primarily geographical. But since the seeds of plants can be borne by the wind or by animals for distances of scores or hundreds of miles, the seeds of such a northern subspecies would not infrequently settle in the territory occupied by a southern one, or vice versa. However, such seeds would not be likely to grow into adult plants, and so to transfer genes from one subspecies to another. This would be prevented by strong adverse selection pressure in the critical seedling stage.

A concrete example of such a situation is given by Baur (1932). *Antirrhinum glutinosum* is found in the Sierra Nevada of southern Spain at altitudes ranging from 800 m to 2,800 m on high mountain slopes which are snow-covered until late in May. Plants taken from the higher altitudes are highly frost resistant and possess a spreading habit which in nature causes them to be closely pressed against the cliffs on which they grow. Although all the plants occurring in nature had this growth habit, and the condition was retained when they were grown under cultivation, their offspring segregated strongly for this condition as well as for frost resistance, indicating heterozygosity of the wild plants for these two highly adaptive characteristics. Two conditions of the habitat of these *Antirrhinum* forms account for this situation. The seedling individuals possessing frost susceptibility and upright growth habit must be quickly eliminated by selection at the high altitudes. The maintenance in the high montane segment of the population of the genes for these strongly disadvantageous characteristics could be accomplished by the continual transfer from lower altitudes of gametes bearing them. Baur observed large numbers of small butterflies borne by the wind from the lower altitudes over the crest of the mountains without damage, and therefore concluded that the transfer of pollen over the necessary distance by these or other insects could occur.

In such examples the two subspecies are kept apart, not by the inability of individuals to cross the geographical barrier separating them, but by the ecological factors which keep them from reaching maturity when they do so. The same arguments apply to related allopatric species.

SPATIAL AND ECOGEOGRAPHIC ISOLATION AS SPECIES-FORMING BARRIERS

By far the commonest type of isolation in both the plant and the animal kingdom is that resulting from the existence of related types in different geographical regions which differ in the prevailing climatic and edaphic conditions. Numerous examples of this type of isolation have already been given in Chapters I and II. If the ranges of the types thus isolated from each other come in contact at any point, and if no other isolating mechanisms exist, then intergrading will occur along these points of contact, and the populations concerned can be regarded as subspecies. But if the related populations are separated from each other by large gaps in which no related forms are found, and if the habitats they occupy are widely different ecologically, then the specific status of these forms is subject to very different interpretations, depending on the concepts held by different biologists.

An example of this situation is found in the genus *Platanus*. The two best-known species of this genus, *P. occidentalis* and *P. orientalis,* occur respectively in the eastern United States and the eastern part of the Mediterranean region. They are markedly different and distinct both in vegetative and in fruiting characteristics, so that Fernald (1931) considers that they represent "sharp specific differentiations." Nevertheless, the artificial hybrid between them, *P. acerifolia,* is vigorous and highly fertile and has perfectly normal meiosis (Sax 1933). The chief isolating mechanism separating *P. occidentalis* and *P. orientalis* is, therefore, the great difference between their natural ranges. This isolation is not merely geographic. The climates prevailing in the eastern United States and the Mediterranean region are so different that types adapted to one of these regions will rarely, if ever, grow spontaneously in the other. Under natural conditions, therefore, these two species could not grow together spontaneously and exchange genes. They are sharply separated by barriers of ecological isolation, even though these are not evident at first glance.

Further studies of hybrids between widely allopatric species will probably reveal that the situation found in *Platanus* is not uncommon in the higher plants. In addition, there are doubtless many examples of generally recognized allopatric species which are sharply discontinuous from each other in morphological characteristics, but occupy regions which have similar climates and are ecologically equivalent. One such example is *Catalpa ovata* and *C. bignonioides.* The former of these two species is found in China and the latter in the eastern United States. These two regions have such similar climates and floras that species native to one of them could probably thrive in the other. Smith (1941) has found that the artificial hybrid between these two species is fully fertile. If, therefore, the geographic barrier between them were removed, they would very likely revert to subspecies of a single species.

The situations represented by *Platanus* and *Catalpa,* as well as many others in which geographic, ecological, and other types of isolation are combined in various degrees, present serious difficulties to the formulation and application of any precise biological species definition. If two sets of populations are ecologically similar and form fertile hybrids, so that they are separated only by spatial isolation, then they cannot be considered as reproductively isolated, and according to the strict application of any of the species concepts given earlier in this chapter they should be placed in the same species, no matter how different and distinct they are in external morphology. On the other hand, if two population systems are sympatric, but rarely or never form hybrids in nature because of their widely different ecological preferences, then they are often considered as distinct species, even though artificially produced hybrids between them are vigorous and fertile (Dobzhansky 1941, pp. 377–378). In order, therefore, to make valid inferences as to the specific status of allopatric, as well as sympatric, population systems, one must determine not only whether they can cross and produce fertile hybrids under the optimum conditions of a cultivated garden plot but, in addition, whether they could coexist in the same territory and hybridize under natural conditions.

Mayr (1942, p. 121) has suggested that in the case of allopatric forms which cannot be hybridized because they cannot be reared under domestication or cultivation, the decision as to whether they are distinct species or merely subspecies can be made by means of inferences. The amount of difference in characters of external morphology between the allopatric forms in question is compared with that between other related forms which are sympatric or at least overlapping in distribution. If this difference is about the same as that found between adjacent and intergrading subspecies, then the isolated, allopatric populations are also regarded as subspecies, but if the difference is greater, they are considered to be distinct species. In order to make such inferences in the case of allopatric populations of plants, one would have to take into account at least five sets of independent variables. These are hybrid sterility, hybrid inviability, cross-incompatibility, seasonal isolation, and ecological isolation. No one of these is closely correlated with morphological differences, as will be evident from numerous examples given later in this chapter. Obviously, therefore, in plants such inferences are very hazardous and unreliable.

These considerations lead us to the conclusion that for a large proportion of plant groups, particularly the hundreds of species of woody plants of tropical distribution, we cannot hope to apply within the foreseeable future any species concept based entirely on reproductive isolation. This leaves only three alternatives open to us. The first is to have two species concepts. One of these, the strictly biological one based on reproductive isolation, would have to be reserved for those groups in which experimental hybridizations have been made or in which the ranges of the species and races overlap to a sufficient extent so that the amount of intergradation or hybridization between them can be determined by studies of naturally occurring populations. The second species concept, which would have to be applied to all groups not suited to experimental studies and with their species and races largely allopatric, would be a much more general and admittedly artificial one. The fact would have to be recognized that the species in these two series of plant groups are not biologically comparable entities. All studies of plant distribution and of the relative amount of species formation in different plant groups would have to take into account the existence of these two different species concepts. This would be a most unsatisfactory situation from every point of view.

The second alternative would be to recognize that at any given moment in the evolutionary time scale, reproductive isolation is important in keeping distinct only those populations which are sympatric or which overlap in their distributions. If subjected to markedly different environments, allopatric populations can without the aid of any additional isolating mechanisms diverge to an extent sufficient to make them entirely distinct from each other in several characteristics. This degree of divergence could be accepted as the attainment of full specific status.

This principle of divergence to the point of sharp discontinuity in several characteristics as a species criterion could be applied to allopatric populations of woody plants in a reasonably precise and systematic fashion by the use of population studies as well as by the usual methods of systematics. In practice, the principle would be applied about as follows. Population A has been recognized by taxonomists as a woody species endemic to Mexico and sufficiently widespread so that millions of trees of it exist. Population B is found in Colombia and Venezuela and is likewise widespread. By studying quantitatively population samples from all parts of the area of distribution of both populations, the systematist could determine with some degree of accuracy the range of variation in each population of all the morphological characteristics by which the two populations might be expected to differ. Once this was done, he could estimate the probability that population B could produce by gene recombination individuals identical in all of their recognizable characteristics with at least some of the individuals present in population A. If this probability were high, or if these similar individuals were found in even a fraction of one percent of the hundreds or thousands studied, then populations A and B would have to be considered subspecies, no matter how different from each other their typical representatives were. But if in two or three genetically independent characteristics the range of variation in population A were entirely separate and distinct from that in population B, so that an individual of B could never be expected to match one of population A, then the two populations could be recognized as distinct species.

The third alternative would be to apply rigidly the criteria set up by Clausen, Keck, and Hiesey (1939), and to recognize as distinct species only those population systems which are separated from each other by internal isolating barriers, so that artificial cross-pollination under cultivation fails to produce hybrids or yields hybrids of reduced viability or fertility. This alternative has the practical difficulties mentioned in connection with the first alternative, but in an accentuated form. As will be pointed out later in this chapter, the final test of fertility of a hybrid in some cases involves raising to maturity progeny of the second generation, and our present knowledge suggests that examples of isolating barriers which do not become evident until the second generation may be particularly common in long-lived woody species, in which a single experiment of this type may occupy a whole lifetime. Furthermore, population systems are being discovered in which two related strains may yield highly sterile hybrids when crossed with one another, but may be able to exchange genes via a third strain, with which both of them can form partly or wholly fertile hybrids. The example of *Oryza sativa,* the cultivated rice, is discussed later in the chapter, and others are known in *Bromus* (Stebbins, unpublished) and *Layia* (J. Clausen, oral communication). The discussions in this chapter of the genera *Aquilegia, Quercus, Ceanothus, Vaccinium,* the family Orchidaceae, and many other groups will make the fact evident to the reader that the strict application of the scheme of Clausen, Keck, and Hiesey would make necessary a drastic revision of the species in many genera as they are now recognized by most working systematists. And the facts mentioned above indicate that whatever might be the theoretical advantages of such revisions, attempting to carry them out under the present state of our knowledge would be totally impracticable and would produce more confusion than order in the taxonomic system as a whole. The great amount of revision which would be necessary to make some groups conform to this system is evident from examples such as that of Osborn (1941), who showed that a hybrid between *Cupressus macrocarpa* and *Chamaecyparis nootkatensis,* produced in cultivation, is highly fertile and yields seedlings which segregate strongly for the parental characteristics. On the basis of this evidence, the suspicion arises that under the scheme of Clausen, Keck, and Hiesey all of the species of *Cupressus* would have to be united with those of *Chamaecyparis* into a single species.

Which of these alternatives is the most desirable would have to be left to the decision of competent systematic workers, trained in modern methods and concepts. Our present state of knowledge does not confirm the opinion, often given by experimental biologists, that only individuals are real and that the species is a purely man-made concept. But it nevertheless is true that there are a number of equally real biological situations to which the traditional concept of the species may be applied. In the opinion of the present writer, the principal task of the experimental taxonomist and the evolutionist should be to study these situations and to spend as little time as possible discussing the definition and application of terms. As our knowledge of the biological facts becomes more complete, we shall gradually achieve a firmer basis, consisting of a large fund of common knowledge, on which to erect our species concept. Until then, the wisest course would seem to avoid defining species too precisely and to be tolerant of somewhat different species concepts held by other workers. The one principle which is unavoidable is that species are based on discontinuities in the genetic basis of the variation pattern rather than on the amount of difference in their external appearance between extreme or even "typical" individual variants.

If we accept this latitude in our species definitions, then we can recognize the existence as species-isolating mechanisms of purely spatial isolation, strictly ecological isolation of sympatric forms, or various combinations of these two isolating factors. And the latter are by far the most common in nature.

ECOLOGICAL SEPARATION OF SYMPATRIC TYPES

Ecological isolation of sympatric species is, according to Mayr (1942, 1947), rather uncommon in animals. In plants, on the other hand, several examples are known. The various species of *Quercus* discussed in Chapter II are excellent ones. Another genus in which ecological isolation plays an important role is *Ceanothus*. This genus of shrubs contains many species in California, some of which present considerable difficulties to the systematist. Hybridization between recognized species is not uncommon both in nature and in gardens where various species are cultivated as ornamentals (McMinn 1942, 1944). All the species have the same number of chromosomes ($n = 12$, McMinn

1942) and similar chromosome complements, meiosis in interspecific hybrids is often quite regular, and many of the hybrids are highly fertile. Nevertheless, some of these species are sympatric over the entire extent of their geographic ranges and remain in general distinct from each other.

Typical examples are *Ceanothus thyrsiflorus*, *C. papillosus*, and *C. dentatus* in the central coast ranges of California. *C. thyrsiflorus* is a tall, vigorous shrub with large leaves and ample panicles of usually pale blue flowers, common on hillsides with relatively good conditions of soil and moisture. *C. papillosus* and *C. dentatus* are smaller and have much smaller leaves with very distinctive shapes and indumentum and small inflorescences of deep blue flowers. They grow on more exposed, drier sites, often with poor or shallow soil. They are more local in their occurrence than *C. thyrsiflorus*, and although the restricted range of *C. dentatus* is entirely included within the geographic distribution of the other two species, no locality is known to the writer which supports both *C. papillosus* and *C. dentatus*.

Evidence that the isolation between *C. thyrsiflorus* and *C. dentatus* is chiefly ecological is provided by their behavior at one locality in which they grow together and in which ecological conditions have been disturbed by the hand of man. On the northwest side of the Monterey Peninsula is a hillside traversed by several shallow gullies and low ridges, on the latter of which the soil is shallow and poor. The normal cover of this hillside is a dense stand of *Pinus radiata*, *P. muricata*, *Arctostaphylos tomentosa*, and *A. hookeri*, among which are mingled scattered shrubs of other species, including *Ceanothus thyrsiflorus* in the gullies and *C. dentatus* on the exposed rock outcrops of the ridges. Along this hillside have been cut several firebreaks or roadways, from which most of the pines and manzanitas (*Arctostaphylos*) have been removed. These have grown up to a luxuriant second growth, in which *Ceanothus* is predominant. On the better soil near the gullies, the species is *C. thyrsiflorus*, while the rocky outcrops on the crests of the ridges support the prostrate, small-leaved species, *C. dentatus*. The flanks of the ridges are occupied by intermediate types, which include every conceivable intergradation between and recombination of the characters of *C. thyrsiflorus* and *C. dentatus*. That these intermediates are of hybrid origin is evident from the fact that seedling progenies raised from them are distinctly more variable than those from typical shrubs of the parental species, and they also show some segregation in the direction of their presumed parents. Under natural conditions, therefore, *C. thyrsiflorus* and *C. dentatus* are distinct and have been so for probably hundreds of thousands of years, while occupying different ecological niches in the same geographic region. The creation of a new habitat by man has at least temporarily broken down the ecological isolation which previously separated them and has therefore reduced greatly their distinctness from each other.

One of the earliest described examples of ecological isolation is that of *Viola arvensis* and *V. tricolor* in Denmark (Clausen 1921, 1922). These two species have different chromosome numbers ($n = 17$ in *V. arvensis* and $n = 13$ in *V. tricolor*), but hybrids between them are fertile and segregate in the F_2 generation to give a great variety of recombination types. In nature, only a very small proportion of the possible recombinations is found in any abundance, and these approach two modes, one of them small-flowered, with yellowish-white petals and pinnate stipules (*V. arvensis*), and the other large-flowered, with blue or violet petals and palmate stipules (*V. tricolor*). The former is strongly calcicolous, while *V. tricolor* is found mainly on acid soils. Intermediate and recombination types can compete successfully only on neutral or faintly acid soil. Since chromosome numbers intermediate between 17 and 13 have not become established in this group, this cytological difference probably reinforces the ecological isolation in keeping the two species separate.

Another good example of ecological isolation is provided by two species of goldenrod of the eastern United States, *Solidago rugosa* and *S. sempervirens* (Goodwin 1937). The former is characteristic of dry, sterile fields over a large area, while *S. sempervirens* occurs in salt marshes along the entire coast. In general, they maintain themselves as distinct, but the hybrid between them, *S. asperula,* is not uncommon. In most localities, only solitary or few hybrid plants are found, but where the transition from a salt marsh to a neighboring field passes through a band of brackish marsh, rather extensive hybrid swarms may occur. Goodwin found that F_1 hybrids between *S. rugosa* and *S.*

sempervirens are fully fertile, but that seed from these plants germinates rather poorly and yields many weak F_2 individuals. In this example, therefore, the two species are kept apart both by ecological isolation and by hybrid sterility effective in the second generation.

The species of *Iris* growing in the Mississippi Delta afford an equally striking and well-analyzed example. The two most common ones are *I. fulva,* a slender, reddish-flowered species inhabiting the clay alluvial soil of natural levees in the shade of the swamp trees, and *I. hexagona* var. *giganticaerulea,* a stouter species with violet-purple flowers which is found in the mucky soil of alluvial marshes, usually growing in full sun (Viosca 1935). When these habitats are disturbed, through the cutting of trees in the habitat of *I. fulva* and the ditching and draining of the marshes occupied by *I. giganticaerulea,* opportunity is provided for hybridization on a large scale and the establishment of the hybrid derivatives in the new habitats formed by these disturbances (Viosca 1935, Riley 1938, Anderson 1949). In *Iris,* as in *Solidago* and *Ceanothus,* the evidence that ecological isolation is the principal barrier between the species concerned is strengthened by the fact that a disturbance of this barrier causes the appearance of large numbers of hybrids and hybrid derivatives, and thereby a great increase of gene flow between the two species.

Marsden-Jones and Turrill (1947) have described a good example of ecological isolation in the flora of Great Britain. *Silene maritima,* a plant of pebble beaches, sea bluffs, scree (talus), and rock faces, is distinguished by several easily recognizable characters from *S. vulgaris,* a common plant of fields and roadsides. When crossed artificially, the two species form vigorous, fertile hybrids in both the F_1 and later generations, so that internal isolating barriers between them are absent. Yet the authors have never observed intergradation or evidence of extensive hybridization between these two species in nature, apparently because their natural habitats are so different that they never have an opportunity to cross. This may be connected with the fact that *S. vulgaris* has entered Britain rather recently, perhaps only since the advent of agricultural man. Before *S. maritima* can be considered completely distinct and isolated from *S. vulgaris,* the absence of intergradation should also be demonstrated for their populations on all parts of the European continent within the range of *S. maritima*. Other examples of sympatric species separated mainly by ecological isolation are *Geum urbanum* and *G. rivale* (Marsden-Jones 1930), *Silene cucubalus* and *S. alpina* (Turrill 1936b), and *Cistus monspeliensis* and *C. salviifolius* (Dansereau 1939).

Ecological isolation as an accessory mechanism, acting to reinforce the discontinuity produced by seasonal isolation or partial hybrid sterility, is a rather frequent phenomenon. Further examples of it will be given later in this chapter.

SEASONAL ISOLATION

Isolation of two populations because of different blooming seasons is occasionally found in plants. One of the best examples is *Lactuca canadensis* and *L. graminifolia* (Whitaker 1944). These two species, although rather different morphologically, have the same chromosome number ($n = 17$), and artificial hybrids between them are fully fertile. In the southeastern United States, where they occur together as common roadside weeds over a large area, they are ordinarily kept distinct through the fact that *L. graminifolia* blooms in the early spring and *L. canadensis* in the summer. They sometimes overlap slightly in their natural blooming periods, and under these conditions hybrid swarms can occasionally be developed.

An example of seasonal isolation which appears to be effective in one part of the range of a species but not in another is that of a Chinese species of the Compositae, tribe Cichorieae, *Ixeris denticulata* (Stebbins 1937). In northern China are found two subspecies: *typica*, which blooms in the fall, and *sonchifolia*, which is a spring bloomer. In this region the two forms are amply distinct in external morphology, and one systematist has placed them in different genera. But in western and southwestern China there occurs another subspecies, subsp. *elegans*, which blooms in the summer. Judging from herbarium specimens, subsp. *elegans* is connected by a series of transitional forms to both *typica* and *sonchifolia*. Both *typica* and *sonchifolia* have the chromosome number $2n = 10$ (Babcock, Stebbins, and Jenkins 1937) and similar chromosome size and morphology. Although they have not been crossed, there is good reason to believe that the hybrid between them would be vigorous and fertile, since *I. denticulata* subsp. *typica* forms fertile hybrids with the more distantly related *I.* (*Crepidiastrum*) *lanceolatum* (Ono and Sato 1935, 1937, 1941).

Seasonal isolation as a means of reinforcing ecological and geographic isolation is well illustrated in certain species of pines of the California coast. *Pinus radiata* and *P. muricata* are found together on the Monterey Peninsula, in the site previously described for *Ceanothus thyrsiflorus* and *C. dentatus.* The two species are amply distinct morphologically; *Pinus radiata* has needles in groups of three, while those of *P. muricata* are in two's, and the color and texture of the needles is very different. The cones of *P. radiata* are relatively large, asymmetrical, and with blunt cone scales, while those of *P. muricata* are smaller, more symmetrical, and with sharp pointed scales. At Monterey, *P. muricata* occupies the poorer sites of the ridge crests, and its stands are therefore completely surrounded by a dense forest of *P. radiata,* which grows in the better soils. Nevertheless, observations of Drs. Palmer Stockwell, H. L. Mason, and the writer indicate that intermediate trees, while not infrequent, form only a fraction of 1 percent of the stand and appear less vigorous and productive than typical *P. radiata* or *P. muricata.* This fact is probably due partly to the poorer adaptation of the intermediate gene combinations, but the possibility of hybridization is greatly reduced by the fact that in normal seasons in this area, *P. radiata* sheds its pollen early in February, while the pollen of *P. muricata* is not shed until April.

P. radiata is kept apart from another species of this group, *P. attenuata,* by a combination of geographical, ecological, and seasonal isolation. The distribution of *P. radiata* is strictly coastal and it occurs in three widely separated localities. That of *P. attenuata* is mostly in the middle and inner coast ranges and in the Sierra Nevada. But near Point Ano Nuevo, in Santa Cruz County, *P. attenuata* grows near the coast and adjacent to *P. radiata.* In this region they appear to be ecologically isolated, since *P. radiata* grows only on the seaward side of the ridges, while *P. attenuata* is confined to their crests on the landward side, out of sight of the ocean. They never grow intermingled, but in one place the forest of *P. radiata* is immediately bordered by an open stand of *P. attenuata.* In this area intermediates have been found, but, as in the previous examples, they form less than 1 percent of the population. Seasonal isolation also enters here. The pollen of *P. attenuata* is shed in April, and therefore much later than that of *P. radiata,* at least in this locality. Hybrids beween these two species produced at the Institute of Forest Genetics, Placerville, have proved vigorous and fertile in both the F_1 and the F_2 generations (Stockwell and Righter 1946).

MECHANICAL ISOLATION

Among the three types of isolating mechanisms which act in the reproductive phase of the parental species, mechanical isolation probably plays the most important role. It is particularly effective in plants with elaborate floral structures adapted to insect pollination, like the milkweeds (Asclepiadaceae) and Orchidaceae. In the former group, the sticky pollen is contained in sacs or pollinia, which are attached in pairs by a clip. This structure is so designed that insects which visit the flowers of *Asclepias* for their nectar invariably pick up one or more of these pollinia on their legs. Pollination is completed by the insertion of the pollinia into slits in the stigma of the flower, which match in size and shape the pollinia of the same species. Nevertheless, as observed by Woodson (1941), the insertion of the pollen clips into the stigmatic slits is a difficult operation, and for this reason only a small proportion of the flowers of *Asclepias* ever become pollinated. This seemingly inefficient mechanism for cross-pollination is in one way very efficient. The various species of this genus differ from each other in the shape of both their pollinia and their stigmatic slits. These differences are large enough so that insertion of the pollinium of one species into the stigma of another is well-nigh impossible. The result is that in spite of the fact that several species of *Asclepias* exist side by side in the eastern United States and many other parts of the world, interspecific hybridization in nature is almost completely absent, and the species are very sharply set off from one another (Woodson 1941). Mechanical isolation, although reducing greatly the reproductive potential of individual plants, maintains the integrity and thereby the adaptive peak represented by the species. The selective disadvantage of a low percentage of fertilized flowers is probably less in

Asclepias than it is in many other genera. Each individual plant is very long-lived and produces large numbers of flowers each year; while the capsule which is formed whenever pollination is effective contains large numbers of seeds. These factors undoubtedly contribute toward the ability of *A. syriaca* and other species to become aggressive weeds, in spite of their apparently inefficient pollination mechanism.

The elaborate floral structures of the orchids probably serve a similar purpose. That the often bizarre and fantastic shapes of these flowers are adapted to securing cross-pollination by insects has long been recognized, and some of them are graphically described in the classic work of Darwin (1862). Although studies of this nature have unfortunately been largely neglected in recent years, one device for insect pollination unknown to Darwin and his contemporaries has been explored. This is the phenomenon of pseudocopulation, in which resemblance of the orchid flower to a female insect attracts the male of that species and causes him to attempt his normal biological function. During this attempt, the insect delivers pollen to the stigma of the orchid flower and receives pollinia from it. According to the excellent review of this subject by Ames (1937), the first example known of pseudocopulation is that of the Australian species, *Cryptostylis leptochila,* which lures the fly *Lissopimpla semipunctata.* Still more interesting and instructive are the examples described by Pouyanne of various species of the European and North African genus *Orphrys,* which are pollinated by male bees belonging to the genera *Andrena* and *Scolia.* The differences between the flowers of these species of *Orphrys* consist partly of superficial resemblances to the females of different species of bees, so that each orchid species attracts only one species of bee. Furthermore, the resemblance is secured in entirely different ways by different species of orchids. For instance, in *Orphrys fusca* and *O. lutea,* which attracts species of *Andrena,* the "abdomen" of the imitation female is oriented so that the male alights with his posterior end adjacent to the stigmatic column, and so receives and delivers the pollen on his abdomen; while *Orphrys speculum* attracts the male of *Scolia ciliata* in such a way that he receives and delivers the pollen with his head (Figure 22).

Pseudocopulation may be a more widespread phenomenon

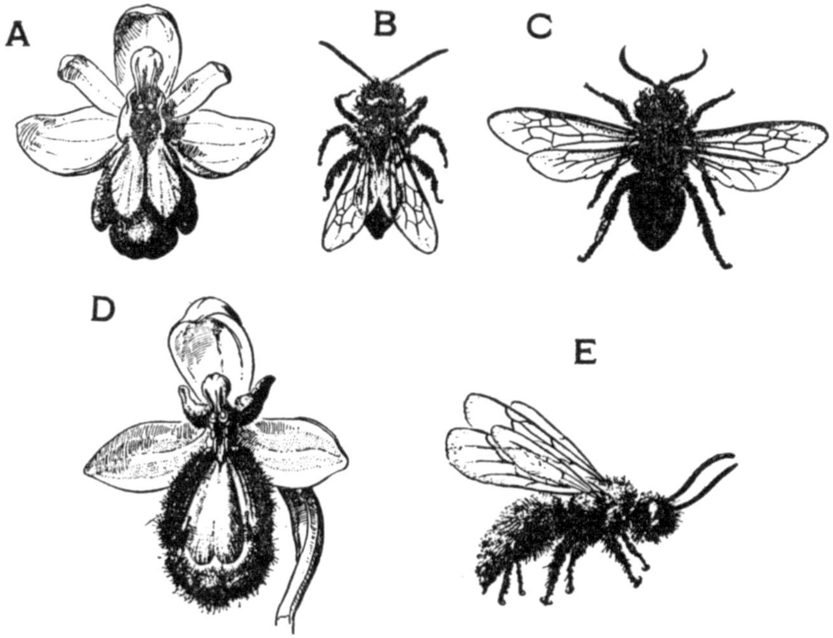

Fig. 22. A, a single flower of *Orphrys fusca;* B and C, males of two species of bees (*Andrena nigroaenea* and *A. trimmerana*) which pollinate it; D, a flower of *Orphrys speculum;* and E, a male bee of the species (*Scolia ciliata*) which pollinates it. From Ames 1937.

than our present knowledge indicates. Dr. W. H. Camp has told the writer of observing the pollination of orchid species in the South American tropics by the males of Syrphid flies. As a method of securing pollination of individual flowers it may not be any more efficient than the more familiar and widespread one of attracting female insects in search of pollen or nectar. But the specificity of this device obviously helps greatly to prevent accidental cross-pollination between different species, and therefore to strengthen reproductive isolation.

In the orchids even more than in the asclepiads the nature of their pollination mechanisms seems to be well adjusted to their mode of life as a whole and to be partly responsible for the pattern of variation and evolution found in the family. Orchids are very long-lived as individual plants, so that failure of seed formation in any one year is no great detriment to them. Furthermore, the number of seeds produced in a single capsule is perhaps the largest in the entire plant kingdom, so that the number of successful pollinations required is less than in most other families. In the temperate regions, with their climatic vicissitudes and comparative poverty of insect life, this system has been only a partial success, and orchids are for the most part rare and comparatively poor in species. But the moist tropics, with their optimum climatic conditions and richness of insect fauna, have proved to be such a favorable environment for the development of these mechanisms for cross-pollination and for reproductive isolation between species that in most of the moister parts of the tropics the orchids lead all other families in the number of their species, and often also in number of individuals.

Recent observations of Mather (1947) have shown that in flowers with an elaborate and distinctive structure and color mechanical isolation can be achieved between two populations even though they are pollinated by the same species of insect. Cultures of two different species of the genus *Antirrhinum,* sect. *Antirrhinastrum, A. majus* and *A. glutinosum,* were grown in alternating blocks of rows, so as to ensure the maximum chance for cross-pollination. The species have long been known to be cross-compatible and *A. glutinosum* is self-incompatible, so that artificial hybrids are obtained with the greatest of ease (Baur 1924, 1932). Nevertheless, in seed harvested from these plots only a very small percentage of hybrids was secured. Following the course of individual bees, Mather found that in any particular flight they nearly always confine their visits to the flowers of one species. After pollinating the flowers of several plants of the species first visited, they not infrequently fly directly over several rows of other species to reach another plant of the one on which they began their rounds.

PREVENTION OF FERTILIZATION

The two other barriers which act in the reproductive phase of the parental species, namely, failure or ineffectiveness of pollen tube growth and failure of fertilization, are rarely the primary cause of reproductive isolation between closely related species. Kostoff (1943, pp. 704–706) lists 68 interspecific combinations in the genus *Nicotiana* which failed because the pollen tubes were unable to reach the ovules. But 50 of these were between species having different chromosome numbers, and the hybrids, had they been obtained, would probably have been weak, sterile, or both. Of the remainder, all but one were intersectional, and similar hybrids in which F_1 plants were obtained always proved to be completely sterile. The cause for this failure of pollen tube growth is often the fact that the style of the maternal species is much longer than that of the species which contributes the pollen, so that the pollen grains are not adapted to traversing the long distance from the stigma to the ovary. In addition, the relatively large, thick pollen tubes produced by some polyploid species have difficulty in penetrating the smaller styles of diploids, so that in some combinations between diploids and tetraploids fertilization fails when the tetraploid is the pollen parent, but succeeds when it is the maternal one.

In the genus *Datura,* Buchholz, Williams, and Blakeslee (1935) found that pollen tubes grow faster in styles belonging to their own species than in foreign ones, and that they often burst in the styles of foreign species. In this genus, however, the species are still more strongly isolated by barriers of hybrid inviability, so that even though pollen tube growth and fertilization are successful, hybrids are rarely obtained (see below). Similarly, *Zea mays* and various species of *Tripsacum* are isolated from each other by the inability of the *Zea* pollen grains and male gametes to function in the ovules of *Tripsacum* and the barrier to the reciprocal hybrid caused by the great length of the styles, or silks, in maize (Mangelsdorf and Reeves 1939). Mangelsdorf secured hybrids between the two genera by artificially reducing the style length of *Zea,* but the high degree of pollen and seed sterility in these F_1 hybrids is perhaps a more potent barrier to gene interchange between these two species than is their initial incompatibility. Difference between the osmotic tension of pollen grains and style tissues is often a barrier to hybridization between species having different chromosome numbers (Watkins 1932, Boyes and Thompson 1937, Schlösser 1936), but in these examples sterility of F_1 hybrids serves as an additional and perhaps a more significant barrier of reproductive isolation between the species concerned.

HYBRID INVIABILITY OR WEAKNESS

The barriers which act by preventing the growth or reproduction of F_1 hybrids are probably the most fundamental and certainly the most widespread of those which separate plant species. They may become manifest at any stage from the beginning of the growth of the hybrid zygote to the maturity of the segregating genotypes of the F_2 generation. The term hybrid inviability may be applied to the inability of the hybrid zygote to grow into a normal embryo under the usual conditions of seed development. It may be caused by disharmony either between the parental sets of chromosomes or between the developing embryo and the endosperm which surrounds it. If the former is the case, then normal mature hybrids cannot be secured under any conditions.

If the growth of the embryo is inhibited by the endosperm, vigorous embryos can often be obtained by dissecting the young embryo from the seed and growing it artificially in a nutrient solution. One of the earliest successful experiments of this nature was that of Laibach (1925) on *Linum perenne* × *austriacum* and its reciprocal. When *L. perenne* is the maternal parent, the embryo reaches maturity, but cannot sprout from the seed. If dissected out artificially, it grows into a vigorous and fertile F_1 hybrid. In the reciprocal cross, *L. austriacum* × *perenne,* hybrids can be obtained by dissecting the very young embryos out of the developing seed and growing them in a nutrient solution. Similar success was obtained by this means in crosses between peach varieties by Blake (1939) and Lammerts (1942), in *Secale cereale* × *Hordeum jubatum* by Brink, Cooper, and Ausherman (1944), and particularly in hybrids of the genus *Datura* by Blakeslee, Satina, and their associates (Blakeslee and Satina 1944, Blakeslee 1945, McLean 1946).

What may be another way of overcoming a barrier of hybrid inviability is described by Pissarev and Vinogradova (1944) in crosses between *Triticum* and *Elymus.* Young embryos of *Triticum vulgare* were dissected out and grafted onto the endosperm in ovules of *Elymus arenarius.* There they finished their growth and germinated from *Elymus* seeds. The mature plants obtained by this method were typical of *Triticum* in appearance, but their reaction to hybridization with *Elymus* differed in that they could be used successfully as maternal parents in the cross *Triticum vulgare* × *Elymus arenarius,* while *Triticum* plants of the same variety grown from normal seed could not function as the pollen or as the female parent in the same hybrid combination. This remarkable result should be repeated with this or other material.

As yet very little is known of the physiological or developmental nature of hybrid inviability, and its genetic basis has been discovered in only a few instances. A well-known example is the interspecific lethal gene found by Hollingshead (1930a) in *Crepis tectorum,* which has no effect on its own species, but causes the early death of the hybrid seedlings in crosses with *C. capillaris, C. leontodontoides,* and *C. bursifolia.* This gene, however, is not the principal agent for isolating the species concerned. The viable hybrids produced when the *C. tectorum* parent does not carry this gene are in every case highly sterile, due to lack of pairing and irregularity of chromosome behavior at meiosis. In the genus *Hutchinsia,* Melchers (1939) has found a similar gene isolating a certain strain of *H. alpina* from other races of this species and from *H. brevicaulis.* In this case the vigorous hybrid derived from strains of subsp. *typica* not carrying the lethal gene is fertile, so that the establishment of this gene in all the individuals of any isolated subpopulation of *H. alpina* would transform such a population into an incipient new species. Similar genes have been found by Brieger (1944a) in *Nicotiana longiflora* × *sanderae.*

In some instances, the hybrid embryo may produce a plant which is too weak to reach maturity or flower normally. A spectacular type of hybrid weakness due to genic disharmony is found in certain hybrids of the genus *Nicotiana,* which form tumors in their vegetative parts (Kostoff 1930, 1943, pp. 613–16, Whitaker 1934b).

In the genus *Epilobium* the extensive studies of Michaelis (1931, 1933, 1938, 1940, Michaelis and Wertz 1935), Lehmann (1931, 1936, 1939, 1942, 1944), and others have revealed the presence of barriers of hybrid weakness and sterility which result from the interaction of certain hybrid genotypes with the plasma of one of the parental species. For instance, the F_1 hybrid *E. luteum* × *hirsutum* is vigorous and fertile when *E. luteum* is the maternal plant, but in the reciprocal cross, in the plasma of *E.*

hirsutum, is weak and sterile. Michaelis, by pollinating the vigorous *E. luteum* × *hirsutum* F_1 with *E. hirsutum* pollen and continuing this type of backcross for 13 generations, obtained a plant which resembled *E. hirsutum* in all of its external characteristics, but had certain cytoplasmic characteristics of *E. luteum.* It formed vigorous F_1 hybrids when pollinated with *E. luteum,* and its cytoplasm resembled that of the latter species in its relatively high permeability to electrolytes and its susceptibility to sudden changes in temperature. Both Michaelis and Lehmann demonstrated numerous genetic differences between different strains of *E. hirsutum,* as well as the related *E. roseum* and *E. montanum,* in their reaction to *E. luteum* and other species. That these differences are genically controlled is shown by the fact that, for instance (Lehmann 1939), an F_1 hybrid between two strains of *E. hirsutum,* when outcrossed to *E. adenocaulon,* gives two types of interspecific hybrids in a 1:1 ratio. Furthermore, some races of *E. hirsutum* give inhibitions of growth when crossed with other races of the same species (Lehmann 1942, Brücher 1943).

The explanation of these results is not clear. That, in contrast to the majority of plant species which have been compared by means of hybridization, the species of *Epilobium* differ markedly in their cytoplasm, is certain. A similar phenomenon appears with respect to the plastids in the genus *Oenothera* (Renner 1936), which likewise belongs to the family Onagraceae. Cytoplasmic and plastid differences of such a striking nature may be a peculiarity of this and some other families of seed plants. The constancy of the plasma and its apparent partial independence of the genotype in *Epilobium* are as yet unique among known examples in the higher plants, although similar phenomena have been found by Wettstein (1937) in hybridizations between species and genera of mosses. Here, however, the reciprocal differences are not inhibitions of growth, but alterations in leaf shape, capsule shape, and other characteristics. There are some other examples of reciprocal differences in hybrids between plant species, but none of these have been investigated with anything like the thoroughness of those mentioned above. Good reviews of this subject are those of Correns and Wettstein (1937) and Caspari (1948).

HYBRID STERILITY

In his thorough review of the subject of hybrid sterility, Dobzhansky (1941) has shown clearly that the failure of hybrids to produce offspring is not connected with any constitutional weakness of the organism as a whole, but rather with specific disharmonies between the parental gene complexes which act during the development of the gonads, at the time of meiosis, or during the later development of the gametophytes or gametes. The earliest classifications of the phenomena of hybrid sterility were based on the time when the sterility becomes manifest. Thus, Renner (1929) distinguishes between gametic and zygotic sterility. The former is the degeneration of the spore tetrads and gametophytes between the time of meiosis and anthesis, and is recognized by the abortion of pollen or ovules, while the latter is the failure of the zygote to develop after fertilization, and is manifest through the abortion of seeds even though the ovules and pollen are normal. Müntzing (1930a) distinguishes between haplontic sterility, which is the gametic sterility of Renner, and diplontic sterility, which may act either in the diploid tissue of the F_1 hybrid before meiosis or in the zygotes of its offspring. Both of these classifications are useful for certain purposes, but the separation of hybrid sterility phenomena into two types, genic sterility and chromosomal sterility (Dobzhansky 1933, 1941, p. 292) is more fundamental. Genic sterility includes all types which are produced by failure of the sex organs to develop up to the point where meiosis can take place or by genically controlled abnormalities of the meiotic process itself, such as abnormalities of spindle formation and genically controlled asynapsis or desynapsis (Clark 1940, Beadle 1930, Li, Pao, and Li 1945, etc., see review in Stebbins 1945). Chromosomal sterility, on the other hand, results from lack of homology between the chromosomes of the parents of a hybrid. When the parental species differ radically from each other in chromosome number, as in the classic example of *Drosera longifolia* × *rotundifolia* (Rosenberg 1909), or when these parents are distantly related to each other, as in the cases of *Nicotiana sylvestris* × *tomentosiformis* (*"Rusbyi,"* Goodspeed and Clausen 1928), *Raphanus sativus* × *Brassica oleracea* (Karpechenko 1927, 1928), and *Gossypium arboreum* × *thurberianum* (Skovsted 1934, Beasley 1942), chromosomal sterility is manifest in the presence of unpaired chromosomes, irregular behavior of the chromosomes at later stages of meiosis, and the passage to the micro- and megaspores of variable numbers of chromosomes. Such obvious irregularities, leading to the possession by the gametes of highly unbalanced chromosomal complements, will, of course, lead to complete sterility of the F_1 hybrid except for the occasional formation and functioning of gametes with the unreduced chromosome number and the consequent production of allopolyploids. With somewhat more closely related species, such as *Lilium martagon album* × *hansonii* (Richardson 1936), *Allium cepa* × *fistulosum* (Emsweller and Jones 1935, 1938, 1945, Maeda 1937, Levan 1936a, 1941b), various *Paeonia* hybrids (Stebbins 1938), and others, pairing of the chromosomes may be nearly complete but show configurations typical of heterozygosity for inversions, translocations, and so on. If, on the other hand, the chromosomes of the parents are similar in number and differ relatively slightly in the nature and arrangement of their gene loci, then the presence of and the reasons for the resulting chromosomal sterility will be far less obvious.

The classic example of this less pronounced type of chromosomal sterility is *Primula verticillata* × *floribunda*. Both parental species here have the haploid chromosome number $n = 9$, and although the F_1 hybrid is completely sterile, its chromosomes usually form at meiosis nine loosely associated bivalents (Newton and Pellew 1929, Upcott 1939). Without further data, one would suspect this to be an example of genic sterility, induced by some type of complementary factor system. But when the somatic chromosome number of this F_1 hybrid is doubled, the resulting allopolyploid, *P. kewensis*, is fully fertile and forms mostly bivalents at meiosis. Although differences in the segregation ratios of genes in a tetraploid as compared to a diploid could account for a considerable difference in their fertility on the basis of the action of complementary gene factors, it could not account for the change from complete sterility to full fertility. The explanation of this change, as suggested by Darlington (1937, pp. 198–199) and Dobzhansky (1941, p. 328), is best made as follows. If we assume that the chromosomes of *P. floribunda* resemble those of *P. verticillata* in respect to their centromeres and at least one of their ends, then they can pair and form chiasmata in an apparently normal fashion. But if differences exist, in the form of small, nonhomologous segments, then this apparently normal pairing actually involves different kinds of chromosomes, and the reduction in chiasma frequency which is found in *P. verticillata* × *floribunda* and similar hybrids is due to the inability of certain segments of their chromosomes to pair and to form chiasmata. The gametes produced by such hybrids are inviable, not because they have deviating chromosome numbers, but because they contain unbalanced, disharmonious combinations of chromosomal material. The production of an allopolyploid by somatic doubling of the chromosome number should in this case result in cells with two sets of exactly similar chromosomes. Pairing and separation of chromosomes in sporocytes consisting of such doubled cells would produce gametes which, in both the number and the genetic content of their chromosomes, would correspond to the somatic cells of the F_1 hybrid and would therefore be perfectly viable and functional. Furthermore, since each chromosome would find in the sporocytes of the doubled tissue a mate exactly like itself, it would pair only occasionally with the partially homologous chromosome derived from the opposite species. In the terminology of Darlington (1937, pp. 198–9), the differential affinity between exactly similar as compared to partly homologous chromosomes would lead to preferential pairing between the chromosomes derived from the same parental species. This explains the fact that although the diploid F_1 hybrid *P. verticillata* × *floribunda* usually forms nine bivalents, the allotetraploid *P. kewensis* forms only 0 to 3 quadrivalents.

The situation can be represented diagrammatically as follows, where F represents any chromosome of *P. floribunda*, V a chromosome of *P. verticillata*, while V_f and F_v represent chromosomes containing mixtures of parental chromosomal material.

Meiosis in parents:	$\frac{F}{F}$ and $\frac{V}{V}$
Constitution of parental gametes:	F and V
Constitution of F_1 somatic cells:	FV
Pairing at meiosis in F_1:	$\frac{F}{V}$, with crossing over

Constitution of F_1 gametes:	F_v and V_f, all or nearly all inviable
Constitution of allopolyploid somatic cells:	FFVV
Pairing at meiosis in allopolyploid:	$\frac{F}{F}, \frac{V}{V}$
Gametes of allopolyploid:	all FV and viable.

It is thus evident that, as Dobzhansky has pointed out, many examples of chromosomal sterility can be detected only by the chromosome behavior and fertility of the allopolyploid derived from the F_1 hybrid. The possibility that sterility in hybrids with apparently normal meiosis might be caused by structural hybridity involving small chromosomal segments was first suggested by Sax (1933) for *Campsis chinensis* × *radicans*. Müntzing (1938) adopted this explanation for the partial sterility of *Galeopsis tetrahit* × *bifida*, as well as for a similar phenomenon in crosses between certain races of *G. tetrahit*. In the latter examples, he obtained strong circumstantial evidence in favor of his hypothesis. Certain interracial F_1 hybrids within *G. tetrahit* form 50 percent of inviable pollen in spite of the presence of 16 pairs of chromosomes at meiosis. In other combinations, the same degree of sterility is accompanied by the occasional presence of two univalents at meiosis, while still other 50 percent sterile interracial F_1 hybrids form at meiosis 14 bivalents and a chain or ring of four chromosomes. This latter configuration is, of course, strong evidence of heterozygosity for interchange of a chromosomal segment. The similar nature of all of the partly sterile interracial hybrids in *G. tetrahit* suggests that their sterility is due to a common cause, structural hybridity, which may sometimes become manifest in the nature of the chromosome pairing, but most often does not.

The widespread occurrence in the higher plants of chromosomal sterility due to heterozygosity for small structural differences is evident from the fact that a large proportion of hybrids between closely related species having the same chromosome number exhibit apparently normal chromosome behavior at meiosis in spite of their partial or complete sterility. Such hybrids have been reported in *Apocynum* (Anderson 1936a), in *Bromus* (Stebbins and Tobgy 1944), in *Ceanothus* (McMinn 1942, 1944), in *Collinsia* (Hiorth 1934), in *Crepis* (Babcock and Emsweller 1936, Jenkins 1939), in *Festuca elatior* × *Lolium perenne* (Peto 1933), in *Ixeris*, subg. *Crepidiastrum* × subg. *Paraixeris* (Ono and Sato 1935, Ono 1937, 1941), in *Lactuca* (Whitaker and Thompson 1941), in *Layia* (Clausen, Keck, and Hiesey 1941), in *Lycopersicum* (Lesley and Lesley 1943, MacArthur and Chiasson 1947), in *Melica* (Joranson 1944), in *Papaver* (Fabergé 1944), in *Phaseolus* (Lamprecht 1941), in *Polygonatum* (Suomalainen 1941), in *Populus* (Peto 1938), in *Setaria* (Li, Li, and Pao 1945), in *Solanum* (Propach 1940, Tatebe 1936), in *Taraxacum* (Poddubnaja-Arnoldi 1939), in *Tradescantia* (Anderson and Sax 1936), in *Tragopogon* (Winge 1938), in *Verbena* (Dermen 1936, Schnack and Covas 1945a,b), in *Elymus* × *Sitanion* (Stebbins, Valencia, and Valencia 1946a), and probably in others. In *Lycopersicum esculentum* × *peruvianum* and some of the *Solanum* hybrids evidence from allopolyploids has shown that their sterility is genic and is retained in the polyploid, while those of *Bromus* (Stebbins 1949b), *Melica, Tradescantia* (Skirm 1942), and *Elymus* × *Sitanion* recovered their fertility upon doubling, as did *Primula kewensis*. Present evidence suggests, therefore, that in the great majority of hybrids between closely related species the decision as to whether the sterility is genic or chromosomal, or both can be made only after allopolyploids have been obtained. The apparently widespread existence of chromosomal sterility due to heterozygosity for structural differences so small as not to materially influence chromosome pairing at meiosis has led the writer to propose a name for this situation: *cryptic structural hybridity* (Stebbins 1945; Stebbins, Valencia, and Valencia 1946a).

The fact must be emphasized here that much of this cryptic structural hybridity is difficult to discover only because in most plants the best stage for studying the gene-by-gene pairing of the chromosomes, the pachytene stage of meiosis, is not clear enough for analysis. In nearly all interspecific hybrids, chromosome pairing is analyzed at the first meiotic metaphase, when the chromosomes are strongly contracted and are associated at only a few points along their length. Even rather large structural differences between partly homologous chromosomes could not be detected at this stage.

The first suggestion as to how cryptic structural hybridity could originate was made by Sturtevant (1938). He pointed out that in some cases an inverted chromosome segment may undergo reinversion, so that for most of the gene loci the original position is restored. But the chances are very small that the breaks which produce the reinversion will occur at exactly the same point as those which produced the original inversion. A chromosome which has suffered inversion and reinversion will therefore differ from its original ancestor by possessing two small segments in an entirely different position. For instance, if the loci on the original chromosomes are ABCDEFGHI, then the original inversion will give the arrangement ABGFEDCHI, and reinversion will very likely produce something like ABHCDEFGI. Now, if the reinverted chromosome is associated in a hybrid with the original one, it will form an apparently normal bivalent, since the displacement of the H segment will be manifest only by two small foldbacks at pachytene, similar to those formed by heterozygosity for small deletions, or it may be completely obscured by nonhomologous pairing. But if chiasma formation and crossing over occurs in any of the regions C to G, two abnormal types of chromosomes will be formed, namely ABHCDEFGHI and ABCDEFGI, the first containing a duplication and the second a deficiency. Such gametes are likely to be inviable or less viable than normal ones, so that sterility in a hybrid heterozygous for one such inversion-reinversion combination will be one half the percent of frequency of chiasmata in the originally inverted segment.

Systems of translocations can occur which will have the same effect as Sturtevant's example, but will produce even more pronounced sterility. If a translocated segment becomes restored to its original chromosome by a second translocation at a slightly different point, pairing between the retranslocated chromosomes and the unaltered ones will be apparently normal, but independent segregation of the slightly unequal bivalents thus formed will automatically lead to the possession by 50 percent of the resulting gametes of a duplication and a deficiency, and therefore of nearly or quite 50 percent sterility of the hybrid. A similar situation, which might arise with even greater frequency in nature, would be produced by pairing between two chromosome sets, both of which were descended from the same original an-

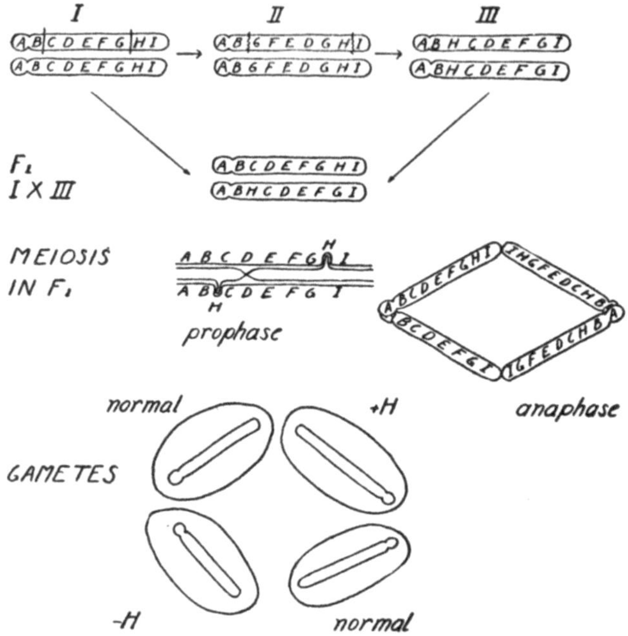

FIG. 23. Diagram showing how partial sterility from cryptic structural hybridity could be produced by means of inversion and reinversion. Further explanation in the text. Original.

cestor and had undergone translocations involving the same two chromosome pairs at slightly different loci. Diagrams of these situations are given in Figs. 23 and 24. It is evident that duplication-deficiency gametes and consequent partial sterility result in this example, not from abnormal pairing within individual bivalents, but from independent segregation of different bivalents. In such hybrids, chromosome pairing even at its most intimate stage, the pachytene of mid-prophase, might appear perfectly normal, since nonhomologous pairing (McClintock 1933) could occur in the arms bearing the translocations.

The hypothetical examples just cited are undoubtedly vastly more simple than most of the conditions of cryptic structural hybridity existing in nature. Müntzing's data (1930a) on the segregation of fertility in the F_2 generation of *Galeopsis tetrahit* × *bifida* indicate that partial sterility in this hybrid is due to a large number of factor differences, each of which may represent a small translocated, inverted, or otherwise displaced chromosome seg-

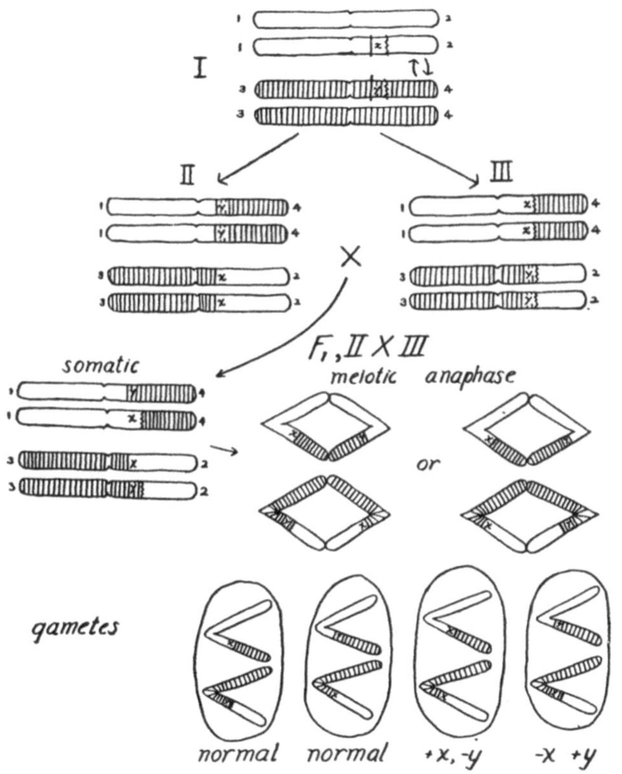

Fig. 24. Diagram showing how partial sterility from cryptic structural hybridity could be produced by means of two independent reciprocal translocations involving the same two pairs of chromosomes. Further explanation in the text. Original.

ment. Similar results have been obtained by Stebbins (1949b) in progeny of hybrids within the complex of *Bromus carinatus* and that of *B. catharticus.* Stebbins, Valencia, and Valencia (1946a) have considered the number of small chromosomal segments for which an F_1 hybrid may be heterozygous before marked chromosomal disturbances will be evident at meiosis. Based on the accurate data of Stadler, McClintock, and others on the sizes of gamete lethal and semilethal deficiencies in maize, they computed that two chromosomes may differ from each other by as many as five or six segments which are large enough to produce lethality or semilethality of the gametes when absent from the gametic complement; and still these two chromosomes would have enough homologous segments in common so that they could pair and form chiasmata with considerable regularity. In most diploid plant hybrids, therefore, fairly regular pairing could take place even though the chromosomal complements of the parents differed from each other by as many as 30 to 50 translocated or inverted segments. And as Sax (1933) has pointed out, heterozygosity for even four or five such segments could be expected to produce a high degree of gametic or haplontic sterility. The frequent presence of good chromosome pairing and at the same time of hybrid sterility of the chromosomal type is therefore to be expected to occur often in hybrids between closely related species.

There is no doubt that many examples of hybrid sterility include a combination of both genic and chromosomal sterility factors. This was first clearly demonstrated by Greenleaf (1941) for the hybrid *Nicotiana sylvestris* × *tomentosiformis*. The diploid F_1 has little or no chromosome pairing and is completely sterile, as might be expected from the fact that its parents belong to different sections of the genus. When the chromosome number of this hybrid was doubled by means of regenerating callus tissue, the resulting allopolyploid had good chromosomal pairing and a high percentage of normal, viable pollen, but set no seed, either with its own or with any other pollen. This was due to the abortion of the embryo sacs at an early age, a condition apparently controlled by certain genetic factors or factor combinations. An example of a different type is *Aegilops umbellulata* × *Haynaldia villosa* (Sears 1941b). Lack of pairing in the diploid F_1 of this intergeneric hybrid is undoubtedly due in large part to dissimilarity between the parental chromosomes and should therefore be termed chromosomal. But the allopolyploid produced from this hybrid remained sterile in respect both to pollen and to seed and showed a high degree of asynapsis, in spite of the fact that each chromosome had a mate with which it was completely homologous. Superimposed on the chromosomal sterility, this hybrid must have had a gene combination causing asynapsis or desynapsis, similar to those which arise as occasional mutations in good species. In the progeny of the allopolyploids between *Triticum aestivum,* on the one hand, and *Agropyron elongatum, A. trichophorum,* or *A. glaucum* on the other, Love and Suneson (1945) found more asynapsis than would be expected on the basis of lack of chromosome homology alone, and Pope (1947) has found a considerable degree of asynapsis in the colchicine-induced allopolyploid of *Triticum aestivum* × *Agropyron elongatum*. Similarly, Stebbins, Valencia, and Valencia (1946b) found genic sterility superimposed upon chromosomal sterility in *A. trachycaulum* ("*A. pauciflorum*") × *Hordeum nodosum*, and Walters (unpublished) found the same in *Agropyron parishii* × *Sitanion jubatum*.

DEGENERATION OF HYBRID PROGENY

The final barrier to gene interchange between species is effective only after the F_2 progeny have been produced. Clausen, Keck, and Hiesey (1940) reported that in *Zauschneria cana* × *septentrionalis* the F_1 plants are vigorous and highly fertile as to seed setting, but the F_2 population consists largely of weak, disease-susceptible individuals, so that intermediates between the two species cannot be maintained. The same is true of *Layia gaillardioides* × *hieracioides* (Clausen, Keck, and Hiesey 1941) and was found by Goodwin (1937a) in *Solidago rugosa* × *sempervirens*, Propach (1937) in *Solanum henryi* × *verrucosum*, Syrach-Larsen (1937) in *Larix Gmelini* × *kaempferi*, and Johnson (1947) in *Populus alba* × *grandidentata*. The F_1 hybrid between *Gossypium arboreum* and *G. herbaceum* is likewise vigorous and fertile (Webber 1935, Hutchinson 1940, Silow 1944) and frequently appears in fields in which its parent species are grown. But F_2 progeny are rarely found; they are apparently too weak to survive. Similarly, the writer has been informed by Dr. S. G. Stephens (oral communication) that while the hybrid between the two New World tetraploid cottons, *G. barbadense* and *G. hirsutum*, is highly fertile, its progeny are so weak that great difficulty is experienced in transferring a gene artificially from one species to the other.

The causes for this degeneration of hybrid segregates may be either genic or chromosomal. It is obvious that if two species differ widely in the physiology of their developmental processes, many recombinations of the various genetic factors affecting these processes will be disharmonious and will produce individuals not adapted to any environment. On the other hand, if chromosomal sterility is operating in the F_1, micro- or megaspores can be produced which are capable of developing into gametophytes and functional gametes in spite of minute duplications or deficiencies, but the accumulation of two or more deficiencies in the zygote may well be fatal to it. Furthermore, McClintock (1942 and oral communication) has shown that different types of tissues have different degrees of sensitivity to the same deficiency. One which can be carried through the pollen grain and pollen tube may be completely lethal when homozygous in the embryo or endosperm. There is obviously no way of distinguishing between genic and chromosomal sterility or inviability in F_2 segregates, and from the practical point of view the distinction is not very important. The highly significant fact about the examples cited in this section is that genetic barriers between species may exist even when they can cross easily and form vigorous, fertile F_1 hybrids. When this is the case, genetic criteria for the separation of species must be explored through the F_2 generation before the existence of isolating mechanisms can be regarded as proved or disproved.

THE INTERRELATIONSHIPS BETWEEN ISOLATING MECHANISMS AND INTERSPECIFIC DIFFERENCES

In the preceding sections reference has often been made to species separated not by one but by combinations of different isolating mechanisms. Too much emphasis cannot be given to such examples; they probably represent the normal condition in nature. Any one of the various isolating mechanisms may be present to a greater or lesser degree as a partial barrier between individuals or races belonging to the same species, and there are relatively few species in the plant or animal kingdoms which are separated by only one type of isolating mechanism. Isolation barriers are usually built up of many different, independent parts. If they are not complete, then the segregating progeny of interspecific hybrids inherit various elements of the original interspecific barrier. They may thus cause its temporary breakdown, or they may build up new isolating barriers upon the foundations of the old one and thereby become the ancestors of new species. This topic will be considered in the following chapter. Here we wish only to emphasize the point that the nature of interspecific isolation barriers precludes the possibility that species usually arise full-fledged at a single step, barring the "cataclysmic" event of polyploidy. The isolating mechanisms responsible for the discontinuity between species, like the morphological and physiological differences between them, are compounded of numerous genetically distinct elements, which in most cases must have arisen and become established in the species populations over a considerable period of time.

Furthermore, the correlation between degree of similarity in morphological characteristics and the effectiveness of the isolation barrier between two species is by no means absolute. To be sure, closely similar species are most often separated by relatively weak systems of isolating mechanisms and can as a rule be hybridized more easily than widely divergent ones. But every systematist who has studied several different, unrelated groups of animals or plants is familiar with the fact that in some groups types which are widely divergent morphologically grade into each other through series of intermediate forms, while in other groups sharp distinctions can be drawn between species which in their external appearance are very similar. Species in unrelated groups are not equivalent to each other either morphologically or ecologically.

A few examples may be cited to bring out this fact. In Chapter II the fact was mentioned that in the genus *Aquilegia* almost any two species can be hybridized and will give vigorous and fertile progeny in the F_1 and F_2 generations. This lack of genetic isolation extends even to the genus *Paraquilegia,* which lacks the spurred petals characteristic of the columbines and therefore in some respects resembles the genera *Isopyrum* and *Anemonella* as much as *Aquilegia.* Earlier in this chapter evidence was cited of extensive intergradation between species of *Ceanothus* which in their typical forms are widely divergent both morphologically and ecologically, and similar examples could be cited from many other genera, such as *Rhododendron* (Sax 1930), *Vaccinium* (Darrow and Camp 1945), *Vernonia* (Steyermark 1939), *Alseuosmia, Coprosma,* and *Veronica* (*"Hebe"*) (Allan 1940). On the other hand, the examples of *Crepis neglecta-fuliginosa* and the *Hemizonia virgata* complex, cited earlier in this chapter, are groups in which isolating mechanisms have developed extensively, accompanied by relatively little divergence in morphological characteristics and ecological preferences.

Further evidence of the relative independence of morphological divergence and the development of hybrid sterility is provided by analyses of the progeny of partly sterile interspecific hybrids. If the factors producing the visible differences between species are the same as or are genetically linked to those responsible for hybrid sterility, then we should expect to find that the relatively fertile F_2 segregates from a partly sterile F_1 would resemble morphologically one or the other of its parents, while those remaining intermediate, or possessing new recombinations of the parental characteristics, should be as sterile as the F_1 or more so. Five examples may be cited which bear on this point.

The first is the ingenious and illuminating study made by Anderson (1936a) and his colleague, R. E. Woodson, on the progeny of the partly sterile hybrid *Apocynum medium* (*A. androsaemifolium* × *cannabinum*). In this study, the geneticist (Anderson) raised the segregating progeny from the hybrid together with that from both parental species and sent specimens of these plants, without information on their origin, to the systematist (Woodson) for identification. The interesting result was that some of the F_2 hybrid derivatives were near enough to one or the other parental species so that they were identified as aberrant forms of them, while others resembled their F_1 hybrid parent so much that they were called *A. medium*. The F_2 plants also varied in fertility, but there was apparently no close correlation between the morphological appearance and the fertility of the plants. Among those considered to be still intermediate (*A. medium*), some were highly fertile, though others were more sterile than the F_1; while those judged to be conspecific with either *A. androsaemifolium* or *A. cannabinum*, though mostly rather fertile, included the most sterile plants of all.

The second example, *Galeopsis tetrahit* × *bifida*, has already been mentioned (Müntzing 1930a). Müntzing analyzed 355 F_2 plants of this cross, scoring them in respect to percentage of apparently normal pollen and to the number of characters in which they resembled either *G. tetrahit* or *G. bifida*. The result is shown in Table 5. Individuals with the value + 5 resemble *G. tetrahit* in five out of the seven characters scored; those with the value 0 are exactly intermediate; and those with − 5 resemble *G. bifida* in five out of seven characteristics. This table shows clearly the complete lack of correlation between sterility or fertility and resemblance to one or the other of the original parents. The visible characters which differentiate the species and the sterility factors which partly isolate them segregate quite independently of each other.

TABLE 5

RELATION BETWEEN POLLEN FERTILITY AND MORPHOLOGICAL CHARACTERS IN *Galeopsis tetrahit* × *bifida*, F_2 [a]

Value figure	Number of plants	Mean pollen fertility (percent)
− 5	2	55–80 (range)
− 4	17	71.90
− 3	33	68.70
− 2	49	67.30
− 1	64	68.30
0	66	64.90
+ 1	61	67.40
+ 2	38	65.25
+ 3	17	60.75
+ 4	5	64.50
+ 5	3	35–75 (range)

[a] Data from Müntzing 1930a.

The third example is the thorough and careful analysis by Terao and Midusima (1939) of a large number of intervarietal hybrids of cultivated rice (*Oryza sativa*). Within this species these workers found that the F_1 hybrids between certain varieties had pollen and seed sterility quite comparable to that of interspecific hybrids. The F_2 progeny of semisterile F_1 hybrids showed a wide range of variation in sterility, as would be expected if multiple factor inheritance were present. The authors state that this segregation for fertility is independent of that for morphological differences, but do not give data to support this statement. However, the lack of correlation between visible intervarietal differences and their intersterility is evident from the fact that some of the most different varieties are nearly or quite interfertile, while others, very similar in appearance, give highly sterile F_1 hybrids. This is shown partly by the authors' chart (Fig. 4, p. 231), and was brought out by a further analysis of their data made by Stebbins (1942a). This analysis revealed the additional fact that some of the most striking examples of morphologically different races which are interfertile and of races which closely resemble each other in appearance but are intersterile are combinations including a race from some center of rice cultivation, such as Hawaii or South America, far from the original home of rice. These varieties are undoubtedly the result of extensive artificial hybridization and selection. This may be responsible for their behavior and suggests that the whole situation in rice has been greatly altered by its long history as a cultivated plant. It is possible that in the wild progenitors of rice, the visible differences and the partial intersterility barriers were correlated with each other, but that artificial selection has produced unnatural recombinations of the genetic factors responsible for these two types of differences.

The last two examples represent apparent exceptions to the lack of correlation between interspecific differences and hybrid sterility. Lamprecht (1941, 1944, 1945a,b) found that from the highly, but not completely, sterile F_1 hybrid between *Phaseolus vulgaris* and *P. coccineus* ("*P. multiflorus*"), lines could be isolated in later generations which, with two exceptions, represented recombinations of all the morphological characteristics separating the two species. These two exceptions (the position of the cotyledons of the germinating seedling, whether hypogeous or epigeous, and the character of the stigma surface) he considered to be characters determined by true species-separating genes. According to his interpretation, genes for the *coccineus* type of cotyledons and stigma cannot survive in *vulgaris* plasma, and, conversely, genes for the *vulgaris* condition of these two characteristics are inviable in the plasma of *P. coccineus*. To explain this situation, Lamprecht developed a new theory about the nature of gene synthesis. This theory needs verification on more carefully controlled material before it can be accepted, and, in view of the example to be presented below, it may not be necessary.

A parallel example is that of *Godetia amoena* and *G. whitneyi*.[1] Hiorth (1942) found that all strains of *G. amoena* found south of the Golden Gate, in central California, were interfertile and constituted the true genetic species, *G. amoena*. On the other

[1] Dr. Harlan Lewis (unpublished) has shown that these species names are incorrectly used by Hiorth and Håkansson. The correct name for the northern species called by them "*G. whitneyi*" is *G. amoena*, while the species which they call *G. amoena* is actually *G. rubicunda*.

hand, all strains classified by systematists as either *G. amoena* or *G. whitneyi* and originating from north of the Golden Gate and northward to British Columbia were likewise wholly or partly interfertile and constituted the species *G. whitneyi*. On the other hand, crosses between strains of *G. amoena* from south of the Golden Gate and *G. whitneyi* invariably form highly sterile hybrids, with a seed fertility of about 3 percent. Both species have the haploid chromosome number $n = 7$, and the F_1 hybrid has good chromosome pairing, usually forming six bivalents and two univalents.

Hiorth found, as did Lamprecht in *Phaseolus*, that, by means of backcrossing, most of the morphological characters peculiar to *Godetia whitneyi* could be transferred to the genetic background of *G. amoena*, and vice versa. A striking exception was the most distinctive difference between the two. Both species usually have a dark colored spot on their petals, although in both there occur genotypes with a recessive allele for the absence of spot. In *G. amoena*, however, the spot is always located at the base of the petal, while in *G. whitneyi* it is in the middle. Hiorth was at first unable to transfer the gene F^x, for central petal spot, from *G. whitneyi* to *G. amoena*, or the corresponding allele F^b, basal petal spot, from *G. amoena* to *G. whitneyi*. F^b plants were obtained in the backcross of the F_1 hybrid to *G. whitneyi*, but these were highly sterile. In one line, however, partly fertile F^b plants were recovered in the fourth backcross generation. These were analyzed cytologically by Håkansson (1947) and were found to have, instead of seven pairs, six pairs plus a chain of three chromosomes, one of them smaller than the other two. Among the progeny from one of these plants was found a fertile plant having seven pairs of chromosomes, which bore the F^b gene. The most plausible explanation of this situation, given by Håkansson, is as follows. The chain of three in the partly fertile F_4 plants consisted of one normal *whitneyi* chromosome, which is pairing at one end with a normal chromosome of *G. amoena*, and at the other with a fragmented chromosome of this species. The gene F^b is located in the normal *G. amoena* chromosome, which also contains factors causing sterility in combination with the chromosomes of *G. whitneyi*. The effect of these sterility factors is counteracted by the presence of the small chromosome from *G. amoena*, which through segregation to the same pole of the two end chromosomes of the chain is often included in the same gamete with the F^b chromosome. The fertile F^b plants with seven pairs originated through crossing over between the chromosome bearing F^b and the normal one from *G. whitneyi,* thus transferring to the *G. whitneyi* genome the F^b factor without the sterility factors originally linked to it.

These experiments suggest that in both *Godetia* and *Phaseolus* the factors causing the morphological differences between the species are different from those responsible for interspecific sterility. The apparently intimate association between certain morphological characteristics and the sterility barriers is one of genetic linkage, which can be broken by crossing over, provided enough offspring are raised.

SOME TYPICAL PATTERNS OF ISOLATING MECHANISMS

Although the complete analysis of the isolating barriers which separate species has been performed as yet in only a relatively small number of genera, nevertheless the data already available show that the principal isolating mechanisms acting to separate the species recognized by the systematist vary considerably from one group to another. This variation, furthermore, is not haphazard, and certain patterns of genetic species differences are becoming recognizable. These patterns appear to bear some relationship to the growth habit and distribution of the groups concerned.

We can recognize first a group of genera in which barriers of incompatibility, hybrid inviability, and F_1 hybrid sterility are relatively weakly developed. The most closely related species are separated largely by ecological or seasonal isolation and by inviability of F_2 segregates. Into this group fall *Pinus, Quercus, Larix, Cistus, Ceanothus, Wyethia, Aquilegia* (which also has some characteristics of the second group), some sections of *Aster* and *Solidago, Platanus, Populus, Rhododendron, Coprosma, Catalpa, Castanea,* and (except for the polyploid complexes) *Vaccinium.* It will be noted that these genera are mostly woody and are all long-lived perennials.

The second group consists of plants with flowers highly specialized for insect pollination and with mechanical isolation strongly developed. Here are found the Asclepiadaceae and the Orchidaceae, mentioned earlier in this chapter. More cytological and genetic studies on these families are, however, needed. In some temperate genera of Orchidaceae, differences in chromosome numbers apparently provide the most important isolating mechanisms (Hagerup 1944).

A third group may be recognized in three genera of the Solanaceae, *Solanum* (diploid species), *Lycopersicon,* and *Datura.* In all of these, hybridization under natural conditions is often difficult or impossible, even when the species are closely similar in external morphology. On the other hand, the F_1 hybrids, when formed, tend to be sterile in spite of good chromosomal pairing and to yield sterile allopolyploids. This indicates the predominance of genic sterility, as well as incompatibility and hybrid inviability.

In the fourth group we may place those genera with an essentially "normal" relationship between amount of difference between species in external morphology, in crossability, in hybrid sterility, and in the chromosomes. Here belong *Layia, Allium, Lilium,* the diploid species of *Nicotiana, Gossypium,* and *Paeonia,* and probably many others. In them, several different types of isolating mechanisms are operating with about equal force.

A fifth group may be recognized in two genera: *Aegilops* (Kihara 1940, Aase 1935, Sears 1941a,b) and *Godetia* (Hiorth 1941, 1942, Håkansson 1941, 1943b, 1946a, 1947), in which an extraordinary amount of chromosomal differentiation has taken place between nearly every species, even those which are morphologically rather similar. In them, nearly every interspecific hybrid has highly irregular chromosome behavior at meiosis, and allopolyploidy is common. In these genera, moreover, closely related species differ markedly in the external morphology of their chromosomes. This group is the direct antithesis of the one first mentioned.

The sixth, seventh, and eighth groups are those in which a large proportion of the barriers between species are due to differences in chromosome number. They may form extensive polyploid series, as in *Rosa, Rubus, Salix, Viola,* and most genera of grasses, or the numbers may be an aneuploid series, as in *Erophila, Carex, Scirpus, Iris,* and *Stipa.* Further barriers may be added by the presence of apomixis, as in *Antennaria, Hieracium, Arnica, Potentilla, Poa,* and *Calamagrostis.* These genera will be discussed in later chapters dealing particularly with polyploidy and apomixis.

These groupings are only approximate, for purposes of orientation, and are certainly not mutually exclusive. Several of the genera mentioned above under one group have some characteristics of other groups. And still other genera exist within the limits of which are found species groups falling into most of the categories mentioned above. The most notable of these is *Crepis* (Babcock 1947). Species of this genus kept apart chiefly by geographical, ecological, or seasonal isolation are *C. pulchra-palaestina* and *C. foetida-thomsonii-eritreensis.* Strong barriers of hybrid incompatibility or inviability between closely similar species are found in the case of *C. alpina-foetida.* The "normal" pattern of the fourth group mentioned above is perhaps the most common one in the genus, but *C. neglecta-fuliginosa* represent a species pair in which great differentiation in the chromosomes has taken place between morphologically very similar species. Polyploidy is present in several different species complexes, an aneuploid series of chromosome numbers is found in the genus as a whole, and apomixis is prevalent among most of the North American species. It is obvious that such genera as *Crepis,* in which many different speciation processes have been taking place in closely related evolutionary lines, are particularly good material for studies of the origin of species.

THE ORIGIN OF ISOLATING MECHANISMS

Although our knowledge of the nature of isolating mechanisms has been considerably increased in recent years, there is as yet no basis on which to erect more than a series of working hypotheses regarding the way in which they originate. The diversity of these mechanisms, as well as the fact that different ones are prominent in different groups of animals and plants, suggests that more than one hypothesis will be necessary to account for the origin of these barriers in the various groups of organisms.

The basic hypothesis held by many zoologists, which has been recognized since before Darwin's time and has recently been elaborated by Dobzhansky (1941, pp. 280–288), Muller (1942), and Mayr (1942, pp. 154–185) is that in sexually reproducing, cross-breeding organisms geographic isolation must precede the formation of any other isolating barrier. The theoretical argument in favor of this hypothesis is as follows. If two populations are occurring in the same habitat and cross-fertilization between their members is possible, then the only type of barrier which will be effective in keeping them separate will be one which will permit the formation of vigorous and fertile offspring in matings within population A or population B, but will cause the weakness or sterility of offspring from matings between A and B. This result can be achieved only through systems of complementary factors, which produce vigorous and fertile individuals when homozygous, but cause weakness and sterility when in the heterozygous condition. If such a barrier were to arise by mutation and selection within a continuous, panmictic population, the individuals in which these mutations first occurred, as well as many of their immediate progeny, would be heterozygous for them, and according to the hypothesis would be weak or partly sterile. This would cause such a strong selection pressure against the isolating factors that their establishment would be impossible. On the other hand, such pairs or larger groups of complementary factors could readily arise in isolated populations, through mutations at different loci. Dobzhansky has suggested the simple situation in which the initial constitution of both populations is aabb, and one of them becomes transformed to AAbb by the occurrence and establishment of a mutation, while the other becomes similarly transformed to aaBB. The single heterozygotes, Aabb and aaBb, are assumed to be viable and fertile, but the double heterozygote, AaBb, is inviable or sterile.

This argument possesses much less force in plants than it does in animals. As recognized by Dobzhansky, it does not apply at all to species which are capable of self-fertilization. In them, the temporary isolation which would be needed for the establishment of complementary factor systems can exist between the individuals of a spatially continuous population. A single heterozygous individual can produce, by selfing, homozygotes for new sterility or incompatibility factors, which can then become established, if favored by selection or random fixation. Furthermore, as pointed out in Chapter V, a considerable proportion of the higher plants which are regularly cross-fertilized are very long-lived and produce a tremendous excess of seeds, so that partial sterility does not place them at a great selective disadvantage. This is particularly true of the numerous species with efficient means of vegetative reproduction. In most of the higher plants, therefore, sympatric speciation is a theoretical possibility.

On the other hand, the available factual evidence points toward the rarity of speciation without previous geographic isolation, that is, of sympatric speciation, in plants just as in animals. The zoological evidence has been well discussed by Mayr (1942, 1947). In those groups of higher plants known to the writer, the relationship known as "Jordan's law" (Jordan 1905) usually holds, as was stated also by Abrams (1905), on the basis of his wide experience with the California flora. That is, the nearest related species to any given species population is found, not in the same area or in a very different one, but either in an adjacent geographic region or in a far distant one with similar climatic and ecological conditions. Fernald (1934) has likewise emphasized the initial importance of geographic isolation.

In support of this opinion, the relationships will be analyzed of some of the species pairs mentioned earlier in this chapter which were stated to be separated by ecological or seasonal barriers rather than by geographical barriers, and which have shown their close relationship by their ability to hybridize with each other. The genus first mentioned is *Quercus*. The species of this genus in eastern North America appear like a group which has differentiated sympatrically during the evolution of the deciduous forests of this region, but this appearance may be simply the result of the long and complex history of these forests and the present lack of geographic isolating barriers. On the other hand, the oaks of the Pacific coast of the United States have a pattern of distribution which agrees perfectly with Jordan's law. Examples are the coast and interior live oaks, *Q. agrifolia* and *Q. wislizenii;* the four species of the "golden oak" group, *Q. chrysolepis, Q. palmeri, Q. vaccinifolia,* and *Q. tomentella;* the three evergreen scrub white oaks, *Q. dumosa, Q. dumosa* var. *turbinella,* and *Q. durata;* and the two large-leaved, deciduous white oaks, *Q. lobata* and *Q. garryana.* The members of these species groups do overlap in their ranges to some extent, but their patterns are such as to suggest strongly that the members of each group were geographically isolated from each other when they became differentiated.

The next example cited of ecological isolation was in the genus *Ceanothus.* Here the evidence suggests the past occurrence of "double invasions" into the central coast ranges (cf. Mayr 1942, p. 173). The three species mentioned were *C. thyrsiflorus, C. papillosus,* and *C. dentatus,* all of which are entirely or largely sympatric. The nearest relatives of *C. thyrsiflorus,* aside from some forms which are certainly to be regarded as subspecies of it, are the rare *C. cyaneus* of southernmost California, the insular endemic *C. arboreus,* and a relatively common subshrub of the north coast ranges, *C. parryi.* This group of species seems to be essentially mesophytic in character and may well have originated as subshrubs in coniferous forests. *C. papillosus,* on the other hand, occurs in its most extreme form (var. *roweanus*) in the chaparral formations of Southern California, and all of its characteristics suggest that it originated under the hot, dry conditions which prevail in these areas. *C. dentatus* shows some resemblances to *C. papillosus* and its relatives, but its prostrate habit and the character of its foliage suggest a closer relationship to a species more typical of the inner central and north coast ranges, *C. foliosus,* which also occurs in relatively dry, chaparral areas. Such a habitat may have been the ancestral one for *C. dentatus.* The present juxtaposition of these three species is best explained as a result of the extensive migrations known to have taken place during the Pleistocene pluvial period, plus the large amount of uplift and faulting which changed completely the topography and probably also the habitats available in the central coast ranges during that epoch.

The ecological and seasonal isolation described in the genus *Pinus* probably has a similar explanation. In this genus, the shedding of pollen is closely associated with the beginning of spring growth in the vegetative shoots. In *P. radiata,* both of these processes begin in late winter, which in central and southern California is the most favorable time for the beginning of plant growth. This species, therefore, is perfectly adapted to the climatic conditions of this region, to which it is endemic, and it almost certainly originated there. On the other hand, *P. attenuata* shows a pronounced winter dormancy, so that both anthesis and the beginning of vegetative growth are delayed until April. This is associated with the fact that the largest, most luxuriant stands of *P. attenuata* are in the interior sections of northern California, and that in some localities it ascends the mountains to an altitude of 6,000 feet (1,800 m). In these regions it is exposed to heavy winter frosts and snowfall, so that its dormancy is under such conditions essential to its survival. These are probably the climatic conditions under which *P. attenuata* became differentiated from the ancestral stock of the closed-cone pines. The other late-blooming species of this group, *P. muricata,* is strictly coastal and not at present exposed to heavy winter frosts. But its main area of distribution is far to the north of the range of *P. radiata,* and it very likely became differentiated on the coast of northern California. These three related species of pines, therefore, probably owe their initial differentiation to ecogeographic isolation and have become associated together only recently, probably during the Pleistocene epoch.

In his discussion of the importance of isolating mechanisms in the evolution of the species of *Crepis,* Babcock (1947a, pp. 147–151) does not state specifically whether or not geographic isolation has usually preceded the formation of other isolating mechanisms. But he points out that of the 182 species which are closely enough related so that they are comparable, 41 percent are completely isolated from their nearest relatives, and many of the remainder more or less overlap, but are largely allopatric. Those which are truly sympatric, such as *C. reuteriana, C. palaestina, C. pterothecoides,* and *C. pulchra,* in the Syria-Palestine region, occur in different ecological niches and are adapted to different growing conditions. Their present association, therefore, may be, as in *Ceanothus,* due to migration from different regions in relatively recent times. Babcock assigns major importance as factors in speciation to geographic isolation and to ecological isolation "brought about through migration, either vertical or horizontal, or both, into new environments." The latter type of isolation corresponds to the modification of geographic isolation which Mayr (1942, pp. 194–198) has discussed in relation to various species of insects and birds. In *Crepis,* therefore, ecogeographic isolation has certainly played a major role in the origin of species, and it is likely that in the majority of cases, if not universally, it has preceded the formation of other isolating barriers.

Although as a general theory the initial importance of geographic or ecogeographic isolation is supported by most of the evidence at present available, the adoption of this theory does not enable us to draw any conclusions about the way in which other isolating barriers arise in geographically isolated populations. And this is the paramount problem in connection with the origin of species. This problem has been brought nearer to solution in recent years by a number of working hypotheses concerning which factual evidence has been or can be obtained. These hypotheses are not mutually exclusive, and in view of the diversity and complexity of genetic isolating mechanisms and of the apparent fact that different ones among them have played the major role in various groups of animals and plants, there is good reason to believe that several different mechanisms have operated to produce the isolating barriers which separate species.

The first hypothesis, which has been specifically or implicitly held by a number of geneticists and has been elaborated particularly by Muller (1940, 1942), is that if two populations are geographically or ecologically isolated from each other for a sufficiently long period of time, differential and divergent processes of gene mutation will inevitably cause the formation of other types of isolating barriers. As a general theory, this hypothesis can no longer be supported, at least in plants. Earlier in this chapter the writer cited the examples of *Platanus* and *Catalpa,* in which species living in different continents and markedly distinct in external morphology are able to form vigorous and fully fertile hybrids. Evidence reviewed in Chapter XIV indicates that such species populations have been separated from each other since at least the middle of the Tertiary period, that is, for thirty million years or more. Even recognizing the fact that a single generation in these species of trees may occupy scores or hundreds of years, one would be inclined to assume that if the origin of internal isolating barriers were inevitable in these isolated population systems, it would have occurred during this period of time. And the two instances cited could be matched by several others. The inevitable effects of genetic divergence produced by mutation and natural selection in geographically isolated population systems cannot account for more than a part of the isolating barriers now existing, and have had relatively little to do with most types of hybrid inviability and sterility. Further evidence in favor of this conclusion is provided by the genetic independence of the factors responsible for hybrid sterility from those producing the visible differences between species, which was discussed earlier in this chapter.

Nevertheless, some isolating barriers are the direct result of evolutionary divergence in isolated populations. This is particularly true of ecological and seasonal isolation, as is evident from the examples cited above in *Quercus, Ceanothus,* and *Pinus.* Barriers caused by inviability of F_2 segregates from interspecific hybrids, and perhaps some examples of inviability of F_1 hybrids, may also be ascribed to this cause. As pointed out by Muller, the mutations responsible for differences in external morphology also cause differences in the rates of developmental processes. Therefore, new recombinations of these mutations may produce disharmonious systems of growth processes and, therefore, weakness or inviability of the hybrid product. The frequency with which this happens cannot be estimated at present, but the evidence presented above, as well as several other considerations (cf. Dobzhansky 1941, p. 284), enables us to say with some certainty that its occurrence is not inevitable.

Two hypotheses have been advanced which postulate natural selection as the guiding force in the establishment of isolating barriers. The first is that suggested by Fisher (1930) and further elaborated by Sturtevant (1938) and by Dobzhansky (1941). The basic assumption of this hypothesis is that, in the words of Dobzhansky (1941, p. 285), "gene recombination in the offspring of species hybrids may lead to formation of individuals with discordant gene patterns, the destruction of which entails a decrease of the reproductive potentials of the species whose members interbreed." The species concerned are viewed as systems of partially isolated and competing subpopulations, with two or more species occurring sympatrically. Interspecific hybridization, wherever it occurs, will produce many sterile or inviable products, and so will reduce the reproductive potential of the two subpopulations participating in it. If, therefore, genes arise in any subpopulation which prevent the occurrence of such interspecific hybrids, such genes will immediately increase the reproductive potential, and so will not only be quickly established by selection within that subpopulation but will also cause that unit to expand and to spread the isolating factors by means of migration pressure.

That this hypothesis cannot be generally applied to plants is evident from two facts. First, numerous examples are now known of crosses between widely different subspecies and partly intersterile species which yield hybrids and hybrid derivatives which not only show no sign of inviability or weakness, but in addition may have a positive selective value under certain environmental conditions. These will be discussed in the following chapter. Here we shall refer only to the example of the F_2 segregates from a hybrid between two physiologically very different subspecies, *Potentilla glandulosa reflexa* and *P. glandulosa nevadensis,* many of which are very vigorous and appear to have a high positive selective value in the three standard environments of Stanford, Mather, and Timberline (Clausen, Keck, and Hiesey 1947, 1948b). Second, equally numerous examples are known, in which barriers of hybrid sterility exist between species which appear never to have been in contact with each other.

Nevertheless, the hypothesis may have a particular application to the origin of certain types of barriers, especially mechanical isolation. The intricate mechanisms for cross-pollination found in such groups as the orchids and the asclepiads are most probably produced by the precise interaction of many different genetic factors. Recombination of these factors in hybrids might be expected to yield many types of disharmonious systems of floral structures, incapable of functioning together. In such groups, therefore, maintenance of the reproductive potential might depend on the evolution of flowers adapted for cross-pollination in a particular way or by particular species of insects, or on the evolution of colors or other recognition marks which would cause the pollinating insect to pass instinctively from plant to plant of one species, avoiding its different relatives, as appears to be the case in *Antirrhinum* (Mather 1947).

Whether the hypothesis of the direct action of natural selection can be extended to account for isolating barriers in plants other than mechanical isolation is doubtful. In animals, it has been applied most frequently to examples of sexual isolation, or instinctively controlled aversion between males and females belonging to different species. For explaining such barriers, it seems to be a most likely hypothesis and fortunately is capable of experimental verification.

Of greater potential importance in plant evolution is the hypothesis that hybrid sterility, particularly of the chromosomal type, may be produced by genetic correlation between the genic or chromosomal differences responsible for this sterility and certain combinations of characteristics which have a positive selective value. Muller (1940, 1942) has suggested the possibility of direct correlations of this nature, which are due to the multiple action of the same genes on development and have been caused by phylogenetic changes in gene function. The reasons why this suggestion cannot be widely applied to plants are the same as those against the hypothesis of the inevitable origin of internal isolating mechanisms between populations isolated for long periods of time. Muller's second suggestion, that structural changes in the chromosomes which could produce chromosomal sterility in hybrids might be correlated with selectively advantageous characteristics through the genetic phenomenon known as position effect, is likewise of limited application to plants. In spite of the large number of chromosomal changes which have been studied in detail, position effect has only rarely been found in the plant kingdom.

A third type of correlation, that due to genetic linkage and the suppression of crossing over in structural heterozygotes, may have widespread significance in explaining the origin of hybrid sterility of the chromosomal type. This hypothesis was first developed by Darlington (1936a,b, 1940) and was discussed further by Muller (1940, 1942). If, for instance, an inversion of a chromosomal segment arises in an individual, then the particular combination of ten or a hundred genes which happens to be located in the inverted segment will be transmitted as a unit to all the offspring of that individual which receive the inversion-bearing chromosome. In the inversion homozygote, crossing over will have no effect on this combination; in the heterozygote, crossing over will be suppressed in the inverted segment, due to inviability of the gametes bearing the crossover chromatids. Now, if the gene combination thus united happens to have a positive selective value or acquires such a value through new mutations, the chromosome bearing it will be spread through the population by natural selection. If this selective advantage is held in only a part of the natural range of the species, then two different chromosomal types will be preserved indefinitely within this species. A similar but even more effective way of building up a diversity of chromosomal types within a species is through the existence of a selective advantage of an inversion heterozygote. For instance, if a particular inversion bears the alleles abcdef, and the combination $\frac{\text{abcdef}}{\text{ABCDEF}}$ has a selective advantage over such combinations as $\frac{\text{abcdef}}{\text{abCDEF}}$, as well as any other type of heterozygous or homozygous combination, then the inversion heterozygotes will be preserved as such, and both chromosomal arrangements will become permanently established in the population.

These situations are now no longer purely hypothetical. The prediction of Darlington (1940, p. 145), that the presence in *Drosophila pseudoobscura* of several different inversion types with characteristic geographic distributions would be explained on this basis, has now been strikingly confirmed by the experiments of Dobzhansky (Wright and Dobzhansky 1946, Dobzhansky 1947a,b). To obtain similar evidence in any plant group is not now possible, because no method is available of studying plant chromosomes in such detail as is possible in the salivary chromosomes of *Drosophila.* But there seems little reason to doubt that such situations exist, particularly in view of the fact that inversion heterozygotes are extremely common in species of *Paris, Paeonia,* and other genera.

Furthermore, translocations of chromosomal segments could become established through their effectiveness in linking groups of genes located on different, nonhomologous chromosomes. This could come about in two ways. In organisms like *Oenothera, Tradescantia,* and the North American species of *Paeonia,* in which pairing and chiasma formation chiefly occur near the ends of the chromosomes and are rare or absent in the regions near the centromere, genes located in the latter regions of two pairs of nonhomologous chromosomes between which an interchange has occurred will automatically become linked, as shown in Chapter XI. But even in plants with random distribution of crossing over, the same result can be achieved via small interstitial translocations or through independent translocations affecting the same two chromosome pairs, as discussed on page 224. In either case there will arise the two interstitially located segments (marked x and y in Fig. 24), which, because they are located on different chromosomes from the segments homologous to them, will usually never be able to cross over and yield viable gametes. Furthermore, the two segments which were transferred by the original interchange will thereafter always be inherited together, since gametes in which they are separated from each other will be inviable. This type of interchromosomal linkage can be achieved only at the expense of the occasional formation of hybrids with 50 percent sterility and a corresponding loss in the reproductive potential of the population, but as pointed out in the previous chapter and as will be further elaborated in the following one, this is not necessarily a serious selective disadvantage in many groups of plants.

The processes discussed so far explain only the establishment in a single species population of several selectively advantageous genotypes having different chromosomal rearrangements, and the partial cleavage of the population which this causes. The ultimate processes are necessarily those which cause the segregation of the chromosomal types into two or more groups, the members of each group being similar enough in their chromosomes so that they are essentially interfertile, and sufficiently different collectively from the members of other groups so that all intergroup matings result in partly or wholly sterile hybrids. Such a segregation is visualized most easily if the species population bearing different chromosomal types with differential selective values becomes broken up into several completely isolated subpopulations, each of which is subsequently exposed to a different kind of selection pressure.

Environmental changes that would promote this type of change in the population structure have occurred repeatedly in the history of the earth. As a typical example may be suggested the probable history of a forest-loving species living in the western United States during the period of mountain building which took place in the latter part of the Tertiary period. In the middle part of this period, during the Miocene epoch, forests of a mesophytic type, containing several different species of deciduous trees, as well as conifers, were widely distributed, and herbaceous plants adapted to this type of habitat could have formed more or less continuous populations over large areas. But the uplift of the Sierra Nevada, the Cascades, and the lesser mountain ranges of the Great Basin cut off the interior of the continent from the moisture-bearing winds from the Pacific and caused the forests to become distributed along the moister, windward sides of the mountain ranges. Forest-loving species thus became restricted to a series of narrow bands, separated by many miles of inhospitable savanna, steppe, or desert country. Furthermore, the blocking off of these ocean winds from the interior undoubtedly caused a great change in temperature relationships, producing much greater winter cold and summer heat in the interior than near the coast. Any forest-loving species which could evolve in response to these changes would therefore not only become segregated into several isolated populations, but in addition the surviving individuals in each population would be forced to evolve in a different direction from that taken by those of other populations. If the ancestral, continuous population had possessed different chromosomal arrangements, some harboring gene combinations adapting the plant to relatively slight temperature changes and others favoring the adaptation to extreme conditions, then the process of isolation and selection would automatically break up the original species into several new species, which would be separated by the internal isolating mechanism of hybrid sterility, as well as by geographic isolation.

The hypothesis just reviewed, therefore, postulates that the possession of different chromosomal arrangements associated in a particular way with adaptive gene combinations may preadapt the species to the formation of new species under the pressure of geographic isolation and differential selection. The most valuable evidence which could be secured in favor of this hypothesis would be the demonstration of the existence of chromosomal types with differential selective values in a species population which appeared to be in the process of breaking up. The information obtained by Hiorth (1942) and Håkansson (1942, 1944b, 1946a) about *Godetia whitneyi* suggests that this species may be such an example.

The next hypothesis, advanced by Wright (1940b), suggests that sterility-producing chromosomal or genic differences which are neutral in their selective value may become established by random fixation in species populations while they are passing through "bottlenecks," or periods when selection has caused them to become much reduced in size. There seems little reason to deny that this could happen, provided the population became sufficiently reduced in size for a long enough period of time. Such a process could explain any internal isolating mechanisms which might be found between species of some oceanic islands living under a relatively uniform environment and of necessity isolated from each other. In continental species, however, there is some question as to whether isolation would often be long enough and the populations small enough so that an effective barrier could be raised. Nevertheless, random fixation could in many cases speed either the direct or the indirect action of natural selection in evolving isolating mechanisms, just as it could in respect to genetic factors affecting the visible differences between species, as discussed in Chapter IV.

The hypothesis that "bottlenecks" have played an important role in the origin of hybrid inviability and sterility is supported by the grouping of genera according to the types of isolating mechanisms found in them, as suggested in the previous section of this chapter. The genera placed in the first group, and characterized by the predominance of ecological and seasonal isolation, as well as inviability of F_2 segregates, are predominantly woody, long-lived, and usually exist in the form of large populations which are freely panmictic. Such trees as pines and oaks tend to dominate their habitats and are less sensitive to small changes in their environment than are most of the smaller herbs. Furthermore, isolation of such populations in time would have to be relatively much longer as compared to most herbs, in order to produce the same period of isolation in terms of generations. It is not surprising, therefore, that in groups of this type the isolating barriers are relatively poorly developed in relation to the morphological and ecological diversity of the individual species, and that they consist chiefly of those barriers which are produced most easily by the direct action of natural selection, without the aid of drastic changes in the size and structure of the populations or of the establishment of different chromosomal types.

On the other hand, the two genera placed in the group at the other extreme, namely, *Aegilops* and *Godetia,* are herbaceous and consist entirely or mostly of annuals. Nearly all of their species occupy pioneer habitats in disturbed ground, and many of them are very local in their distribution. Others, which are now widespread as weeds, were probably also much restricted before the advent of man. Such groups have undoubtedly gone through many changes in population size and have been subjected to many alterations in their environment, to which they may have been particularly sensitive. They are therefore groups in which the processes hypothesized above as most important in the origin of internal isolating barriers would be expected to proceed most rapidly. In addition, certain species groups in other genera which have been used in this chapter as examples of the strong development of barriers of hybrid inviability and sterility, such as *Crepis neglecta-fuliginosa* and certain species of *Hemizonia* and *Layia,* are likewise annuals living in disturbed, unstable habitats. On this basis, we should expect that the presence in a genus of many polytypic species containing large numbers of morphologically and ecologically differentiated subspecies would be an indication that in this genus the most effective speciation processes were proceeding at a relatively slow rate. Rapid speciation would be indicated by the presence of many relatively constant, closely similar species with restricted and adjacent distributions, but separated by well-developed barriers of hybrid inviability or sterility.

The final hypothesis concerning the origin of internal isolating mechanisms was suggested by the present writer (Stebbins 1942a, 1945). This is that new isolating mechanisms may be compounded from old ones through the segregation of fertile derivatives from partly sterile interspecific hybrids. This hypothesis will be discussed in the following chapter, as one of the most important effects of hybridization. Experimental evidence now available and still being obtained suggests that isolating mechanisms can arise in this manner, and that in some plant genera this method may have played a rather important role. It can be particularly effective in originating new barriers of chromosomal sterility in species groups with predominant self-fertilization.

Our present knowledge of isolating mechanisms, therefore, supports the statement given at the beginning of this chapter, that descent with modification and the origin of species are essentially different processes. Furthermore, since six or more different hypotheses which have been suggested to account for the origin of interspecific isolating mechanisms can all be supported by some evidence in the case of certain particular groups of animals or plants, we can conclude that the evolutionist must deal, not with a single process, the origin of species, but with several different processes, the origins of species.

CHAPTER VII

Hybridization and Its Effects

THE ROLE of hybridization in evolution has been one of the most controversial topics in the whole field of evolutionary study. Some authors, particularly Lotsy (1916, 1932) and Jeffrey (1915) have assigned a dominant role to this process. Other botanists, particularly systematists occupied chiefly with identification, classification, and the compiling of generic monographs and local floras, have been reluctant to recognize the existence of more than a very small number of hybrids in nature, and have assigned relatively little importance to these. Many zoologists, also, such as Mayr (1942), failing to detect evidence of hybridization in their material, have minimized its evolutionary effects.

The evidence to be presented in this chapter indicates that the true situation, at least as far as the higher plants are concerned, lies somewhere between these extremes. Even if we use the term hybridization in its broadest sense, namely, the crossing of any two genetically unlike individuals, the greatest possible importance we could assign to it would be as a third major evolutionary process, recombination, with an importance not exceeding that of mutation and selection. If we restrict the term to its most commonly accepted usage, namely, crossing between individuals belonging to reproductively isolated species, its importance must be considerably less. On the other hand, careful studies of numerous groups of higher plants from the cytogenetic as well as the systematic point of view have shown definitely that in many of them interspecific hybrids are rather common in nature. Furthermore, these hybrids have frequently given rise to offspring of later generations which have considerably modified the pattern of variation in the groups to which they belong. In fact, the accumulating evidence may make possible the generalization that nearly all of the plant genera which are "critical" or intrinsically difficult of classification owe their difficulty largely to either the direct effects of interspecific hybridization or the end results of hybridization accompanied by polyploidy, apomixis, or both, as discussed in Chapters VIII, IX, and X.

THE FREQUENCY OF HYBRIDIZATION IN PLANTS AND ANIMALS

The statement has recently been made (Mayr 1942, p. 122, Turrill 1942a) that interspecific hybrids are much less common in animals than they are in plants. This is undoubtedly true for certain groups of animals. Two reasons may be assigned for this. In the first place, the higher animals, both arthropods and vertebrates, possess a most effective type of isolating mechanism which by its very nature cannot exist in plants. This is sexual isolation, which consists of an instinctive aversion on the part of males for females of another species (Dobzhansky 1941, pp. 261–267). In some groups, such as fishes (Hubbs and Miller 1943), sexual isolation may have its maximum efficiency only when the species are occurring in their natural environment and are represented by a large number of individuals, and it may break down in extreme habitats, where males find available few or no females of their own species. But Levene and Dobzhansky (1945) have shown that males of *Drosophila pseudoobscura* will not mate with females of *D. persimilis* any more frequently when such females are the only ones available to them than they will when they can make the choice between females of *D. persimilis* and those of their own species. Sexual isolation may therefore become developed until it is just as permanent and absolute a barrier as hybrid sterility. Its primary effect will be to lower greatly the frequency with which F_1 interspecific hybrids occur in nature.

The second factor which increases the frequency of hybrids and hybrid derivatives in many groups of plants is the great longevity of their individuals and, more particularly, the efficient methods of sexual reproduction. These points have been discussed in Chapter V, so that here we need only emphasize the fact that in plant groups in which the individual genotypes can be preserved for great numbers of years and can be spread over large areas, the selective disadvantage of a relatively low pollen and seed fertility is much less than in organisms of which the individual genotypes have a relatively limited life span. In such groups, therefore, the selective advantage of occasional hybridization between species, that is, the ability to produce radically new adaptations to new environmental conditions which may arise, may outweigh the disadvantage incurred by the sterility of such hybrids. On this basis, of course, we should expect to find more examples of natural hybridization in perennial groups than in annual ones, and the latter should be characterized by the presence of sharper species boundaries. Accurate data on this point are not yet available, but it should be noted that of the 16 or more examples of hybridization discussed in this chapter only two, *Helianthus* and *Zea*, involve annual species. The greater frequency of polyploidy, which is often associated with hybridization, in perennial herbs than in annual herbs is discussed in the following chapter.

In spite of these differences, which reduce the frequency of hybrids in animals as compared to plants, interspecific hybridization may not be as uncommon in animals as is usually believed. In plants, every living individual of a species can be observed and its morphological characteristics studied. Although in certain favorable localities natural hybrids occur at a frequency of one to several percent, such localities represent only a small fraction of the total distribution of the parental species. There are probably very few plant groups in which the ratio of natural F_1 hybrids to individuals of the pure species is more than one in ten or one in a hundred thousand. In most animals the critical examination of tens or hundreds of thousands of individuals of a species is so laborious or impractical that it is rarely carried out. Because of this fact, rare hybrid individuals may never be discovered. Furthermore, since relatively few animal species can be bred and hybridized in captivity, the identity of suspected hybrids cannot often be verified experimentally. The experimental work with artificial hybrids and particularly with hybrid derivatives of plants has shown that in their morphological characteristics they often differ considerably from the appearance which one might predict on a priori grounds. It is possible, therefore, that in museum collections there exist a considerable number of specimens of animal species hybrids or hybrid derivatives of which the identity is not recognized. Finally, the processes of meiosis and gamete formation can be much more easily studied in and are known for a much larger number of species of plants than of animals. If these processes are not intimately known, the existence of barriers of partial hybrid sterility between closely related species may not be detected. In animals, an individual is often considered fertile if it produces any offspring at all, but in plants individuals which produce only 5 to 10 percent of the normal number of seeds and have been found to possess the abnormalities of meiosis usually characterizing interspecific hybrids are judged to be partially sterile hybrids between valid species. It is thus possible that in animals, some of the so-called subspecies which are connected by occasional intermediate forms actually are closely related species, separated by barriers of partial sterility. That this may be true in *Peromyscus maniculatus,* for instance, is shown by the fact that Cross (1938) has found the somatic chromosome number 52 in subsp. *hollisteri* and the number 48 in six other subspecies. In the higher plants, groups having such different chromosome numbers would be judged to constitute different species. Another example is the genus *Platysamia,* in which Sweadner (1937) has shown that the females of hybrids between entities which otherwise would be recognized as subspecies are almost completely sterile.

Natural interspecific hybrids have been found relatively frequently in fishes (Hubbs and Hubbs 1932, Hubbs and Kuronuma 1942, Hubbs, Walker, and Johnson 1943), in certain groups of toads (Blair 1941), and in some mollusks, such as the genus *Cerion* (Bartsch 1920). In these groups sexual isolation seems to be relatively poorly developed as compared to the warm-blooded vertebrates and the higher insects, and the individuals are fairly long-lived. We may expect, therefore, that patterns of speciation in marine invertebrates, as well as in many groups of fishes and amphibians, will be more nearly like those in the higher plants than are the patterns found in most warm-blooded vertebrates and insects.

All these considerations suggest that, while hybridization is certainly less common in animals than in plants, and is correspondingly less important as a factor in evolution, its influence in certain groups may be considerable. The points brought up in this chapter cannot be stated categorically to apply to plants alone.

DEGREES OF HYBRIDITY

Although the terms "hybrid" and "hybridization" are usually applied to crosses between individuals belonging to different species, this is by no means the only usage of the term. As Darlington (1940) has pointed out, there are various types of hybrids and hybridity, some of them within the taxonomic species, and some of them between species. From the genetic viewpoint, interspecific hybridization is only a special case of a much more widespread phenomenon. Moreover, in its effects, as well as its evolutionary importance, it has much in common with other types of hybridization.

The simplest type of hybridity is present in all individuals of a sexually reproducing, cross-fertilized species. It has been pointed out repeatedly earlier in this book that genetic heterozygosity is the normal condition in nature. Individuals are normally heterozygous or "hybrid" for a large number of different allelomorphic gene loci, and they produce variable, segregating offspring, whether through selfing or crossing. The difference in this respect between intersubspecific or interspecific hybrids, on the one hand, and the so-called "pure" individuals characteristic of a subspecies or species, on the other, is entirely quantitative, not qualitative. This fact will become evident to anyone who raises side by side the seedlings of a known or suspected interspecific hybrid and of a plant of its parental species in any cross-fertilized genus, such as *Quercus* or *Ceanothus*.

Hybrids may also be formed between members of different, partially isolated subpopulations having different gene frequencies. Such hybrids will as a rule be more heterozygous than the individuals of a cross-fertilized population and will serve a different purpose in evolution. As Wright (1931, 1940b) has pointed out, the most efficient type of population structure for the promotion of evolutionary diversification consists of the division of a large population into several small subpopulations, partly isolated from each other. Under these conditions, random fixation will tend to establish in each subunit characteristics which may be of no immediate selective value, but may enable the subpopulation to explore new "peaks" of adaptation by developing gene combinations with a new type of adaptive value. These valuable new genes or gene combinations may be transferred from one subpopulation to another by occasional migration followed by hybridization. This type of crossing, therefore, is an essential part of the "migration pressure," which is responsible for the flexibility of Wright's system of partially isolated subpopulations.

Hybridization between members of different subspecies differs only in degree from that between partly isolated subpopulations of the same subspecies. Such hybridization can be expected on a large scale whenever two subspecies or ecotypes, occupying different ecological niches, are able to come in contact with each other. The genetic consequences of such hybridization in the F_2 and later generations will be segregation and recombination of the genetic factors responsible for adaptation to different ecological conditions. Its evolutionary consequences will depend on the conditions under which the hybridization takes place. If the two hybridizing subspecies are old, well-established ones, the chances are that they represent gene combinations so well adapted to the environments they occupy that any new combinations created by hybridization will almost certainly be less favorable and therefore selected against in the original environments. In the absence of new habitats, therefore, hybridization between subspecies will lead to the production of many ill-adapted genotypes, and therefore to evolutionary disintegration rather than to progress.

On the other hand, the situation will be entirely different if the hybridization takes place in an unstable, rapidly changing environment. If enough hybrids are formed, then out of the great array of segregates which they will produce in later generations, some are likely to be better adapted to the new environments than are any individuals of the parental subspecies. Examples of this process going on in nature will be given later in this chapter.

As was pointed out in Chapter VI, species do not necessarily differ more from each other in characteristics of external morphology than do subspecies, although this is usually the case. On this basis, some interspecific hybrids may be no more heterozygous for the genetic factors responsible for such morphological differences than are hybrids between extreme variants belonging to the same species. The really outstanding difference between interspecific hybrids and those within the species is that the former bridge a natural barrier or system of barriers of reproductive isolation. If these barriers consist entirely of ecological, seasonal, or mechanical isolation, then the F_1 hybrid is fully fertile and is in no way different from hybrids between subspecies. Such interspecific hybrids in *Platanus, Ceanothus, Catalpa, Quercus,* and other genera have been mentioned in previous chapters. More often, however, hybrid sterility is present, reducing to a greater or lesser degree the reproductive capacity of the F_1 hybrid. Even this, however, is not a unique property of hybrids between species. Semisterile hybrids between individuals of the same species have long been known in *Oenothera, Campanula* (Darlington and Gairdner 1937), *Galeopsis* (Müntzing 1930a, 1938), and many other genera. And in maize, *Crepis* (Gerassimova 1939), and other genera, strains of a species which form almost completely sterile hybrids with other strains of the same species have been produced artificially by subjecting the plants to X rays and then selecting progeny containing appropriate types of structurally altered chromosomes. Usually, the sterility of an interspecific F_1 is greater than that of the intraspecific hybrids mentioned above, but this is not necessarily so. The distinguishing feature of interspecific hybrid sterility is that it is always present and differs relatively little depending on which individuals of the parental species are used for crossing. Sterility barriers within species, on the other hand, are relatively local in distribution, so that if two individuals happen to be partially isolated from each other by such a barrier, they can nevertheless exchange genes indirectly through outcrossing with other individuals of the same species. Interspecific hybrids, therefore, are even in their most significant characteristic, sterility, different quantitatively rather than qualitatively from some hybrids within the species. This is to be expected on the basis that, as hypothesized in Chapter VI, species in most instances arise gradually from subspecies.

Sterile interspecific hybrids may be roughly divided into two groups: those which are capable of producing some viable pollen and seed, through selfing, intercrossing between F_1 individuals, or backcrossing to the parental species, and those which are completely sterile, except for the occasional production of allopolyploid derivatives. In the terminology of Clausen, Keck, and Hiesey (1939), the first are hybrids between ecospecies, and the second, between cenospecies. The two groups, however, are by no means sharply separated from each other. Many hybrids exist which are almost completely sterile, but do produce seed at the rate of one in a thousand, one in ten thousand, or one in a million ovules. Such hybrids in the genus *Paeonia* were called to the writer's attention by Dr. A. P. Saunders, after he had observed many of them for periods of several years (Saunders and Stebbins 1938). If hybrid sterility usually has the multifactorial basis postulated for it in the preceding chapter, whether these factors are genes or small chromosomal rearrangements, we should expect to find a larger proportion of species separated by barriers producing a very low degree of fertility, than by those causing partial fertility of intermediate degrees. This point can be illustrated by an oversimplified example. If two species, A and B, are separated by a barrier consisting of several factors, each of which by itself causes the death of one half of the gametes in an F_1 hybrid, then the pollen fertility of the F_1 will be 50 percent if the parents differ by one such factor, 25 percent if they differ by two, 12.5 percent if they differ by three, and $100/2^n$ percent if they differ by *n* such factors. With an increasing number of factors, therefore, this ratio forms a logarithmic curve which approaches, but never reaches, zero. In many genera, some modification of this curve, produced by the fact that different factors have different effects on F_1 sterility, is the probable distribution of sterility values for F_1 hybrids between species separated by barriers of hybrid sterility of increasing intensity.

The evolutionary possibilities of hybrids which can produce even a very few viable offspring on the diploid level are obviously very different from those which are either completely sterile or produce only rare polyploids or apomictic derivatives. The first type of hybrid will therefore be discussed in the remainder of this chapter, and the second type in the two following chapters.

SOME GENERAL PRINCIPLES CONCERNING HYBRIDIZATION

Before the aftereffects of interspecific hybridization are discussed in particular, some important general characteristics about hybridization must be reemphasized. In the first place, although the F_1 progeny of an interspecific cross are usually as much like each other as are the different individuals of the parental species, the offspring in the F_2 and later generations are extremely variable, due to Mendelian segregation of the genetic factors responsible for the interspecific differences. The extent of this variability cannot be appreciated without first-hand experience of it. If the reader has never seen a large F_2 progeny of a hybrid between two widely different subspecies or closely related species, he should study carefully some well-described and illustrated example of one, such as Clausen's (1926) of *Viola arvensis* × *tricolor,* Müntzing's (1930a) of *Galeopsis tetrahit* × *bifida,* or Clausen, Keck, and Hiesey's (1947) of *Layia glandulosa* subsp. *typica* × subsp. *discoidea.* The striking fact about many of these progenies is not only their variability but also the presence of variants which look as if they have entirely "new" characteristics and represent recombination types whose occurrence could never have been predicted from a study of the original parents of the cross. In some of these progenies, such as those of *Apocynum* (Anderson 1936a) and *Quercus* (see Chapter II), types close to the original parents can be recovered in the F_2 generation, but in most others this is not possible unless a very large number of individuals is raised.

The second of these generalizations is that although segregation in the F_2 of an intervarietal or interspecific hybrid produces a very large number of recombination types, these are by no means a random sample of the total array of possible recombinations of the phenotypic characteristics of the parents. Correlations between groups of parental characteristics are always evident, so that the recombinations found represent a series of oscillations about a central axis, which is a condition of intermediacy or a greater or lesser approach to one or other of the parental species in all of the interspecific differences simultaneously. This fact has been brought out most strikingly by Anderson (1939) in his analysis of the cross between *Nicotiana alata* and *N. langsdorffii.* These two species are radically different in every visible characteristic, but many of these differences are of a similar nature. For instance, the leaf tips, calyx lobes, and corolla lobes are all relatively elongate and acute in *N. alata,* and short and blunt in *N. langsdorffii. N. alata* is larger and coarser in all of its parts; its corollas are not only larger than are those of *N. langsdorffii* but the lobes are still larger in relation to the size of the corolla (Fig. 25). As shown in Fig. 25, the corolla shapes found in the F_2 population do not

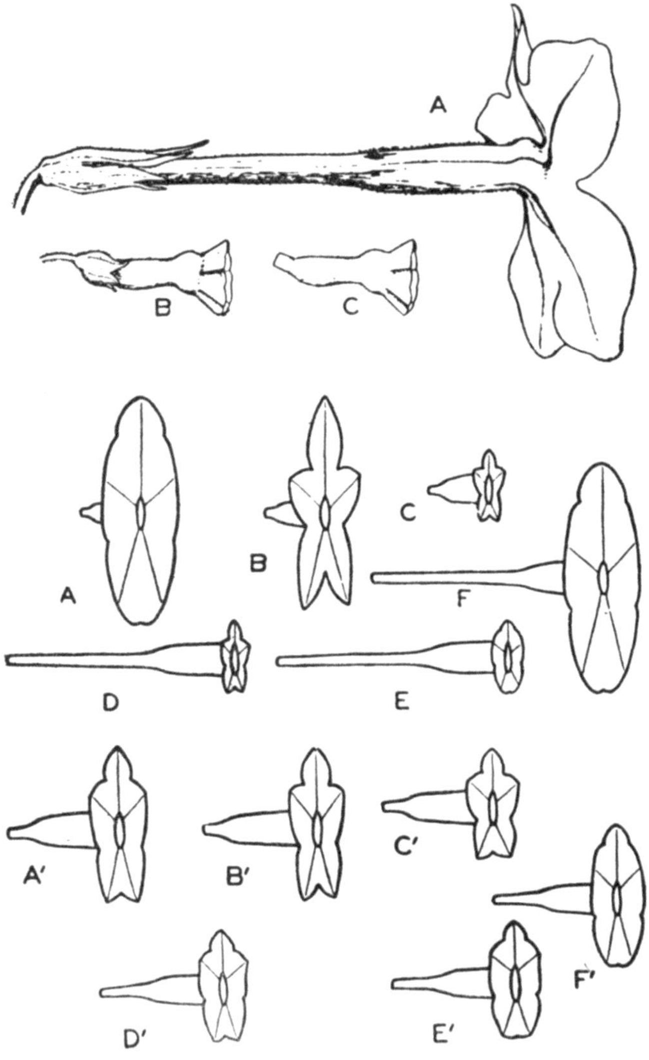

FIG. 25. At top (A), typical flowers of *Nicotiana alata* and (B and C) *N. langsdorffii*. Middle (A–E), extreme recombinations to be expected with complete recombination of tube length, limb width, and lobing of *N. alata* and *N. langsdorffii*. Below, actual extremes of recombination found in 147 plants of the F_2 generation of *N. alata* × *langsdorffii*. The letters A′ to E′ correspond to A to E in the middle figure. From Anderson 1939.

include such extreme recombinations as a corolla tube similar to that of *N. langsdorffii* together with lobes like those of *N. alata.* And Fig. 25 shows that when scored in respect to an aggregate of the parental differences, the individuals of the F_2 population group themselves for the most part near a line connecting the two parental species.

Two explanations may be given for this situation. In the first place, many of the parallel differences between the parental species are the result of developmental correlation in the action of the genes responsible for these differences, as has been emphasized in Chapters III and IV. In *Nicotiana alata* × *langsdorffii,* Anderson and Ownbey (1939) and Nagel (1939) have produced evidence suggesting that these differences are associated with genically controlled differences in hormone activity. The second explanation, as pointed out by Anderson (1939), is that genetic linkage will be more strongly evident in F_2 and F_3 populations if the character differences are governed by multiple factors, as is most frequently the case in interspecific hybrids, than if they are controlled by single factors. If, as in the *Nicotiana* hybrid cited, the haploid chromosome number is 9, then some linkage is almost certain to occur between any two characters each of which is governed by nine or more genes. Such linkages will eventually be broken up in later generations, but they will have a great effect on the distribution of variants in the immediate progeny of an intersubspecific or an interspecific hybrid.

The third generalization about the partially sterile offspring of natural interspecific hybrids in the wild is that they have a much greater chance of producing offspring from the abundant viable pollen of the numerous plants of the parental species which surround them than from the scant, poorly viable pollen produced by themselves or by the few other F_1 individuals which may be present. For this reason, the offspring of most natural interspecific hybrids are far more likely to represent backcross types than true F_2 segregates. This situation can be expected to become accentuated in later generations by natural selection. The combination of characters represented by the parental species have been tried over many hundreds or thousands of generations and have shown themselves to be adaptive; the chances that any new combination will prove equally adaptive are relatively small.

Hence, the nearer a backcross segregate approaches to one or the other of the parental species in characters of adaptive value, the greater its chance of survival. Since selection will have relatively little effect on recessive genes, these might be expected to be transferred relatively easily from one species to another by means of hybridization and backcrossing. In any case, the net effect of interspecific hybridization in nature between partially interfertile species would be expected on a priori grounds to be the reversion of the hybrid offspring toward one or other of the parental species. This effect would be particularly strong in the case of old, well-established species living in a stable habitat, and would be likely to be counteracted only if great disturbance of this habitat caused the original species to be ill-adapted to the new conditions.

The final generalization is the direct outcome of the previous one. In its ultimate evolutionary importance, hybridization depends directly on the environment in which it takes place. Hybridization between well-established and well-adapted species in a stable environment will have no significant outcome or will be detrimental to the species populations. But if the crossing occurs under rapidly changing conditions or in a region which offers new habitats to the segregating offspring, many of these segregates may survive and contribute to a greater or lesser degree to the evolutionary progress of the group concerned. This point has been particularly emphasized by Wiegand (1935) and Anderson (1948, 1949).

INTROGRESSIVE HYBRIDIZATION AND ITS EFFECTS

With these principles in mind, we can now examine some concrete examples of interspecific hybrids and their progenies in nature. According to expectation, we find that the majority of such progenies consist of the products of backcrossing, rather than of true F_2 and later generation segregates. An example which has been thoroughly studied is that of *Iris fulva* and *I. hexagona* var. *giganti-caerulea* (Riley 1938, Anderson 1949). These two species occur sympatrically in the Gulf Coast region of the southern United States, particularly in the lower Mississippi Delta. They are very different in external morphology, and *I. hexagona* var. *giganti-caerulea* is more closely related to *I. brevicaulis* of the northern Mississippi Valley than it is to *I. fulva*. The F_1 hybrid between *fulva* and HGC (these two abbreviations were adopted by Anderson for the sake of convenience, and so are used here) is recognizably intermediate and partly fertile. The isolating barrier between the two species consists partly of hybrid sterility, but ecological isolation may play an equal or even more important role. *Fulva* occurs in wet clay soils, mostly along the edges of rivers and drainage ditches, in partial shade, while HGC grows in the mucky soil of tidal marshes, in full sun. Where farming activity has caused the clearing of woodland and the partial drainage and pasturing of swamps, new, intermediate habitats are available; these contain a many-colored array of hybrids and hybrid derivatives of the brick-red flowered *fulva* and the variegated, blue-flowered HGC. Nevertheless, the populations do not contain a complete blend of the two species, but, most commonly, individuals which are more nearly like HGC and contain certain characters or character combinations which suggest *fulva*. Apparently, hybridization followed by backcrossing and selection of backcross types has caused certain genes and gene combinations from *fulva* to pass across the barrier separating the two species and to become incorporated into the genic complement of HGC. This transference of genetic material across an incompletely developed interspecific barrier, usually via a partially sterile F_1 hybrid, by means of repeated backcrossing and selection of well-adapted backcross types, has been termed by Anderson and Hubricht (1938a) *introgressive hybridization* or *introgression*. It has been treated in monographic fashion by Anderson (1949).

The detection of introgressive hybridization in species populations depends on two characteristics of this process. In the first place, it can occur only in that part of the geographic range of a species which overlaps the distribution of closely related species, and then only when the habitat provides an ecological niche for the establishment of the introgressive types. If, therefore, the variation pattern of a species is being altered by introgressive hybridization, this pattern should contain more variability in regions where the ranges of two related species overlap than where either species is growing by itself. Also, this variability should be greater in newly opened and much-disturbed habitats than in old, stable ones. The second significant characteristic is that introgression, like all types of intervarietal and interspecific segregation, follows the principle of correlation between different characteristics, as discussed in the previous section. If hybridization and subsequent introgression are taking place between species A and B, then the variation pattern of species A should be increased in the direction of species B in and near the regions where A and B are found together and where their habitats have been disturbed in relatively recent times. Furthermore, the variant, introgressive individuals of species A should not possess different characteristics of species B recombined at random with those of species A. Each individual should vary in the direction of species B in several of the characteristics distinguishing the two species, although obviously any particular characteristic would be expressed to different degrees in different individuals. The manner in which the variation pattern of a species is altered by hybridization and introgression is shown diagrammatically in Fig. 26.

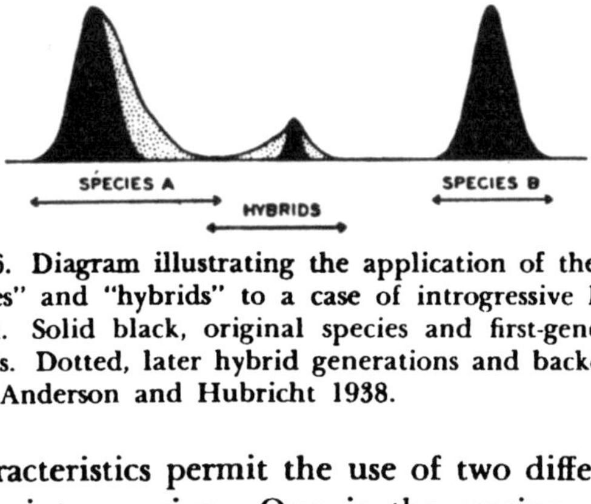

FIG. 26. Diagram illustrating the application of the terms "species" and "hybrids" to a case of introgressive hybridization. Solid black, original species and first-generation hybrids. Dotted, later hybrid generations and backcrosses. From Anderson and Hubricht 1938.

These characteristics permit the use of two different methods for detecting introgression. One is the scoring of herbarium specimens of a species with respect to certain individual characteristics which might be expected to be altered by introgression. If introgression is present, the character selected should be more variable in that part of the range of a species which coincides with the range of its relatives and should vary in the direction of the species with which introgression is suspected. This method was used by Anderson and Hubricht (1938a) for *Tradescantia occidentalis* in relation to *T. canaliculata,* for *T. canaliculata* in relation to *T. subaspera,* and for *T. bracteata* in relation to *T.*

canaliculata. The data, although admittedly very scanty, suggest that the variation patterns of *T. occidentalis* and *T. bracteata* have been modified by introgression from *T. canaliculata* (see Table 6). The latter species, on the other hand, has been unaltered by hybridization with *T. subaspera.* Another study of introgressive hybridization on the basis of individual characteristics as found in herbarium specimens is that of Epling (1947b) on *Salvia apiana* and *S. mellifera.* In the latter work, statistical comparison of population means in respect to two characteristics, one of the inflorescence and one of the corollas, failed to show any significant difference between colonies of *S. mellifera* close to known occurrences of the F_1 hybrid between *S. mellifera* and *S. apiana,* and other colonies of *S. mellifera* far outside of the range of *S. apiana.* In this instance, therefore, frequent interspecific hybridization is not accompanied by introgression of genes across the barrier formed by partial sterility of the F_1 hybrid.

The second method of detecting and estimating the extent of introgressive hybridization is through the use of the *hybrid index* (Anderson 1936c). As stated in Chapter I, this index is a modification of the statistical principle of discriminant functions. It makes use of the principle that, in contrast to the variability produced by the occurrence and segregation of individual mutations, variation caused by hybridization and introgression is characterized by correlations between characteristics which otherwise are genetically independent of each other.

The procedure is as follows. To each characteristic an index number is assigned, which is always 0 for the condition typical of one species, and may vary from 1 to 6, depending on the character, for the condition typical of the other. Thus, the characters are weighted according to their importance as diagnostic characteristics of the species concerned. This weighting is, of course, subjective, but must be based on the worker's intimate knowledge of the species concerned. Greater weight is attached to characteristics which are known to be relatively constant in regions where each of the parental species occurs by itself. If artificial hybrids and F_2 offspring have already been produced between the two species concerned, the weighting can take into account the probable number of genes which differentiate the two parental species with respect to each character. It is obvious that since the effectiveness

Table 6

Comparisons of Herbarium Material of *Tradescantia canaliculata*, marked "Can.," *T. bracteata* outside the Range of *T. canaliculata*, marked "Bract.," and *T. bracteata* within the Range of *T. canaliculata*, marked "Bract. (Can.)" [a]

	Node Number								Leaf Number										Internode		Tuft		
	2	3	4	5	6	7	8	9	5	6	7	8	9	10	11	12	13	14	Increase	Decrease	None	Weak	Strong
Bract.		11	1						1	7	3	1							6	6	4	7	1
Bract. (Can.)	2	9	9	2					1	8	8	4	1						12	10	1	3	19
Can.	2	12	18	19	14	5	3	1		3	10	22	12	10	2		2	3	56	18		26	42

[a] From Anderson and Hubricht 1938.

of the hybrid index depends largely on correlation due to genetic linkage, greater weight should be attached to those characteristics which are controlled by multiple factors than to those which are governed by a single gene.

A typical example of this assignment of index values is given in Table 7. From this table it is evident that individuals typical of *Iris fulva* will receive the total index of 0, those typical of *I. hexagona* var. *giganti-caerulea,* that of 17, and the hybrids and their segregating and backcross derivatives, various intermediate scores. The larger the number of characters used, the larger is the number of genetic correlations reflected in the index values; and in the case of characters governed by multiple factors, the index values usually reflect the true genetic situation most closely when the largest possible number of intermediate conditions is recognized for each character and a maximal range is assigned to each index value. On the other hand, care should be taken to avoid including two different characteristics, such as the size and propor-

TABLE 7

CHARACTERS AND INDEX VALUES OF *Iris fulva* AND *I. hexagona* VAR. *giganti-caerulea* [a]

I. fulva	Index value	*I. hexagona* var. *giganti-caerulea*	Index value
1. Tube of perianth (hypanthium) yellow	0	Hypanthium green	2
2. Sepals orange-red	0	Sepals blue-violet	4
3. Sepal length 5.1–6.4 cm	0	Sepal length 8.6–11.0 cm	3
4. Petals narrowly obovate	0	Petals cuneate-spatulate	2
5. Anthers extruded beyond limbs of styles	0	Ends of anthers about 1 cm below ends of style limbs	2
6. Appendages of style branches small, barely toothed	0	Appendages of style branches large, deeply lacerate-toothed	2
7. Crest of sepals absent or very small	0	Crest of sepals present	2
Total index value	0		17

[a] Data from Riley 1938.

tions of the leaves and the same characters of the sepals and petals, which might be governed partly or wholly by the same genetic factors and therefore show developmental correlation.

Fig. 27 shows the distributions of total index values in a colony of typical *Iris fulva,* one of typical *I. hexagona* var. *giganti-caerulea,* and two which contain plants typical of HGC plus

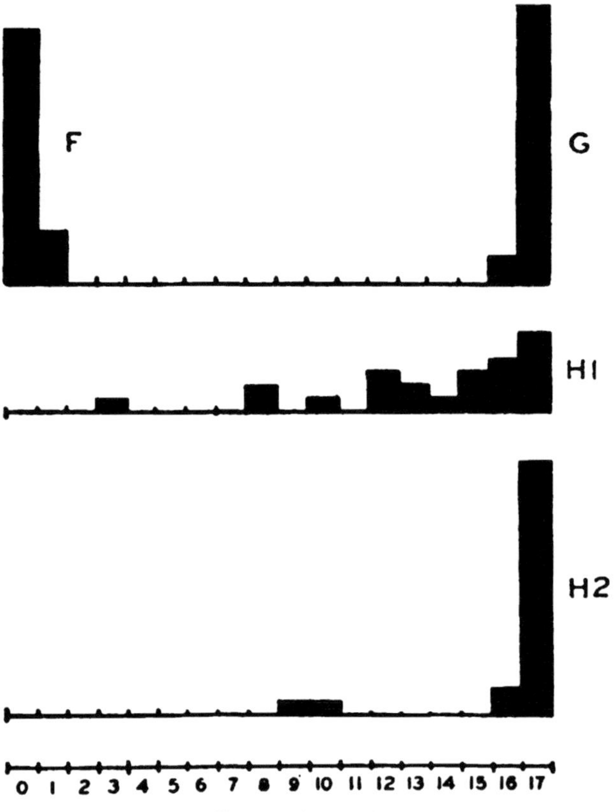

FIG. 27. Frequency distributions of total index values of 23 individuals in a typical colony of *Iris fulva* (F), one typical of *I. hexagona* var. *giganti-caerulea* (G), and two containing hybrid types (H1 and H2). From Riley 1938.

hybrids and their derivatives. In each colony, 23 plants were scored. Although the four colonies were only 500 feet or less from each other, the typical colonies grew in habitats characteristic of their respective species, while those containing hybrid types grew in an old stream bed, now heavily pastured.

A number of examples are now available of the use of this index to detect hybridization and introgression. That of Riley (1939) on *Tradescantia* demonstrates the passage of genes across a sterility barrier formed by the difference in chromosome number between a diploid species (*T. paludosa*) and two tetraploid species (*T. canaliculata* and *T. hirsutiflora*). Goodwin (1937a) on *Solidago rugosa* × *sempervirens*, Wetmore and Delisle (1939) on *Aster multiflorus* × *novae-angliae*, and Heiser (1947, 1948) on *Helianthus annuus* × *bolanderi* and *H. annuus* × *petiolaris* have supported their studies of natural populations with parallel analyses of artificial hybrids and their derivatives. In *Helianthus* the introgressive types have spread as weeds over a considerable area in California and are beginning to assume the character of newly formed subspecies, or ecotypes. Other population studies which have demonstrated the occurrence of hybridization and introgression are those of Dansereau (1941b, 1943) on *Cistus*, of Dansereau and Lafond (1941) and Dansereau and Desmarais (1947) on *Acer nigrum* × *saccharophorum*, and of Stebbins, Matzke, and Epling (1947) on *Quercus ilicifolia* × *marilandica*. Heiser (1949) has made a thorough review of the literature in this field.

In nearly all of the examples cited in this section, the species concerned are to a certain extent "weedy" in character and the introgressive types have become established in habitats much disturbed by the activity of man. On the other hand, the two examples cited by Epling (1947a,b) of hybridization in nature without subsequent introgression, namely, *Salvia apiana* × *mellifera* and *Arctostaphylos mariposa* × *patula*, involve nonweedy species occurring in undisturbed habitats. A similar example is that of *Pinus brutia* and *P. halepensis* in Greece (Papajoannon 1936). These two strictly indigenous species hybridize extensively in two small districts, but populations only a few miles away and elsewhere show no evidence of the effects of hybridization or backcrossing. In a totally different environment, namely, the tropical rain forest of Brazil, Seibert (1947) and Baldwin (1947) have noted that the species of *Hevea* occurring in undisturbed virgin forests are relatively constant and distinct from each other, while in the vicinity of towns there are hybrid swarms involving various ones of the species usually recognized by systematists. Darrow and

Camp (1945) have also emphasized the importance of man's activity in increasing the number of hybrids and hybrid derivatives in the genus *Vaccinium,* while great emphasis has often been given to this factor in the origin of the highly complex situations which exist in genera such as *Crataegus* (Wiegand 1935), *Rubus* (Brainerd and Peitersen 1920), and *Amelanchier* (Wiegand 1935).

There is little doubt, therefore, that the majority of the examples of hybridization and introgression which can be found in plant populations at the present time are associated with the disturbance of old habitats and the opening up of new ones through human activity.

Indirect evidence of a different type for the importance of disturbance of the habitat in permitting the establishment of introgressive types is provided by Dansereau's study of × *Cistus florentinus* (*C. monspeliensis* × *salviifolius*). This hybrid has been produced artificially by Bornet (Gard 1912) and has 90 percent of obviously abortive pollen. It is a very common hybrid, reaching a frequency of 10 percent in the large stands of *Cistus* found in parts of Italy, and it has crossed back to both parents. The data of Dansereau, although based on only three characters — form of leaves, texture of leaves, and number of flowers per inflorescence — suggest that *C. salviifolius* is everywhere more variable than *C. monspeliensis,* and that much of this variation is in the direction of the hybrid with *C. monspeliensis.* It is likely, therefore, that the products of backcrossing from the F_1 to *C. salviifolius* have more often been successful than have backcross types involving *C. monspeliensis.* In other words, hybridization and introgression have promoted a flow of genes from *C. monspeliensis* to *C. salviifolius,* but not in the reverse direction. Although the ranges of these two species coincide for the most part, *C. monspeliensis* prefers relatively dry, sunny sites, while *C. salviifolius* usually occurs in more mesic habitats, particularly in the shade of oaks and pines. The disturbance of the Mediterranean regions by human activity during the past centuries has consistently and progressively destroyed the forested areas containing habitats of the latter type and has opened up sites suitable for *C. monspeliensis.* Typical plants of the latter species, therefore, have been as well adapted to these newly opened habitats as are genotypes modified by introgression. But *C. salviifolius* has constantly found itself exposed to habitats drier than those to which it is best adapted, and in the colonization of these new sites it has profited by the acquisition from *C. monspeliensis* of genes for greater resistance to drought.

Baker (1948) has studied a somewhat different type of example in *Melandrium dioicum* (*Lychnis dioica*) and *M. album*. The former species is indigenous in the forested areas of Europe, including Great Britain; *M. album* is a field weed, indigenous in the Near East, which has been spreading with human cultivation ever since the Neolithic Age. They hybridize easily, and the F_1 hybrids are rather fertile, although some sterility is present in both F_1 and F_2 generations. Baker was able to distinguish between ecologically neutral characters, such as length of calyx teeth, color of petals, and fertility of pollen, and those of adaptive significance. Depending on the degree to which the original forests are intact or cut down and replaced by plowed fields, a region contains pure *M. dioicum,* pure *M. album,* or populations of various hybrid and introgressive types. Baker recognized three stages of invasion and replacement of *M. dioicum* by *M. album*: first, the introduction into each species of ecologically neutral characters possessed by the other; second, the establishment in ecologically modified areas of various intermediate types; and, third, the disappearance of *M. dioicum* except for ecologically neutral characters which have become incorporated into the germ plasm of *M. album*. The strains of this weed introduced into North America all appear to be introgressive forms of *M. album,* containing some genes from *M. dioicum*.

In view of the complexity of adaptation, as discussed in Chapter IV, the present writer is somewhat doubtful of the ability of an observer to distinguish between adaptive and nonadaptive characteristics in hybrid derivatives. This somewhat reduces the general applicability of Baker's scheme of stages. In his example, the plants considered by him to be "pure" *M. album* may themselves differ from the indigenous eastern Mediterranean form of the species in possessing genes of *M. dioicum* acquired through hybridization and introgression in the remote past. Physiological characteristics acquired in this manner may have aided in adapting them to the climate of northern Europe.

Valentine (1948) has described a situation involving *Primula vulgaris* and *P. elatior* in Britain, in which hybrids and hybrid derivatives appear to have persisted in about the same proportion in a population for many years, although they may have been produced originally by disturbance of a natural area. This suggests that hybridization and introgression can reach a stable equilibrium, if ecological conditions originally disturbed later become relatively constant.

We cannot conclude, however, that hybridization and introgression are geologically recent phenomena in the history of plant populations. Throughout the history of the world there have been great environmental disturbances, caused by natural fires, landslides, volcanic eruptions, floods, the rise and fall of inland seas, glaciations, and similar agencies. Each of these has given opportunities for the occurrence and spread of hybrids and their derivatives among some plant groups. The detection of introgression resulting from these past disturbances is, however, beset with considerable difficulties. The F_1 hybrids from which the introgressive types arose have in most cases disappeared completely, and in many cases the parental species are likely to have altered the geographic distribution which they possessed at the time of the disturbance. The best evidence for hybridization in the remote past can be obtained from a study of naturally occurring allopolyploids of hybrid origin, as discussed in the next two chapters.

The best example known to the writer providing evidence of the occurrence of interspecific hybridization and introgression in

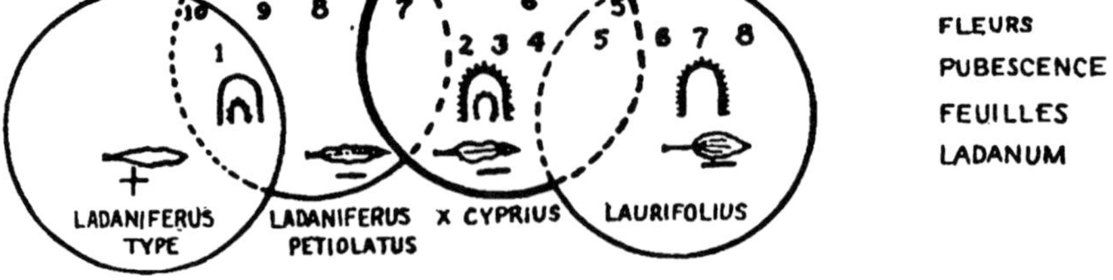

FIG. 28. Ideographs showing the number of carpels (upper row of figures), number of flowers per inflorescence (lower row of figures), pubescence of leaves and sepals, shape and margin of leaves, and presence or absence of ladanum in *Cistus ladaniferus*, *C. laurifolius*, and their F_1 hybrid. From Dansereau 1941.

the remote past and the subsequent establishment of the introgressive types as a new variety with a wide geographic range, is the work of Dansereau (1941) on *Cistus ladaniferus* and *C. laurifolius*. These two species, the only members of the section

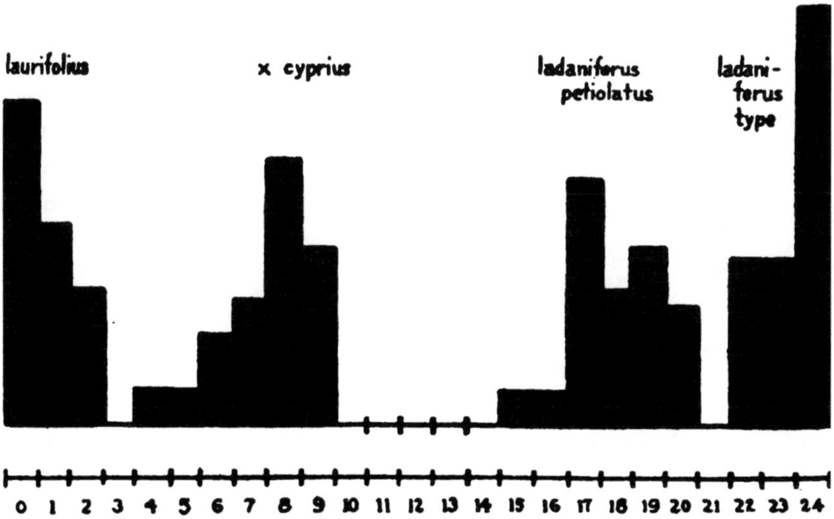

FIG. 29. Diagram showing index values and their frequencies obtained from the five characters shown in Figure 28. From Dansereau 1941.

Ladanium, occur together over most of Spain, Portugal, and southern France, and occasional F_1 hybrids are found throughout this region. This hybrid also was produced artificially by Bornet (Gard 1912), and has 80 percent of sterile pollen. Figure 29 shows that in respect to index values compounded from the seven characters illustrated in Fig. 28, namely, number of carpels, number of flowers, pubescence of sepals, length of petioles, margin of leaves, width of leaves, and presence or absence of ladanum, the variation patterns of the two species are completely discontinuous with each other and with the intermediate one formed by the F_1 hybrids (× *C. cyprius*).

Typical *C. laurifolius* occurs in central Italy and western Asia Minor without *C. ladaniferus* (Fig. 30), while in Corsica and North Africa is found its subspecies *atlantica,* which differs in being smaller in certain of its parts (Fig. 28). The range of typical *C. ladaniferus* does not extend beyond that of *C. laurifolius,* but its variety *petiolatus* is abundant along the coast of North Africa, in Morocco and in Algeria. All the differences which distinguish this variety from typical *C. ladaniferus* are in the direction of *C. laurifolius*. This is good evidence in favor of the hypothesis of Dansereau, that var. *petiolatus* originated from past hybridization

FIG. 30. Map of the geographical distribution of *Cistus ladaniferus*, *C. laurifolius*, their hybrids, and the hybrid derivative *C. ladaniferus* var. *petiolatus*. From Dansereau 1941.

between typical *C. ladaniferus* and *C. laurifolius*. One might suggest an alternative hypothesis, namely, that var. *petiolatus* is the primitive ancestor from which the two other types diverged, but this hypothesis seems less likely in view of the present geographic distribution of the three entities. The present center of distribution and variability of the genus *Cistus* as a whole is the Iberian Peninsula (Dansereau 1939), which is the home of *C. ladaniferus typicus* rather than of var. *petiolatus*.

Another example which suggests the occurrence of hybridization in the remote past is that of the sugar maples, *Acer saccharophorum* and *A. nigrum* (Dansereau and Lafond 1941, Dansereau and Desmarais 1947). The degree of intermediacy described in this example, however, suggests that the two "species" concerned are merely subspecies or ecotypes, which coincide to a large extent in their geographic distribution, but are separated by ecological barriers. Anderson and Hubricht (1938b) found them to be distinct in southern Michigan, while the greatest assortment of "pure" and "hybrid" colonies located near each other was found by Dansereau in the glaciated region of southern Quebec.

The studies made by Fassett (1944a,b, 1945a,b,c) on three closely related North American species of *Juniperus* show how the effects of contact between related types may be entirely different depending on the environmental conditions at the point of contact. *J. virginiana,* of the eastern United States, *J. horizontalis,* of the northeastern United States and Canada, and *J. scopulorum,* of the Rocky Mountains, are in general allopatric, and might well be considered well-marked subspecies rather than species, all of them being perhaps subspecies of the Old World *J. sabina.* Where two of these three forms come together at the margins of their ranges, the following different phenomena are found. *J. virginiana* and *J. scopulorum* intergrade like typical subspecies. They form hybrid swarms in southwestern North Dakota, and "introgressive" types of *J. virginiana,* containing some characteristics of *J. scopulorum,* are found hundreds of miles to the southward and eastward, in Nebraska and Kansas. These two forms have obviously been in contact for a long time, and in their region of contact, intermediate types of habitat are widespread. *J. scopulorum* and *J. horizontalis* occur together in northeastern Wyoming, and each can be found in its purest form a short distance from the other. In one locality, the east side of the Big Horn Mountains, however, a stand of *J. scopulorum* growing across the road from typical *J. horizontalis* contained many introgressive individuals. On the western side of these mountains a large stand of pure *J. scopulorum* is accompanied by smaller groups of semiprostrate individuals which in other respects correspond to that species, but may have originated through backcrossing from ancient hybrids with *J. horizontalis,* which is absent from the vicinity.

J. virginiana and *J. horizontalis* meet in several regions, two of which Fassett has carefully studied: the driftless area of Wisconsin and the coast of southern Maine. In the former area he found one locality containing only pure *J. virginiana* and *J. horizontalis* growing side by side, another in which pure *J. virginiana* was accompanied by a semiprostrate form of possible hybrid derivation, and still another occupied by both species plus a hybrid swarm of intermediate types. The latter condition was the usual one in southern Maine. These diverse situations prevailing at different localities of contact of these three entities might be due to the fact that they themselves are sufficiently diverse genetically so that the forms growing in one region cross easily with members of another "species," while those found in another region do not; or the abundance or paucity of hybrids and their derivatives in any particular region may be due to the particular selective forces at work in that environment. Additional more detailed analyses of situations such as this, supplemented by experimental work, should contribute greatly to our understanding of the nature of species barriers and the significance of hybridization.

One of the boldest hypotheses, involving extensive introgression, is that of Mangelsdorf (Mangelsdorf and Reeves 1939, Mangelsdorf and Cameron 1942, Mangelsdorf 1947) regarding the origin of certain varietal characters in maize. These characters, particularly cylindrical ears; grains in few, straight rows; hard, stiff cobs; and small, flat kernels, are believed to have entered the germ plasm of maize through introgression from hybrids with species of the related grass genus *Tripsacum*. This hybridization is believed to have yielded as a by-product another more or less constant species, *Zea (Euchlaena) mexicana,* or teosinte. Evidence in favor of this hypothesis is the concentration of "tripsacoid" characteristics in the numerous varieties of maize found in the lowlands of southern Mexico and Guatemala, which is the center of variability of the genus *Tripsacum,* and the correlation of these morphological characteristics with a cytological peculiarity most strongly developed in the species of *Tripsacum,* namely, the presence of heavily staining knobs on the ends of several of the chromosomes at mid-prophase. The evidence provided by artificial hybridization is in some ways favorable and in others unfavorable to the hypothesis. Mangelsdorf could produce the F_1 hybrid only by a special technique which involves shortening the styles or silks of maize. There is considerable doubt that it occurs naturally at present. The hybrid is highly sterile, but does set some seed with maize pollen. Fertile types extracted in later generations are mostly plants with 20 chromosomes and morphologically typical of maize, but since maize and *Tripsacum* chromosomes can occasionally pair and exchange segments, transfer of *Tripsacum* genes into the maize germ plasm is possible and occurred in Mangelsdorf's experiments. Nevertheless, this occurrence is so rare that in order to produce the "explosive" increase in variability which Mangelsdorf attributes to introgression from *Tripsacum* into the maize of Guatemala and Mexico, the occurrence of a large number of spontaneous hybrids would have been necessary. Such hybridization would be a very unlikely occurrence with maize and *Tripsacum* as strongly cross-incompatible as they are now. On the other hand, the data carefully gathered and critically evaluated by Mangelsdorf (1947) show that teosinte differs from maize genetically in a peculiar way. The genetic differences are confined to certain chromosomal segments with definite locations; but each segment, rather than containing its own particular complement of genes, contains and shares with the other segments some genes which influence each of the diverse and unrelated morphological characteristics differentiating teosinte from maize. It is difficult to see how such a situation could have arisen except through hybridization.

A possible solution to this dilemma might consist in alternatives to the Mangelsdorf hypothesis which do not involve recent hybridization between modern maize and any contemporary species of *Tripsacum*. The basic haploid chromosome number of *Tripsacum*, $x = 18$, is almost certainly a polyploid one, and the whole genus may well have an allopolyploid origin. The nine- or ten-paired ancestors of modern *Tripsacum* are very likely now extinct, but they may have existed in Mexico or Guatemala before these countries were as extensively disturbed by human cultivation as they now are. They may have included forms which could hybridize more readily with primitive maize than can any modern species of *Tripsacum*.

Even if the hypothesis of the hybrid origin of teosinte is rejected as too improbable, there remains a strong possibility that the characteristics peculiar to the "tripsacoid" maize varieties of Guatemala and Mexico arose through hybridization and introgression. If, as seems likely from the evidence of Mangelsdorf and his associates, as well as of Brieger (1944a,b), maize originated in South America, and if teosinte existed as a wild species before the introduction of cultivated maize into Guatemala and Mexico, then this introduction must have been accompanied by extensive hybridization between the two species, and selection of valuable backcross, introgressive types would have been made by the aboriginal cultivators. Hybrid corn is now the most valuable and highly developed form of maize; and hybridization almost certainly played an important role in the evolution of this most interesting crop plant.

From the preceding discussion we may conclude that introgressive hybridization is in many ways similar to evolutionary divergence through mutation, recombination, and selection. One important difference is that the genes which take part in this process enter the germ plasm of the species, not through mutation, but through transfer from another species across a barrier of reproductive isolation. A second is that not single genes, but groups of them, are added to the genetic complement of the species. Nevertheless, the similarities between evolutionary change through introgressive hybridization and that of the more usual type are great enough so that they can be directly compared. A group of closely related, incompletely isolated species, or ecospecies in the sense of Turesson and Clausen, Keck, and Hiesey (1939), can be likened to the type of population structure characterized by Wright (1931, 1940) as that which makes for the most rapid evolutionary progress; that is, a large population subdivided into many, partly isolated smaller ones. In this case, however, the isolation is not spatial, as in Wright's model, but is provided by the reproductive isolating mechanisms which separate the species. The isolation between the subpopulations is in time rather than in space, since F_1 hybrids occur only occasionally. Each species has in its past evolution "climbed" a different "adaptive peak," so that even though these species are closely related and sympatric, they will probably occupy a different ecological niche in their community or exploit their environment in a somewhat different way. This is true of all of the species pairs mentioned above. Therefore, introgressive hybridization between such related species represents the crossing of genes from one "adaptive peak" to another and makes possible the formation of gene combinations capable of climbing new "peaks." In its action it is therefore essentially similar to migration pressure in the partially subdivided population model of Wright. This analogy is entirely in accord with the known conditions under which introgression is most evident, namely, when new "adaptive peaks" in the form of unoccupied environmental niches are available to the population.

HYBRIDIZATION AND THE ORIGIN OF NEW TYPES

Introgressive hybridization, whatever may be its importance in modifying and amplifying the variation pattern of certain individual species, is nevertheless by its very nature not a way of producing new morphological or physiological characteristics, and therefore of progressive evolution. It merely produces convergence between previously more distinct species. There is, however, evidence that in some instances hybridization can result in the appearance of types which are actually new. These may represent various degrees of divergence and distinctness from their parental populations. The least remarkable, but perhaps the most frequent, are new races or subspecies which may arise from hybridization between preexisting subspecies of the same species, provided that a new and intermediate habitat is available to them.

Examples of the origin of such races are not numerous and by their very nature are hard to establish through observation of wild populations. Although the recognition of an F_1 hybrid between members of two adjacent subspecies or species is not a very difficult matter and can be verified experimentally with relative ease in many groups of plants, the identity of segregates in later generations may be much more difficult to recognize and to verify. Furthermore, considerable familiarity with the climatic and physiographic history of a region is needed before a habitat can be recognized as relatively new.

One example which nevertheless seems to be of this nature is *Potentilla glandulosa* subsp. *hanseni* in the Sierra Nevada of California (Clausen, Keck, and Hiesey 1940). As is shown by their chart (Table 1) and as brought out in the text (p. 44), this subspecies is intermediate in a whole series of morphological characteristics between subspp. *reflexa* and *nevadensis*. It also occupies a habitat, the mountain meadows at middle altitudes, which is intermediate between that occupied by the two last-mentioned subspecies. These meadows are moister and cooler than the warm dry slopes which in the same region are the habitat of subsp. *reflexa,* and on the other hand are considerably warmer than the subalpine and alpine habitat of subsp. *nevadensis*. Furthermore, both subsp. *reflexa* and subsp. *nevadensis* range far beyond the area in the central Sierra Nevada occupied by subsp. *hanseni*. Finally, the mountain meadows in which subsp. *hanseni* is found are new habitats created by the disturbance of topography resulting from the Pleistocene glaciation. They represent either filled-in lake beds or outwash plains which have poor drainage because of their flat surfaces. The most plausible hypothesis, therefore, is that in late glacial or early postglacial times subsp. *hanseni* was produced by hybridization between subspp. *reflexa* and *nevadensis*, and that it then entered the meadow habitat newly available to it. Clausen, Keck, and Hiesey (1947) have shown that a great array of segregates are produced in the F_2 generation of the cross between these two subspecies, and that many of these are very well adapted to conditions in the meadow at Mather, which is occupied by subsp. *hanseni*. Since the authors have not published any comparison between these segregates and subsp. *hanseni* in morphological characteristics, it is not known how nearly this subspecies has been reproduced artificially.

Another probable example of this nature is *Vaccinium corymbosum*, the common high-bush blueberry of the glaciated regions of northeastern North America (Camp 1945). This polymorphic entity, occupying a habitat known to be relatively recent, is believed on morphological grounds to have resulted from hybridization between four other "species" which occur south of the glaciated territory. The arguments for recognizing these five entities as distinct species, and for the complex phylogeny which Camp postulates, are rather involved and not supported by enough data to be very convincing.

A still more striking example of the origin of a new race through hybridization is that of *Abies borisii-regis*, carefully analyzed by Mattfeld (1930). The common fir of central Europe, *A. alba*, is constant and typical in the northern and western part of the Balkan Peninsula, extending southward to the northern boundary of Greece. In the mountains of central and southern Greece it is replaced by *A. cephalonica*, which is likewise constant and typical throughout the main part of its range. But in northern Greece there occurs a series of intermediate forms, which at their northern limit most resemble *A. alba* and grow with trees typical of that species, and at the southern limit of their distribution resemble and accompany typical *A. cephalonica*. Trees similar to these intermediate forms are the only ones found on the Athos Peninsula, in northeastern Greece, as well as in parts of

Macedonia and in the Rhodope Mountains of Bulgaria. The fir forests of the latter regions are isolated by distances of 60 to 100 miles (100 to 160 km) from those of the Grecian peninsula. The intermediate form from the Rhodope Mountains was named by Mattfeld as a distinct species, *A. borisii-regis,* but the evidence presented by him indicates to the present writer that it, as well as *A. alba* and *A. cephalonica,* should be treated as races of a single polytypic species.

Mattfeld has considered three possibilities regarding the nature of *A. borisii-regis*: first, that it might be considered an intermediate race, corresponding to and of parallel origin with *A. alba* and *A. cephalonica*; second, it might be an original, heterozygous and genotypically rich population from which *A. alba* segregated and migrated northward, and *A. cephalonica* similarly segregated in the south; and, third, that *A. borisii-regis* represents a series of products of ancient hybridization between *A. alba* and *A. cephalonica,* plus derivatives of segregation and backcrossing.

The first possibility is highly improbable because of the nature of the intermediate populations. In no area are they a constant, easily recognizable entity, as are *A. alba* and *A. cephalonica;* rather, each mountain range possesses a different complex of intergrading and recombining forms, which have in common only the characteristic that they show the various diagnostic characters of *A. alba* and *A. cephalonica,* and no others, combined in different ways and with various degrees of intermediacy. The greatest amount of variability is in northern Greece, in the populations which are continuous with those of the other two races; but the isolated populations of Athos and the Rhodope Mountains also show much evidence of segregation and recombination.

The second possibility is rejected by Mattfeld because no characteristics can be seen in these variable, intermediate populations except for those of *A. alba* and *A. cephalonica.* If northern Greece were the original gene center for this complex of *Abies,* it should certainly contain some genes and genotypes which did not become segregated into the populations of *A. alba* and *A. cephalonica.*

This leaves the third possibility, that of ancient hybridization, as the most likely. Its likelihood is strengthened by paleontological, and particularly geological, evidence. *A. alba* and its close relative of Asia Minor, *A. nordmanniana,* are characteristic elements of the Colchian flora, fossil remains of which indicate that it was widespread throughout central Europe in the latter part of the Tertiary period. *A. cephalonica,* on the other hand, is not only strikingly different morphologically from *A. alba* and *A. nordmanniana* but in addition it occupies a different floristic province, that of the Grecian-Asia Minor Mediterranean flora. This flora appears to have developed in isolation from the Colchian flora during the latter part of the Tertiary period, since the two have very little in common. The most likely hypothesis, advanced by Mattfeld, is that *A. alba* and *A. cephalonica* were well isolated from each other during the Miocene and Pliocene epochs, and that hybridization began with the southward migration of *A. alba* in response to the cooling of the climate at beginning of the Pleistocene glacial, or "diluvial," period. *A. borisii-regis,* therefore, is probably descended from a series of hybrid swarms which have existed for several hundred thousand years. On the Grecian peninsula, their variability is continually being increased by influx of genes from the parental races, but on Athos, the Rhodope Mountains, and probably on the island of Thasos, they are forming intermediate races that are relatively constant and true breeding, although they are still more heterozygous than the parental races.

Other examples will doubtless become available when more groups have been studied critically with this possibility in mind. Further evidence on the ease with which crossing between subspecies can lead to new ecotypes or subspecies could be obtained by experiments on the artificial establishment in new environments of the products of intersubspecific hybridization. The writer is at present conducting such experiments in the genus *Bromus,* but the degree of their success cannot as yet be estimated.

The examples already mentioned are new types only in the sense that they are entities which are recognizably different from their parental populations and occupy new habitats; morphologically they do not contain any new characteristics. But the origin from hybridization of races or species with characteristics that are new, in the sense that they could not have been predicted on the basis of examination of the parental types, has been reported several times. The simplest of such cases are those involving types of gene recombinations well known to geneticists, that is, the interaction of different allelomorphic series of simple Mendelian factors. A good illustration is found in the work of Brainerd (1924) and Gershoy (1928, 1932, 1934) on the progeny of natural interspecific hybrids between eastern North American species of *Viola*. Although the principal character differences between these species give simple segregations, and many of the F_2 and F_3 individuals show reversion to one or the other of the parental species, nevertheless a considerable proportion of these offspring are considerably modified from the condition found either in the parents or in the F_1, and some of these modified types will eventually breed true. An illustration of such possible "new" types in respect to leaf shape is given in Fig. 31, showing the leaves of *V. pedatifida, V. sagittata,* their F_1 hybrid, and various F_2 segregates. Brainerd's comment on a situation similar to this is as follows (1924, p. 165).

In these various ways there has arisen in the numerous progeny of the hybrid under discussion a considerable diversity of foliage, such as would present insoluble difficulties to a taxonomic student who did not know that these diverse forms all came from one individual, by close-fertilized reproduction, in the short period of three or four years. The extreme differences are such as would warrant the making of several distinct species, according to the hasty methods of ordinary practice.

In comparing the various segregating types illustrated by Brainerd with the stable and constant species of *Viola* found in the eastern United States, one becomes struck by the possibility that various ones of the unusual leaf forms, such as those found in *V. palmata, V. brittoniana, V. triloba, V. stoneana,* and *V. viarum,* have been derived by the stabilization of hybrid derivatives. There is considerable reason for suggesting that the number of stable, recognizable species in the subsection *Boreali-Americanae* has been considerably increased by hybridization. Brainerd (1924) cites an example of the possible birth of such a new species in the naturally occurring offspring of *V. affinis* × *sagittata*.

Similar "new" types appear to be segregating from natural hybrids of *Iris* in the Mississippi Delta region of the southern United States (Viosca 1935, Foster 1937, Riley 1938, Anderson 1949). A large number of these have been named as species by

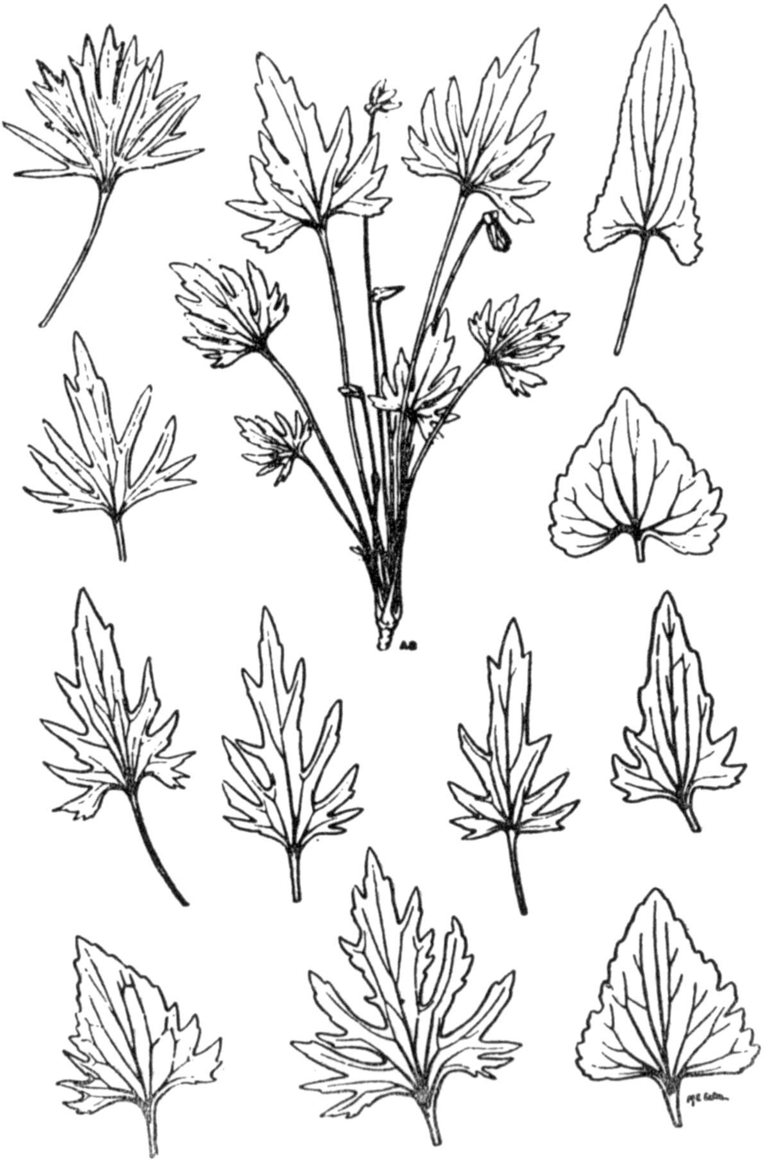

FIG. 31. Leaf of *Viola pedatifida* (upper left), leaf of *V. sagittata* (upper right), plant of the F_1 hybrid *V. pedatifida* × *sagittata* (upper middle), and leaves of nine different F_2 segregates from this hybrid. From Brainerd 1924.

the taxonomic "splitters." In the subgenus *Eubatus* of the genus *Rubus*, there is a diversity of forms in the eastern United States. The data of Brainerd and Peitersen (1920) and Peitersen (1921) on segregation of seedling progenies from putative hybrids found in the wild and corresponding to some of the species recognized by systematists, although far too scanty, suggest that many of these "species" are unstable hybrids or hybrid derivatives. In this subgenus, however, the variation pattern has been greatly complicated by polyploidy and apomixis, as will be discussed in Chapter X.

Some evidence at hand suggests that the recombination of genetic factors in the offspring of interspecific hybrids may sometimes lead to new types radically different from those found in either parent. Lotsy (1915) illustrates some striking examples of this nature in the progeny of *Antirrhinum glutinosum* crossed with a peloric form of *A. majus*. Hagedoorn and Hagedoorn (1921) cite the example of Vilmorin's hybrid between *Argemone mexicana* and *A. platyceras*, from which several strongly aberrant types segregated in the F_2 generation. Some of these were different from either parent in such fundamental characteristics as the number of sepals or of carpels. Similar aberrant types appeared in the F_1 hybrids of *Paeonia lactiflora* ("*P. albiflora*") and various members of the complex of *P. anomala* (Saunders and Stebbins 1938). Less extreme new types are reported by Clausen (1926) as segregates from hybrids between *Viola arvensis* and *V. tricolor*. Griggs (1937) has emphasized the significance of such types in the origin of cultivated plants. Their importance in evolution depends, of course, upon whether they can ever become established in nature. The examples cited above of *Argemone* and *Paeonia* are teratological and not fitted to any environment, while Clausen has remarked that the new types in *Viola* disappear in favor of typical *V. arvensis* when cultivation is abandoned. Nevertheless, the possibility exists that under particular environmental conditions some such new types may become established and may contribute to evolutionary progress. The detection of an existing species or subspecies as a new derivative of past hybridization is well-nigh impossible, since by definition such new types would not be recognizably intermediate between their parental species.

Another way in which hybridization may result in evolutionary progress is through its stimulation of the mutation rate. In Chapter III the example was cited of the increased mutability of the petal spot gene of *Gossypium hirsutum* when transferred by means of interspecific hybridization to the genetic background of *G. barbadense* (Harland 1937). Furthermore, if Sturtevant's and Shapiro's hypothesis of the gradual reduction of mutation rates through selection of mutation suppressors is correct (see page 99), then interspecific hybridization followed by segregation and recombination would be a way of eliminating these suppressors and of enabling a previously stable evolutionary line to progress rapidly in response to a newly changing environment. Unfortunately, evidence for increased mutation rates following hybridization is difficult to obtain because new mutations occurring in the strongly segregating progeny of an interspecific hybrid are in most cases impossible to detect. Such evidence will be obtained most easily in genera like *Gossypium,* which contains several species that can be crossed and produce partially fertile hybrids, and in which a number of different gene loci have been identified. Giles (1940) has produced evidence that gross structural changes in the chromosomes occur three times as frequently in the hybrid *Tradescantia canaliculata* × *humilis* than in its parents. He suggested that this difference might be due to disturbance of the normal coiling cycle of the chromosome threads or chromonemata in the chromosomes of the hybrid.

THE EFFECT OF HYBRIDIZATION ON INTERSPECIFIC ISOLATING MECHANISMS

Offspring produced from partly sterile interspecific hybrids often retain the sterility of their F_1 progenitors, but in many instances have been known to become more fertile. The most spectacular and widely known of such examples of recovery of fertility are those in which the chromosome number of the hybrid has been doubled, and a constant, fertile, allopolyploid species has been produced. These will be discussed in the following chapter. Less widely known and studied, but of equal or greater evolutionary importance, are examples of the recovery of fertility and the establishment of constant, true-breeding lines from highly sterile interspecific hybrids without doubling of the chromosome number. Such increase in fertility has been reported by Brainerd

(1924) and Clausen (1926, 1931) in *Viola*, by Ostenfeld (1929) in *Polemonium mexicanum* × *pauciflorum*, by Müntzing (1930a) in *Galeopsis tetrahit* × *bifida*, by Anderson (1936a) in *Apocynum androsaemifolium* × *cannabinum* and (unpublished) in *Nicotiana alata* × *landsdorffii*, by Winge (1938) in *Tragopogon pratensis* × *porrifolius*, by Lamprecht (1941) in *Phaseolus vulgaris* × *coccineus*, by Zakharjevsky (1941) in *Triticum durum* × *timopheevi*, by Hiorth (1942) and Håkansson (1946a, 1947) in *Godetia amoena* × *whitneyi*, and has been found by the writer in the progeny of several hybridizations in the genera *Bromus* and *Elymus*. Constant, true-breeding lines have been extracted from such fertile segregates in *Galeopsis, Tragopogon, Phaseolus, Triticum, Viola*, and *Godetia*.

The evolutionary importance of such fertile hybrid derivatives depends both on their morphological characteristics and on the degree of fertility or sterility which is found when they are crossed back to their original parents. Three situations can exist with respect to these fertility relationships. First, the fertile hybrid derivative may form fertile hybrids with one of its parental species, and sterile ones with the other. In this case it represents either the complete recovery of the genic complement of one parent or the establishment of an introgressive type. This is true of the lines established in *Tragopogon* and some of those in *Phaseolus*. Second, the line might conceivably form fertile hybrids with both of the parental species, and so represent the obliteration of the isolating barrier between them. This situation, however, has never been reported so far as the writer is aware, and on the basis of our knowledge of genetic isolating mechanisms would not be expected unless the fertility of the F_1 hybrid were itself relatively high. The third possibility, and the one which is by far the most important from the evolutionary point of view, is that the new stabilized, fertile line would form partly sterile F_1 hybrids in crosses with either of its parental species. This is apparently true of some of Lamprecht's derivative lines in *Phaseolus*, although his data are not as complete as might be desired. It may also be true in *Viola, Nicotiana, Godetia*, and *Triticum*, but the appropriate hybridizations have not been made, or at least have not been reported. In an earlier paper (Stebbins 1942a) the writer suggested that if, as now seems even more evident, barriers of hybrid sterility are made up of many different genetic factors, either genes or small chromosomal segments, a considerable proportion of those fertile types which have been extracted by segregation, without doubling of the chromosome number, from partly sterile interspecific hybrids should form partly sterile hybrids with both of their parental species. Furthermore, the larger the number of genetic factors which contribute to the original sterility, the smaller is the number of fertile lines which can be extracted in a given number of generations, but the larger is the proportion of these fertile types which will be thus isolated from both of their original parents. If we use a simplified model of complementary sterility factors, we can make the following calculations by the use of the expansion of binomials as required for genetic experiments of this nature. If the parents differ by one pair of complementary sterility factors producing 50 percent gametic sterility in the F_1, then 50 percent of the F_2 offspring will be fully fertile and will form fertile hybrids with one or the other of the original parents. If the parents differ by two such factor pairs, and the F_1 is 25 percent fertile, then 25 percent of the F_2 offspring will be fully fertile, but only half of them will form fully fertile hybrids with one or other of the original parents. The other half of these fertile F_2 derivatives will consist of two types, each of which will form partly sterile hybrids with both of the original parents, and in addition will be partly intersterile with each other. The general formula may be expressed as follows. If n is the number of independently segregating factor pairs which separates the parental species, and each factor pair causes when heterozygous the death of 50 percent of the gametes, then the percent of individuals in the F_2 generation which will be fully fertile is $100 \times 1/2^n$. But the percentage of these fertile types which will be partly intersterile with their parents and therefore a potential new species is $100 \times (1 - 1/2^{n-1})$. With a difference between the parents of four factor pairs, this last figure is 87.5 percent, and if the number of factors is six and the gametic fertility of the F_1 hybrid is reduced to 1.56 percent, then the percent of fully fertile types in F_2 is also 1.56 percent, and nearly 97 percent of these will be partly intersterile with their parents. Even though, as stated earlier, this simple accumulation of similar sterility factors is probably never actually followed in the evolution of sterility barriers in nature, nevertheless it is probably approached nearly enough so that the results calculated on this basis will be approximated to a considerable degree in the progeny of many interspecific hybrids.

The testing of this hypothesis in any group of plants obviously requires a long series of experiments extending over a considerable period of time. Even more work would be required to prove conclusively that any particular wild species has originated in this fashion, since one would first have to identify its parental species by means of morphological studies and would then have to obtain F_1 hybrids and later-generation progenies on a large scale in order to obtain a sufficiently large proportion of the multitude of segregant types so that some of them could be expected to be similar to or at least approximating the particular segregants which had become selected out of the progeny of the natural hybrid leading to the species in question. Nevertheless, strong circumstantial evidence has been produced by Epling (1947a) that the Californian species *Delphinium gypsophilum* arose in this fashion. It is intermediate morphologically between *D. recurvatum* and *D. hesperium* and occupies a natural range in the inner coast ranges of central California which is climatically and edaphically intermediate between that of the two other species. All three species are diploids, with eight pairs of chromosomes, except that *D. gypsophilum* also has a tetraploid race. The F_1 hybrid between *D. hesperium* and *D. recurvatum* resembles *D. gypsophilum* morphologically, but differs from it in having a large amount of aborted pollen and a very low seed set under open pollination, while *D. gypsophilum* grown under the same conditions is fully fertile as to both pollen and seed. More important, the progeny obtained from crossing this F_1 with wild *D. gypsophilum* have a higher fertility than those derived from backcrossing the same F_1 to its immediate parents. Apparently, therefore, the artificial F_1 hybrid between *D. hesperium* and *D. recurvatum* resembles the naturally growing species *D. gypsophilum* in its fertility relationships, as well as external morphology, and the most likely hypothesis is that the latter species has been produced by segregation from such a hybrid. The fact that it retains its intermediate character would be explained on the assumption that the morphological differences between *D. hesperium* and *D. recurvatum* are governed by multiple factors.

THE SIGNIFICANCE OF HYBRIDS IN AGRICULTURE AND FORESTRY

From the evidence presented in this chapter, it is apparent that under favorable environmental conditions some phases of evolution can be greatly speeded up by interspecific hybridization. The parallel between organic evolution in nature and the improvement of cultivated plants through artificial breeding and selection on the part of man is close enough so that we might expect such hybridization to be important in plant breeding as well. Furthermore, the value of this procedure should vary depending on the character of the plant under cultivation, the use to which it was put, and the environmental conditions under which it was grown. This is just what we find. Interspecific hybridization has been carried out by plant breeders on an empirical basis for centuries, and in some groups, such as roses and tulips, it has played a major role in the origin of the types now being cultivated. And in recent years some of these plant hybridizers, like the Vilmorins in France, Burbank in the United States, and Michurin in Russia, have achieved spectacular results and wide popular acclaim through the use of this method. Furthermore, the usefulness of hybridization has proved great in some crops and relatively slight in others, depending on various factors. This fact can be brought out by a brief comparison of the amount of hybridization which has figured in the ancestry of modern varieties of field and forage crops, orchard crops, truck crops, and horticultural ornamentals.

In these four classes of cultivated plants, hybridization has been most important in garden and greenhouse ornamentals. The majority of the widely grown varieties of perennial herbs and shrubs, such as roses, tulips, hyacinths, narcissus, pansies, orchids, primroses, chrysanthemums, irises, delphiniums, and rhododendrons, are of hybrid origin, and during the past hundred years plant hunters have been exploring every corner of the globe for new species of these genera which might be combined with the well-known ones to produce an ever-increasing array of novelties. It is in this field that the daring, imagination, and keen eye of the professional hybridizer are most richly rewarded.

Next in importance are the orchard crops, both fruits and berries. Many of these, such as apples, cherries, plums, peaches, grapes, various types of berries, citrus fruits, and bananas, have been much hybridized, while in others, such as pears, walnuts, olives, avocados, and papayas, all the numerous varieties now being grown have probably originated by selection from a single original species. Except in the genus *Rubus*, the hybridization has been restricted for the most part to crosses between closely related species belonging to the same section or subgenus, while in ornamentals hybrids between members of different subgenera or even genera have been more frequent.

In the field and forage crops, such as cereal grains, fiber crops like cotton and flax, tobacco, and forage crops like alfalfa and the clovers, hybridization between reproductively isolated species has been still less frequent. Several of them, such as wheat, cotton, and tobacco, are allopolyploids and therefore are derived originally from hybrids between widely different species, as will be discussed in the following chapter. But this hybridization, even if it took place after the parental species had been brought into cultivation, was spontaneous, not guided by the hand of man. The recent improvement of these crops by breeding has mainly been through intervarietal crossing and selection within the genetic species. Interspecific hybridization has in many instances contributed much to this improvement, but usually the crosses have been between closely related species, like the emmer and the *vulgare* or *aestivum* species of wheat, and the sea island and the upland cotton. Wider crosses, such as wheat with rye or quack grass (*Agropyron*), figure prominently in recent literature, but the polyploid derivatives of such crosses are all still in the experimental stage. Furthermore, the greatest use of interspecific hybridization in these crops in recent years has been not through the selection of new types from their segregating progeny, but through the transfer of individual characters from one species to another through a careful combination of hybridization, backcrossing, and selection. The majority of the characteristics of the more desirable parental variety is combined with one particular character, such as resistance to disease, insect pests, drought, or early maturing, derived from the other (Briggs 1938, Hayes and Immer 1942). In both purpose and method, this technique appears like a highly refined version of the introgressive hybridization which occurs so frequently in nature.

Finally, hybridization between distinct species has been least important in the truck crops or vegetables. In nearly all of the most important ones — carrots, beets, turnips, radishes, asparagus, celery, spinach, lettuce, peas, snap beans, lima beans, and tomatoes — the cultivated varieties have all been derived by selection from a single diploid species or from a complex of interfertile taxonomic "species." In some of them, like the cultivated *Brassica oleracea,* including cabbage, cauliflower, broccoli, and Brussels sprouts, and *Cucurbita pepo,* including most types of garden squashes and pumpkins, the varieties which have been selected from a single species may be so widely different that their specific identity would hardly be suspected without careful study. In these crops, even hybridization and backcrossing for the purpose of transferring single characters has not been widely used, and the results of such hybridization are still largely in the experimental stage. The vegetables which form an exception to this rule are the tubers, like the potato and the sweet potato.

These differences between the classes of crops in respect to the importance of hybridization can be explained on the basis of two assumptions: first, that hybridization is of much greater value in plants which are usually propagated by vegetative means and less valuable in those reproduced primarily by seed; and, second, that the value of hybridization between widely different species is comparatively high in crops in which quantity of yield is the principal objective of breeding and falls lower as the demands for quality of the crop become more and more exacting.

The first assumption would explain the greater importance of hybridization in ornamentals and orchard crops as compared to field and truck crops. In these first two categories a large proportion of the most important varieties, even those of nonhybrid origin like the seedless table grapes and the double varieties of the common Chinese peony, produce no viable seeds and are propagated exclusively by vegetative means, via grafts, root divisions, or cuttings. Others which do produce seeds, like apples, pears, and peaches, are so heterozygous because of previous intervarietal or interspecific hybridization that seedling progeny are for the most part segregates with undesirable characteristics of yield or quality. Hybridization is more valuable in such crops because a long and expensive program of selection and genetic purification is not needed in them. Any single valuable hybrid individual, once obtained, can immediately become the progenitor of a new variety and can be perpetuated indefinitely.

Here we find an analogy to the significance of hybridization in organic evolution. Earlier in this chapter the greater importance was indicated of hybridization in plants with efficient means of vegetative reproduction. This is because the hybrid product, even though highly sterile, can persist for a long time if it is well adapted to its environment, and it may eventually yield more fertile derivatives.

The differences in the demands for quality would explain most of the differences in the importance of hybridization in ornamentals as compared to orchard crops, and in field crops as compared to vegetables. In garden flowers and shrubs, quality is very important, but the demand is not so much for a particular quality as for an aesthetic value which can be achieved in a number of different ways. In fact, novelty is of itself often desirable, and this is, of course, the one characteristic of quality which is obtained more easily by hybridization than by any other method. On the other hand, the demands for quality in fruits and fruit products are exacting and conservative, in that the traditional flavors are those usually desired. Furthermore, certain additional characteristics of quality are essential in those fruits which are regularly shipped long distances, dried, or canned. To obtain as a result of wide interspecific crossing even individual genotypes which will meet all these demands on quality is no easy task, and the breeder cannot rely on luck to nearly as great an extent as he can with many types of ornamentals.

Similar considerations hold in a comparison between field and truck crops. In certain of the former, such as wheat, cotton, and flax, the demands on quality are exacting, but in others, particularly those like field corn, barley, oats, and alfalfa, which are used principally as feed for animals, they are much less so. Most of the vegetable crops, on the other hand, have originally been selected out of a large number of potentially edible types of roots, leaves, or stems because they possess certain qualities of flavor, tenderness, or succulence, and the first consideration of the breeder dealing with such crops is to maintain these qualities. In addition, vegetables, like fruits, must under modern conditions possess shipping or canning qualities to be widely useful. If such qualities are to be maintained in a crop reproduced entirely or largely by seed, the breeder cannot afford to break them down completely by introducing into the stock a whole complement of widely different genes obtained from a distantly related species.

Here again we have an analogy to organic evolution. If the adaptation of the species to its environment is of such a broad, general type that many different gene combinations are equally adaptive, then the chances are very good that some of the segregates from interspecific hybridization will have a selective value, and this process may figure prominently in evolution. But if certain of the adaptations possessed by the species are very exact and specific, like the elaborate flower structure of orchids and milkweeds, then the products of hybridization will almost certainly be nonadaptive, and this process will take a relatively small part in evolution.

In recent years, a type of plant breeding has been undertaken which differs radically from the breeding of typical cultivated plants and more nearly approaches the conditions found in organic evolution. This is the breeding of trees for reforestation and of forage grasses and other plants for the revegetation of uncultivated pastures and range lands. This differs from the breeding of cultivated crops chiefly in the fact that the finished product cannot be provided by man with the optimum conditions for growth, but must be able to compete successfully under more or less natural conditions. In forest trees, three additional characteristics affect the breeding process. In the first place, the amount of time and space needed for growing to maturity a single crop is much greater than that needed for annual seed crops, but, like the latter, only one harvest is possible for each planting. This puts a relatively high premium on the quality of the seed and should make it economically profitable to spend large amounts of time and money in securing the best possible seed. Second, the length of generations in forest trees makes impossible the establishment and maintenance of genetically pure stocks after hybridization. Finally, as Righter (1946) has pointed out, genetic uniformity is not necessarily a desirable quality in stands of forest trees, which occupy a relatively heterogeneous environment, and in which the destruction through natural causes of 90 percent of the original stand of seedlings is actually beneficial.

All of these differences favor the use of interspecific hybrids in forest-tree breeding. The same amount of time and money used in securing a large amount of hybrid seed would be a relatively high proportion of the total cost of raising the crop in an annual plant and a much lower one in forest trees. And the heterozygosity or partial sterility of F_1 hybrids is a relatively slight disadvantage, since breeding and artificial selection from such hybrids is impractical, and genetic uniformity is not necessarily a desirable characteristic. Additional characteristics which increase the value of interspecific hybridization in forest-tree breeding are that the demands on quality in a lumber- or pulp-producing tree are relatively general and not very exacting, and, finally, that in most genera of trees, as pointed out in the preceding chapter, hybridization between widely different species often yields vigorous and partly or wholly fertile offspring.

As a result, interspecific hybridization has played a prominent role in the breeding of forest trees. This will be evident from a study of the reviews of this subject by Syrach-Larsen (1937), Johnson (1939), Smith and Nichols (1941), and Richens (1945), in which many examples are cited. A short description of the principal work in two of the most important genera, *Populus* and *Pinus,* will serve to illustrate the nature of this work. In poplars, the experiments of several workers (Stout, McKee, and Schreiner 1927, Wettstein 1933, 1937, Johnson 1942) have shown that the F_1 progeny of some hybrid combinations, both between and within species, may be more vigorous under conditions of cultivation than either parent. This hybrid vigor, however, is by no means universal. It is not found in some hybrids between species belonging to different sections, such as *P. tremula* × *nigra* (Wettstein 1933) and *P. balsamifera virginiana* × *grandidentata* (Stout and Schreiner 1934). In other hybrids its manifestation may be very irregular. Johnson (1942) found that the average vigor of hybrid progenies between *P. alba* and either *P. grandidentata* or *P. tremuloides* was greater than that of their parents, but the variability of the seedling lots was in every case very large, and there was much overlapping. He found that there was in general no inverse correlation between vigor of the seedlings and pulping quality of the wood, so that hybrid vigor, when obtained, carries a practical advantage in this genus. Furthermore, since vegetative propagation through slips or cuttings is practicable in poplars, the best method of increasing yield in this genus may be to select particular parental trees which give exceptionally vigorous hybrids, to obtain a large number of F_1 seedlings, perhaps over a period of years, and to reforest with series of cuttings from several different vigorous F_1 genotypes.

In pines, as in poplars, many interspecific crosses yield hybrids more vigorous than their parents (Righter 1946). Vegetative reproduction is impracticable in this genus, but Righter has shown that large quantities of F_1 seed can be produced relatively economically. In pines, therefore, reforestation with hybrids will in every case make use of the seedlings obtained directly from crossing. The mean vigor of the progeny will be relatively important, and the occurrence of particularly vigorous individual genotypes will be less important than it is in poplars. Righter has suggested that in some instances, in which previous progeny tests have shown that many vigorous trees can be obtained from the F_2 generation, reforestation of an area planted previously to an F_1 population of an interspecific hybrid or a series of different hybrids can be secured by leaving a few vigorous seed trees. The high quality of the strongly segregating F_2 progenies might be maintained by natural selection plus artificial culling of undesirable segregates. If such reforestation practices are carried out, botanists of the future should be able to follow spectacular experiments on speeding up the rate of evolution on a stupendous scale.

In the revegetation of uncultivated pasture or range lands, the nearest possible approach is made to the conditions of organic evolution in nature, since natural selection can here be only modified by controlling the amount and time of grazing and by other management practices, while frequent artificial reseeding is often so impracticable that successful natural reseeding is a particularly desirable characteristic of the forage crop to be sown. Since livestock will eat and thrive on forage plants of a great variety of types, the principal characteristics which the desirable strains should have are vigor, competitive ability, and resistance to drought, cold, and other extreme conditions. Such characteristics can often be found in interspecific hybrids.

The use of wide crosses in this manner is discussed by Love (1946). He found that when sample plots were seeded to a mixture of three species of *Stipa* and the plants allowed to mature, a number of natural interspecific hybrids were found after the first season in which the species were able to cross-pollinate. These hybrids were more vigorous than their parents, and under the dry conditions prevailing at Davis, California, had the additional advantage of remaining green later in the spring and becoming green earlier in the fall. Although sterile, they are very long-lived, and as long as the parental species were present in sufficient number, might be expected to be replaced periodically. Their presence, therefore, might be a permanent asset to a pasture seeded originally to a mixture of these or any other groups of related species capable of forming vigorous F_1 hybrids. Interspecific hybridization followed by polyploidy may have an even wider application in the revegetating of uncultivated pasture and range lands, as will be discussed in the following chapter.

CHAPTER VIII

Polyploidy I: Occurrence and Nature of Polyploid Types

THE PHENOMENON of polyploidy, or the existence in genetically related types of chromosome numbers which are multiples of each other, is one of the most widespread and distinctive features of the higher plants and was one of the earliest of their cytogenetic characteristics to become extensively studied and well understood. One of De Vries's original mutations of *Oenothera lamarckiana*, mut. *gigas*, was early recognized to be a tetraploid (Lutz 1907, Gates 1909), and the artificial production by Winkler (1916) of a tetraploid form of *Solanum nigrum* through decapitation and the regeneration of callus tissue was perhaps the first example of the direct production in a laboratory experiment of a new, constant genetic type. Meanwhile, explorations of chromosome numbers in various plant genera were revealing the presence of series consisting of multiples of a basic number, as in *Drosera* ($2n = 20$ and 40, $x = 10$, Rosenberg 1909)[1]; *Dahlia* ($2n = 32$, 64, $x = 16$, Ishikawa 1911); *Chrysanthemum* ($2n = 18$, 36, 54, 72, 90, $x = 9$, Tahara 1915); *Triticum* ($2n = 14$, 28, 42, $x = 7$, Sakamura 1918, Sax 1922); and *Rosa* ($2n = 14$, 28, 42, 56, $x = 7$, Täckholm 1922). Then, Digby (1912) discovered that the spontaneous occurrence of a constant fertile type, *Primula kewensis*, from a sterile interspecific hybrid, *P. verticillata* × *floribunda*, was associated with the doubling of the chromosome number, but she failed to grasp the significance of this phenomenon. Subsequently Winge (1917) formulated his hypothesis of hybridization followed by polyploidy as a method for the origin of

[1] In this book, the somatic number of any particular form will be designated as $2n$ and its gametic number as n, regardless of its degree of polyploidy, unless it is an unbalanced type with an uneven somatic number. The most probable basic haploid or gametic number for any genus or for any polypoid series that forms part of a genus is given the letter x, and the somatic numbers of unbalanced types like triploids, pentaploids, etc., are designated as $3x$, $5x$, $7x$, etc.

species, but he was apparently unaware of *P. kewensis*, the only example supporting it. This hypothesis was experimentally verified by Clausen and Goodspeed (1925) through the artificial synthesis of *Nicotiana digluta* from *N. tabacum* × *glutinosa*, and several other artificially produced allopolyploid species were reported in quick succession, the most spectacular of them being that combining the radish and the cabbage, *Raphanobrassica* (Karpechenko 1927, 1928). A few years later, Müntzing (1930b) reported the first example of the artificial synthesis by this method of a well-known Linnaean species, *Galeopsis tetrahit*, from a hybrid between two others, *G. pubescens* × *speciosa*. Strong circumstantial evidence was then produced by Huskins (1931) that one of the only valid wild species known to have originated in historic times under human observation, *Spartina townsendii*, is an allopolyploid derived from the cross *S. alterniflora* × *stricta*. Finally, the colchicine technique of doubling chromosome numbers, developed through the cytological discovery of Dustin, Havas, and Lits (1937), followed by its practical application by Blakeslee and Avery (1937), has provided cytogeneticists with a most valuable tool for experimental studies of polyploidy. For a complete list of literature on the history and application of this technique, see Eigsti 1947.

Meanwhile, the realization that many of our most valuable crop plants, such as wheat, oats, cotton, tobacco, potato, banana, coffee, and sugar cane, are polyploids was supplemented by decisive evidence concerning the actual parentage of some of them, as in the examples of *Triticum aestivum* (McFadden and Sears 1946), *Gossypium hirsutum* (Beasley 1940a, Hutchinson and Stephens 1947, Stephens 1947), and *Nicotiana tabacum* (Goodspeed and Clausen 1928, Kostoff 1938a, Greenleaf 1941, 1942). The role of polyploidy in the origin of new types being created by the plant breeder was recognized in the case of the loganberry (Crane and Darlington 1927, Crane 1940a, Thomas 1940b), nessberry (Yarnell 1936), veitchberry (Crane and Darlington 1927), the cultivated blueberries (Darrow, Camp, Fischer, and Dermen 1944, Darrow and Camp 1945), and "perennial wheats" (Wakar 1935a,b, 1937). The final step in the use of polyploids in connection with plant breeding and organic evolution, namely, the establishment as a spontaneous, natural series of populations of a new species produced artificially by this method, is also not far distant.

DISTRIBUTION OF POLYPLOIDY IN THE PLANT KINGDOM

Polyploidy is known to a greater or lesser degree in all groups of plants. Although it has been studied very little in the thallophytes, multiple series of chromosome numbers have been reported in some genera of algae, such as *Cladophora, Chara,* and *Lomentaria* (Tischler 1931, 1936, 1938). In the fungi, it is apparently rare or lacking. Among the bryophytes, mosses are classical examples of polyploidy, both natural and artificially induced (Wettstein 1924, 1927). It is found in all the major divisions of the vascular plants, namely Psilopsida (*Psilotum*), Lycopsida (*Selaginella*), Sphenopsida (*Equisetum*), Filicinae (most genera), gymnosperms (*Sequoia, Podocarpus, Ephedra*), and angiosperms.

The distribution of polyploidy among the various groups of vascular plants is very irregular and shows no obvious relationship to the phylogenetic position of the group. The only two surviving genera of the primitive class Psilopsida, *Psilotum* and *Tmesipteris,* have such high chromosome numbers that there is a very strong probability that they represent the last relics of ancient, highly developed polyploid series. Similar very high chromosome numbers exist in nearly all genera of the ferns, particularly the Polypodiaceae. And typical polyploid series have been found within some species of this family, notably *Polypodium vulgare* (Manton 1947). There is good reason to believe, therefore, that polyploidy has played a particularly important role in the evolution of the family Polypodiaceae and may be responsible for many of its taxonomic complexities.

On the other hand, polyploidy is particularly uncommon among the living gymnosperms. It is unknown in the cycads or *Ginkgo,* and the only known polyploid species of conifers are *Pseudolarix amabilis* (Sax and Sax 1933), *Sequoia sempervirens* (Dark 1932, Buchholz 1939), *Juniperus chinensis* var. *pfitzeriana* (Sax and Sax 1933), and probably some species of *Podocarpus* (Flory 1936). It is apparently more common in the Gnetales (Florin 1932). Among the angiosperms in general, the proportion of polyploid species was roughly estimated by the writer, when preparing a review of the frequency of polyploidy in another connection (Stebbins 1938b), to be about 30 to 35 percent. But these species are very irregularly distributed. Widespread families which are nearly or entirely devoid of polyploidy are Fagaceae, Moraceae, Berberidaceae, Polemoniaceae, and Cucurbitaceae. Families characterized by a high frequency of polyploidy, often involving genera as well as species, are Polygonaceae, Crassulaceae, Rosaceae, Malvaceae, Araliaceae, Gramineae, and Iridaceae. In addition, the high aneuploid series of chromosome numbers found in the Cyperaceae and the Juncaceae are probably a modified form of polyploidy. In other families, neighboring genera differ widely in the frequency of this phenomenon. In the Salicaceae, it forms the major basis of the variation pattern in *Salix,* but is very rare in *Populus.* It is highly developed in *Betula,* but almost unknown in other genera of the Betulaceae. It is common in *Dianthus* (Caryophyllaceae), but uncommon in *Silene;* nearly absent in *Aquilegia,* sporadic in *Anemone,* but predominant in *Thalictrum* (Ranunculaceae); rare in *Lactuca* and *Prenanthes,* but common in *Hieracium* (Compositae); rare in *Lilium,* but common in *Tulipa* (Liliaceae). In some genera, such as *Potentilla, Primula, Solanum, Pentstemon,* and *Crepis,* some species groups are strictly diploid, while others form extensive polyploid series. Most genera of angiosperms contain some polyploidy, most often in the form of tetraploid or hexaploid species, either scattered through the genus or concentrated in some of its sections.

A classification of the genera of angiosperms on the basis of the percentage of polyploid species found in them (Stebbins 1938b) revealed the fact that, on the average, the highest percentages of polyploids are found in perennial herbs, a smaller proportion in annuals, and the lowest percentages in woody plants. The genera included in this survey are mostly those of temperate climates, since very few tropical genera are well enough known for this purpose. Data on tropical groups may show different relationships, particularly between perennial herbs and woody plants. Some suggestions as to the explanation of these differences in the incidence of polyploidy will be offered in the following chapter.

DIRECT EFFECTS OF POLYPLOIDY

The evidence to be presented in this chapter indicates that polyploidy as it occurs in nature is most often associated with hybridization, either between species or between different subspecies or races of the same species. If this assumption is correct, any conclusions about the physiological and ecological effects of chromosome doubling *per se* that are based on studies of naturally occurring polyploids must be considered hazardous. If, therefore, we are to understand the significance of polyploidy in nature, we must first examine the effects of artificially induced polyploidy on pure species of various types. Studies of this sort have been made by a number of workers and have been reviewed by Noggle (1946). Some of the more comprehensive studies are those of Blakeslee (1939), Randolph (1941a), Straub (1940), Pirschle (1942a,b) and Larsen (1943).

A survey of these studies brings out the fact that the only generalization which may be safely made about the morphological and physiological effects of polyploidy is that these depend greatly on the nature of the original genotype. The popular conception, that polyploidy usually produces *gigas* types, which are larger than their diploid ancestors, is now known to be true only in special instances, particularly if the original diploid is strongly heterozygous, as is true of the progenitor of the first known *gigas* tetraploid, that of *Oenothera lamarckiana.* To be sure, polyploidy increases cell size in the meristematic tissues, but the eventual size of individual organs and of the plant as a whole depends as much on the amount of cell elongation and on the number of cells produced during growth as it does on the initial size of the cells. Both of these latter processes may be adversely affected by the polyploid condition. In general, the *gigas* effects of polyploidy are seen most often and are expressed most strongly in organs with a determinate type of growth, such as sepals, petals, anthers, few-seeded fruits, and seeds. In compound, many-seeded fruits, like those of the tomato, tetraploidy may cause a reduction in size, because of the partial sterility of the tetraploids and the consequent reduction in the number of seeds produced. In addition to the occasional *gigas* effect on individual parts of the plant, and more rarely on the plant as a whole, the following effects of chromosome doubling occur often enough to deserve consideration, although none of them is universal.

An increase in size of the individual cells is perhaps the most widespread effect of polyploidy. It often makes possible the use of measurements of certain cells of the plant, particularly the guard cells of the stomata and the mature pollen grains, to suggest the diploid or the polyploid condition of plants represented only by herbarium specimens, or in which for other reasons actual counting is not practicable. These measurements cannot be used indiscriminately and are valueless unless the sizes of stomata and pollen grains are known in individuals of the same or closely related species in which the chromosomes have actually been counted. Nevertheless, if these basic facts are known, and if the chromosomes of all of the diploid and the polyploid species of a group are of about the same size, then measurements of the size of stomata and pollen grains can often provide valuable supplementary evidence on the relative geographic distribution of closely related diploids and polyploids, and can also give suggestions of where exploration in the field is likely to yield desired diploid or polyploid members of a particular species complex.

Nevertheless, the work of Tobler (1931) and Barthelmess (1941) on mosses shows how the effects of chromosome doubling on the size of mature cells can vary with the nature of the original diploid genotype. Using the well-known method of regenerating gametophytes from the stalk of the spore capsule, Tobler obtained diploid gametophytes from each of 54 segregating haploid individuals derived from a cross between two races of *Funaria hygrometrica,* one of which had cells twice as large as those of the other. The range in cell volume among the diploid plants was about the same as that among their haploid progenitors, but the mean volume for the diploids was about 3.5 times that for the haploids. There was, however, no significant correlation between cell size in the haploid individuals and in their diploid derivatives. In some genotypes, doubling had almost no effect, while in others it increased the volume to eight times that found in the haploid. Furthermore, there was a significant negative correlation (— 0.65) between the initial size of the cells and the degree to which this was increased by doubling. Genotypes with initially small cells were affected much more than those with large ones. The results of Barthelmess (1941) on *Physcomitrium piriforme* were essentially similar.

A number of secondary effects are associated with the primary effect of polyploidy on cell size. The first of these is on water content. Since in most mature plant cells the cytoplasm is appressed to the wall, and the central part of the cell is occupied by the vacuole, the relation of water to protoplasm depends partly on the ratio between volume and surface in the cell. Furthermore, since any increase in cell size causes a corresponding increase in the ratio of volume to surface area, such increases are likely to increase the water content of the cell relative to the amount of protoplasm, and therefore to lower the osmotic tension. This tendency was demonstrated experimentally by Becker (1931) in mosses, by Schlösser (1936) in a strain of tomato, and by Hesse (1938) in *Petunia*. But Fabergé (1936) found no significant difference in water content between the cells of diploid and tetraploid tomatoes, while Fischer and Schwanitz (1936) record that in polyploids of *Anthoceros* cell size is reduced and osmotic tension is increased. Since many factors other than size affect the osmotic properties of cells, the tendency for polyploidy to reduce osmotic tension can be expected to be less general than the effect on cell size. The effect of polyploidy on cell size and cell elongation is probably responsible for the differences between diploids and autotetraploids in the content of various substances, such as protein, chlorophyll, cellulose, auxin, and various vitamins, as discussed in the reviews of Randolph (1941a) and Noggle (1946).

Another series of effects of chromosome doubling depends on changes in growth rate. In general, the growth rate of autopolyploids is slower than that of their diploid progenitors, although the degree of retardation may vary greatly with the nature of the original genotype. This causes autotetraploids to flower relatively later and in some instances to flower over a longer period of time. Another effect on growth is the reduction in the amount of branching, which occurs very frequently, but not universally (Hesse 1938, Pirschle 1942a,b, Levan 1942b). In many grasses this effect results indirectly in a reduction of the size of the plant as a whole, since the number of basal shoots or tillers per plant is significantly reduced, as in *Stipa lepida* (Stebbins 1941b).

The least consistent of the effects of chromosome doubling are those on the shape of the individual organs of the plant. This is as expected, since these shapes are the product of the interaction of tendencies affecting growth rate as well as cell size, and the individual processes may be differentially affected by polyploidy. The most consistent effect, since it is the one most directly affected by cell size, is on the thickness of leaves and other appendages. Randolph, Abbe, and Einset (1944) have found that the leaves of tetraploid maize are consistently thicker than those of the diploid, and that this increase is about the same for each variety tested. Similar differences in leaf thickness have been reported by most other workers on autotetraploids. The leaves and other organs are usually shorter and broader in autotetraploids, but this effect is by no means universal (Pirschle 1942a). In *Stipa lepida* (Stebbins 1941b) progenies of sister diploid plants differed in this respect, the autotetraploids having in one instance significantly broader leaves and in another significantly narrower leaves than their diploid ancestors.

Of equal and perhaps greater significance from the evolutionary point of view than those on the morphology and physiology of the plants are the effects of polyploidy on fertility and genetic behavior. The most conspicuous and universal of these is the reduction of pollen and seed fertility in autopolyploids as compared with their diploid ancestors. Like other effects of chromosome doubling, this varies greatly with the genotype of the diploid. Some tetraploid varieties of maize set 80 to 95 percent of seed (Randolph 1935, 1941a), and 75 to 80 percent fertility is found in colchicine-produced autotetraploids of the grass species *Ehrharta erecta* (Stebbins 1949b). At the other extreme, Einset (1944, 1947a) found only 5 to 15 percent of seed setting in autotetraploids of cultivated lettuce, while Beasley (1940b) found almost complete sterility in an autotetraploid of *Gossypium herbaceum*. Most autotetraploids lie somewhere between these extremes.

The opinion originally held about the cause of sterility in autopolyploids, namely, that it is due to the formation and irregular segregation of multivalent associations of chromosomes (Darlington 1937), has had to be revised in recent years. Randolph (1941a) concluded that sterility in autotetraploid maize is largely controlled by specific genes or gene combinations and is chiefly physiological in nature. Sparrow, Ruttle, and Nebel (1942) found that differences between the fertility of autotetraploids derived from different varieties of the snapdragon *Antirrhinum majus* were not correlated with the differences in the frequency of multivalents at meiotic prophase and metaphase, but did show a positive correlation with the frequency of lagging chromosomes and other abnormalities seen at later stages of meiosis. Similar results were obtained by Myers and Hill (1942) and Myers (1943) in comparisons of meiotic abnormality and fertility in different strains of *Dactylis glomerata*, a wild species which behaves cytologically like an autotetraploid, and by Myers (1945) in autotetraploid *Lolium perenne*. Einset (1944, 1947a) found that a small proportion of the high sterility in autotetraploid lettuce is due to abortion of pollen grains and ovules, but that most of it is caused by the failure of pollen grains to complete growth down the styles and by the inhibition of fertilization. Westergaard (1948) found in the species complex of *Solanum nigrum* great differences in seed production between artificially produced autopolyploids of different diploid species, as well as between allopolyploids from different interspecific hybrids. Since these differences could not be explained on the basis of the meiotic behavior of the chromosomes, he concluded that fertility and sterility in allopolyploids, as well as in autopolyploids, are to be explained in genetic terms rather than on a cytological basis.

These data all suggest that the principal causes of sterility in autotetraploids, as well as in some allopolyploids, are a series of disharmonies produced at various stages of the sexual cycle, of which disturbances of the spindle and other external features of meiosis are among the most important. The effects of abnormal pairing between groups of three and four chromosomes completely homologous to each other are of relatively minor importance. This point of view is strengthened by the fact that in autogamous plants crossing between races tends to increase the fertility of the tetraploids, as in *Antirrhinum* (Sparrow, Ruttle, and Nebel 1942) and *Lactuca* (Einset 1947), and that similar increases can be obtained in allogamous plants by selection within the heterozygous material, as in *Fagopyrum* (Sacharov, Frolova, and Mansurova 1944).

Polyploidy also affects the incompatibility relationships of some self-incompatible ("self-sterile") species. If the diploid is strongly self-incompatible, as in *Brassica* and *Raphanus* (Howard 1942), *Oenothera organensis* (Lewis 1943), and *Taraxacum kok-saghyz* (Warmke 1945), the autotetraploid is likely to be equally so. But if some self-compatibility is present at the diploid level, this may be greatly increased in the polyploid, as in *Petunia* (Stout and Chandler 1941), *Allium nutans* (Levan 1937a), *Pyrus* spp. (Crane and Lewis 1942, Lewis and Modlibowska 1942), and *Trifolium repens* (Atwood 1944). Lewis (1943, 1947) has described carefully the effect of polyploidy on genetic mechanisms for self-incompatibility and has concluded that competition between pairs of different self-sterility (S) alleles is chiefly responsible for the increased compatibility.

The evidence from natural polyploids suggests that in many instances self-incompatibility is maintained at the polyploid levels, at least in the flowering plants. In the Gramineae, which have been the subject of many studies on self-incompatibility (Beddows 1931, Smith 1944), there are many polyploid species which are self-incompatible to at least such a degree that they are rarely or never inbred in nature. Well-known examples are *Bromus inermis, Festuca arundinacea, F. ovina,* and *F. rubra, Dactylis glomerata, Agropyron repens,* and *A. cristatum.* On the other hand, most of the polyploid species in this family which are self-compatible or strictly autogamous, such as *Bromus secalinus, B. mollis, B. rigidus, Hordeum nodosum, Avena fatua,* and *A. barbata,* appear to have arisen from self-compatible or autogamous diploids. Lewis and Modlibowska (1942), however, record that tetraploid species of *Tulipa* and *Hyacinthus* are self-compatible, while corresponding diploid species are self-incompatible. Natural polyploidy has probably had some effect on self-incompatibility, but this has certainly been much less than the alterations of this condition which have been produced by gene mutations.

Another effect of polyploidy is on the genetics of segregation, due to the presence of duplicated genes. The early study of this subject by Haldane (1930) has now been followed by several others, which are reviewed by Little (1945). The principal effect in all cross-bred populations is, of course, the increase of the proportion of individuals heterozygous at any locus and the decrease in frequency of homozygosity, so that the effects of recessive genes, and particularly gene combinations, are rarely realized in the phenotype. This is due to the fact that homozygotes must possess four similar genes at any locus rather than only two, and three degrees of heterozygosity are possible with respect to each gene locus. Segregation ratios are further complicated by the degree to which the chromosomes form multivalent associations, their chiasma frequencies, and the way in which the multivalents segregate. The details of these segregations are significant chiefly to plant breeders, who can work with genetically pure material.

The final and most important effect of polyploidy is the genetic barrier which is immediately erected between a polyploid and its diploid progenitor. Autopolyploids are usually rather difficult to cross with related diploids, and the F_1 progeny of such matings are highly sterile triploids. Polyploidy, therefore, is one way, and perhaps the only way, in which an interspecific barrier can arise at one step, and thus give an opportunity for a new line to evolve independently and to diverge from the parental type.

POLYPLOIDY AND HYBRIDIZATION

The fact that polyploidy is the best way of producing constant, fertile species from sterile interspecific hybrids has been recognized ever since the experimental confirmation of Winge's hypothesis through the synthesis of *Nicotiana digluta.* Typical allopolyploids or amphiploids, which resemble diploids in the regularity of chromosome pairing at meiosis and in the constancy of their genetic behavior, have now been synthesized in forty or more instances, and the actual or probable allopolyploid character of scores of wild species has been determined. At least half of the naturally occurring polyploids are probably strict allopolyploids, in that they have originated from F_1 hybrids between two ancestral species which were so distantly related to each other that their chromosomes were essentially nonhomologous. This does not include species which are diploid with respect to the basic number of their genus, but of which this generic number may be of polyploid derivation, as in the subfamily Pomoideae of the Rosaceae.

The remainder of natural polyploids, including between fifty and a hundred species which have been analyzed to date, and probably several hundred or even several thousand yet to be investigated, have certain cytogenetical properties which are often associated with autopolyploidy. In particular, they resemble related diploid species more or less closely in external morphology, and their chromosomes form a greater or lesser number of multivalent associations at meiosis, indicating some duplication of chromosomal material. We cannot conclude from this evidence, however, that such polyploids have all been derived from pure, fertile diploid species rather than from interspecific hybrids. Both the morphological and the cytological evidence on polyploids of known origin warns us against such inferences.

There are now several examples which show that even extreme allopolyploids, derived from hybridization between species having widely different chromosomes, may resemble one or the other of their parental species so closely that they have not been recognized as distinct by systematists. One of the most striking of these is *Nasturtium microphyllum.* This species was originally regarded by Manton (1935) as an autotetraploid form of the ordinary watercress, *N. officinale,* chiefly on the grounds of its external morphology. When, however, the true autotetraploid of the latter species was produced artificially, it was found not only to be entirely different in appearance from the wild plant but in addition to form sterile hybrids with it (Howard and Manton 1940). Furthermore, the hybrid between the natural tetraploid and typical *N. officinale* was found to possess 16 bivalent and 16 univalent chromosomes, indicating that one of the two genomes present in the former is completely nonhomologous with that of the latter. The natural tetraploid was therefore described as a new species, *Nasturtium uniseriatum* (Howard and Manton 1946), and its authors now believe it to be an allotetraploid derived from a hybrid between *N. officinale* and some species of a different but related genus, *Cardamine.* Airy-Shaw (1947) has shown that the tetraploid species had already been recognized as *Nasturtium microphyllum.*

A similar example is *Madia citrigracilis* (Clausen, Keck, and Hiesey 1945a). On morphological grounds, this species was first thought to be a form of *M. gracilis,* but later its haploid chromosome number was found to be $n = 24$, while that of *M. gracilis* is $n = 16$. Finally, *M. citrigracilis* was synthesized artificially via the triploid hybrid between *M. gracilis* and the very different diploid species *M. citriodora* ($n = 8$). Since the triploid F_1 has practically no chromosome pairing, and *M. citrigracilis* regularly forms 24 bivalents, the latter ranks as a true allopolyploid, as does also *M. gracilis.* Furthermore, there is another hexaploid, even more similar to *M. gracilis,* as well as a third species, *M. subspicata,* which is diploid, but is nevertheless so similar to *M. gracilis* in external appearance that the two were thought to belong to the same species until they were analyzed cytologically. Typical *M. gracilis* may be an allotetraploid of *M. subspicata* and some other species. In the complex of *M. gracilis,* therefore, we have a diploid, a tetraploid, and two hexaploid forms which on the basis of external morphology were judged by experienced systematists thoroughly familiar with the genus to belong to one species, but which actually include allopolyploids derived from hybridization between at least three, and perhaps four, original diploid species so distantly related that their chromosomes were

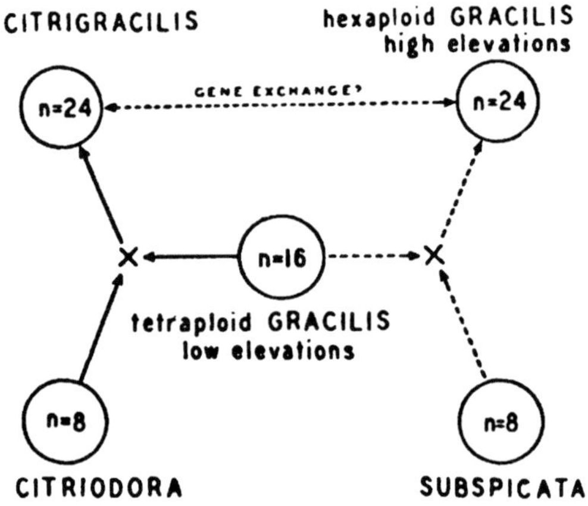

FIG. 32. *Madia citrigracilis* and allies. Relationships arrived at by experiment indicated by solid lines; those by circumstantial evidence, by broken lines. From Clausen, Keck, and Hiesey 1945a.

almost completely nonhomologous with each other (see Fig. 32).

The two preceding examples show that even typical allopolyploids may be placed by systematists in the same taxonomic species as one of their parents if their cytogenetic condition has not been analyzed. Such polyploids, however, nearly always possess morphological characteristics by which they may be recognized, once these are made clear. This is definitely true of the two examples mentioned above. If, however, a polyploid has originated from hybridization between two closely related species, with chromosomes partly homologous to each other, some derivatives of this polyploid may be hardly distinguishable from autopolyploids of one or the other parental species. And if, through backcrossing, such a polyploid acquires a preponderance of genes derived from one or the other of the parental species, it may fall entirely within the range of variation of the latter.

A striking example of such a situation was produced artificially by Mehlquist (1945) in the genus *Dianthus*. He pollinated the tetraploid *D. chinensis* ($2n = 60$), the garden pink, with pollen from the very different-looking diploid species *D. caryophyllus* ($2n = 30$), the ordinary greenhouse carnation. From this he obtained, in addition to a large number of sterile triploid hybrids, a partly fertile tetraploid hybrid containing a set of 30 chromosomes from a normal gamete of *D. chinensis* plus 30 chromosomes ap-

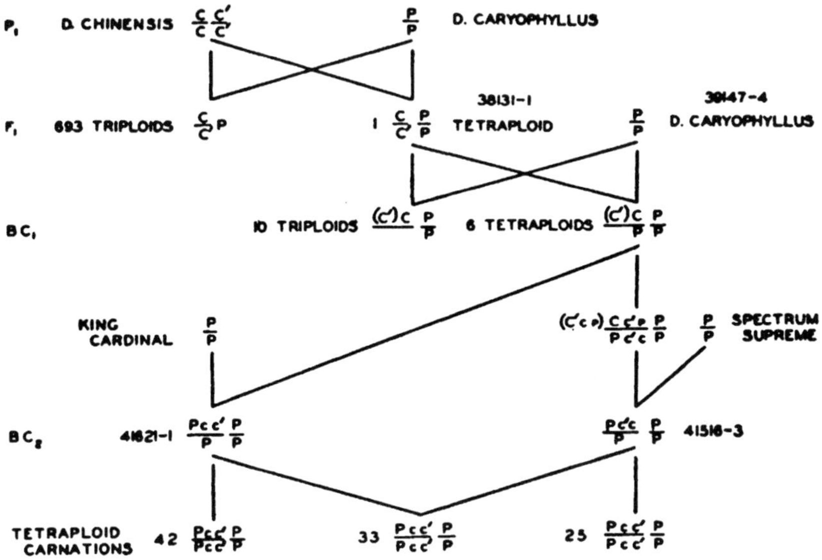

FIG. 33. Diagram showing the pedigree of a tetraploid carnation which, though morphologically similar to *Dianthus caryophyllus*, must contain some genes derived from *D. chinensis*. Further explanation in the text. From Mehlquist 1945.

parently derived from an unreduced gamete of *D. caryophyllus*. Such a tetraploid is essentially an allotetraploid and produces gametes containing one genome from *D. chinensis* and one from *D. caryophyllus*. Repeated backcrossing for two generations of this tetraploid to *D. caryophyllus* produced a series of plants which were vigorous and fertile and appeared like autotetraploid plants of the carnation. They had thick, stiff leaves, larger flowers, and other similar characteristics, but the influence of genes from *D. chinensis*, which, as shown in the pedigree, Fig. 33, must have been present, was not detectable. As in the case of many individuals which are the product of introgressive hybridization, their origin could not have been discovered by studying their external appearance.

That this type of polyploid is often produced in nature seems inevitable. The examples of introgressive hybridization in *Tradescantia* discussed by Anderson, Riley, and others (see Chapter VII) frequently involve tetraploids. It is not improbable that many of the apparently autopolyploid individuals of *T. canaliculata, T. occidentalis,* and other species of this genus (Anderson and Sax 1936, Anderson 1937b) are actually derived from such introgression and contain genes from one of the other species. This possibility is strengthened by the observation of Giles (1941) that the triploid hybrid between the diploid *T. paludosa* and tetraploid *T. canaliculata* is morphologically very similar to the latter. On theoretical grounds, one would expect that genotypes having morphological and physiological characteristics similar to a well-established wild species would usually have a higher selective value than entirely new hybrid combinations, and therefore that natural selection would tend to favor backcross derivatives from allopolyploids whenever these could be produced and had a reasonably high fertility. This seems to be true in the genus *Galium.* Fagerlind (1937) records that autopolyploids belonging to different species of this genus frequently hybridize when they grow together in the same locality, but that truly intermediate types are much less common than forms which closely resemble one species, but possess a few characteristics of another.

The foregoing discussion shows that in many instances one cannot decide on the basis of external morphology alone whether a given polyploid contains only genes derived from a single interfertile species population, or whether it has obtained through hybridization genes from two or more species. For such decisions, the external morphology of the chromosomes, as well as their behavior at meiosis, often provides valuable evidence, but this also is by no means infallible. Some authors, as Bergman (1935a) in *Hieracium,* and Sørenson and Gudjónsson (1946) in *Taraxacum,* have suggested that if three or more morphologically similar sets of chromosomes can be recognized in the somatic divisions of the root tips, this provides strong evidence in favor of the autopolyploid nature of the species concerned. But studies of meiosis in interspecific hybrids of *Paeonia* (Stebbins 1938a) and many other genera have shown that chromosome sets which look almost identical with each other in the ordinary root tip preparations may actually differ by large and numerous structural rearrangements. In genera like *Taraxacum,* in which species that form completely sterile hybrids may have chromosomes so similar to each other that they can pair perfectly in the F_1 (Poddubnaja-Arnoldi 1939a), conclusions based on the external morphology of the chromosomes may be particularly misleading.

Similar difficulties accompany the interpretation of the nature of polyploidy on the basis of chromosome association at meiosis. Müntzing and Prakken (1940) have pointed out that in some polyploids of experimental origin and known to contain three or more identical chromosome sets or genomes, the chromosomes form mostly pairs at meiosis. Furthermore, the number of trivalent and quadrivalent configurations formed may differ between autopolyploids produced from different strains of the same species, as in the snapdragon *Antirrhinum majus* (Sparrow, Ruttle, and Nebel 1942). In polyploids derived from undoubted interspecific hybrids, such as *Primula kewensis* (Upcott 1939), *Lycopersicum peruvianum-esculentum* (Lesley and Lesley 1943), and several other examples listed elsewhere by the writer (Stebbins 1947a), a smaller or larger number of multivalent associations may be formed. This is to be expected from the fact that, as stated in Chapter VI, most F_1 diploid hybrids between closely related, but distinct, species have a high degree of chromosome pairing in spite of their sterility. Furthermore, since this pairing indicates that such hybrids possess a considerable number of homologous segments in common, one would expect polyploids produced from them to have many gene loci present four times. Therefore, as Dawson (1941) has pointed out, tetrasomic ratios, which sometimes have been assumed to be a diagnostic of autopolyploids, may be expected in polyploids derived from some types of sterile interspecific hybrids.

The only conclusions which can be drawn from the preceding facts are as follows. In the first place, no single criterion can be used to decide whether a given natural polyploid has arisen from a fertile species or a sterile interspecific hybrid. Before even any hypothesis can be erected concerning the nature of polyploidy in a given species, it must be thoroughly studied from the cytogenetic as well as the morphological point of view, and all of its immediate relatives must be equally well understood. Before any final decisions can be made, the form in question should be hybridized with its putative diploid ancestor or ancestors, or, better yet, it should be resynthesized. In the second place, no classification of naturally occurring polyploids which recognizes only two contrasting categories, autopolyploids and allopolyploids, can hope to express even roughly the pattern of cytogenetic variation which polyploidy produces, no matter on what criteria these categories are based. Such an oversimplification does more to hide the complexities of origin and relationship which exist in such groups than to clarify them.

The difficulty resides not only in the fact that various intermediate conditions connect typical autopolyploidy with typical allopolyploidy. In addition, the four criteria which are generally used to distinguish between these two types, namely, morphological resemblance, chromosome behavior, presence or absence of tetrasomic segregation, and the fertility or sterility of the diploid from which the polyploid was derived, do not run parallel to each other. It is quite possible for a polyploid to be judged an autopolyploid on the basis of one or more of these criteria, and an allopolyploid on the basis of the remaining ones. For this reason, the writer has presented elsewhere (Stebbins 1947a) an amplified classification of polyploid types, in which four principal categories have been recognized. There is no doubt that these categories are not sharply distinct from each other and are connected by a series of borderline cases. This is inconvenient to those whose primary purpose is classification, but nevertheless the categories recognized represent modal types, with characteristics of their own, so that their recognition provides a firmer basis for understanding the true character and origin of the various types of natural polyploids. A summary of this classification will therefore be presented in the following section.

TYPES OF POLYPLOIDS AND THEIR CHARACTERISTICS

The four types of polyploids recognized by the writer (1947a) are *autopolyploids, segmental allopolyploids, true* or *genomic allopolyploids,* and *autoallopolyploids.* The first two occur in nature predominantly or entirely at the level of triploidy or tetraploidy; true allopolyploidy can occur at any level from tetraploidy upwards, while autoallopolyploidy is confined to hexaploidy and higher levels of polyploidy. The term amphiploid, coined by Clausen, Keck, and Hiesey (1945a), is suggested as a collective term to cover all types of polyploids which have arisen after hybridization between two or more diploid species separated by barriers of hybrid sterility. It therefore includes segmental allopolyploids, true allopolyploids, and autoallopolyploids plus aneuploids which have arisen from hybridization between two species belonging to an aneuploid series with lower numbers, as in *Brassica* (Nagahuru U, 1935, Frandsen 1943) and *Erophila* (Winge 1940). A diagram showing the interrelationships of these categories is presented in Fig. 34.

Typical autopolyploids have now been produced artificially from a large number of species, mostly of cultivated plants. Their morphological and physiological properties have already been

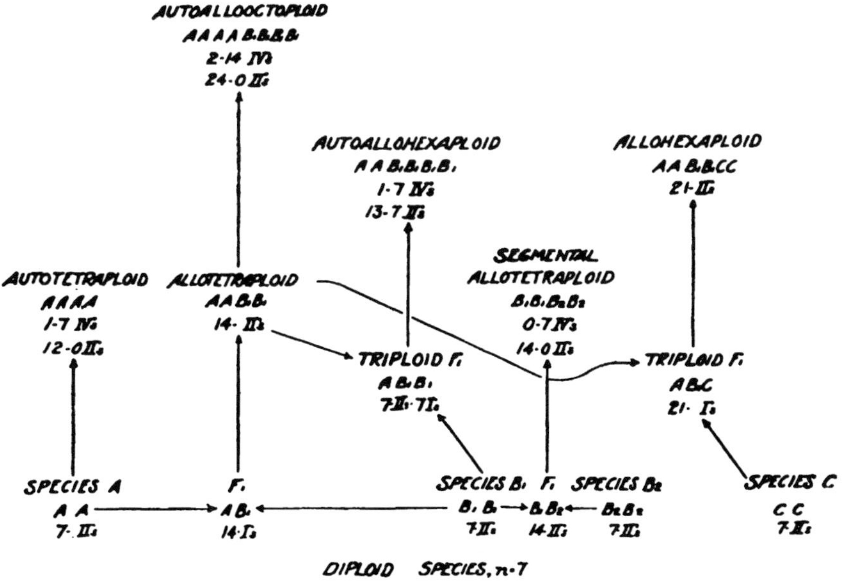

Fig. 34. Idealized diagram showing the interrelationships, genome constitution, and mode of origin of typical autopolyploids, allopolyploids, segmental allopolyploids, and autoallopolyploids. Original.

discussed. In general, the resemblance between such polyploids and their diploid progenitors is remarkably close. Such differences as exist in size and shape are considerably less than those ordinarily found between different subspecies or ecotypes of the same species. The physiological differences, likewise, although they would permit autopolyploids to grow in different situations from those occupied by their diploid progenitors, are relatively small compared to those between many ecotypes having the same chromosome number.

As has been emphasized by Clausen, Keck, and Hiesey (1945a), as well as by the present writer (Stebbins 1947a), many examples of natural polyploids once thought to be autopolyploids are now known or suspected to be of hybrid origin, and the number of polyploids which appear to have arisen in nature from a single diploid type is relatively small. One of the clearest examples is *Galax aphylla* (Baldwin 1941). This species belongs to a monotypic genus of the eastern United States which has no close relatives. The tetraploid differs in little besides its more sturdy character and thicker leaves. It has about the same geographic distribution as the diploid, but its range is somewhat wider. Similar examples are *Sedum ternatum* (Baldwin 1942a) and *S. pulchellum* (Baldwin 1943, Smith 1946). A probable autotriploid of this character is that of *Fritillaria camschatcensis* (Matsuura 1935). This plant is completely sterile, reproducing only by offsets from its bulbs. It occurs in lowlands throughout the island of Hokkaido, Japan, while its fertile diploid ancestor is confined to mountain summits. *F. camschatcensis* occurs also in Kamchatka, the Aleutian Islands, and the Pacific coast of North America south to the state of Washington, and is the only member of the section Liliorhiza found in the Old World (Beetle 1944). The chromosome number of the plants growing outside of Japan is not known. A related species of California, *F. lanceolata,* also has autotriploid and autotetraploid forms (Beetle 1944). These likewise are characterized by their robust stature, but so far as is known they occur only within the range of the diploid form.

In a number of instances, natural tetraploids are related to a series of diploid forms which differ morphologically from each other and have different geographical distributions and ecological preferences but form fertile hybrids on the diploid level, and therefore, unless other types of isolating mechanisms are present, may be considered ecotypes or subspecies of the same species rather than separate specific entities. Such tetraploids are essentially autopolyploids, but may deviate in morphological characteristics, as well as ecological preferences, from any single wild diploid form. These differences are due to the presence of new gene combinations, as well as to the tetraploid condition. The best-known example of such a situation is *Biscutella laevigata* (Manton 1934, 1937). This tetraploid is widespread in the glaciated regions of the central Alps, but it has a series of diploid relatives which are scattered through central, southern, and southwestern Europe, from Austria to central France, mostly in lowland sites or in mountain chains not covered by the Pleistocene glaciation. Many of these diploids have been recognized as species, but the more conservative systematic treatments group them with *B. laevigata,* usually as varieties or subspecies. Clausen, Keck, and Hiesey have suggested that the isolating barrier produced by the difference in chromosome number should be recognized, but that all the diploids should be placed in one species, for which the oldest name is *B. coronopifolia.* Manton obtained some evidence indicating that hybrids between these diploid entities are fertile. She notes that typical *B. laevigata* is rather similar to a diploid form known as *B. minor,* from the unglaciated portion of the Austrian Alps, but that it possesses the ability to produce stolons or root buds, a character found in certain diploid forms of western Europe, particularly *B. arvernensis* and *B. lamottii.* It is likely, therefore, that the present tetraploid contains genes derived from several of the different diploid forms, and that this fact, as well as its tetraploid condition, is responsible for its wide distribution and range of ecological tolerance. Whether it originated from ancient hybrids between two or more different diploid forms, which may have grown together during one of the interglacial periods of the Pleistocene, or whether its present condition is the result of hybridization between tetraploids which were derived from different ones of the modern diploid species or subspecies cannot be told, and is perhaps not of great importance.

A similar case is *Dactylis glomerata.* This tetraploid pasture grass has long been thought to be derived by autopolyploidy from the forest-loving diploid of central and northern Europe, *D. aschersoniana*, and Müntzing (1937) suggested that both the morphological and the ecological differences between the two species are due solely to the effects of doubling the chromosome number in *D. glomerata*. But Myers (1948) obtained a diploid plant from Iran which in its narrow, glaucous leaves, contracted inflorescences, and strongly ciliate glumes possesses in an extreme form the characteristics of *D. glomerata* which distingush it from *D. aschersoniana*. It resembles forms identified as *D. hispanica*, *D. juncinella*, and *D. woronowii*. Hybrids between this diploid and *D. aschersoniana* are fully fertile and resemble *D. glomerata* in size, date of maturity, and other characters.

Intervarietal autopolyploids of this sort may be found to be not uncommon when more polyploids are analyzed with this possibility in mind. Suggested ones are *Allium schoenoprasum* (Levan 1936b), *A. paniculatum-oleraceum* (Levan 1937b), *Polygonatum commutatum* (Eigsti 1942, Ownbey 1944), *Cuthbertia graminea* (Giles 1942), *Eriogonum fasciculatum* (Stebbins 1942c), and some of the various polyploids of *Vaccinium* (Camp 1945, Darrow and Camp 1945). One would expect this type of autopolyploid to segregate in the direction of its ancestral diploid subspecies, except in habitats which favored new combinations of genes. Randolph (1941a) has pointed out that the smaller amount of segregation in intervarietal tetraploids than in diploid hybrids of similar origin should favor the persistence of hybrid vigor in all combinations in which this phenomenon appears. For several reasons, therefore, we should expect intervarietal or intersubspecific polyploids to be the commonest type of autopolyploids in nature.

The second type of polyploid is that termed by the writer (Stebbins 1947a) a *segmental allopolyploid*. It may be defined as a polyploid containing two pairs of genomes which possess in common a considerable number of homologous chromosomal segments or even whole chromosomes, but differ from each other in respect to a sufficiently large number of genes or chromosome segments, so that the different genomes produce sterility when present together at the diploid level. This definition includes a great range of forms. At one extreme are types like *Zauschneria californica* (Clausen, Keck, and Hiesey 1940), which is derived from two species (*Z. cana* and *Z. septentrionalis*) capable of crossing to form a fertile F_1 hybrid, but one which produces many inviable offspring in the F_2 generation. Near the other extreme are *Primula kewensis* (*P. floribunda-verticillata,* Newton and Pellew 1929, Upcott 1939), *Saxifraga adscendens-tridactylites*[2] (Drygalski 1935), and probably *Triticum durum-timopheevi* (Zhebrak 1944a,b, 1946). These are descended from diploid hybrids with a certain amount of pairing at meiosis, but with considerable meiotic irregularities. The polyploids form mostly bivalents, and, though somewhat more variable, behave not very differently from typical allopolyploids.

The writer (Stebbins 1947a) has listed more than twenty-five allopolyploids of this type which have been produced artificially in recent years. Many of these, like *Crepis foetida-rubra* (Poole 1931), *Layia pentachaeta-platyglossa* (Clausen, Keck, and Hiesey 1945a), and *Allium cepa-fistulosum* (Jones and Clarke 1942), have been so sterile in both the original and later generations that they would have been complete failures under natural conditions. But others, like *Primula kewensis, Tradescantia canaliculata-humilis* (Skirm 1942), and *Nicotiana glauca-langsdorffii* (Kostoff 1938b), have either been fertile from the start or have yielded highly fertile and in some instances constant types after a number of generations of selection.

Natural polyploids of this type have not as yet been recognized in many instances. Both in their close morphological resemblance to diploid forms and in the presence of multivalent configurations of chromosomes at meiosis, they simulate autopolyploids and have most often been confused with them. The actual nature of a segmental allopolyploid is uncertain until both of its parents have been identified by appropriate hybridization experiments, although external morphology may in some instances provide strong clues. The clearest examples of natural segmental allopolyploids, in addition to *Zauschneria,* are *Galium mollugo-verum* (*G. ochroleucum*), which originated through hybridization between two autopolyploids (Fagerlind 1937), and the cultivated species *Aesculus carnea* (*A. hippocastanea-pavia,* Upcott 1936).

Another well-known polyploid which probably is of this type is

[2] For the sake of uniformity, artificial allopolyploids, except those whose names are well established in the literature, will be designated by the names of their two parental species, in alphabetical order and separated by a hyphen.

Solanum tuberosum, the common potato. Its somatic number is $2n = 48$, which makes it tetraploid, since the basic number for *Solanum,* as well as a large number of other genera in the Solanaceae, is $x = 12$. Both multivalent configurations and tetrasomic ratios of gene segregation have been found in this species (Cadman 1943, Lamm 1945). But Ivanovskaya (1939) found that a haploid plant was not only completely sterile, but in addition formed occasional univalents and 7 percent of bridge-fragment configurations at meiosis, which phenomena are much less common in typical diploid plants. She therefore concluded that *S. tuberosum* is a polyploid derived from a hybrid between two species having chromosomes similar enough to pair, but not enough to permit free interchange of genetic materials. Propach (1940) found that hybrids between diploid species of the section *Tuberarium* of *Solanum* are rather difficult to obtain, but that all of the F_1 plants secured, whether fertile or sterile, had good chromosome pairing. Lamm (1945) found that the number of multivalents in *S. tuberosum* is significantly less than that in an artificial autotetraploid of the diploid *S. rybinii,* and that the number of chromosome pairs with satellites seen at somatic mitosis is two, just as in diploid species of this section. He concluded that "polyploidy in *S. tuberosum* is mainly autopolyploidy with slight differentiations in the allopolyploid direction." Cadman (1943) found associations of six and eight chromosomes in strains of *S. tuberosum.* Since only bivalents are formed in its diploid relatives, these higher associations may be due to heterozygosity for segmental interchanges. *S. tuberosum,* therefore, has the cytological and genetic properties to be expected in a segmental allopolyploid, but the case cannot be considered definite until one or both of its diploid ancestors have been identified.

One of the most definite examples of a natural segmental allopolyploid is the tetraploid *Delphinium gypsophilum* (Lewis and Epling 1946, Epling 1947a). As was described in Chapter VII, this species exists in both a diploid and a tetraploid form, and the diploid is both morphologically and cytogenetically very similar to the F_1 hybrid between *D. hesperium* and *D. recurvatum.* This hybrid is highly sterile, although it produces some seed, but chromosome pairing at meiosis is nearly normal. It is likely, therefore, that tetraploid *D. gypsophilum* was produced directly from the F_1 or F_2 hybrid between *D. hesperium* and *D. recurvatum* and therefore meets all the requirements for a segmental allopolyploid. It is somewhat more widespread in geographic distribution than its diploid counterpart.

Another group of probable segmental allopolyploids are the tetraploid species of *Paeonia.* All of them form a greater or lesser number of multivalents at meiosis (Dark 1936, Stebbins 1938a and unpublished) and on these grounds, as well as their superficial resemblance to certain diploid species, Barber (1941) and Stern (1946, 1947) concluded that they are autopolyploids. But, as pointed out elsewhere by the writer (Stebbins 1948), the various tetraploids show morphological resemblances to more than one of the diploid species, and Stern in some of his species descriptions recognizes the presence of intergrading between tetraploids which he relates to different diploid species. Furthermore, some of the tetraploids, like *P. peregrina* and *P. officinalis,* have only one or two multivalents at meiosis, and in some sporocytes may form ten bivalents. In plants with large chromosomes like those of *Paeonia* and with random chiasma formation, the presence in a plant of four completely homologous genomes should cause it to form a high proportion of quadrivalents. In fact, even some triploid hybrids between widely different species, like *P. albiflora* × *tomentosa,* may form as many as four or five trivalents. The cytological evidence, therefore, agrees with that from external morphology in suggesting that the tetraploids of *Paeonia* are mostly segmental allopolyploids.

Another probable example of a segmental allopolyploid is *Lotus corniculatus.* Dawson (1941) found that this tetraploid forms, almost entirely, bivalents at meiosis, but because he found tetrasomic inheritance for certain genetic characters he concluded that this species is an autopolyploid of the closely similar diploid species *L. tenuis.* Tome and Johnson (1945), however, secured an artificial autotetraploid of *L. tenuis* and found that it was morphologically different from *L. corniculatus* and would not hybridize with it. Careful analysis of other tetraploids believed to be autopolyploids may very well show that many of them are also segmental allopolyploids.

The most important genetic difference between segmental allopolyploids and typical or genomic ones is the ability of the former to segregate in respect to some of the characteristics by which their ancestral species differ from each other. This results directly from their ability to form multivalents at meiosis. Such multivalents represent two types of chromosome pairing. Each quadrivalent in a newly formed segmental allotetraploid contains two completely homologous chromosomes, derived from the same species or even the same diploid individuals, and two partly homologous ones, derived from different species, but possessing in common a greater or lesser number of homologous segments. The genomes of this type of polyploid may be designated as $A_1A_1A_2A_2$, and the individual chromosomes as $1A_11A_11A_21A_2$, $2A_12A_12A_22A_2$ . . .$nA_1nA_1nA_2nA_2$, where n is the basic haploid number, A_1 and A_2 being the parental genomes. Pairing between two $1A_1$ chromosomes is designated as *homogenetic association,* and between $1A_1$ and $1A_2$ as *heterogenetic association.* The writer (Stebbins 1947a) has discussed elsewhere the derivation of these terms and their relationship to the parallel, but by no means synonymous, terms auto- and allosyndesis. Autosyndesis refers to the pairing of chromosomes derived from the same parental gamete of a particular plant, regardless of the similarity or difference from each other, while allosyndesis refers similarly to pairing between chromosomes derived from different parental gametes, such as normally occurs both in diploids and in polyploids. Depending on whether the plant is a newly formed polyploid, a hybrid between two autopolyploids, or a long-established polyploid, autosyndesis may be either homo- or heterogenetic association.

The frequency of heterogenetic association as compared to homogenetic association depends on the relative degree of affinity between the completely homologous as compared to the partly homologous chromosomes, or their *differential affinity* (Darlington 1937, pp. 198–200). This degree of affinity is determined largely by the size of the homologous segments as compared to the nonhomologous segments in the partly homologous chromosomes, but the absolute chromosome size, the genetically determined chiasma frequency, and the distribution of chiasmata, whether at random or localized, all play important roles. In some plants with small chromosomes, like *Primula kewensis* and *Lotus corniculatus,* differential affinity is marked, and the homologous chromosomes usually exhibit preferential pairing, so that most of the sporocytes contain largely or exclusively bivalents. But in *Paeonia* and in *Crepis foetida-rubra* (Poole 1931), both of which have considerably larger chromosomes, multivalents are regularly formed, and preferential pairing is at a minimum.

On the basis of their results in *Layia* and *Madia,* Clausen, Keck, and Hiesey (1945a) have concluded that the heterogenetic association in segmental allopolyploids, which they characterize as (p. 72, Table 12) inter-ecospecific amphiploids with the identity of the parental genomes lost in recombination, will always lead to segregation of disharmonious gene combinations, and thus to the eventual failure of the line. They cite as examples of this failure *Layia pentachaeta-platyglossa, Aquilegia chrysantha-flabellata* (Skalinska 1935), the examples of Poole and of Hollingshead (1930b) in *Crepis,* and *Primula kewensis.* They recognize that in *P. kewensis* there is as yet little evidence of increasing sterility, but they cite the amount of genetic variation in different lines of this species reported by Newton and Pellew (1929) and Upcott (1939) as a sign that it would not be likely to remain as a stable species in nature. On the other hand, they do not refer to the earliest example of a segmental allopolyploid which eventually yielded, after six to seven generations, relatively stable and fertile lines, namely, *Nicotiana glauca-langsdorffii* (Kostoff 1938b). A more recent series of examples is the large number of allopolyploids produced by Zhebrak (1944a,b, 1946) between the various tetraploid wheats of the emmer series (*Triticum durum, T. dicoccum,* etc.) and the tetraploid *T. timopheevi.* The cytology of these polyploids was not investigated, but since Lilienfeld and Kihara (1934) found 12 out of a possible 14 bivalents to be the most common amount of pairing in diploid hybrids of this combination, and since the chromosomes are large and have random chiasma distribution with a normal frequency of more than two chiasmata per bivalent, multivalents would be expected in these tetraploids. Zhebrak found a certain amount of segregation in respect to the differences between the parental species, although no forms closely resembling either of the parental species were recovered after several generations of selfing. On the other hand, these polyploids were fertile when first formed, and have remained so.

Gajewski (1946) obtained evidence on the relation between

chromosome homology and fertility in a segmental allopolyploid between *Anemone silvestris* and *A. multifida.* The original production of this allopolyploid by Janczewski, in 1892, was probably the first example ever observed of a fertile, constant type obtained from a sterile interspecific hybrid, although the cytological basis for the increase in fertility was, of course, not recognized by Janczewski. The F_1 hybrid between *A. silvestris* ($2n = 16$) and *A. multifida* ($2n = 32$) forms three to seven pairs of chromosomes at meiosis and is highly sterile, but its allopolyploid forms one quadrivalent in 50 percent of the sporocytes, in addition to bivalents and occasional univalents, and in the second generation varied in fertility from 22 to 70 percent of normal seed setting. This fertility was somewhat increased in the following generation, and the data indicate that fully fertile lines will eventually be produced by selection. Gajewski presents a chart based on the relation between chromosome pairing in the F_1 and fertility of the derived allopolyploid of 19 different interspecific combinations recorded in the literature, and this indicates that there is a negative correlation between the amount of pairing in the F_1 hybrid and the degree of fertility in the derived allopolyploid. But in his study of 20 hybrids and their derived allopolyploids in *Aegilops* and in *Triticum,* Sears (1941) found no such correlation, and if Sears's allopolyploids are plotted on Gajewski's chart, all evidence of correlation disappears. Chromosome pairing is only one factor affecting the fertility or sterility of allopolyploids; the degree of similarity of the pairing chromosomes, their size, their chiasma frequency, and various genically or plasmatically controlled sterility factors must also be considered (see page 306).

On theoretical grounds, the hypothesis presented in Chapter VI, that chromosomal sterility is due to heterozygosity for translocations and inversions of small chromosomal segments, and that these segments are inherited independently of the morphological and physiological differences between the parental species, leads to the postulate that segregation and selection over a long enough period of time should eventually lead to the production of fertile, genetically stable types derived from segmental allopolyploids. The length of time needed for this to happen could be estimated only on the basis of some knowledge of the number of chromosomal differences between the parental species. Since this is not available at present for any example, no estimates of this nature can be made.

Segmental allopolyploids, therefore, have two distinctive properties. The first is the ability for genetic segregation, in respect to both the morphological differences between the parental species and the chromosomal differences which formed the sterility barrier between these parents. Selection pressure in favor of increasing fertility will affect the segregation of these chromosomal differences in two ways. If the chromosomes of the parental species were largely similar and differed by only one or two small nonhomologous segments, selection would favor the elimination of these segments and evolution in the direction of autopolyploidy. But if the chromosomes differed initially by numerous or large segments, but nevertheless possessed enough homology so that they could occasionally form multivalents in the original, "raw" allopolyploid, selection would favor the further differentiation of these chromosomes by means of mutation and further chromosomal rearrangements and the elimination of multivalent formation. Thus, segmental allopolyploidy is an unstable condition, which through segregation and chromosomal alteration, guided by selection for fertility, will evolve in the direction of either auto- or typical allopolyploidy. However, the direction of this evolutionary trend will be according to the constitution of individual chromosomes, not of the entire complement of the polyploid. In many segmental allopolyploids, some of the parental chromosomes may be so similar that selection will favor elimination of structural differences and the formation of four similar chromosomes, while others of the initial parental sets may have been so strongly differentiated that further differentiation and the elimination of multivalent formation will be favored.

The stable derivatives eventually produced from the raw segmental allopolyploid may therefore be expected to have one of three different types of constitution. First, they may be completely autopolyploid in respect to the structural make-up of their chromosomes, although still intermediate between their original parental species in external morphology and in ecological preferences. Second, they may become cytologically allopolyploid, so that they rarely or never form multivalents or exhibit heterogenetic association. Finally, they may come to possess some chromosomal types present in the form of four complete structural homologues, and others of the same original complement so strongly differentiated that they form only bivalents. This third type will be permanent, stable segmental allopolyploids. They will regularly form a small number of multivalents and will segregate for some of the morphological differences between the parental species, but they will nevertheless be highly fertile and able to maintain themselves under natural conditions.

The second property of segmental allopolyploids is that they can be expected to form partly fertile hybrids through backcrossing with autopolyploid derivatives from either of their parental species. By means of hybridization and introgression they may greatly obscure or completely obliterate the morphological and genetic barrier which existed between these parental autopolyploids. This fact is probably largely responsible for the usual treatment of this type of polyploid in the systematic literature; they are usually placed in the same taxonomic species as one of their diploid ancestors, and their existence may cause conservative botanists to combine both of their ancestral diploids into the same species.

The third type of polyploid, the typical or genomic allopolyploid, is a much better known type of plant. Well-known artificial examples are *Nicotiana glutinosa-tabacum* (*N. digluta,* Clausen and Goodspeed 1925), *Brassica oleracea-Raphanus sativus* (*Raphanobrassica,* Karpechenko 1927, 1928), and *Secale cereale-Triticum aestivum* (*Triticale,* Müntzing 1939). Natural allopolyploids among wild species whose parentage has been proven by synthesis of the same or a similar polyploid from diploid ancestors are *Galeopsis tetrahit* (*G. pubescens-speciosa,* Müntzing 1930a,b), *Rubus maximus* (*R. idaeus-caesius,* Rozanova 1934, 1938, 1940), *Nicotiana rustica* (*N. paniculata-undulata,* Goodspeed 1934, Eghis 1940), *Brassica napus* (*B. campestris-oleracea,* Nagahuru U, 1935), *B. juncea* (*B. campestris-nigra,* Frandsen 1943), *Madia citrigracilis* (Clausen, Keck, and Hiesey 1945a), and *Bromus arizonicus* (*B. haenkeanus-trinii,* Stebbins, Tobgy, and Harlan 1944, Stebbins 1949b). Cultivated species of which the allopolyploid origin has been similarly demonstrated are *Nicotiana tabacum, Gossypium hirsutum,* and *Triticum aestivum* (see page 372). In addition, more or less convincing indirect evidence has been obtained for the parentage of the following natural allopolyploids: *Spartina townsendii* (*S. alterniflora-stricta,* Huskins 1931), *Pentstemon neotericus* (*P. azureus-laetus,* Clausen 1933), *Iris versicolor* (*I. setosa-virginica,* Anderson 1936b), *Prunus domestica* (*P. divaricata-spinosa,* Rybin 1936), *Poa annua* (*P. exilis-supina,* Nannfeldt 1937, Litardière 1939), *Nicotiana arentsii* (*N. undulata-wigandioides,* Goodspeed 1944), *Artemisia douglasiana* (*A. ludoviciana-suksdorffii,* Clausen, Keck, and Hiesey 1945a, Keck 1946), *Oryzopsis asperifolia* and *O. racemosa* (*O. holciformis vel aff.-O. pungens vel. aff.,* Johnson 1945), and *Bromus marginatus* (*B. aff. laevipes-stamineus,* Stebbins and Tobgy 1944, Stebbins 1947c). In many other examples one of the parental species of an allopolyploid has been identified, and a still larger number of species has been shown to have the chromosome number and behavior which one would expect in a typical allopolyploid.

Typical allopolyploids are derived from hybridization between two or more distantly related species, of which the chromosomes are so different that they are unable to pair in the diploid hybrid, or form only a small number of loosely associated bivalents. The hybrids which are capable of giving rise to allopolyploids are usually completely unable to give diploid progeny, so that interchange of genes between their parental species is impossible. The species ancestral to allopolyploids, therefore, usually belong to different cenospecies, according to the definition of Clausen, Keck, and Hiesey (1939, 1945a). In their systematic position as based on characteristics of external morphology, they may belong to the same section of a genus, as with some examples in the genera *Nicotiana* and *Madia,* but more often they belong to different sections, subgenera, or even genera.

A typical allopolyploid, therefore, contains two or more sets of very different genomes, and may be given the formula AABB, AABBCC, and so on, where each letter represents a set of chromosomes of the basic haploid number for the genus. The only type of pairing which normally occurs is that between similar chromosomes of the same genome, or homogenetic association. This causes allopolyploids to breed true to their intermediate condition and to segregate relatively little. Furthermore, hybrids formed by backcrossing an allopolyploid to either diploid parent or to their autopolyploid derivatives are usually partly and sometimes completely sterile. A typical allopolyploid, therefore, in contrast to a segmental allopolyploid or an autopolyploid, not only is often fully fertile and constant from the start; in addition it is strongly isolated from and as a rule morphologically discontinuous with its nearest relatives. The result of this condition is that most of the polyploid species complexes which are difficult problems for the systematist contain segmental allopolyploids, autopolyploids, or both; typical allopolyploids most often stand out as clearly marked species.

Nevertheless, such polyploids, although they resemble diploids in many respects, still possess a number of qualities connected with their polyploid origin. The most significant of these is the presence of duplicated genetic material, which enables them to withstand losses of chromosomes or chromosome segments which would be fatal to diploid organisms. This ability is most strikingly manifest in wheat (*Triticum aestivum*), in which Sears (1944b) has been able to produce 21 different viable nullisomics, each lacking a different one of the 21 chromosome pairs present in normal *T. aestivum*. Sears showed, furthermore, that genetic material could be duplicated in chromosomes completely incapable of pairing with each other. Plants monosomic for chromosome II are morphologically very similar to those monosomic for chromosome XX and in both cases are definitely weaker than normal plants. But a plant which has 42 chromosomes and is monosomic or nullisomic for chromosome II and trisomic or tetrasomic for chromosome XX is nearly normal in every respect. Nevertheless, although they compensate each other in their influence on viability and fertility, chromosomes II and XX are completely unable to pair at meiosis, even when they have no other mates.

In another allopolyploid which has been carefully analyzed, *Nicotiana tabacum,* Clausen and Cameron (1944) have produced and studied the phenotypic characteristics of monosomics lacking one member of each of the 24 chromosome pairs normally present in the species. Each of these monosomics has its recognizable distinctive features, and the viability of gametes lacking a chromosome differs greatly depending on the particular chromosome concerned. In *Gossypium hirsutum,* on the other hand, the amount of duplication of genetic material appears to be so slight that monosomics are difficult to obtain and rarely perpetuate themselves even in artificial cultures (M. S. Brown, oral communication). Allopolyploids, therefore, may differ considerably from each other in the amount of genic duplication they possess, even though they are much alike in their normal genetic and cytological behavior.

Another property of many allopolyploids which results from the presence of some duplicated genetic material is the ability of their chromosomes to undergo occasional heterogenetic association, and so to segregate with respect to some of the characteristics which differentiated their parental species, or for characteristics which have originated by mutation since the origin of the polyploid. The progeny resulting from such segregation appear like sudden mutants. They have been most carefully studied in certain cereals, where they are responsible for the "fatuoid" variants in cultivated oats (*Avena sativa*) and the "speltoids" in wheat (*Triticum aestivum*). Huskins (1946) has reviewed carefully the extensive literature on these forms and has marshaled an abundance of evidence to show that they have actually resulted from this type of pairing, rather than from hybridization or from the addition or subtraction of whole chromosomes, as other authors have believed. The detection of variants of this nature is very difficult except in well-known and genetically constant species like wheat and oats, but there is little doubt that occasional heterogenetic association is an important source of variation in many wild allopolyploids. Huskins (1941) has made the apt suggestion that in most polyploid species chromosomal changes are a more important source of variation than gene mutation.

In an allopolyploid which is well adapted to its environment, chromosomal variation is likely to lead to deleterious variants, and selection will favor mutations which make the chromosomes more different from each other. The original, or "raw," allopolyploid is likely to become progressively "diploidized," until its behavior more nearly resembles that of a diploid species. R. E. Clausen (1941) has determined accurately the nature of this diploidization for certain chromosomes of *Nicotiana tabacum,* by comparing cultivated strains of this species with allopolyploids newly produced from its putative parents, *N. sylvestris* and *N.*

tomentosiformis. These "raw" allopolyploids have in duplicate certain factors, such as the dominant allele to the mammoth factor, the normal allele to a factor responsible for asynapsis, and another to a recessive white-seedling character, which are all present only singly in *N. tabacum*. Since the chromosomes of the "raw" allopolyploid pair perfectly with those of *N. tabacum* in the F_1 hybrid, the elimination of duplicate alleles during the evolution of *N. tabacum* appears to have been either by mutation or by the loss of very small chromosomal segments.

Although typical allopolyploids can be formed only through hybridization between species the chromosomes of which are so different from each other as to be nearly or quite incapable of pairing, by no means all polyploids from such hybrids are successful allopolyploids. If the parental species are separated by factors for genic as well as for chromosomal sterility, the allopolyploid combining their genomes will be as sterile as their diploid hybrid. For instance, *Nicotiana sylvestris-tomentosiformis*, although it has normal chromosome pairing and good pollen, is completely female sterile because of abortion of the embryo sac at the two- to four-celled stage (Greenleaf 1941, 1942). Nearly all of the allopolyploids between *N. sylvestris* and members of the *N. tomentosa* complex, if they are produced through doubling of the somatic tissue of the diploid F_1 hybrid, have the same type of sterility (Clausen 1941). But Kostoff (1938a) was able to produce a fertile allopolyploid of this combination by using unreduced gametes of the F_1 hybrid, and through the intermediate stage of a triploid. On the basis of evidence obtained from hybrids between Kostoff's allopolyploid and the one produced by somatic doubling, Greenleaf (1942) concluded that during one of the stages in the formation of Kostoff's allopolyploid, the female sterility factor was eliminated by heterogenetic association and crossing over.

An allopolyploid exhibiting a very different type of genic sterility is the one produced by Sears (1941b) between *Aegilops umbellulata* and *Haynaldia villosa*. The F_1 hybrid of this combination had a mean value of only 0.3 pairs at meiosis and was completely sterile, as might be expected from the remoteness of the relationship between its parents. But the allopolyploid, produced by somatic doubling through colchicine treatment of the F_1 hybrid, was also highly sterile and had 6 to 26 out of a possible

28 unpaired univalent chromosomes, in spite of the fact that each chromosome possessed a potential mate with which it was completely homologous. Li and Tu (1947) have found a similar situation in the colchicine-produced allopolyploid, *Aegilops bicornis-Triticum timopheevi.* The best explanation of this situation is, as Li and Tu have suggested, that the parental species differ by certain genetic factors affecting timing or some other process essential to chromosome pairing at meiosis. At present, no criteria are available by which one may tell whether such factors are at work in an F_1 hybrid, and so predict whether or not the allopolyploid will be sterile. More intensive study of various examples of artificial allopolyploids of all types will be needed before satisfactory generalizations can be made concerning the probable success or failure of the allopolyploids of any type, based on the study of parental species and the diploid hybrid.

The fact that artificially produced "raw" polyploids of all types have for the most part proved only partly fertile lends great importance to ways in which fertility may be increased in later generations. That such an increase can take place is evident from the observations of Sando (1935) on *Triticum turgidum-Haynaldia villosa,* of Katayama (1935) on *Triticum dicoccoides-Aegilops ovata,* of Oehler (1936) on *Triticum durum-Aegilops triuncialis,* of Kostoff (1938b) on *Nicotiana glauca-langsdorffii,* and of Gajewski (1946) on *Anemone multifida-sylvestris.* The two sets of factors which probably cause this increase in fertility are briefly mentioned by Kostoff and are discussed by Gajewski. These are, first, alterations in chromosomal structure which would eliminate heterogenetic association, with its consequent formation of gametes containing duplications and deficiencies of chromosomal segments; and, second, gene mutations suppressing or counteracting the genetic physiological disharmonies of meiosis which are responsible for asynapsis of completely homologous chromosomes, with the consequent formation and irregular distribution of univalents. Both of these factors are probably operating in different polyploids, and their relative importance most likely depends on the nature of the particular "raw" polyploid concerned. Fagerlind (1944a), based on the observations of Wettstein (1927) and Wettstein and Straub (1942) on the moss genus *Bryum,* has suggested that in perennial species, physiological adjustment leading to greater fertility can occur during the lifetime of a single plant, presumably through somatic mutations. The determination of the relative importance in polyploids of different types of these factors for increasing fertility is one of the major tasks of plant breeders making use of artificial polyploids, and obviously it is of great importance to an understanding of plant evolution.

Although the majority of allopolyploids are either tetraploids or hexaploids, and so combine the genomes of only two or three different ancestral species, a number of higher polyploids of this type are known. For instance, one of the most common species of grasses of western North America, *Bromus carinatus* and its relatives, is an allopolyploid containing four different genomes (Stebbins and Tobgy 1944). Another species belonging to a different subgenus, *B. trinii,* is an allohexaploid containing three different genomes, and the F_1 hybrid, *B. carinatus* $\times$ *trinii,* forms mostly univalents at meiosis, with 5 or 6 bivalents as the maximum amount of pairing. The allopolyploid produced from this hybrid has 98 chromosomes as the somatic number, and at meiosis it forms either 49 bivalents or 47 bivalents and one quadrivalent (Stebbins 1949b). This polyploid is vigorous and fertile, and the first stage of establishing it in nature as a spontaneous species is proving successful. Here, therefore, is an example of a probably successful allopolyploid which contains seven different genomes, acquired by a succession of hybridizations probably extending over a period of millions of years and involving seven different ancestral diploid species.

The fourth type of polyploid can exist only at the level of hexaploidy or higher, and combines the characteristics of the two preceding ones. This is the type called by Kostoff (1939c) an *autoallopolyploid.* He cites the example of *Helianthus tuberosus,* a hexaploid with $2n = 102$ chromosomes, which produces with the diploid *H. annuus* ($n = 17$) a tetraploid hybrid forming 34 bivalents at meiosis. On this basis, Kostoff has assumed that *H. tuberosus* has the genomic formula $A_tA_tA_tA_tB_tB_t$ and *H. annuus,* the formula B_aB_a, the genomes designated by the letters A and B being entirely different from each other, but those with the same capital letter having enough chromosomal segments in common so that they can pair normally. Thus, in respect to the

A genome, *H. tuberosus* is either autopolyploid or segmental allopolyploid, and we should expect to find in it both multivalents and tetrasomic inheritance. But the presence in this species of the genome designated B makes it partly an allopolyploid, and at least one interspecific hybridization must have been involved in its origin.

Two other well-known polyploids which are probably of this type are *Phleum pratense* and *Solanum nigrum*. The former species was judged by Gregor and Sansome (1930) to be an allopolyploid, an opinion shared by Clausen, Keck, and Hiesey (1945a). But Nordenskiöld (1941, 1945), after intensive study, considered it to be an autopolyploid containing the diploid complement of *Phleum nodosum* trebled, while Myers (1944) noted cytological characteristics indicating autopolyploidy. But since two different haploid plants of *P. pratense*, one studied by Nordenskiöld and one by Levan (1941a), typically form 7 bivalents and 7 univalents, their genomic formula must be AAB, and that of the diploid AAAABB. The evidence of Nordenskiöld suggests that the A genome is very likely that of *P. nodosum*, and the B genome, that of the diploid *P. alpinum*. The tetraploid *P. alpinum*, which Gregor and Sansome believed to be ancestral to *P. nodosum*, was shown by Nordenskiöld (1945) to be an allopolyploid containing the B genome and a still different one, belonging to some diploid species as yet unidentified.

Solanum nigrum was considered by Jorgenson (1928) to be an allopolyploid, but Nakamura (1937) believed it to be an autopolyploid, since he found quadrivalents at meiosis in this species and noted a strong resemblance to a diploid species from southern Japan described by him as new, but actually conspecific with *S. nodiflorum* Jacq. The haploid of *S. nigrum* ($n = 36$) forms approximately 12 bivalents and 12 univalents (Jorgenson 1928), so that it is most likely an autoallopolyploid containing four genomes from *S. nodiflorum* or some other species closely related to it, and two from some diploid species as yet not identified. Both in this example and in the previous one, the resemblance of the autoallohexaploid to one of its diploid ancestors is so strong that the two have been placed in the same species by most systematists. This is likely to be true of most autoallopolyploids which contain two or more genomes derived from one species and only one genome from another. It is likely that a considerable proportion of the polyploids at the hexaploid level or higher which are believed to be autopolyploids are actually of this constitution.

Another type of autoallopolyploid can be produced by doubling the chromosome number of a typical allotetraploid, to produce an octoploid with the genomic formula AAAABBBB. A typical example is the autopolyploid produced by Clausen (1941) from *Nicotiana tabacum*. Among wild species, *Rubus ursinus* ($2n = 56$) is probably of this nature, since the hybrid between this species and the diploid *R. idaeus* forms 14 bivalents and 7 univalents (Thomas 1940a,b). According to Fedorova (1946), *Fragaria grandiflora, F. chiloensis,* and *F. virginiana* are similar. Other high polyploid species are known to form multivalents and to resemble closely certain diploids or lower polyploids. Typical of these are *Agropyron elongatum,* in which a form with $2n = 70$ chromosomes resembles one with $2n = 14$ (Wakar 1935, Simonet 1935); *Pentstemon neotericus,* an octoploid ($2n = 64$), which is probably derived from a closely similar hexaploid, *P. azureus,* and a diploid species, *P. laetus* (Clausen 1933); and *Rubus lemurum* (Brown 1943), with $2n = 84$ chromosomes, which is closely similar to and apparently forms partly fertile hybrids with the octoploid *R. ursinus,* mentioned above. Most of the high polyploids in the genus *Chrysanthemum* (Shimotomai 1933) are probably of this nature, as are those described by Callan (1941) in *Gaultheria* and *Pernettya.* In fact, there is good reason to believe that most of the polyploids at the octoploid level or higher contain some duplicated chromosomes or chromosome segments, but the behavior of artificial autopolyploids indicates that few if any of these higher polyploids can be strict autopolyploids. It is very difficult, and perhaps not of major importance, to make the distinction between those which combine autopolyploidy with allopolyploidy and the higher polyploids which are partly segmental and partly typical allopolyploid in constitution.

When extensive duplication of chromosomal material exists, as it does in such polyploids, regular behavior of the chromosomes at meiosis is not essential to the production of viable gametes, since many different combinations of various numbers of chromosomes can function. Love and Suneson (1945) found that an

F_1 hybrid between *Triticum macha* and *Agropyron trichophorum*, both of which have $2n = 42$, gave rise to a fertile F_2 plant with 70 chromosomes. The best explanation is that an unreduced gamete with 42 chromosomes united with a partially reduced one having 28. Plants with 84 and other numbers of chromosomes might be expected to arise from this same F_1, so that it is potentially the progenitor of a number of different evolutionary lines, each with a different chromosome number and capable of becoming a different species. In the genus *Saccharum* Bremer (1928) and Grassl (1946) have found at high polyploid levels a great variety of different chromosome numbers, ranging from $2n = 60$ to $2n = 120$. Many of these numbers are found in the recently produced "noble canes," and are therefore the result of plant breeding in recent times. Most of these forms, whether euploid or aneuploid, are reasonably fertile.

At these higher levels of polyploidy, therefore, various combinations of the auto- and allopolyploid condition are probably the most common situation. In some favorable instances analysis of these polyploids and the discovery of their diploid ancestors may be possible. But in the majority of them, their origin is too complex to be analyzed in its entirety, and at least one of the diploid ancestors may be extinct.

POLYPLOID PERMANENT HYBRIDS: THE *Rosa canina* COMPLEX

A most unusual type of autoallopolyploid is found in the roses of northern Europe belonging to the *Rosa canina* complex. Many years ago Blackburn and Harrison (1921), as well as Täckholm (1922), found that most forms of this group are pentaploids with 35 somatic chromosomes, though there are forms with 28 and 42, but that in any case they form at meiosis only 7 bivalents, and 14, 21, or 28 univalents. The behavior of these univalents in the meiosis leading to pollen formation is entirely different from that in megasporogenesis in the ovules. In the former divisions, the univalents are usually eliminated, so that the functional pollen grains have only 7 chromosomes, all derived from the bivalents. In megasporogenesis all of the univalents are included in the functional megaspore, so that this cell and the resulting embryo sac and egg cell have, in the *caninae* roses with 35 chromosomes as the somatic number, 28 chromosomes. The union at fertilization of 28 + 7 restores the normal somatic number. The constancy of individual microspecies in this complex, as well as the large number present, led Täckholm (1922) to believe that apomixis exists in this group, and some experiments on emasculation and hybridization seemed to confirm this suspicion. But Blackburn and Harrison believed these roses to be sexual and ascribed their peculiar variation pattern and cytological behavior to a type of balanced heterogamy. This opinion was shared by Darlington (1937), Fagerlind (1940b), and particularly by Gustafsson (1931a,b, 1944).

The careful cytogenetic work of Gustafsson (1944) and Gustafsson and Håkansson (1942) has provided the basis for an illuminating and highly probable hypothesis about the constitution of this group and the reasons for its anomalous and intricate variation pattern. Gustafsson found that hybrids between two common members of this complex, *R. canina* and *R. rubiginosa,* are strongly matroclinous, as are also hybrids between either species and a relatively distantly related diploid, *R. rugosa.* Nevertheless, the true hybrid nature of F_1 plants produced from these crossings was evident from certain features of their external morphology, and more particularly from the behavior of the chromosomes at meiosis. This is precisely the reverse of what is found in a series of hybrids between normal species. The related forms, *R. canina* and *R. rubiginosa,* yield F_1 hybrids with approximately 7 bivalents, although cells with 3 to 6 bivalents, as well as those with multivalents, are not uncommon. But the F_1 hybrids of the wide intersectional crosses between *R. canina* or *R. rubiginosa* and *R. rugosa,* with the latter as the pollen parent, have a much higher number of bivalents, mostly 11 to 14. The only possible explanation of this fact is that at least 4 to 7 of these bivalents are due to autosyndesis of chromosomes derived from the maternal gamete. Both *R. canina* and *R. rubiginosa,* therefore, are considered by Gustafsson to possess an "internal autotriploidy," which makes possible pairing between two sets of chromosomes derived from the same maternal gamete, provided that one of these does not have exactly similar mates contributed by the paternal gamete.

After considering several possible genomic formulae, Gustafsson has concluded that the most likely one for the somatic constitution of the pentaploid *canina* roses is $A_1A_1A_2CD$, while

R. rugosa is CC, and the F_1 hybrids are A_1A_2CCD. Since the chromosomes of A_1 and A_2 are only partly homologous and may have been derived from different original diploid or tetraploid species, the "internal autotriploidy" would be segmental allotriploidy according to the terminology used in this chapter. The degree of similarity between A_1 and A_2 varies according to the species of *caninae* studied. In hybrids between *R. canina* and either *R. rugosa* or *R. rubiginosa* the number of bivalents is somewhat smaller than in either *R. rubiginosa* × *rugosa* or *R. rubiginosa* × *canina,* indicating that a greater amount of autosyndesis is possible between chromosomes belonging to *R. rubiginosa* than between those of *R. canina* × *rubiginosa* and its reciprocal, which results in greater sterility of hybrids with *R. canina* as the maternal parent, and provides additional evidence in favor of Gustafsson's hypothesis of autosyndesis.

Fagerlind (1945), while verifying Gustafsson's hypothesis of internal auto- or segmental allopolyploidy, has shown that this condition extends beyond triploidy and, in at least some of the *canina* roses, may involve complete or partial homologies between all of the sets present. He crossed a tetraploid member of the *canina* group, *R. rubrifolia,* reciprocally with *R. rugosa.* With *R. rubrifolia* as mother, the tetraploid F_1 strongly resembled that species, and as in Gustafsson and Håkansson's hybrids produced 7 to 14 bivalents. But the reciprocal hybrid, *R. rugosa* × *rubrifolia,* while diploid as expected, nevertheless consistently produced 7 bivalents. This shows that the pairing or "A" set of *R. rubrifolia* is homologous to the haploid genome of *R. rugosa.* Since Fagerlind has stated from preliminary information on hybrids with *R. rugosa* as the maternal parent that the pairing sets of different roses of the *canina* group differ widely from each other in their genic content, Gustafsson's formula may yet hold for the species investigated by him, although it is obviously invalid for *R. rubrifolia.* On the other hand, the hypothesis of Fagerlind, that all the genomes in the *canina* roses are potentially capable of pairing with each other, but are usually kept from doing so by genetic factors, seems more likely at present.

Fagerlind has stated that new microspecies are formed in this group chiefly by hybridization between existing forms followed by segregation of the genes present in the pairing genomes, as well as by mutations occurring in the nonpairing genomes. Gustafsson and Håkansson, on the other hand, emphasize the importance of heterogenetic association between chromosomes belonging to different genomes as a source of variation. This type of association would be expected most often as autosyndesis in an F_1 hybrid between two microspecies having somewhat dissimilar chromosomes. All three of these processes have probably contributed to the multitude of microspecies found in this group.

This situation in the *canina* roses provides a most interesting parallel to that in the majority of the complex-heterozygote races of *Oenothera,* discussed in Chapter XI. Both are permanent heterozygotes or hybrids and are heterogamous, in that their male gametes contain different chromosomal material from their female ones. Both probably owe their success and aggressiveness to the permanent possession either of hybrid vigor or of favorable combinations of genes. Both consist of a large number of relatively constant biotypes or microspecies, which owe their constancy partly to autogamy and partly to a great restriction of the amount of pairing and crossing over which can take place, in relation to the number of gene loci present. In both groups, new biotypes or microspecies can arise either through crossing between preexisting ones or through an exceptional type of pairing and crossing over. The chromosomes of both groups contain some elements which regularly pair and cross over and others which are normally unpaired. In *Oenothera,* the unpaired regions are the differential segments, in those portions of the chromosomes near the attachment constrictions or centromeres; while in the *canina* roses they are the univalents. There is, however, an important difference in the genic content of the pairing portions as compared to the nonpairing portions of the chromosome complements in the two groups. In *Oenothera,* the differences between races appear to lie mostly in the differential or nonpairing segments, while in the *canina* roses, if the hypothesis of Gustafsson and Håkansson is correct, differences must exist in both the pairing and the nonpairing chromosomes. That the univalents of these roses are active and with a full genic content, rather than devoid of genes like the B chromosomes of maize and *Sorghum,* is evident from the fact that monosomic types lacking one of them have a lowered vigor and fertility as well as differences in external morphology from their normal sibs.

Gustafsson and Håkansson have shown conclusively that the septet theory of Hurst (1925, 1928, 1932), according to which all of the polyploid species of *Rosa* contain genomes derived from only five original and sharply differentiated diploid species groups, cannot explain the constitution which they have found in the *canina* group, at least in the form in which Hurst presented it. On the other hand, they show that these tetraploid, pentaploid, and hexaploid roses could have arisen in a number of different ways from hybridization between normal diploid, tetraploid, and hexaploid species, all of which are well known in the genus. That they have had at least partly a common origin seems clear, since their peculiar method of stabilizing the results of a very irregular type of meiosis has, so far as is known, arisen nowhere else in the plant kingdom. It is not necessary, however, to assume that all the modern forms originated from a single hybridization, since new types could also arise from crosses between the permanent hybrids and other normal diploid or tetraploid species.

THE POLYPLOID COMPLEX

In many groups of plants, polyploids of some or all of the types already mentioned exist together with their diploid progenitors. And if the diploids are related closely enough to each other so that they can produce segmental allopolyploids as well as autopolyploids, exchange of genes is permitted on the tetraploid level between entities which on the diploid level are almost completely isolated from each other by chromosomal sterility. The presence of autoallopolyploids at the hexaploid or higher levels involving two or more genomes of one species and one or more of another will result in the existence of forms which are morphologically very similar to some of the segmental allopolyploids at the tetraploid level, and like them are intermediate between two or more of the original diploid species, but are isolated from these tetraploids because of the difference in chromosome number. By these means can arise the type of variation pattern designated by Babcock and Stebbins (1938) the ***polyploid complex***. Such a complex may be visualized as a series of distinct pillars, representing the diploid forms, which support a great superstructure of intermediate polyploids (Fig. 34). The systematic complexity of such groups is obvious. The smaller complexes containing only two or three original diploids are usually classified as a single species by conservative systematists; the larger ones, which may contain as many as ten diploid species plus various recombination types at the polyploid levels, are the most notable examples of "critical" or difficult genera, of which satisfactory classifications are difficult or impossible to make.

The example discussed principally by Babcock and Stebbins was that of the American species of *Crepis,* in which the polyploids are apomictic. This type will be discussed in Chapter X. A sexual example which the writer also had in mind (cf. Stebbins 1939) is that of the Mediterranean species of *Paeonia.* This consists of five or six distinct diploids plus a number of auto- and segmental allopolyploids which combine their characteristics, as discussed in a previous section. Two others, which are considerably smaller, are the genus *Zauschneria* (Clausen, Keck, and Hiesey 1940, 1945a) and the species complex of *Eriogonum fasciculatum* (Stebbins 1942c). In the former, there are three relatively localized diploid species, *Z. cana,* of coastal southern California, *Z. septentrionalis,* of northwestern California, and *Z. garrettii,* of Utah. The common tetraploid form of coastal California, *Z. californica,* is intermediate between *Z. cana* and *Z. septentrionalis,* but grades imperceptibly into forms which appear like autopolyploids of one or the other species. In the Sierra Nevada and the mountains of Southern California are found broad-leaved forms, *Z. californica latifolia,* which the authors believe to be autopolyploids of *Z. septentrionalis,* but some of these resemble *Z. garrettii* in both their external morphology and ecological preferences. At the tetraploid level, therefore, is found the same range of morphological characteristics and climatic adaptation as exists in the diploids. But whereas the diploids include only three relatively uniform and distinct types representing three extremes of this range of variation, the tetraploids form a continuous series throughout the whole range, with intermediate types by far the most common. The situation in *Eriogonum* is similar, except that here only two "diploid" (perhaps originally tetraploid) species are involved. A much larger and more intricate, but otherwise similar, polyploid complex is the subgenus *Cyanococcus* of *Vaccinium* (Darrow and Camp 1945, Camp 1945).

In their discussion of the polyploid complex, Babcock and Stebbins emphasized the importance of studying the distribution and the morphological characteristics of the entire complex before making any generalizations concerning its origin or its evolutionary tendencies in relation to climatic and ecological trends. The importance of such complete knowledge has become more and more evident as additional polyploid complexes have been reported in the literature. In both this and the next chapter, numerous examples are mentioned of erroneous conclusions about the effects of polyploidy in a particular group. These statements were made on the basis of a partial knowledge of the group and had to be revised when it was more completely known.

CHAPTER IX

Polyploidy II: Geographic Distribution and Significance of Polyploidy

THE FACT that polyploid species may have different geographic distributions and ecological preferences from those of their nearest diploid relatives has been noted by a number of authors. There has, however, been no unanimity of opinion on the effects of polyploidy on distribution, and a number of different tendencies have been noted. The principal ones of these are as follows.

Hagerup (1932) first brought out the concept that polyploids are more tolerant of extreme ecological conditions than their diploid relatives, and he illustrated it with examples taken from the desert flora of Timbuktu, in northwestern Africa. The most striking of these examples is in the grass genus *Eragrostis,* in which Hagerup found three species: a diploid annual of lake margins, *E. cambessedesiana;* a tetraploid perennial, *E. albida,* found in somewhat drier places; and a robust perennial octoploid, *E. pallescens,* found on the very dry sand dunes. Hagerup believed these to form an autopolyploid series, based on Hitchcock's study of their external morphology. But the hazards of determinations of the type of polyploidy on such evidence have already been pointed out, and these are particularly great in a genus like *Eragrostis,* which is even for the grasses very poor in diagnostic characteristics for the separation of species. Furthermore, the large number of African species of this genus is almost unknown cytologically and completely unknown in respect to the behavior of the chromosomes at meiosis. *Eragrostis,* therefore, would seem at present to be a particularly unfavorable example on which to base any generalizations about polyploidy and distribution or ecological preferences. Based on the example of *Empetrum,* which contains a diploid, *E. nigrum,* in northern Europe, and a tetraploid, *E. hermaphroditicum,* in Greenland, Hagerup suggested that polyploidy might be associated with the severe cold of arctic regions. But in later papers (1933, 1940), the same author cited two examples, *Vaccinium uliginosum* and *Oxycoccus microcarpus,* in which diploid representatives of the species as ordinarily recognized by systematists have a more northerly distribution than tetraploids referred to the same species. All three of these examples are believed by Hagerup to be autopolyploid series, but in each one he studied only the European and Greenland representatives of groups which have their centers of distribution on the North American continent, and in none of them was cytological or genetic evidence obtained as to the nature of the polyploidy. All, therefore, suffer from incompleteness of the data, but, in the example of *Oxycoccus,* Camp (1944) has produced some evidence that the tetraploid is either an allopolyploid or at least an intersubspecific autopolyploid which owes its characteristics to gene combinations derived from two ecologically different types, the high arctic *O. microcarpus* and the temperate American *O. macrocarpus. Campanula rotundifolia* (Böcher 1936) is another European tetraploid which has a diploid relative in the high arctic regions.

The hypothesis of Hagerup, that the frequency of polyploids in regions with a severe climate is due to the direct action of the environment in inducing failure of meiosis in the sex cells, was examined by Gustafsson (1948) through a study of the frequency of polyploids in a particularly severe locality, the Dovre massif of the Norwegian mountains, in which the snow does not melt until late summer. Here the proportion of polyploids among sexual species was actually lower than the average for the Scandinavian mountains, and not appreciably higher than in lowland plants of this country.

That polyploid is favored by existence along the seacoast was concluded by Shimotomai (1933), on the basis of his studies of *Chrysanthemum* in Japan, and by Rohweder (1937), after studying the Darss-Zingst area, in northern Germany. But in the genus *Artemisia,* Clausen, Keck, and Hiesey (1945a) have shown that in the complex of *A. vulgaris* in western North America the coastal species, *A. suksdorfii,* is diploid, the widespread desert species, *A. ludoviciana,* is tetraploid, while the hexaploid *A. douglasiana* grows in the most favorable habitats of all three species, namely, the valleys and coastal hills of California and Oregon. Tischler (1937) found a high percentage (65 percent) of polyploids in the halophytic flora of certain islands in the North Sea, and from that concluded that polyploidy is associated with adaptation to the severe conditions of saline habitats. This conclusion was apparently supported by the data of Wulff (1937a) on the halophytic flora of Schleswig-Holstein, but Tarnavschi (1939) found only 26 percent of polyploids among 38 halophytic species of the east coast of Rumania. Sokolovskaya and Strelkova (1940) found 50 percent of polyploids in the alpine flora of the Caucasus, but in the corresponding floras of the Pamirs and the Altai Mountains of central Asia, which have more extreme climates, the percentages of polyploids were, respectively, 65 percent and 85 percent. From this they concluded that polyploidy is causally related to the tolerance of these extreme climatic conditions. But Bowden (1940), after obtaining the chromosome numbers of a considerable number of species of flowering plants from various countries, which were growing under cultivation in Virginia, found no correlation between the incidence of polyploidy and the winter hardiness of the species.

One of the most widely known studies of the relation between polyploidy and distribution is that of Tischler (1935), in which he reported that among four different European floras studied by him the lowest percentage of polyploidy was found in the most southerly one, that of Sicily, with ony 31 percent of polyploids. Schleswig-Holstein, in northern Europe, was found to have 44 percent of polyploids, while two more northerly localities, the Faroe Islands and Iceland, were reported as having still higher percentages of polyploids, the figures being 49.4 percent and 54.5 percent, respectively. In a more recent publication (Tischler 1946), the percentage of polyploids on the Greek Islands of the Cyclades was found to be about the same as in Sicily (34 percent). Tischler's results have been criticized on the ground that many of his counts were based, not on plants found in the region for which the species was scored as diploid or tetraploid, but on records in the literature for the species concerned. Such records are hazardous, because many species have diploid and polyploid races with different distributions. However, both Tischler and Wulff (1937b) later made counts on numerous species from

Schleswig-Holstein, and their more complete data were reported, along with those for many other localities in northern Europe as well as the Faroes, Iceland, and Spitzbergen, by Löve and Löve (1943). The data for the extreme conditions and very poor flora of Spitzbergen were first reported by Flovik (1940). The values reported by Löve and Löve were higher for all localities than those given by Tischler, and the percentage recorded for Spitzbergen, 77 percent, was by far the highest of all. But the gradient of increasing polyploidy northwards was the same. The statistical significance of all their differences was found by Löve and Löve to be very high when floras from different latitudes were compared. All these authors showed that the percentage of polyploids is from 19 to 32 percent higher among monocotyledons, which in these floras consist mainly of Gramineae and Cyperaceae, than it is among dicotyledons.

The apparently greater percentage of polyploids in higher latitudes and higher altitudes than in regions with milder climates is attributed by both Tischler and Löve and Löve, as well as by Müntzing (1936), to the greater hardiness of the polyploids, although Löve and Löve suggest also that the greater tolerance of autopolyploids to the photoperiodic conditions of long days may be an additional factor. The assumption of a greater hardiness in polyploid species is criticized by Bowden (1940), Nielsen (1947), and Gustafsson (1947c, 1948), chiefly on the grounds that when diploids are compared with their most nearly related auto- or allopolyploids the greater hardiness of the polyploids is by no means consistently present. The same fact was noted by Clausen, Keck, and Hiesey (1940, 1945a) for the western American species of the genera *Zauschneria*, *Achillea*, and *Artemisia*. Gustafsson has pointed out sources of error in comparisons of percentages of polyploidy based on studies of entire floras. The percentage of polyploids varies greatly from one family to another, being particularly high in the Gramineae, the Cyperaceae, and the Rosaceae. Hence, if a particular flora happens to possess an unusually high percentage of species of one or more of these families, it will automatically have a higher percentage of polyploids than floras which have lower percentages of them. Furthermore, as was demonstrated by Müntzing (1936) and Stebbins (1938b), the highest percentage of polyploids is to be found in perennial herbs, while annuals and woody plants have lower percentages of polyploidy. High percentages of polyploidy are, therefore, to be expected in floras containing a great preponderance of herbaceous perennial species, as is the case in most cool temperate and subarctic floras (Raunkiaer 1934), while, conversely, floras like those of the Mediterranean region, which are rich in woody plants and annuals, can be expected to contain a relatively low percentage of polyploids.

These sources of error can be overcome only by studying the problem of polyploidy and distribution not from the point of view of entire floras, which may contain a very large or a small percentage of diploid species which are nearly or quite incapable of forming successful polyploids regardless of the climatic conditions to which they are exposed, but through analysis of a series of species groups containing polyploids which are known or suspected on the basis of good evidence to be derived directly from the diploid species with which they are compared. Until recently this approach has been impractical because of the small number of polyploid complexes which were sufficiently well known. Before a group can be safely subjected to distributional analysis, the entire world distribution of its members must be known, and direct or indirect evidence on the actual or probable chromosome numbers of its members must be available at least for representative localities throughout its range.

By searching the literature on polyploidy, the writer has assembled 100 examples of such groups (genera, subgenera, or sections of genera) which are at least well enough known so that they can be used tentatively for the type of comparison needed. The list is too long to reproduce here, but some features of it may be mentioned, and the principal conclusions drawn from it will be summarized. In the first place, all the examples are of groups confined to the temperate and arctic floras of the northern hemisphere. No tropical or Southern Hemisphere groups are known well enough for this purpose. Nearly one half (46) of the groups occur to a large extent in the regions covered by the Pleistocene glaciation. Twenty-one are Holarctic in distribution, 56 are North American, and 23 are of the Old World. Nearly all of the latter are European, while of the North American examples two are essentially northern, 21 eastern, and 33 western. The sample, therefore, is by no means a random one, and it reflects both the amount of cytotaxonomic study which has been done in the various regions and the degree to which the present writer is familiar with their floras.

The principal results are as follows. In 60 of the 100 examples, the polyploids as a group have a wider distribution than their diploid ancestors, in seven the area of diploids and polyploids is about equal, and in 33 the polyploids occupy a smaller area. There exists a tendency, therefore, for polyploids to have wider geographic ranges than their diploid ancestors, but this tendency is by no means universal. There was no difference in this respect between the groups found in glaciated regions and those outside of the areas covered by the Pleistocene ice. When the geographic position of the diploids was compared with that of the polyploids, the largest number of examples (31) was found to be those in which, as in *Zauschneria* and *Biscutella,* the diploids are distributed about the periphery of the range of the group as a whole, and the polyploids occupy the central portion. However, the number of examples (28) in which the diploids are centrally located, as well as of those (27) in which the diploids are more southern and the polyploids more northern in distribution, are both only slightly, and probably not significantly, smaller. In seven examples the diploids have a more northerly distribution than the polyploids, while in the remaining seven the distribution is of another nature or not noticeably different between diploids and polyploids.

These results indicate that no general rules can be formulated to govern the relation between the distribution of diploids and tetraploids, at least at present. The higher percentage of polyploids in high northern latitudes can be ascribed to a combination of the following causes. First, the floras of these latitudes contain an exceptionally high percentage of perennial herbs with efficient means of vegetative reproduction, or of hemicryptophytes, in the terminology of Raunkiaer. Because of their growth habit such plants are more favorably disposed toward polyploidy, as will be brought out in a later section. Second, the majority of these areas have been subjected to very drastic changes in climatic and edaphic conditions in relatively recent times, accompanying the successive advances and recessions of the Pleistocene ice sheets. Polyploids, when allopolyploid or when derived from crossing between races or subspecies of a species, are likely to possess wide ranges of tolerance of climatic and edaphic conditions, as has been demonstrated experimentally for many interspecific and interecotypic hybrids. They are thus ideally suited to the colonization of areas newly opened to plants, and therefore floras of such areas might be expected to contain higher percentages of polyploids than those of older, more stable regions. Finally, it is likely that in some groups, particularly the grasses and the sedges, many of the polyploids may actually be more resistant to severe climatic conditions than their diploid ancestors (Flovik 1938). The evidence at present, however, indicates that this tendency is by no means general.

In an earlier publication (Stebbins 1942c) the suggestion was made that polyploids, as new genetic types, are likely to be particularly well adapted to the colonization of newly available areas, and examples illustrating this tendency were cited. These are *Iris versicolor* in the glaciated areas of eastern North America (Anderson 1936b), *Biscutella laevigata* in the glaciated portions of the central Alps (Manton 1934, 1937), *Crepis* in the portions of the northwestern United States covered by lava flows during the latter part of the Tertiary period (Babcock and Stebbins 1938), and the *Eriogonum fasciculatum* complex in the portions of coastal California subjected to elevation and faulting during the Pleistocene epoch. Others are the tetraploids of *Tradescantia* in the newly disturbed areas of the central United States (Anderson and Sax 1936, Anderson 1937b), those of *Vaccinium,* subgenus *Cyanococcus,* in the glaciated portions of eastern North America (Camp 1945), those of certain species of *Galium* in the glaciated portions of northern Europe (Fagerlind 1937), and the tetraploids of *Paeonia* in central Europe (Stern 1946, 1947). Among the 100 examples discussed in the preceding paragraph, 55 showed indications that the tetraploids had occupied newer territory than the diploids, 4 had apparently the reverse distribution, and in 41 examples no difference could be noted between the age of the areas occupied by the diploids and those of the tetraploids. In many of these, however, such evidence could probably be obtained by sufficiently careful study.

Melchers (1946) has recently proposed a scheme of genic recombination and selection in autopolyploids which would convert them from the ill-adapted types which have resulted from nearly all experiments on the artificial production of polyploids to the highly successful polyploid species found in nature. Experimental testing of this scheme should be possible in the near future. Nevertheless, its application is limited to autopolyploids, which, according to the evidence presented in Chapter VIII, are a relatively small proportion of the successful natural polyploids.

The tendency of polyploids to occupy newly available habitats might lead one to expect a relatively high percentage of polyploidy among the weedy species which have taken over the new habitats made available by man. But Heiser and Whitaker (1947) found no significantly greater percentage of polyploidy in the weeds of California as compared to the nonweedy species. Their results are explained largely by the fact that the ruderal habitats occupied by weeds are adapted primarily to annuals, in which the percentage of polyploidy is lower than it is in perennial herbs. When the weedy annual species of two families, the Gramineae and the Compositae, were compared with the nonweedy annuals of the same families, the weeds were found to include a higher percentage of polyploids in both families, although this difference was most evident in the Gramineae.

The fact should be emphasized that polyploidy is by no means the only way in which plant species can adapt themselves to the occupation of new habitats. If at the time when they produce polyploid derivatives, diploid species still possess a wealth of ecotypes adapted to different environments, the diploids may along with the polyploids expand their distribution areas as new habitats become available. This has been true of such diploids as *Rosa blanda, R. woodsii,* and other roses of the section *Cinnamomea* (Erlanson 1934), of *Vaccinium angustifolium* and other diploid species of blueberries (Camp 1945), of the diploid forms of *Crepis acuminata* and *C. atribarba* ("*C. exilis,*" Babcock and Stebbins 1938), of *Polygonatum pubescens* (Ownbey 1944), and of *Hieracium umbellatum* (Turesson 1922b, Bergmann 1935a). The difference in geographic distribution between diploids and polyploids can nevertheless be maintained even under these conditions, because of the tendency of the tetraploids to occupy different new habitats from those taken over by the diploids.

The present state of our knowledge concerning the effect of

polyploidy on plant distribution may therefore be summarized about as follows. Polyploids usually have different geographic distributions from their diploid ancestors, and are likely to be particularly frequent and diverse in regions newly opened to colonization by plants, except when these are suited primarily to annual species. Their areas of distribution tend to be larger than those of their diploid ancestors, but this tendency has many exceptions. In some instances they show greater tolerance of extreme environmental conditions than their diploid ancestors, but this tendency is not as marked as some authors have believed. The most valuable new information on this subject will be obtained, not from statistical studies of the frequency of diploids and polyploids in the entire flora of various regions, but from careful analysis of an increasing number of individual species groups, with attention paid both to the morphological characteristics and the interrelationships of the diploids and polyploids, as well as to their particular climatic and ecological preferences.

POLYPLOIDY AS EVIDENCE FOR FORMER DISTRIBUTIONAL PATTERNS

The tendency for some polyploids to have wider ranges of ecological tolerance than their diploid ancestors may in some instances cause them to survive after these ancestors have become extinct. As examples of such relict polyploids may be cited *Sequoia sempervirens,* the Pacific coast redwood; *Lyonothamnus floribundus;* and the living species of the genera *Fremontia, Psilotum,* and *Tmesipteris* (Stebbins 1940a). In other instances, the diploid ancestor or ancestors may have disappeared from the region occupied by the polyploid, but be still living in other parts of the earth. If this is the case, allopolyploids may provide valuable evidence on the past distribution of species groups and eventually of entire floras. As has already been pointed out, carefully conducted experiments of a suitable nature can make possible the identification of the actual parents of an allopolyploid or intersubspecific autopolyploid, or at least of their nearest living diploid relatives. If two diploid or low chromosome types demonstrated to be the ancestors of a particular allopolyploid are now found in regions remote from each other, one must assume that at some former period in the earth's history they had a different geographic distribution and occurred together, so that they could hybridize to form the allopolyploid. With this information as a starting point, and with some conception of the geological history and fossil flora of the regions in which the species occur, significant inferences can often be made as to the time and place when these hybridizations took place, and therefore of the distribution of certain elements of the flora during past epochs of the earth's history.

Several examples of this nature have been given elsewhere (Stebbins 1947b,c). One of the best documented is *Iris versicolor* (Anderson 1936b). This species, the common blue flag of northeastern North America, is a high polyploid with $2n = 108$ chromosomes, and it almost certainly originated from hybridization

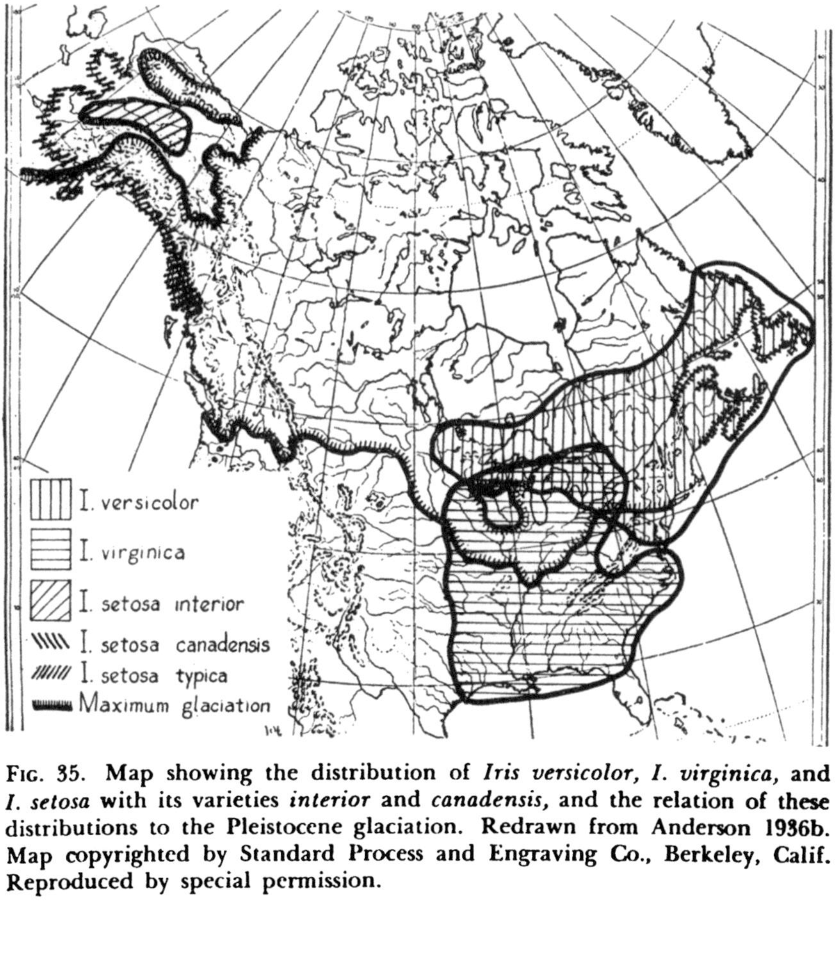

FIG. 35. Map showing the distribution of *Iris versicolor, I. virginica,* and *I. setosa* with its varieties *interior* and *canadensis,* and the relation of these distributions to the Pleistocene glaciation. Redrawn from Anderson 1936b. Map copyrighted by Standard Process and Engraving Co., Berkeley, Calif. Reproduced by special permission.

between *I. virginica* var. *shrevei* ($2n = 72$) and *I. setosa* var. *interior* ($2n = 36$) or some very similar form. At present, these two ancestors are separated from each other by the entire distance of 5,400 km from Minnesota to central Alaska (Fig. 35). Since *I. versicolor* occurs at present with *I. virginica,* but not with *I. setosa,* the plausible assumption is made by Anderson that at some time in the past the range of *I. setosa* var. *interior* extended southeastward to the north central part of the United States, where it overlapped with that of *I. virginica* var. *shrevei.* There the hybridization took place which resulted in the formation of *I. versicolor.* Since the latter species is found almost entirely within the territory covered by the Wisconsin ice, it probably originated during the Pleistocene epoch. *I. setosa* var. *interior* probably occurred along the moraine of the retreating Wisconsin glaciers and was able to hybridize with the plants of *I. virginica* which advanced up the Mississippi Valley in response to the warming of the climate in late glacial times. A similar distribution of the polyploid *Oxycoccus quadripetalus* in relation to its diploid ancestors *O. microcarpus,* of northwestern North America and Eurasia, and *O. macrocarpus,* of the east central United States, led Camp (1944) to suggest a similar origin for this polyploid species. *O. quadripetalus,* however, has attained a much wider distribution than *Iris versicolor,* since it occurs in the arctic regions of both hemispheres. This may very well be due to the superior means of seed dispersal afforded by its fleshy berries.

In *Bromus carinatus* and its relatives we have a group of allopolyploids which must have originated from hybridization between diploid species of the section *Bromopsis,* now found chiefly in North America, and hexaploid species of the section *Ceratochloa,* which are at present confined to South America (Stebbins 1947b,c). Pairing of the chromosomes in appropriate hybrids indicates that if the genomic formula of the former diploids is expressed as LL, and that of the hexaploids as AABBCC, then the octoploids of the *B. carinatus* complex have the formula AABBCCLL (Stebbins and Tobgy 1944, Stebbins 1947c). Since these octoploids must have originated in North America, their hexaploid ancestors existed on this continent at some time in the past. Interpretation of the fossil grasses described by Elias (1942) from late Tertiary deposits of the North American Great Plains suggests that the nearest living relatives of these fossils are species of the genus *Piptochaetium* now confined to South America, and indicates that other species which at present are indigenous only to temperate South America may have existed in North America in the latter part of the Tertiary period, and that the hexaploids of *Bromus,* sect. *Ceratochloa,* were among these species.

Two allopolypoids which suggest the former existence in North America of a species group now entirely Eurasian are *Oryzopsis asperifolia* and *O. racemosa.* These species have the haploid number $n = 23$ and are morphologically intermediate between North American species of the subgenus *Euoryzopsis,* having $n = 11$, and certain Eurasian representatives of the subgenus *Piptatherum,* which have $n = 12$ (Johnson 1945). The evidence summarized by Chaney (1936, 1947) on the woody plants known as fossils from the mid-Tertiary deposits of the western United States suggests that these Asiatic woodland species existed in North America during the middle or early part of the Tertiary period and that the allopolyploid species of *Oryzopsis* arose at that time.

The possibility that the New World allotetraploid species of *Gossypium* originated in North America in the early part of the Tertiary period through hybridization between one group of diploids now found only in South America and another now confined to Asia has been discussed elsewhere (Stebbins 1947b). Hutchinson and Stephens (1947) have summarized the evidence proving as conclusively as such things can be demonstrated that the New World cottons are allopolyploids of the Old World *G. arboreum* or its near relative and an American diploid closely similar to or identical with *G. raimondii,* of Peru. They have assumed that *G. arboreum* was brought from India to the New World by aboriginal man and that the hybridization and allopolyploidy which gave rise to the modern cultivated cottons took place under cultivation in South America. The difficulties of this hypothesis are discussed elsewhere (Stebbins 1947b). The possibility of the early Tertiary origin of the allotetraploid cottons is at least strong enough to warrant a search for similar examples among strictly wild species groups of the semiarid tropics.

A group of European tetraploids which gives evidence of past hybridization between diploid species groups now widely separated from each other is that of the species of *Paeonia* related to *P. officinalis.* Evidence is given elsewhere (Stebbins 1938a, 1939, 1948) that this species and some of its relatives, particularly *P. peregrina,* arose as allopolyploids between *P. cretica, P. daurica,* and other southeastern European diploids on the one hand, and primarily Asiatic diploids related to *P. anomala* on the other. That these Asiatic and European species could have existed together in northern Europe during the latter part of the Tertiary period is suggested by the composition of the fossil flora of this age described by Reid and Reid (1915) from the lower Rhine basin. It contains a mixture of species still found in parts of Europe and of others now strictly Asiatic. From these examples one can see that the evidence from allopolyploidy for past distribution patterns of herbaceous plant species agrees well with that provided by fossil floras, whenever the latter are available.

FACTORS DETERMINING THE ORIGIN AND SPREAD OF POLYPLOIDY

From the evidence presented in the preceding sections certain factors can be summarized which favor the development of polyploidy in plant groups. These are of two types, internal and external. The internal factors consist chiefly of the manner of growth and the reproductive potentialities of the diploid species, as well as the genetic relationships of the existing diploid species to each other. Müntzing (1936) and the writer (1938b) have brought out the fact that polyploidy is more frequent in perennial herbs than it is in annuals. Fagerlind (1944a) and Gustafsson (1947c, 1948) have produced further data to support this conclusion, and Gustafsson has pointed out, furthermore, that among perennials those species with effective means of vegetative reproduction, such as stolons, rhizomes, bulbs, and the "winter buds" of *Potamogeton* and other aquatic plants, are more likely to be polyploid than are species without such structures. Both Müntzing and Gustafsson interpret this evidence to mean that the origin of the perennial habit, as well as of the accessory methods of vegetative reproduction, is a direct consequence of the polyploidy. But against this interpretation are the following facts. In the first place, a very large number of autotetraploids has now been produced from annual diploids belonging to a wide variety of plant genera. While some of these autotetraploids, like those of *Zea*

(Randolph 1935), have a subperennial habit, the majority of them have remained strictly annual. And there is no evidence at all that the induction of polyploidy has caused the appearance of rhizomes or any similar methods of vegetative reproduction where they did not exist in the diploid forms. On the other hand, in many of the genera showing the characteristic correlation between polyploidy and efficient vegetative reproduction, such as *Biscutella, Fragaria, Rubus, Antennaria, Arnica, Artemisia,* and *Hieracium,* diploid species exist in which vegetative reproduction is just as well developed as it is in the tetraploids (Fagerlind 1944a). In many instances vegetative reproduction and the perennial habit are undoubtedly intensified by the onset of polyploidy, but they are rarely if ever originated by this condition.

The undoubted correlation between frequency of polyploidy and the development of perenniality and vegetative reproduction is better explained by assuming that when such characteristics are already present in diploid species, polyploidy can arise and become established more easily than is possible in annuals or in perennials without effective means of vegetative reproduction. Such an assumption is fully warranted by the fact that in most instances the origin of polyploidy involves a period of partial sterility which may last for several generations before the polyploids have become stabilized (Fagerlind 1944a). In autopolyploids this sterility "bottleneck" is at the polyploid level. Most of the artificially induced autopolyploids are partly sterile, as was stated in Chapter VIII, and highly fertile types can be obtained from them most easily through intercrossing between autopolyploids of the same species derived from different diploid genotypes, and selection for high fertility from this heterozygous material. Fertile autopolyploids might occasionally be expected from diploid interracial hybrids, but the latter are, of course, relatively uncommon. Potential allopolyploids must face the "bottleneck" of partial sterility at the diploid level, and in the case of segmental allopolyploids this recurs in the early stages of polyploidy.

The fact that perennials with efficient means of vegetative reproduction are better equipped than annuals to pass through such "bottlenecks" is obvious and needs no further elaboration. A further significant fact is that plants with efficient means of vegetative reproduction are better equipped to take advantage of the greater range of tolerance of extreme environmental conditions which is known to exist in many interracial and interspecific hybrids (Clausen, Keck, and Hiesey 1947). Such hybrids, if they are capable of spreading by means of stolons, rhizomes, tubers, or similar devices, may drive out their parents through direct competition, even though their ability to form viable seed is vastly inferior. Both auto- and allopolyploidy may therefore be expected most frequently in perennials with efficient means of vegetative reproduction.

Gustafsson (1948b) has also postulated that cross-fertilization, or allogamy, promotes the origin of polyploidy, while autogamy restricts it. It is certainly true that many allogamous groups have high percentages of polyploidy, while, conversely, there are many autogamous groups which lack it. It is possible, however, that this apparent correlation between allogamy and polyploidy is indirect and spurious. In Chapter V a correlation was demonstrated between allogamy and the presence of efficient means of vegetative reproduction, while a large proportion of annuals was shown to be autogamous. In order, therefore, to demonstrate a true correlation between allogamy and a high frequency of polyploidy, data should be obtained from species with the same growth habit. There is some indication that, if such data are obtained, the expected correlation will not be found. In species of Gramineae with the caespitose perennial habit, for instance, high polyploidy has been developed in many species groups of such genera as *Bromus, Danthonia,* and *Stipa,* which are largely autogamous through the presence of cleistogamous flowers. Also, among the annual Compositae, tribe Madinae, polyploidy is most highly developed among the autogamous species of the genus *Madia,* but is rare or lacking among the allogamous species of *Layia* and *Hemizonia.*

Clausen, Keck, and Hiesey (1945a) have suggested that certain types of interrelationships between the species of a group favor the establishment and spread of allopolyploidy. According to them, a large number of polyploids arising from interspecific hybrids with intermediate degrees of chromosome homology, that is, those characterized as segmental allopolyploids by the present writer, will be unsuccessful because of partial sterility or genotypic instability. That this is not entirely true is evident from the successful artificial production of such polyploids in various genera (see page 323), as well as the various natural examples cited in Chapter VIII. Nevertheless, the proportion of segmental allopolyploids which are unsuccessful is definitely higher than that of allopolyploids or amphiploids derived from widely different species. Hence, the development of allopolyploidy in a genus is greatly favored by the presence in it of many species which are sufficiently closely related to each other so that they can produce vigorous F_1 hybrids, but strongly enough differentiated so that their chromosomes are nearly or entirely incapable of pairing with each other. Conversely, the presence of many species having chromosomes partially homologous with each other restricts the ability of the group to form extensive polyploid series. Evidence from artificially produced examples indicates that autopolyploidy is rarely, if ever, successful at higher levels than tetraploidy, so that in general any one species can produce by autopolyploidy only a single other genetically distinct type.

As a rule, therefore, extensive polyploid complexes are produced only after some of the speciation processes in a genus on the diploid level have reached a certain stage of maturity, so that at least some of the species have diverged widely from the others with respect to their chromosomes. Since, as will be pointed out in Chapter XIV, rates of evolution vary widely from one group to another, this maturity may be reached in a short period of chronological time or may never arrive, depending on the group concerned. The genus *Aquilegia* is cited by Clausen, Keck, and Hiesey as one in which species differentiation has not reached the point favorable to allopolyploidy, although it belongs to one of the most primitive families of angiosperms, the Ranunculaceae, and has a distribution pattern which suggests at least a mid-Tertiary age. Whether the absence of polyploidy in *Aquilegia* can be explained on this basis is doubtful, since in genera like *Zauschneria* and *Vaccinium,* of which the species are just as poorly differentiated cytologically on the diploid level, both auto- and allopolyploidy are well developed. A better example might be the Old World species of *Crepis,* in which natural polyploids are rare, and the ones produced artificially from diploid species (*C. foetida-rubra, C. capillaris-tectorum*) have been unsuccessful. The only extensive group of polyploids in *Crepis* has overcome the sterility handicap by means of apomixis, as mentioned in Chapter X. The genus *Madia* may be cited as one which is perhaps relatively young, since it consists mostly of highly specialized annuals adapted to relatively new habitats in coastal California, but which has nevertheless reached the stage of maturity in species differentiation which makes allopolyploidy successful.

The chief external factor favoring the establishment of polyploidy is the availability of new ecological niches. Unless these are at hand, the new polyploids will be forced to face competition with the already established diploids, and in such struggles they are almost certain to be at a disadvantage, particularly during the first few generations of their existence. Polyploidy, therefore, may be looked upon as a process which is most effective as a means of enabling species groups which have reached a certain stage of depletion of their biotypes, and of sharp divergence of specific entities, to adapt themselves to new environmental conditions which arise relatively suddenly. It is much less important in stable environments and in diploid species which are still widespread and rich in ecotypic differentiation.

These facts may afford the best explanation of the rarity of polyploidy among woody plants of the temperate zone (Stebbins 1938b). These plants have all of the features which various authors have considered favorable to polyploidy: long life, frequent vegetative vigor, and small chromosomes. Furthermore, the basic haploid numbers of their genera are frequently very high, such as $x = 21$ in *Platanus*, $x = 23$ in *Fraxinus* and other Oleaceae, and $x = 41$ in *Tilia*. These numbers suggest that polyploidy is possible in such genera and may have been one of the important evolutionary processes active while they were being differentiated from related genera, as discussed in the following section. The absence of well-developed polyploid series in them may indicate merely that polyploids have been at a selective disadvantage during their recent evolutionary history. Such a hypothesis is favored by the fact that many of the woody genera which are most abundant in the regions covered by the Pleistocene glaciation, such as *Salix* and *Betula*, have strongly developed polyploid series.

POLYPLOIDY AND THE ORIGIN OF HIGHER CATEGORIES

Polyploidy is now widely recognized as one of the principal methods for the formation of new species among the higher plants. The species originating by this process, however, are for the most part very similar to their diploid ancestors in external morphology and in ecological preferences, or else they contain recombinations of the characteristics found in these ancestors. The long-continued evolution needed to differentiate genera, families, orders, and phyla appears to have taken place chiefly on the diploid level, or at least on the homoploid level in groups which had a polyploid origin. The evidence in favor of this assumption, which was first clearly expressed by Levitzky and Kuzmina (1927) in their study of the genus *Festuca,* by Avdulov (1931) in his cytological survey of the family Gramineae, and by Manton (1932) in her similar study of the Cruciferae, is that in most families polyploidy occurs sporadically in various genera, with the polyploids showing few if any characteristics not also found in diploid species, while the greatest scope of morphological and ecological variation, including types which form connecting links between genera, is most often at the diploid level.

Nevertheless, there is some evidence that many genera and even subfamilies or families of seed plants have had a polyploid origin. This evidence must of necessity be indirect, since these higher categories are all so old that the ancestral species which gave rise to them are almost certainly extinct, so that the polyploidy cannot be repeated experimentally, as it can in the case of many modern polyploid species. The three types of evidence which have been employed are, first, evidence from external morphology and chromosome numbers; second, evidence from "secondary pairing"; and, third, evidence from the number of satellites and nucleoli in the somatic complement.

The last two lines of evidence have been subject to frequent criticism. Secondary pairing was first clearly described and illustrated by Ishikawa (1911) and Lawrence (1931) in *Dahlia variabilis,* which is unquestionably a high polyploid with $2n = 64$ chromosomes. But its use for determining the polyploid origin of higher categories dates from the study of Darlington and Moffett (1930, Moffett 1931) on the origin of the tribe Pomoideae of the family Rosaceae. In both of these groups, the small chromosomes form only bivalents at meiosis, but these bivalents are arranged in groups of two or three, rather than being equally spaced on the metaphase plate. Darlington and Moffett found that in the diploid species of Pomoideae studied by them the most frequent pattern of arrangement of the 17 bivalents was in three groups of 3 bivalents and four groups of 2, and they interpreted this to mean that the tribe Pomoideae originated from diploid species with $x = 7$ by doubling of the entire set, plus the addition of a third partial set consisting of 3 of the original 7 bivalents ($7 + 7 + 3 = 17$). The presence of $x = 7$ as the basic haploid number of many genera in the family, particularly within the tribe Rosoideae, was used as supporting evidence for this hypothesis.

The concept of secondary association and interpretations based on it have been frequently criticized. Sax (1931b, 1932) found that the pattern of secondary association described by Darlington in *Pyrus* could not be recognized in other genera of the tribe, and he suggested that the Pomoideae are hypertetraploids derived from primitive Prunoidae with $x = 8$ ($8 + 8 + 1 = 17$). Clausen (1931b), based on his studies of aneuploidy and polyploidy in *Viola*, sect. *Melanium,* expressed the need for caution in interpreting the origin of basic chromosome numbers, and he suggested that secondary association might arise as a result of segmental interchange, which could cause the occurrence of duplicated segments on otherwise nonhomologous chromosomes. Heilborn (1936) believed that secondary association is an artefact, based on a tendency for chromosomes of similar size to attract each other as they congregate on the metaphase plate. His conclusions, though they might have some validity in the species of *Carex* studied by him, which contain chromosomes differing greatly from each other in size, would be inapplicable to *Dahlia, Pyrus,* and other genera in which such differences do not exist. Catcheside (1937), after a careful study of secondary association in *Brassica oleracea,* including measurements and statistical analysis of the distance between the bivalents, concluded that the phenomenon is a real one, and suggested that *B. oleracea* ($n = 9$) is a hyperploid derived from an ancestor with the haploid number $x = 6$. Sakai (1935) and Nandi (1936) both studied secondary association in *Oryza sativa* ($n = 12$), and they concluded that this species is a hypertetraploid derived from ancestral species with $x = 5$.

One of the weaknesses of the interpretations of phylogeny based on secondary association is that most of them are at variance with conclusions reached on the basis of comparative morphological and cytological studies of the groups concerned. As between the opinion of Darlington, that the Pomoideae are derived from the Rosoideae, and that of Sax, that they are more nearly related to the Prunoideae, the latter agrees much more nearly with the morphological evidence. The Rosoideae contain a number of characteristics and tendencies not found at all or relatively uncommon in the Pomoideae: tendency toward the herbaceous or scandent habit; tendency toward yellow flowers; and carpels numerous and containing only one ovule. On the other hand, the Prunoideae more nearly resemble the Pomoideae in habit, leaf shape, inflorescence, and character of sepals and petals than does any other tribe of the family Rosaceae, and these two tribes possess in common the substance amygdalin, which is not found in the Rosoideae or in any other tribe of the family except for the two mentioned above. There are, however, certain morphological and anatomical characteristics in which most genera of the Pomoideae resemble the Spiraeoideae more than they resemble the Prunoideae, such as the presence of five carpels with several or numerous ovules (Juel 1918). On morphological grounds, therefore, the most probable inference as to the origin of the Pomoideae is that they arose as amphidiploids between primitive or ancestral members of the tribes Spiraeoideae and Prunoideae. Since the basic number most commonly found in the Spiraeoideae is $x = 9$ (Sax 1931b, 1936), while that in the Prunoideae is $x = 8$ (Darlington 1928, Sax 1931b), the basic haploid number of the Pomoideae, $x = 17$, is directly explained by this hypothesis.

The conclusion which Catcheside (1937) reached on the basis of secondary association in the genus *Brassica,* namely, that the original basic number for the genus is $x = 6$, does not agree with the conclusions reached by Manton (1932) on the basis of her comprehensive morphological and cytological survey of the family Cruciferae as a whole. She found that $x = 6$ occurs only occasionally in this family, and not at all in *Brassica* or its near relatives. Furthermore, in every instance of the occurrence of this basic number, it is found in a species representing the specialized end point of an evolutionary series and appears to have been evolved by reduction from $x = 7$ or $x = 8$, as will be discussed in Chapter XII. Manton concluded that the most probable original basic number for *Brassica,* as for many other genera of Cruciferae, is $x = 8$.

Similarly, the conclusion of Sakai and Nandi, that the original basic chromosome number of *Oryza* is $x = 5$, is at variance with the evidence produced by Avdulov (1931) in his thorough and comprehensive survey of the family Gramineae. Avdulov showed that the basic number $x = 12$ is found not only in *Oryza* and most other genera of the tribe Oryzeae but also is by far the commonest one in the primitive genera of grasses placed by Avdulov in his series Phragmitiformes, as well as in the tribe Bambuseae. The genus *Ehrharta,* which in many respects appears more like an ancestral prototype of the Oryzeae than does any other living genus, also has $x = 12$. On the other hand, species with five pairs of chromosomes are very uncommon among the Gramineae, and this number is not found in any primitive genus, nor in any genus related to *Oryza*.

So far as this writer is aware, no critical study exists in which evidence from secondary association has been correlated with that from comparative morphology and cytology of a number of different species and genera and agreement has been found between the different lines of evidence. At present, therefore, secondary association can be considered an actual phenomenon and one which in many instances suggests the polyploid nature of a species or genus, but one which may be considerably modified by segmental interchange, duplication of chromosome segments, and other phenomena not at all related to polyploidy. It is therefore not a reliable index of the exact basic haploid number possessed by the original ancestors of a group.

Equally or more unreliable is the evidence from the number of satellites and nucleoli. The use of this evidence has been advocated most strongly by Gates, who has published a comprehensive review of the subject (Gates 1942). In some instances the possession of four satellites and nucleoli is evidence of tetraploidy, but many polyploids never have more than two of these structures. Furthermore, there are examples, such as *Leontodon incanus* and

L. asperrimus, which have only four pairs of chromosomes, two of which bear satellites, and consequently often form four nucleoli at the mitotic telophase (Bergman 1935b).

For these reasons, the only safe inferences which can be made about whether a particular basic number is of ancient polyploid origin are those based on thorough and well-correlated studies of the comparative morphology and cytology of a large number of intimately related species and genera. Even when all possible evidence has been obtained, questions such as these are open to a certain amount of doubt, and the final conclusions must be regarded as indicating which of several possible situations has the greatest degree of probability.

There is, nevertheless, evidence that polyploidy played a role in the origin of genera among the earliest of the angiosperms. Whitaker (1933) found the haploid number $x = 19$ not only in *Magnolia* and *Liriodendron,* but also in *Trochodendron, Tetracentron,* and *Cercidiphyllum,* woody genera of the primitive order Ranales which bear only relatively distant relationships to each other (Nast and Bailey 1945). In other woody genera of this order, namely, *Illicium, Schizandra, Kadsura,* and *Michelia,* the basic number is $x = 14$, and $x = 12$ is characteristic of one family of woody Ranales, the Lauraceae (Darlington and Janaki-Ammal 1945), while $x = 7$ is found in the genus *Anona* of another such family, the Anonaceae (Asana and Adatia 1945). Although the Magnoliaceae and their relatives obviously did not arise by allopolyploidy between any forms resembling the living Lauraceae and Anonaceae, nevertheless there exists a strong possibility that the commonest basic numbers in the primitive woody Ranales were $x = 12$ or 6, and $x = 7$. If this were true, the living Magnoliaceae and their relatives would have to be looked upon as very ancient allopolyploids which have long outlived the extinction of their diploid ancestors. The Anonaceae and the Lauraceae could be regarded as families in whose ancestors polyploidy did not become so firmly established, and which underwent much more progressive evolution on the diploid level.

Anderson (1934) has suggested that the woody Ranales and through them the angiosperms as a whole originated as allopolyploids from 12-paired ancestors related to the conifers or Ginkgoales and 7-paired ancestors related to the Gnetales. On grounds of external morphology this suggestion is so highly improbable as to be hardly worthy of serious consideration.

Another primitive group in which polyploidy may have played an important role in the differentiation of genera is that of the subfamilies Mimosoideae and Caesalpinoideae of the Leguminosae. In both tribes, the commonest haploid numbers for several genera are $x = 12$, $x = 13$, and $x = 14$ (Senn 1938a, Darlington and Janaki-Ammal 1945). One of the largest genera of the tribe, *Cassia,* also has many, perhaps a majority of, species with these as the haploid numbers, but other species exist with $x = 8$ and $x = 7$ (Jacob 1940), suggesting that the last two are probably the basic numbers for the genus. In another genus of the Caesalpinoideae, *Cercis,* the haploid numbers found are $x = 7$ and $x = 6$. All this evidence suggests that the latter are the original basic numbers of the ancestral genera of Caesalpinoideae and Mimosoideae, and that the basic numbers found in most of the modern genera are of polyploid derivation.

In a genus of the Leguminosae, subfamily Papilionoidae, namely, *Erythrina,* the basic number almost certainly indicates an ancient polyploidy. Atchison (1947) has found either $2n = 42$ or $2n = 84$ in all of 33 species of this genus investigated. Since the commonest basic numbers found in other genera of the tribe Phaseoleae, to which *Erythrina* belongs, are $x = 11$ and $x = 10$, the basic number, $x = 21$, of this genus is probably derived by allopolyploidy. Since *Erythrina* is distributed in the tropics of both the Old and the New World, its polyploidy may be very ancient (see Chapter XIV).

Some well-known genera and subfamilies of temperate regions of which the basic number is probably of ancient polyploid derivation are *Platanus* ($x = 21$, Sax 1933), *Aesculus* ($x = 20$, Upcott 1936), *Tilia* ($x = 41$, Dermen 1932), and the Oleaceae, subfamily Oleoideae ($x = 23$, Taylor 1945). The Salicaceae probably represent such a family, since their basic haploid number is $x = 19$ (Darlington and Janaki-Ammal 1945). Many other similar examples of ancient woody genera could be cited. In an earlier publication (Stebbins 1938b), the writer pointed out that basic chromosome numbers are on the average higher in woody genera than they are in herbaceous ones, although polyploid series within genera are less frequent. At that time the suggestion was offered that these relatively high basic numbers might have been the original ones for the angiosperms, and that the lower numbers characteristic of herbaceous genera were derived through stepwise reduction in the basic number, as is known to occur in many genera, such as *Crepis,* and will be discussed in Chapter XII. The more recent evidence, however, suggests that $x = 6$, 7, and 8 are more likely to have been the original basic numbers for the flowering plants, and that $x = 12$, 13, 14, and all higher basic numbers are for the most part of ancient polyploid derivation (Stebbins 1947a).

This hypothesis has important implications for the phylogeny and the evolutionary history of the flowering plants. Evidence presented earlier in this chapter indicates that polyploids are most successful as invaders of newly opened habitats. Our knowledge of the fossil floras of the Tertiary period indicates that ever since the beginning of this period the genera and families of woody angiosperms have existed in approximately their present form (see Chapter XIV). The polyploidy which accompanied their origin must therefore date from the Cretaceous period or earlier. One of the most likely hypotheses, therefore, is that the new habitats which offered themselves to the newly formed polyploids of the woody angiosperms were those made available by the decline and extinction of large numbers of gymnosperms during the early and middle part of the Cretaceous period. Once these habitats were filled, the opportunities were over for rapid bursts of evolution and the "cataclysmic" origin of species and genera among woody plants adapted to mesophytic conditions, so that after this time only isolated examples of this type of evolution occurred among such plants.

If this hypothesis is correct, students of angiosperm phylogeny should look for traces of ancient allopolyploidy resulting from hybridization between species which were the ancestral prototypes of many of our modern families. The reticulate pattern of evolution which is the typical one when allopolyploidy has occurred should characterize the interrelationships between families and orders of flowering plants. If this is actually the case, an explanation might be at hand for the fact that botanists have had much more difficulty in deciding on the true interrelationships between families and orders of flowering plants than have students of insects or vertebrates when dealing with similar groupings in their organisms.

It should be emphasized, however, that this view of polyploidy as having played a major role in the origin of families and orders of flowering plants does not imply that progressive evolution is furthered by this process. In every example in which its immediate effects have been analyzable, polyploidy has appeared as a complicating force, producing innumerable variations on old themes, but not originating any major new departures. In Chapter XIII, the fact will be emphasized that the number of families of angiosperms is much larger than that of major evolutionary trends which can be traced in the group. This multiplication of families, containing various recombinations of a relatively small number of primitive and specialized morphological characteristics, may have been greatly aided by polyploidy. But neither the basic chromosome numbers of angiosperm genera nor any other type of evidence supports the concept that such progressive trends as sympetaly, epigyny, or any other of the trends in floral specialization which form the basis for the natural system of families and orders of flowering plants has been caused or promoted by polyploidy.

POLYPLOIDY IN PLANTS AND ANIMALS

One of the most interesting and puzzling features of polyploidy is the small role which it has played in animals as compared to its dominant role in the higher plants. Vandel (1938), after reviewing the chromosome numbers in the animal kingdom, reached the conclusion that polyploid species of animals are known only in hermaphroditic groups, such as pulmonate mollusks, and in parthenogenetic forms, such as rotifers, the genus *Artemia* among crustaceans, and the moth, *Solenobia*. Since the review of Vandel, three outstanding examples of actual or probable polyploidy among natural animal species have been described, namely, in the ciliate Protozoa (*Paramecium*, Chen 1940a,b), in the order Curculionidae, or weevils, among the insects (Suomalainen 1940a,b, 1945, 1947), and in the family Salmonidae among fishes (Svärdson 1945). The first two of these three examples agree with previously known ones in that the ciliate Protozoa are essentially hermaphroditic and all the polyploid Curculionidae are parthenogenetic. But the family Salmonidae appears to represent the first example of a sexual polyploid complex, with all of its attendant difficulties of classification of species and genera, which has been detected in the higher animals.

The classic theory to explain the scarcity of polyploidy in bisexual animals, first offered by Muller (1925), is that polyploidy upsets the balance of sex determination to such an extent that regular segregation of sex factors is impossible in polyploids. This hypothesis was based on the fact that triploid individuals of *Drosophila melanogaster* are either females or intersexes, but never normal functional males. Early evidence from the plant kingdom tended to support this hypothesis. Thus, Wettstein (1927) found that polyploids produced artificially from dioecious species of mosses are monoecious or hermaphroditic, while natural polyploid species in *Bryum* and other genera are the same, a fact recently confirmed for the genus *Mnium* by Heitz (1942). Similarly, Hagerup (1927) found that the diploid *Empetrum nigrum* of Europe is dioecious, while the tetraploid *E. hermaphroditicum* of Greenland is hermaphroditic, with perfect flowers.

But the fact was realized somewhat later that typical polyploid series exist in genera of dioecious flowering plants, and that the polyploid members of these series in most cases show as regular a segregation of male and female types as do the diploids. This was pointed out for *Salix* by Müntzing (1930b), and it is also true of *Rumex acetosella* (Löve 1944), of certain polyploid but normally sexual species of *Antennaria* (Bergman 1935c, Stebbins unpublished), and of the genera *Buchloe* and *Distichlis* among the grasses (Avdulov 1931, Stebbins and Love 1941).

Direct evidence that dioecism and polyploidy are not incompatible with each other was produced almost simultaneously by Warmke and Blakeslee (1940) and by Westergaard (1940) from their studies of autotetraploid plants of *Melandrium dioicum* (*Lychnis alba* or *L. dioica*). This species has long been known to possess a typical X — Y sex-determination mechanism, with the male as the heterozygous sex. The autotetraploids of the constitution XXXX were, of course, females, but all other types (XXXY, XXYY, and XYYY) were entirely or essentially males, quite contrary to their counterparts in *Drosophila*. Because of this fact, the establishment of lines containing only XXXX and

XXXY types was easily accomplished, and these lines were as typically dioecious as their diploid ancestors.

These results show that the existence of functional bisexual polyploids depends on the presence in the group of a sex-determining mechanism in which the chromosome peculiar to the heterozygous sex (the Y or Z chromosome) possesses strongly dominant sex-determining factors. If these factors are only weakly dominant, or if, as postulated by Bridges for *Drosophila,* the Y or Z chromosome is neutral and the nature of the sex depends on the balance between the autosomes and the X or W chromosomes, then the sex-determining mechanism cannot work on the polyploid level. This shows that Muller's principle holds for some and perhaps the majority of bisexual animals, but could hardly be expected to be responsible for the scarcity of polyploidy in the animal kingdom as a whole.

A further reason for the scarcity of polyploidy in animals is provided by the studies of Fankhauser (1945a,b) on the morphological characteristics and the development of polyploid individuals of certain species of amphibians. In such species as *Triturus viridescens* and *Eurycea bisliniata,* eggs obtained from normal fertilization and raised artificially without special treatment regularly produce a small percentage of triploid or tetraploid individuals. Adult polyploid animals, although they can be reared in captivity, are unknown under natural conditions, and no polyploid species are known of *Triturus* or any other genus of amphibians. The artificially raised adult salamanders contain both typical males and sterile females with underdeveloped ovaries and are always weaker than normal individuals.

The causes for the weakness of artificially raised polyploid salamanders lie partly in the disturbances produced by polyploidy in the size and number of cells in various vital tissues and organs. Most of them, including the brain and the spinal cord, have fewer cells than do the corresponding organs of diploids. It is obvious that this condition would affect metabolic processes like digestion and nervous reaction. Expressed in general terms, the statement can be made that in addition to the incompatibility between polyploidy and certain types of sex-determining mechanisms, a second important barrier to the success of polyploidy in animals, as was first suggested by Wettstein (1927), is the disharmony in the complex processes of animal development produced by this sudden alteration of the genotype. Polyploidy is successful in plants because their developmental processes are so much simpler (see Chapter V, p. 182).

The scarcity of polyploidy in animals has another important implication for students of this phenomenon in plants. Long-continued trends of specialization and the production through adaptive radiation of families and orders adapted to particular modes of life have been much more characteristic of evolution in animals than in plants. The fact itself is additional evidence in favor of the conclusion expressed in the previous section of this chapter, namely, that polyploidy, although it multiplies greatly the number of species and sometimes of genera present on the earth, retards rather than promotes progressive evolution.

THE PRACTICAL SIGNIFICANCE OF POLYPLOIDY

The practical value to plant breeders of an intimate knowledge of the natural processes of evolution has been previously emphasized earlier in this book, particularly in connection with ecotype differentiation and interspecific hybridization. Polyploidy is not only one of the best known of evolutionary processes; in addition, it is the most rapid method known of producing radically different but nevertheless vigorous and well-adapted genotypes. For these reasons, it is the large-scale evolutionary process most suitable for development as a technique in plant breeding.

Nevertheless, attempts to use polyploidy as a plant breeding technique have up to the present not proved entirely successful. This is partly due to the fact that methods of producing polyploids in large quantities have been known for only twelve years, six of which were disturbed by a great war and the inevitable retardation of agricultural research which it caused. But in addition, evidence obtained from the large-scale production of artificial polyploids has shown that many conceptions previously held about polyploidy are only partly true. Before artificial polyploidy can be used by plant breeders with the greatest possible efficiency, both its particular advantages and its limitations must be thoroughly understood. Finally, the nature of artificial polyploids has demonstrated the fact that, although polyploidy is not necessarily a slower technique than the older method of intervarietal hybridization and selection, it is also no more rapid. Newly produced artificial polyploids are rarely if ever of economic value; they must be adapted to the needs of the breeder by means of selection and testing just as do newly produced hybrids. In view of this fact, and of the fact that the production and introduction to agriculture of a new variety of such a crop as wheat requires about fifteen years, it is not surprising that as yet no new variety of crop plant produced through artificial polyploidy has found its way into large-scale agriculture.

The use of polyploidy as a tool in plant breeding was not possible until a satisfactory method was obtained for producing polyploids on a large scale. This was achieved by the colchicine technique, as mentioned in Chapter VIII (p. 299). Other chemical substances have also been found to double the chromosome number of plants, but, as shown in the careful experimental comparison by Levan and Ostergren (1943), none of them is as efficient as colchicine.

The numerous polyploids produced with the aid of colchicine include both autopolyploids and allopolyploids of various types. As mentioned in the preceding chapter, the effects of induced autopolyploidy are various, but in some instances the polyploids, when first produced, have possessed economically desirable qualities. In ornamental flowering plants, these qualities have been a greater durability and a heavier texture (Emsweller and Ruttle 1941), while in crop plants they have consisted largely of higher concentrations of valuable substances such as vitamins and various types of proteins, of increased size of seeds, and in some forage crops, such as red clover (*Trifolium pratense),* of a greater yield of dry matter (Randolph 1941, Levan 1942b, 1945, Noggle 1946). These have, however, been balanced in most instances by various undesirable qualities, chiefly slower growth and lower fertility. The reduced seed fertility is the chief drawback of autopolyploids of cereal crops and others in which the fruits are chiefly used. The only such crop in which autopolyploidy appears to have effected an immediate improvement is buckwheat (*Fagopyrum esculentum,* Sacharov, Frolova, and Mansurova 1944).

As mentioned in the preceding chapter, the reduced fertility of autopolyploids is nearly always greater when their diploid progenitors are homozygous. For this reason, successful autopolyploids can never be expected from pure lines of self-pollinated diploids. The best possibility for obtaining desirable autopolyploid varieties of such plants lies in the production of polyploids from many different diploid lines with various desirable characteristics, with subsequent hybridization and selection from this initially diverse material. In fact, Levan (1945) has pointed out that since polyploidy radically affects the balance of multiple factors acting on a single character, this proper balance, even in the case of allogamous plants, must be restored by hybridization and selection from a wide range of genetic material at the tetraploid level.

As Wettstein (1927), Levan (1942b), and many other workers have found, the chromosome numbers of a species can under no conditions be doubled indefinitely without deleterious results. Although some autotetraploids are equal to or superior to their diploid progenitors in vigor or other desirable qualities, autooctoploids are nearly always so abnormal as to be sublethal, although this effect is much more pronounced in some plants than it is in others. For each species, there is an optimum chromosome number, which may be diploid, tetraploid, or hexaploid, but is rarely higher. As yet no way is known of predicting from the characteristics of a diploid what its optimum degree of ploidy may be. The writer's experience in the genus *Bromus* (Stebbins 1949b), as well as the series of chromosome numbers known in such genera as *Buddleia*, which reaches $2n = 300$ (Moore 1947), and *Kalanchoe* (Baldwin 1938), which reaches about 500, indicate that these limits are much more flexible in allopolyploids.

If we judge by the frequency with which the different types of polyploidy appear among crop plants, we must consider allopolyploidy to be more important to agriculture than autopolyploidy. Six of the major crop plants may be autopolyploid species, but available evidence for most of these indicates that they are just as likely segmental allopolyploids, and some of them may be typical allopolyploids. These are the potato (*Solanum tuberosum*, Lamm 1945), coffee (*Coffea arabica*, Krug and Mendes 1941), banana (*Musa* vars., Cheeseman and Larter 1935, Wilson 1946, Dodds and Simmonds 1948), alfalfa (*Medicago sativa*, Ledingham 1940), peanut (*Arachis hypogaea*, Husted 1936), and sweet potato (*Ipomoea batatas*, King and Bamford 1937). On the other hand, there are five leading crops and several minor ones in whose ancestry allopolyploidy has certainly played a major role. The origins of tobacco and cotton have been discussed earlier in this book. The other leading crop plants which are allopolyploids or partly of allopolyploid derivation are wheat, oats, and sugar cane (Bremer 1928, Grassl 1946), while less important allopolyploid crop plants are plums, apples, pears, loganberries and other commercial berries of the genus *Rubus,* and strawberries.

The commercial wheats form one of the most striking examples of allopolyploidy in the plant kingdom, and they show what may be expected from this process in the future development of crop plants. The tetraploid wheats of the emmer or *durum-dicoccum* series were apparently derived from the wild tetraploid species *Triticum dicoccoides* of Syria (Aaronsohn 1910). The formula given these wheats by Kihara (1924) is AABB, suggesting an allopolyploid origin, a theory supported by the fact that they regularly form 14 bivalents at meiosis. The A and B genomes are, however, closely related genetically, and under some circumstances chromosomes belonging to these two genomes may pair with each other. The ancestral *T. dicoccoides,* therefore, may have originally been a segmental allopolyploid which became diploidized in response to natural selection for high fertility. The hexaploid wheats, which comprise the typical bread wheats (*T. aestivum* or "*T. vulgare*"), contain the A and B genomes, although in some varieties these are considerably modified, as is evident from irregularities in pairing in hybrids between certain hexaploid wheats and the tetraploids (Love 1941). In addition they have a third genome, C, the chromosomes of which are nearly or entirely incapable of pairing with those belonging to A and B.

The identification of this C genome is one of the most interesting chapters in the history of polyploidy and reveals the occurrence in prehistoric times of one of the greatest miracles in agriculture. The hypothesis that this genome is derived, not from *Triticum* in the strictest sense, but from the related genus or subgenus *Aegilops* has been suggested by many authors. With the arrival of the colchicine technique, Thompson, Britten, and Harding (1943) were able to produce an allopolyploid from the hybrid between *Triticum dicoccoides* and *Aegilops speltoides.* This artificial allopolyploid for the most part had 21 pairs of chromosomes at meiosis, was reasonably fertile, and its hybrids with forms of *Triticum aestivum* were partly fertile, although they had rather irregular meiosis. Later, McFadden and Sears (1946) produced a somewhat similar allopolyploid, using as parents *T. dicoccoides* and *A. squarrosa.* This proved to be highly fertile, and in external morphology it closely resembled the *Triticum spelta* form of *T. aestivum.* F_1 hybrids between the artificial allopolyploid and *T. spelta* formed 21 bivalents at meiosis and were fully fertile, showing that this newly created allopolyploid, *T. dicoccoides-A. squarrosa,* had essentially the chromosomal constitution of the bread wheats. There is little doubt, therefore, that the typical bread wheats (*Triticum aestivum*) originated from hybrids between the emmer or hard wheats and the weedy grass *Aegilops squarrosa.*

A. squarrosa is a small, useless weed with hard glumes or husks, long divergent awns, or beards, and a fragile, shattering inflorescence, which grows in abundance on the borders of wheat fields in many parts of the Near East, from the Balkans to Armenia, Iran, and Afghanistan. The original hybridization and doubling probably took place in southwestern Asia, between cultivated emmer-type wheat and plants of *A. squarrosa* infesting its fields (McFadden and Sears 1946).

The origin of the *vulgare* type wheats illustrates two points about natural allopolyploidy which are of the greatest importance to plant breeders. In the first place, the diploid parents of agriculturally valuable allopolyploids need not themselves possess outstanding qualities. As in the case of inbred lines of maize, and perhaps to an even greater extent, it is the combining ability that counts. Secondly, the origin of successful natural allopolyploids involves a succession of rare accidents. Therefore, given human purpose, a thorough knowledge of the new qualities desired and of the genetic make-up and interrelationships of the diploid species available, and a not inconsiderable amount of patience and perseverance, plant breeders should be able to create allopolyploids better suited to the needs of mankind than any which nature has produced, and to do this in a vastly shorter period of time.

Any program of breeding for improved new allopolyploid types must include not only a careful study of the interrelationships of the species concerned and of the most suitable methods for hybridization and doubling the chromosome number but, in addition, a full realization of the limitations of this method. As has been emphasized in Chapter VIII, a large proportion of vigorous interspecific hybrids yield unsuccessful allopolyploids, partly because of irregular chromosome pairing at the polyploid level and partly because of the presence of some other type of sterility. As yet, no generalizations can be made by which the success or failure of an allopolyploid can be predicted with any degree of certainty, even after the F_1 diploid hybrid ancestral to it has been obtained, but some guiding suggestions can be made, based on the experience of various workers in this field.

In the first place, the increase in fertility of a doubled hybrid is due principally to one cause, namely, a change in the manner of pairing of the chromosomes due to the presence in the allopolyploid of two completely homologous sets. If the sterility of the diploid hybrid is due to some cause other than the lack of similarity between the chromosomes, the allopolyploid will be no more fertile than its undoubled progenitor. F_1 hybrids with abortive flowers, with anthers that fail to produce normal archesporial tissue and microspore mother cells, or with carpels that do not form normal ovules cannot be expected to yield successful allopolyploids. As mentioned in the preceding chapter, there are various additional types of sterility, which become evident only after the chromosome number of the hybrid has been doubled. In some instances of genic sterility, as in the fertile *Nicotiana sylvestris-tomentosiformis* line produced by Kostoff (1938a, Greenleaf 1941), this type of sterility can be circumvented. But in general, the breeder must expect that a considerable proportion of the allopolyploids he produces will be unsuccessful because of sterility factors which cannot be eliminated.

One type of sterility which can be controlled in at least some instances is that found in the allopolyploid between wheat and rye (*Triticale*, or *Triticum aestivum-Secale cereale*). Strains of this allopolyploid have now been in existence for over fifty years, and although many show desirable qualities of luxuriance, hardiness, and large size of grain, none has as yet become sufficiently fertile and productive to succeed as a commercial crop. Müntzing (1939), after an analysis of four different strains of *Triticale* and of various hybrids between them, concluded that the original cause of this sterility, as well as of the fact that some strains of this allopolyploid segregate weak individuals, lies in the fact that rye is normally a self-incompatible, cross-pollinated plant. Normal rye plants are all or nearly all heterozygous for lethal or semilethal genes, so that when artificially inbred they yield a large majority of weak and partly sterile individuals. The allopolyploid strains which have originated from wheat × rye hybrids by the functioning of unreduced gametes, as is true of most of the strains now available, also contain rye genomes heterozygous for lethal and semilethal genes, which are forcibly inbred because of the dominance of factors for self-fertility present in the wheat genomes. The segregation of weak and sterile types is therefore to be expected. Müntzing (oral communication) has suggested that this difficulty may be overcome by using as the original parent a strain of rye rendered relatively homozygous and vigorous by inbreeding and rigid selection over several generations. The same difficulties and considerations will apply in every example of an allopolyploid which combines the genomes of a self-compatible and a self-incompatible species. Among these may be mentioned the perennial wheats produced from hybrids between *Triticum* and *Agropyron*. It is possible that better success with these allopolyploids may be obtained by first breeding the *Agropyron* parent for constancy, vigor, and fertility.

Since much of the sterility in allopolyploids results from the formation of multivalents and other types of abnormal pairing, reduction in fertility can be expected if the diploid hybrid which gives rise to the allopolyploid has any amount of chromosome pairing, although, as stated in Chapter VIII, there is no correlation between the amount of pairing in the F_1 hybrid and the degree of sterility found in its derivative allopolyploid. Furthermore, the sterility originating from this cause may in many instances be overcome by selection in later generations, although this is not always true. At present, the cytogeneticist does not yet have enough knowledge of the causes of sterility to enable him to make more than the tentative suggestions given above for the guidance of plant breeders. More information on the nature of hybrid sterility is badly needed.

Allopolyploids or amphiploids can be used in plant breeding in three different ways. The first is the relatively modest aim of transferring across a barrier of interspecific sterility a valuable character which is carried by a single genetic factor. An outstanding example is the transfer of the necrotic lesion type of resistance to the tobacco mosaic disease from *Nicotiana glutinosa* to *N. tabacum* by means of the amphiploid *N. digluta* (Holmes 1938, Gerstel 1943, 1945a). In this instance, backcrossing of *N. digluta* to *N. tabacum,* followed by selfing, produced plants with the entire chromosomal complement of *N. tabacum* except for the single chromosome bearing the desired factor. The success of this method depends not only on the fertility of the amphiploid but, in addition, on the viability and the fertility of the alien substitution race (Gerstel 1945a), which contains a whole chromosome pair foreign to the species of crop plant in which it resides. It is likely that the method will be successful in economic species which are themselves polyploid, but not usually in diploid crop plants.

The second use of amphiploids is the incorporation into a valuable crop plant of some desirable characteristic which is controlled by many genetic factors. Perhaps the most outstanding example of this is the use of the amphiploids between wheat and various species of *Agropyron* to produce perennial wheat. The extensive literature on this subject cannot be reviewed here; the reader may be referred to the papers by Peto (1936), White (1940), Love and Suneson (1945), and Sule (1946). Amphiploids between *Triticum* and *Agropyron* have been produced independently in Russia, Canada, and the United States. The earliest, those of the Russian worker Tsitsin, have yielded more favorable combinations of perenniality, high productivity, and desirable grain quality than have the Canadian and the American amphiploids. Unfortunately, they have not been subjected to careful cytological studies by modern methods, so that their chromosomal constitution is almost unknown. Sule (1946) has suggested that winter wheats, which already possess winter hardiness of the vegetative parts, are the most suitable wheat parents of such amphiploids, and he has indicated that the best method of improvement is through the intercrossing of different amphiploids, as well as of strains derived from backcrossing to wheat. The degree of stability and constancy of these Russian amphiploid derivatives has not been carefully described in publications, and the degree of regularity of their chromosome behavior is not known. But Love and Suneson (1945) found a high degree of chromosomal irregularity and inconstancy in fertile derivatives from amphiploids involving *Triticum durum*, as well as *T. macha* and *Agropyron trichophorum*. They have not been able to obtain constant, highly fertile, and productive strains which combine perenniality and the desirable grain quality of the bread wheats. The difficulty of obtaining such types lies partly in the complex genetical basis of the perennial condition, but perhaps more important is the irregular chromosome behavior in backcross types from the original amphiploid to wheat. Since none of the initial amphiploids containing the complete chromosome set of *Agropyron* in addition to that of wheat have the required quality of grain, backcrossing is necessary in order to produce useful perennial wheats. Although perhaps not impossible, the goal of producing a constant perennial wheat with productivity and quality equal to that of the best of the present annual varieties has not been reached and will require a large amount of time, patience, and skill on the part of plant breeders.

McFadden and Sears (1947) have suggested that in wheat the transfer of desirable genes from one polyploid species to another can be most efficiently carried out by the use of entire genomes; a procedure which they term "radical wheat breeding." For instance, if the breeder wishes to transfer genes for disease resistance from the tetraploid *Triticum timopheevi* to the hexaploid *T. aestivum*, he should not do this via direct hybridization between the two species, but should first obtain an allohexaploid between *T. timopheevi* and *Aegilops squarrosa*, since the latter species contains the C genome which is present in *Triticum aestivum*, but absent from *T. timopheevi*. This synthetic allohexaploid would hybridize more easily with *T. aestivum* than would *T. timopheevi*, and the F_1 would have a more regular meiosis and would be more fertile. Backcrossing this F_1 to a desirable variety of *T. aestivum* should permit the incorporation into this variety of any genes or chromosome segments from *T. timopheevi* that might be needed. Theoretically, this method should be more efficient than direct hybridization. Nevertheless, since hybridization and allopolyploidy often lead to quite unexpected results, the method should be tried out on a relatively large scale before being recommended without qualification.

The third use of amphiploids is as entirely new crop plants with qualities different from any now known. These are not likely to be useful in such crops as cereal grains, cotton, and tobacco, in which standards of quality have been established and stabilized for a very long time and must be met by any new variety which is to be useful to the processors of these crops. On the other hand, in the newer types of crop plants, particularly in plants used for animal feed and for forage, in which vigor, productiveness, and adaptability to various environmental conditions are the chief desirable qualities, there is room for many new amphiploid types.

Among the newer crops which have risen by amphiploidy in recent years, perhaps the most notable are the various berries related to the blackberry and the raspberry, of the genus *Rubus*. A careful account of these, containing a good review of the literature, is given by Clausen, Keck, and Hiesey (1945a). Two typical examples are the nessberry, an allotetraploid derived from the diploid species *R. trivialis* and *R. strigosus,* and the loganberry. The latter is apparently a typical allopolyploid, although its origin was of an unusual nature. Thomas (1940b) has shown that it must have originated from a hybrid between the Pacific coast dewberry, *R. ursinus,* and the cultivated raspberry, *R. idaeus,* and that a diploid pollen grain of the latter species must have functioned to produce it. Since *R. ursinus* is an autoallopolyploid, with the genomic formula AAAABBBB, and *R. idaeus* may be given the formula II, the F_1 hybrid of the type described above had the formula AABBII and was therefore at once constant and fertile.

The use of allopolyploids as improved types of forage plants is just beginning. Armstrong (1945) has suggested that the *Agropyron-Triticum* allopolyploids will be more useful in Canada as forage plants than as perennial grains. Clausen, Keck, and Hiesey (1945b, and unpublished) have found that amphiploid apomicts produced by hybridizing widely different species of *Poa* have unusual vigor and adaptability to a wide range of ecological conditions. Similar characteristics have been found by the writer (Stebbins 1949b) in various amphiploids in the genus *Bromus* and in the tribe Hordeae. The habitats for which these new types of forage plants are designed are on theoretical grounds ideally suited to colonization by newly formed amphiploids. They are in some respects, particularly the climate and the underlying soil formation, essentially similar to habitats which existed before the arrival of man. But man's activity in clearing trees and brush, in introducing large numbers of grazing animals, and in allowing the natural cover to become greatly depleted, has so altered the original conditions that the plants constituting the original vegetation are unable to grow as vigorously and abundantly as they formerly did, even under the most careful system of management of the land. The restoration of the world's depleted range and pasture lands can be accomplished to a great extent by proper management and by reseeding with old species known to be adapted to the area concerned. But the progressive plant breeder, if he can produce plants particularly adapted to the new conditions which will prevail on such restored and well-managed lands, should be able to increase greatly their productivity, and so make more worth while the effort and expense required for proper management practices. Amphiploids, because they involve the potential achievement in a short time of an entirely new range of adaptability, should be the ideal type of plant on which to base such a program of breeding. Furthermore, controlled experiments, designed to follow the adaptation of these newly created amphiploids to their new habitats, should enable evolutionists to observe selection in action under almost natural conditions. The use of amphiploids for revegetation of depleted grazing lands, therefore, should provide a medium by which a program of agricultural improvement can be combined with fundamental experimental studies on the dynamics of evolution.

CHAPTER X

Apomixis in Relation to Variation and Evolution

THE TERM APOMIXIS is a general one, covering all types of asexual reproduction which tend to replace or to act as substitutes for the sexual method. The more familiar term parthenogenesis, though better known to most biologists, applies to only a part of the apomictic phenomena found in plants. In most animals higher than the Protozoa, the Porifera, and the Coelenterata parthenogenesis is the only possible method of regular asexual reproduction. It appears in various groups of animals, such as rotifers, crustaceans, Hymenoptera, and aphids, but is by no means as widespread as it is in plants. This is in accordance with the principle, developed in Chapter V, that the relatively simple structure and development of plants, as well as their tendency for indeterminate growth, increases the selective value in them of genetic systems which deviate from continuous sexual reproduction and allogamy.

Careful analyses of the apomictic phenomena in a number of plant genera, and more or less casual surveys of apomixis in many more, have shown that numerous methods of apomixis exist in the plant kingdom, and that their genetic nature and evolutionary origin are both multiform and complex. The earlier reviews of the literature on apomixis by Rosenberg (1930) and the writer (Stebbins 1941a) have now been superseded by the vastly more thorough, complete, and discerning review and analysis of Gustafsson (1946, 1947b,c). This chapter, therefore, will be based chiefly on this review, particularly in respect to the factual material included, and for more detailed information on every phase of this subject, as well as for a complete list of literature, the reader is advised to consult Gustafsson's work.

METHODS OF APOMICTIC REPRODUCTION

On the part of botanists not specializing in this field one of the chief barriers to an understanding of the apomictic phenomena has been the highly complex terminology it has developed, as well as the differences of opinion which have arisen among specialists as to the correct usage of terms. There are two reasons for this. In the first place, the phenomena themselves are complex and in many ways quite different from any which are regularly encountered in organisms reproducing by normal sexual methods. Secondly, the early research on apomixis was carried out before the morphological nature and the genetic significance of the various parts of the normal sexual cycle were fully understood by most botanists, so that many terms were used uncritically and were defined in ways that included erroneous interpretations of the phenomena they purported to describe. We owe chiefly to Winkler (1920, 1934) the first classifications and terminology of the apomictic phenomena which show correctly their relationship to corresponding sexual processes. In particular, Winkler first stressed the all-important point that since the normal sexual cycle of the higher plants includes two entirely different and equally essential processes, meiosis and fertilization, which are separated from each other by the entire period of growth and development of the gametophyte, these two processes can be expected to be influenced by entirely different environmental and genetic factors. Every harmonious apomictic cycle must therefore provide either a single substitute for both of these processes or a separate substitute for each of them, with coordination of these two substitutes. Thus, parthenogenesis through development of the egg cell without fertilization is not possible unless previously some process has occurred which has circumvented the reduction of the chromosomes at meiosis and has produced an embryo sac and egg cell with the diploid chromosome number. On the other hand, the mere occurrence of various altered developmental processes which lead to diploid gametophytes and egg cells does not ensure the occurrence of parthenogenesis and apomictic reproduction, as has been shown in a number of organisms.

Nevertheless, the terminology of Winkler is not complete, since some of the modifications of meiosis which lead from an archesporial cell in the ovule to a diploid embryo sac have not been recognized. For these reasons, Stebbins and Jenkins (1939), Fagerlind (1940a), and Stebbins (1941a) proposed amplified terminologies, in which names were given to as many as four different ways of circumventing meiosis and producing a diploid embryo sac. Gustafsson (1946) has pointed out that Fagerlind's karyological classification is difficult to apply because of the numerous intermediate situations which exist between the various recognized modifications of meiosis. In addition, he shows by the clarity of his lengthy exposition of the apomixis phenomena that a relatively simple terminology is sufficient for all practical purposes.

Gustafsson does not include in his classification of the apomixis phenomena those isolated occurrences of haploid parthenogenesis which have been reported from time to time in plants as well as in animals. Since these occurrences are known only in laboratory cultures and are without evolutionary significance, this decision is correct, and the chart presented by the writer (Stebbins 1941a, Fig. 1) may therefore be simplified to that extent (Fig. 36).

The classification of Gustafsson includes two principal types of apomixis, vegetative reproduction and agamospermy. The former may be considered as apomixis whenever the normal sexual processes are not functioning or are greatly reduced in activity. This may come about through a variety of causes, both phenotypic and genotypic, as was discussed in Chapter V. Under such conditions, structures like stolons, rhizomes, and winter buds, which normally act only as accessory methods of reproduction, may assume the entire reproductive function, so that the species under these conditions is essentially apomictic, although the same genetic types may under other conditions be normal sexual organisms. Gustafsson cites a number of species, such as *Elodea canadensis, Stratiotes aloides, Hydrilla verticillata,* and various members of the family Lemnaceae which in northern Europe reproduce wholly by such asexual means, but elsewhere are normally sexual.

More striking, and perhaps of broader significance, are those methods of vegetative reproduction in which the propagules occur within the inflorescence and replace the flowers. This phenomenon, often termed vivipary, is well known in such genera as *Polygonum (P. viviparum), Saxifraga, Allium, Agave,* and some

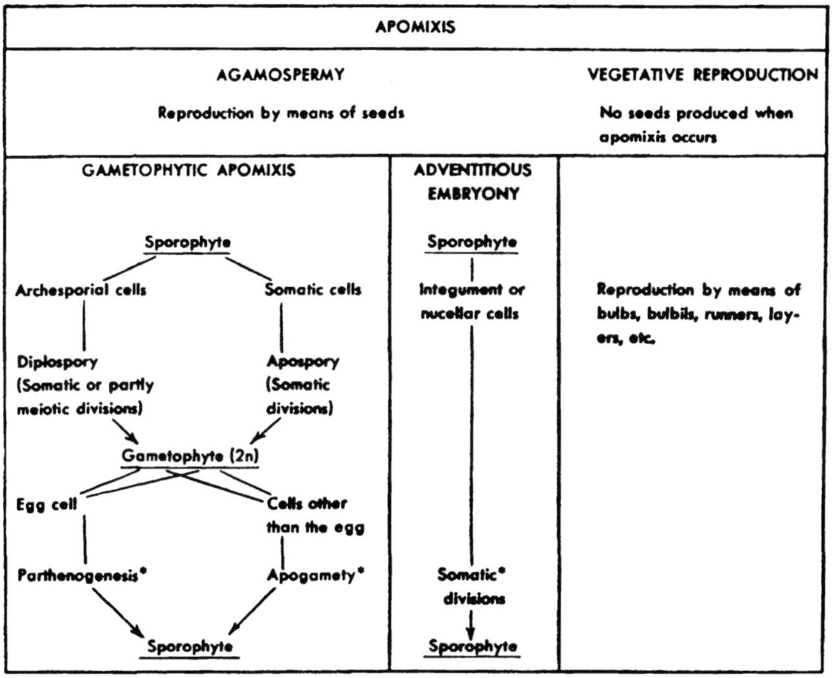

*The processes at this level can take place either autonomously or by pseudogamy, i.e., under the influence of pollen tubes or endosperm development.

Fig. 36. Chart showing the interrelationships of the processes of apomixis found in the higher plants. Modified from Gustafsson 1946.

genera of grasses, particularly *Poa* and *Festuca.* Species or races with this mode of reproduction may fail entirely to produce seed, and therefore from the genetic point of view may be quite comparable to the obligate apomicts which reproduce by seed.

Agamospermy includes all types of apomictic reproduction in which embryos and seeds are formed by asexual means. The essential feature of all types of agamospermy is that meiosis and fertilization are circumvented, so that the embryo (or embryos) which develops in the seeds is usually identical with its maternal parent in both chromosome number and genic content. The simplest method is adventitious embryony, in which embryos develop directly from the diploid, sporophytic tissue of the nucellus or ovule integument, and the gametophyte stage is completely omitted. This phenomenon is best known in various species of the genus *Citrus,* but it is also found in *Alchornea ilicifolia* and *Euphorbia dulcis* (Euphorbiaceae), in *Ochna serrulata* (Ochnaceae), in *Eugenia jambos* (Myrtaceae), in *Sarcococca pruniformis* (Buxaceae), in *Opuntia aurantiaca,* and in species of *Hosta* and *Allium* (Liliaceae) and of *Nigritella, Zygopetalum,* and *Spiranthes* (Orchidaceae). References are given by Stebbins (1941a) and Gustafsson (1946). For some reason, adventitious embryony seems to be relatively frequent in species native to warm temperate or tropical climates.

More common are methods of agamospermy which include from the morphological viewpoint a complete sporophyte-gametophyte-sporophyte cycle of alternation of generations, but in which diploid gametophytes arise as a result of some method of circumventing meiosis. These were collectively termed *agamogony* by Fagerlind (1940a) and Stebbins (1941a), a term which, as Gustafsson pointed out, had been used previously by Hartmann in a totally different sense. Fagerlind (1944b) then suggested the substitute term *apogamogony,* which Gustafsson rejected as illogical. He himself refers frequently to those types of agamospermy which are not adventitious embryony by the rather long and awkward phrase "apospory and diplospory followed by parthenogenesis." Since a collective term is obviously needed, the suggestion is here made that the term *gametophytic apomixis,* that is, apomixis via a morphological gametophyte, which sometimes can function sexually, is both descriptive and appropriate.

There are two principal ways of circumventing meiosis: apospory, in which a diploid embryo sac is formed directly from a cell of the nucellus or of the inner integument by a series of somatic divisions; and diplospory, in which the embryo sac arises from a cell of the archesporium, but in which meiotic divisions are either entirely omitted or are so modified that pairing and reduction of the chromosomes does not occur. These various modifications and the omission of meiosis have frequently been designated according to the plant in which they were first found (*Antennaria* scheme, *Taraxacum* scheme, etc.), but Fagerlind (1940a), Stebbins (1941a), and Battaglia (1945) have used special terms for them. Which of these methods of terminology is the more desirable, or whether any special terms are needed for these phenomena, is a matter outside of the present discussion of apomixis in relation to evolution. Whatever the method of formation of diploid gametophytes, the end result, from the evolutionary point of view, is essentially the same. Figure 36 shows a modification of the simplified terminology recognized by Gustafsson to describe the phenomena included in agamospermy, or apomixis with seed formation.

Diploid gametophytes formed by apospory or by diplospory may give rise to embryos through the multiplication of either the egg cell (parthenogenesis) or some other cell (apogamety). The former phenomenon is by far the most common one in the seed plants, but apogamety is frequent in ferns. Both processes, as well as adventitious embryony, are in some plants autonomous, but in others require pollination and the development of pollen tubes. Such apomicts are said to be pseudogamous. Like sexual forms, they will not seed unless pollination has occurred, but the plants formed from such seed are maternal in character. The effect of pollination appears to vary with the species concerned. In most pseudogamous species, such as *Allium odorum* (Modilewski 1930), *Potentilla collina* (Gentcheff and Gustafsson 1940), *Poa* spp. (Håkansson 1943a, 1944a), and *Parthenium incanum* (Esau 1946), the embryo begins development autonomously and even precociously, but the endosperm will not develop unless it is fertilized. Pollination and partial fertilization are therefore essential for continued growth of the embryo. Gustafsson nevertheless cites some examples, particularly the species of *Ranunculus* investigated by Häfliger, in which fertilization of the endosperm possibly does not take place. When parthenogenesis is autonomous, development of the embryo, the endosperm, or both are very often so precocious that at the time when the flowers open and the stigmas are ready to receive pollen the formation of the embryo and seed has already started.

One of the most conspicuous features of many apomictic plants is the disturbance of the meiotic divisions of both megasporogenesis in the ovules and microsporogenesis in the anthers. This is present occasionally in plants which reproduce by vegetative means or by adventitious embryony, but it is not characteristic of them. In plants possessing apospory ("somatic apospory" of Fagerlind and of Stebbins 1941a), meiosis is likewise either normal or characterized by slight abnormalities, such as a small number of univalent and multivalent chromosomes, in both microsporogenesis and the formation of the megaspore tetrads and embryo sacs which become replaced by the products of apospory. This comparative regularity may be connected with the fact that most of the aposporous apomicts have pseudogamous embryo development. On the other hand, in some of the American species of *Crepis,* which combine apospory with autonomous development of embryos and seeds, the archesporial tissue of the anthers develops so abnormally that the pollen mother cells degenerate before they reach the stage of meiosis (Stebbins and Jenkins 1939).

The most striking abnormalities of both micro- and megasporogenesis are found in those plants which combine some form of diplospory (including "generative apospory" of Stebbins 1941a) with autonomous development of embryos and seeds. These abnormalities have been intensively studied in *Hieracium, Taraxacum, Antennaria, Calamagrostis,* and some species of other genera. This work was reviewed by Stebbins (1941a) and even more carefully by Gustafsson (1947b). The most frequent abnormalities are failure of chromosome pairing, failure of contraction of the chromosomes, retardation of meiosis, and precocity of the meiotic divisions. These abnormalities, which were found thirty years ago by Rosenberg (1917), were considered by Ernst (1918) as strong evidence for his hypothesis that apomictic forms are of hybrid origin, and therefore that apomixis is caused by hybridization. Further knowledge of these phenomena, however, has made such an opinion untenable, although some of the abnormalities may be a direct or indirect result of past hybridization between species.

The most conspicuous feature of these abnormalities is that they are often, and perhaps usually, different in the male and the female organs of the same species. In *Antennaria fallax,* for instance (Stebbins 1932b), the occasional male intersexes found in nature have microspore meiotic divisions which are essentially regular, with a high degree of bivalent formation and normally contracted chromosomes. But in the ovules, meiosis is either entirely omitted or may in some ovules occur in a very abnormal fashion, with practically no chromosome pairing. In such an example, it is quite clear that the failure of pairing has nothing to do with the lack of homology between dissimilar chromosomes, but is connected with some disturbance in the normal development of the ovular and archesporial tissue. Meiosis in the anthers of various *Hieracium* apomicts has been studied by several workers, and a series of abnormalities has been described; but these are not duplicated in the ovules. Finally, Nygren (1946) has found that in *Calamagrostis lapponica* meiotic divisions occur in the anthers, although these are irregular and characterized by a varying number of univalents, while in the ovules the divisions are essentially mitotic; but in *C. chalybaea,* the reverse is the case, since the divisions of the pollen mother cells are essentially mitotic, while those of the megaspore mother cells are more like meiosis, although the formation of restitution nuclei causes them to produce diploid megaspores and gametophytes. In *Hieracium,* the reversion toward mitosis in the pollen mother-cell divisions is accompanied by great precocity in the time when they begin, while the same change in the ovules is associated with a delay in the onset of the divisions. In *Calamagrostis,* mitosislike divisions in the pollen mother cells may be associated either with a delay in their onset or with precocity.

The complexity and multiformity of these disturbances of meiosis, as well as the differences between anthers and ovules of the same plant, can be explained only if we assume that the normal course of meiosis requires the carefully synchronized occurrence of a series of developmental processes, and that it can be upset in a number of different ways by various environmentally and genetically controlled disturbances of one or more of these processes. They provide a wealth of material for studies of the differences between mitosis and meiosis and their causes, which still has by no means been exploited to its full extent.

One of the results of this diversity in the nature of meiotic abnormalities in apomicts is that not infrequently some ovules of a plant may produce diploid embryo sacs through failure of meiosis, while in other ovules of the same plant successful completion of the meiotic divisions will yield haploid embryo sacs and egg cells. The former ovules contain diploid egg cells capable of parthenogenetic development, usually through pseudogamy; while the latter will not form seeds unless fertilized. Such plants will therefore produce some completely maternal progeny through apomixis and other offspring by the normal sexual process. They are termed *facultative apomicts.* In plants which form diploid embryo sacs by means of apospory, facultative apomixis may also occur because in certain ovules the aposporic embryo sac either does not develop at all or develops so slowly that it does not crowd out the haploid one derived from the megaspore.

GENETIC BASIS OF APOMICTIC PHENOMENA

A number of hybridizations have now been performed between sexual types and related facultative or obligate apomicts of the same genus. Segregation in later generations from these crosses has shown in every case that the apomixis of the species is genetically controlled, but the basis of inheritance has been different in the various groups studied. The simplest situation is that found by Levan (1937b) in the predominantly apomictic *Allium carinatum,* a species in which the flowers are largely replaced by bulbils which reproduce it vegetatively. A cross between *A. carinatum* and the sexual species *A. pulchellum* showed that bulbil production in the former species is largely controlled by a single dominant gene, although the expression of the character may be greatly modified by other genes and by various environmental factors. Gustafsson's brief analysis of the meager data on the inheritance of adventitious embryony in *Citrus* has led him to the suggestion that in this group the inheritance is more complex, and apomixis is recessive to sexuality.

The most extensive genetic data on apomixis have been obtained in the complexes of *Poa pratensis* and *P. alpina* and in the genus *Potentilla.* In both of these groups the mechanism of apomixis is aposporous or diplosporous (in *Poa alpina*) gametophyte development followed by pseudogamous or (in some species of *Potentilla*) autonomous development of the embryos. The genetic basis in both is complex, as revealed through several publications by Müntzing and by Akerberg on *Poa,* and by Müntzing, Christoff and Papasova, and Rutishauser on *Potentilla* (see Gustafsson 1947b,c, for complete list of references). In *Poa,* crosses between facultatively apomictic and sexual biotypes, as well as between different facultative apomicts, yield a great predominance of sexual plants, showing that apomixis is recessive to sexuality (Müntzing 1940). Furthermore, since progeny from such crosses yield some plants which possess apospory, but not regular pseudogamous embryo development, and others which have little apospory, but yield a high percentage of haploid progeny through pseudogamous parthenogenesis (Håkansson 1943a), the various processes of the apomictic cycle are in this genus controlled by different genes, and successful apomictic reproduction is "due to a delicate genetic balance" between a series of factors.

The results in *Potentilla* are similar when diploid apomicts are crossed with each other or when apomicts with a low chromosome number are crossed with sexual forms. On the other hand, crosses between two high polyploid apomicts or between a polyploid apomict and a sexual form with a low chromosome number yield mostly apomictic progeny. Several genes for apomixis, therefore, may in the case of *Potentilla* dominate over a smaller number for sexuality. In *Rubus,* as in *Poa,* crosses between two facultative apomicts give sexual progeny.

In the genus *Parthenium,* the apomicts are pseudogamous. Although no hybrids between sexual and apomictic biotypes have yet been reported, evidence has been obtained to indicate that the two essential processes of apomixis — development of gametophytes with the unreduced chromosome number and of embryos without fertilization — are genetically independent of each other. This evidence consists of the behavior of a particular strain, No. A42163, described by Powers (1945). Meiosis fails and unreduced embryo sacs are formed in 90 percent of the ovules, but the eggs of these gametophytes cannot develop without fertilization as well as pollination. As a result, the offspring of this plant and its progeny consist of 65 to 90 percent of morphologically aberrant, dwarf triploids. The parental plant has 72, the progeny, 108 chromosomes. These triploids, in their turn, produce only unreduced gametophytes and egg cells, and progeny with still higher chromosome numbers (about 140) and more aberrant morphology. In crude language, this line polyploidizes itself out of existence.

Dominant factors for apomixis have been reported in *Hieracium* by Ostenfeld (1910) and Christoff (1942), but their evidence is considered doubtful by Gustafsson, since it deals with a cross, *H. auricula* × *aurantiacum,* involving a diploid sexual and a polyploid apomictic species. In general, the sum of evidence from

Hieracium suggests a genetic basis for apomixis similar to that in *Potentilla,* in which crosses between two polyploid facultative apomicts are likely to yield apomictic progeny.

Summing up, we may say that as a rule the apomictic condition is recessive to sexuality, although increasing the chromosome number often increases its tendency toward dominance. However, this recessiveness is not usually that of a single gene, but is due to the fact that a successful apomictic cycle is produced by the interaction of many genes, so that any cross between an apomictic form and a sexual one, or between two apomicts having different gene combinations controlling their apomixis, will cause a breaking up of these gene combinations and therefore a reversion to sexuality, sterility, or some abnormal genetic behavior. It is only in the relatively simple types of apomixis, like vegetative reproduction and adventitious embryony, that simple genetic behavior can be expected.

APOMIXIS, HYBRIDIZATION, AND POLYPLOIDY

The first hypothesis developed to explain apomixis was that of Ernst (1918), who believed that it is caused by hybridization between species. His evidence consisted mainly of the fact that disturbances of meiosis and of other developmental processes in the development of the sex cells are widespread in apomictic forms, and often simulate the same phenomena found in interspecific hybrids. As was pointed out earlier in this chapter, this evidence by itself is by no means conclusive. Supporting evidence produced by Ernst centered in the great polymorphism in apomictic groups and the difficulties of delimiting species in most of them. The latter situation has now been confirmed in many groups besides those known to Ernst, and in some of them, such as *Rubus, Poa, Crepis, Parthenium,* and *Antennaria,* convincing evidence has been obtained for the hybrid origin of many of the apomicts. Nevertheless, there is no evidence at all that hybridization by itself can induce apomixis. Hybrids between different sexual species of *Rubus* (Peitersen 1921), *Antennaria* (Stebbins 1932a), *Potentilla* (Christoff and Papasova 1943), *Taraxacum* (Poddubnaja-Arnoldi 1939a,b, Koroleva 1939, Gustafsson 1947b), and other genera have in no instance shown any clear indications of apomictic reproduction, even though their parental species may be closely related to known apomictic forms.

A similar situation exists in respect to the relationship between apomixis and polyploidy. Many groups of plants are known in which the diploid species are exclusively sexual and the polyploids, largely apomictic. This correlation is not, however, complete; all of the different types of apomixis are now known to occur in diploid forms. Diploid species with vegetative apomixis by means of bulbils or similar structures ("vivipary") are known in *Allium, Agave,* and *Lilium,* although in all these genera apomictic polyploids occur with equal or greater frequency. In a larger number of genera with this type of apomixis (*Polygonum viviparum, Ranunculus ficaria, Cardamine bulbifera, Saxifraga* spp., various species of *Festuca, Poa, Deschampsia,* and other Gramineae), all, or nearly all, the apomicts are polyploid.

Among agamospermous apomicts, those with adventitious embryony contain a relatively high proportion of diploids. *Citrus* is the classic example of this situation, but others are *Nothoscordum bivalve, Alnus rugosa, Sarcococca pruniformis,* and some species of *Eugenia.* In the latter genus, polyploids with adventitious embryony occur also, but *Ochna serrulata* and *Nigritella nigra* are the only polyploid species with this type of apomixis which are not known to have diploid relatives also possessing it.

Among the numerous groups of apomicts with gametophytic apomixis only three are known in which some of the apomictic forms are diploid. These are *Potentilla,* particularly *P. argentea* (A. and G. Müntzing 1941), the *Ranunculus auricomus* complex (Häflinger 1943, cited by Gustaffsson 1947b), and a form of *Hieracium umbellatum* (Gentscheff, cited by Gustafsson 1947b). In both *Potentilla* and *Hieracium,* the polyploid apomicts far outnumber the diploid ones, and polyploid apomicts also occur in *Ranunculus.* On the other hand, the list of apomictic groups given by Stebbins (1941a) includes 24 of this type, in which the apomicts are exclusively polyploid. Additional groups of this nature which may now be cited are *Parthenium* (Rollins, Catcheside, and Gerstel 1947, Esau 1944, 1946, Stebbins and Kodani 1944), *Rudbeckia* (Battaglia 1946a,b, Fagerlind 1946), *Paspalum* (Burton 1948), and probably *Crataegus* (Camp 1942a, b), although the latter genus is badly in need of careful study.

On the other hand, evidence obtained recently makes very unlikely the assumption that polyploidy directly initiates apomictic reproduction. Autopolyploids of *Taraxacum kok-saghyz* produced by Kostoff and Tiber (1939), by Warmke (1945), and by the writer (unpublished) show no signs of apomixis, although all of the triploid and tetraploid "species" which are closely related to *T. kok-saghyz* and live in the same region are apomictic (Poddubnaja-Arnoldi and Dianova 1934). Similarly, Gardner (1946) found complete sexuality in a high polyploid form of *Parthenium argentatum,* although closely related strains of the same species are apomictic.

The usual close association between apomixis, on the one hand, and polyploidy, interspecific hybridization, and polymorphy, on the other, must be explained on an indirect rather than a direct basis. Hybridization promotes apomixis in two ways. In the first place, the combinations of genes necessary to initiate an apomictic cycle are probably put together most easily by hybridization, either between species or between different forms of the same species. A genetical model illustrating how this could happen is presented by Powers (1945). He conceives of a minimum of three pairs of genes, as follows: Gene pairs *A* vs. *a* control, respectively, normal meiosis and failure of chromosome reduction; *B* vs. *b* control fertilization of the egg (whether haploid or diploid) and failure of fertilization; while *C* vs. *c* control failure of the egg to develop without fertilization vs. autonomous (usually precocious) development of embryo and endosperm. Apomixis will occur regularly only in the triple recessive homozygote, *aabbcc.* Gustafsson points out that pairs *B, b* and *C, c* control processes closely integrated with each other, and which might be expected to be controlled by the same genes. Since the action of any of the three alleles *a, b,* or *c* would be lethal or strongly deleterious to the individual unless the other two were present, such alleles would be likely to survive in natural populations only if they were recessive. Either inter- or intraspecific hybridization would be essential to put together such gene combinations.

Conditions similar to those which would be expected in two of the single recessive homozygotes postulated by Powers are known as isolated instances in a few plant species. In *Scolopendrium* (Andersson-Kotto 1932) and *Leontodon hispidus* (Bergman 1935b) recessive mutant forms are known which produce gametophytes by apospory, but these cannot produce new sporophytes without fertilization. According to the scheme of Powers, these would have the combination aaBBCC. The form of *Parthenium argentatum,* described in the previous section, which produces only triploid offspring is probably also of this constitution, as are the occasional aposporous, but not parthenogenetic, individuals which have been found in *Oxyria digyna, Antennaria dioica, Coreopsis bicolor,* and *Picris hieracioides* (cf. Stebbins 1941a). Håkansson (1943a) studied a sexual plant of *Poa alpina* derived from a cross between a sexual biotype and an apomictic biotype, and he found in it a tendency to produce up to 15 percent of haploid embryos by pseudogamous parthenogenesis. Hagerup (1944) found that in *Orchis maculata* haploid embryos frequently arise, due to precocious division of the egg cell. Both these forms can be considered to possess the combination AABBcc.

Powers's hypothesis thus seems to be the most plausible one yet developed for the genetic explanation of apomixis, although the situation in many apomicts may be even more complex. Each of the separate processes, and particularly the failure of meiosis, may in some instances be affected by more than one gene.

The second way in which hybridization promotes apomixis is through the greater vigor and tolerance of a wide range of ecological conditions possessed by hybrid genotypes of which the parents differ greatly in their ecological preferences. This has been demonstrated clearly by Clausen, Keck, and Hiesey (1945b) in the case of such crosses as *Poa pratensis* $\times$ *P. lapponica* or *P. ampla,* as well as in hybrids between widely different sexual ecotypes in *Potentilla glandulosa* (see page 108). The immediate selective advantage of a mechanism tending to preserve and reproduce such well-adapted genotypes is obvious.

After reviewing all the evidence in relation to polyploidy and apomixis, Gustafsson has reached the plausible conclusion that, although apomixis can be induced in diploids by favorable gene mutations, the action of many of these apomixis-inducing genes is stronger on the polyploid level than it is on the diploid level. The best evidence for this is obtained from ferns, particularly the work of Heilbronn (1932) on *Polypodium aureum.* The sporophyte of this species normally produces haploid spores and gametophytes by means of regular meiosis. Such normal gametophytes, like those of most ferns, cannot develop further unless the egg is fertilized to form a new sporophyte. But diploid gametophytes may also be formed aposporously by regeneration from pieces of leaves. These, like their haploid counterparts, produce eggs which are capable of fertilization and yield tetraploid sporophytes. But in addition they can under certain environmental conditions produce diploid sporophytes directly from their vegetative tissue, a process known as apogamety (see diagram, Fig. 36). Tetraploid gametophytes, produced aposporously from tetraploid sporophytes, are almost incapable of producing sex organs, and they reproduce regularly by apogamety. A somewhat similar tendency has been described by Springer (1935) in the moss *Phascum cuspidatum.*

That this property of increasing apomixis with higher levels of polyploidy is not a general one is evident from the comparative studies of Beyerle (1932), which have demonstrated that sporophytes of different fern species differ greatly from each other in their capability for regeneration. Wettstein (1927) has in addition found that the tetraploid, octoploid, and even 16-ploid gametophytes of *Funaria, Bryum,* and other genera of mosses are normally sexual and quite different from those of *Phascum cuspidatum.*

In the flowering plants, Levan (1937b) has produced a polyploid form similar to the bulbilliferous species *Allium oleraceum* by crossing two varieties of the diploid, nonbulbilliferous *A. paniculatum,* and he is of the opinion that the polyploidy was directly responsible for the bulbillifery. Turesson (1930, 1931b) noted that in *Festuca ovina* the degree of vivipary increases with increase of the chromosome number, but in other species of Festuca, such as *F. rubra, F. californica,* and *F. arundinacea,* hexaploid, octoploid, and even decaploid forms exist which show no sign of vivipary. There is good evidence, therefore, that polyploidy can reinforce the action of genes favoring such types of apomixis as vivipary or other forms of vegetative reproduction.

Adventitious embryony is the one type of apomixis which seems to be more common in diploid than in polyploid plants. We should expect, therefore, that the action of genes for this type of apomixis would not be favored by polyploidy. But, in the case of gametophytic apomixis, we should expect a particularly strong reinforcing action of polyploidy on genes for its component processes. There is, however, relatively little evidence for this from the flowering plants. Gustafsson does not cite any examples from the angiosperms of the reinforcement by polyploidy of a genetic tendency for nonreduction and the formation of diploid gametophytes. In *Hieracium hoppeanum,* Christoff and Christoff (1948) have obtained evidence for a tendency away from nonreduction after artificial doubling of the chromosome number. The form investigated is a pentaploid with the somatic number 45, and like all apomicts of *Hieracium* subg. *Pilosella* it produces unreduced gametophytes by apospory. A form with 90 chromosomes obtained from colchicine treatment also reproduces in this manner to a large extent, but it yields in addition offspring with 45 chromosomes, which resemble the undoubled form. Apparently embryo sacs with the reduced chromosome number have been formed, and eggs from these develop parthenogenetically. Embryological studies showed that in both typical *H. hoppeanum* and the doubled form, the products of meiotic division of the archesporial cell degenerate and are replaced by embryo sacs derived from the integument and the nucellus. In the doubled form, however, some of the nucellar cells expand and go through meiosis, thereby producing embryo sacs with the reduced chromosome number.

Evidence that a tendency toward pseudogamous parthenogenesis may be induced by increasing the chromosome number was produced by Håkansson (1943a, 1944) in *Poa alpina.* He obtained hybrids between a sexual race of this species with $2n = 24$ chromosomes and an apomict having $2n = 38$. Some of the F_1 plants, resulting from the fertilization of unreduced egg cells, had 41 to 43 chromosomes. These plants, although they always produced by meiosis embryo sacs with the reduced chromosome number, nevertheless yielded a high proportion of haploid offspring through parthenogenesis. Their sister plants with the normal complement of 30 chromosomes obtained from haploid gametes of the parental forms were not parthenogenetic at all. Håkansson and Gustafsson assumed that this situation could not be explained on a genic basis, since the F_1 offspring capable of parthenogenesis receive twice as many genes from their sexual parent, but the same number of genes from their apomictic parent as do those which are purely sexual.

In artificial autotetraploids of some purely sexual angiosperms eggs with the reduced chromosome number may develop parthenogenetically. This was found by Randolph and Fischer (1939) in maize and by Warmke (1945) in *Taraxacum kok-saghyz*. In every instance, the number of parthenogenetic diploids was very small, but nevertheless the fact is demonstrated that polyploidy may promote the autonomous development of the egg cell into an embryo.

Fagerlind (1944a) has suggested that in some, perhaps the majority of, agamic complexes the apomixis first arose on the diploid level. The present association between polyploidy and apomixis is believed to be secondary, and due to the fact that polyploids, particularly of unbalanced chromosomal types, can reproduce themselves more efficiently when apomictic than when sexual. This may be a partial explanation of the correlation in some groups, but it seems hardly applicable to such complexes as *Antennaria, Rubus, Potentilla,* and *Poa,* which contain a considerable number of sexual polyploids.

SOME TYPICAL AGAMIC COMPLEXES

Whatever may be its causes, the connection between apomixis, hybridization, and polyploidy is so intimate that it cannot escape the observation of anyone who studies thoroughly a group of apomicts and their closest sexual relatives. Species groups in which these three sets of processes have been operating possess a characteristic variation pattern, not unlike that described in Chapter VIII for polyploid complexes, but more intricate, and with fewer discontinuities between entities which can be recognized as species. Such groups were characterized by Babcock and Stebbins (1938, Stebbins and Babcock 1939) as *agamic complexes,* on the basis of their study of a typical example, the American species of *Crepis.* This complex remains the only one which has been studied in its entirety from both the systematic and the cytological point of view. But in other agamic complexes, particularly *Poa, Potentilla, Rubus, Parthenium, Taraxacum,* and *Hieracium,* portions of them have been studied much more intensively than *Crepis* from the cytogenetic as well as from the taxonomic point of view, and variation patterns have been revealed which deviate considerably from those found in *Crepis.* The best conception of the variation pattern in such complexes can therefore be obtained by summarizing first the situation in *Crepis,* and then indicating the ways in which various other agamic complexes depart from it.

The species of *Crepis* native to North America form a polyploid series based on the haploid number $x = 11$. Eight of them are diploid, but one, *C. runcinata,* is very different from the others and has no polyploid apomictic relatives. The agamic complex, therefore, is based on seven primary diploid species, *C. pleurocarpa, C. monticola, C. bakeri, C. occidentalis, C. modocensis, C. atribarba* ("*C. exilis*"), and *C. acuminata.* With the exception of the last two, these are all much restricted in distribution, and appear as relicts, as shown on the map, Fig. 37. Although their ranges overlap to a considerable extent, they have very different

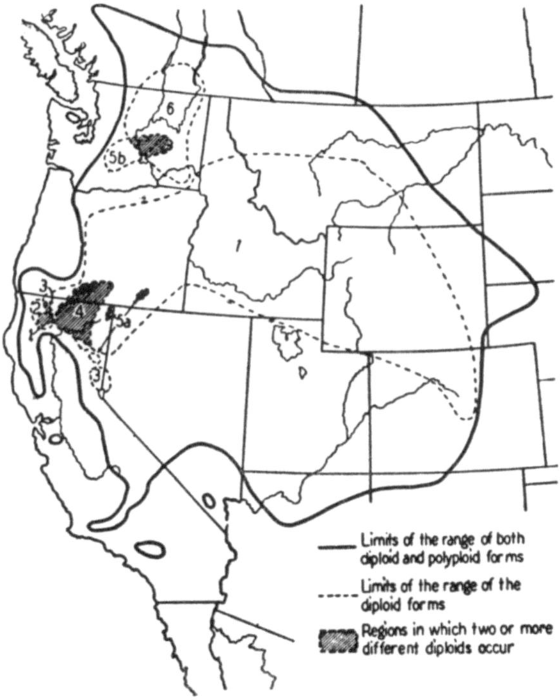

FIG. 37. The relative areas of distribution of diploid sexual forms of *Crepis* species in western North America and of their polyploid apomictic derivatives. From Babcock and Stebbins 1938. Base map copyrighted by Rand McNally Company. Reproduced by special permission.

ecological preferences, and no localities have been found containing two diploid species growing side by side. For this reason, no diploid hybrids between them are known. Since these species are extremely difficult to grow under cultivation, genetic work on them has not been attempted.

These sexual diploids comprise only a minute fraction of the total population of *Crepis* in western North America. Far more numerous are the polyploids, which have somatic chromosome numbers of 33, 44, 55, 77, and 88. All these polyploids which have been tested prove to be apomictic, although some of them are facultative apomicts. They include a certain percentage of forms which resemble closely each of the seven diploids, but the large majority show in their external morphology various combinations of the characteristics of two or more diploids. They thus appear to be allopolyploids, an assumption which is borne out by their cytology, so far as it is known (Stebbins and Jenkins 1939). This hypothesis is also supported by the fact that whenever the ecological and climatic preferences of an apomict have been analyzed, they have been found to fall between those of the diploid species between which the apomict is intermediate morphologically. In geographic distribution, the apomicts as a whole exceed the range of their diploid ancestors, but not to a great extent, because of the wide range of one diploid species, *C. acuminata.* This is the only species of which the nearest apomictic relatives have a narrower geographic distribution than the diploid form. On the other hand, the highly restricted diploids, *C. occidentalis* and *C. modocensis,* possess polyploid apomictic derivatives which are found throughout most of the range of the complex, although most of these seem to contain some genes from *C. acuminata, C. bakeri,* or both.

The most important point about the variation pattern in this complex as a whole is that while each of the seven diploids possesses certain distinctive morphological characteristics not found in other diploid species, none of the polyploids possesses characteristics which cannot be explained as a result of recombinations of those found in the diploid species, plus the effects of polyploidy. Until all the diploids were recognized, the relationships between the different apomicts could be perceived only dimly, and superficial resemblances caused the grouping together in the same systematic category of forms with very different relationships and ecological preferences. But recognition of the seven primary diploids permitted the grouping of the apomicts around them, either as apparent autopolyploids or as allopolyploids combining the different characteristics of various diploids. There thus appeared a variation pattern which resembles that shown in Fig. 34 except that it is more complex. The diploids can be thought of as seven pillars, which by themselves are sharply distinct from each other and so widely different that they appear to have affinities with Old World species which are placed by Babcock in different sections. They are connected by a much larger superstructure of polyploid apomicts, which represent all sorts of intergradations and recombinations of their characteristics.

Another feature of this pattern as observed in *Crepis* is that its diversity is very great in the regions occupied by the sexual species, but much less in the peripheral areas where only apomicts are found. In the latter regions, the same biotype or apomictic clone can often be recognized over distances of scores or even hundreds of miles, while in the vicinity of the sexual types each separate colony seems to possess its own cluster of distinctive apomicts. The apomicts are distributed in just the manner which would be expected if they had been formed by hybridization and polyploidy involving the sexual species and had radiated outward from the areas which the latter occupy.

From the systematic and the phytogeographic points of view, *Crepis* has two advantages over most of the known agamic complexes. The total range of its complex is confined to one geographic area and life zone, namely, the semiarid cold steppe or arid transition zone of the northwestern United States and adjacent Canada. This makes it much easier to comprehend as a single unit than each of the widespread Holarctic or even bipolar complexes, such as *Poa, Rubus, Antennaria, Taraxacum,* and *Hieracium.* Secondly, it is so recent that the diploid species which gave rise to it are all still in existence and are growing in the same regions as their nearest relatives among the polyploid apomicts. In this writer's opinion, many of the apparent divergences from *Crepis* in the variation pattern of other agamic complexes will be found to disappear when these are as fully known.

Although many agamic complexes have variation patterns essentially similar to that in *Crepis,* others deviate from this genus in certain important respects. These are, first, the number of sexual species on which the complex is based; second, the extent to which the range of variation in the complex is covered by modern sexual species; third, the relative frequency of facultative as compared to obligate apomixis; and, fourth, the presence of polyploid sexual species as well as diploid apomictic species.

So far as is known, the agamic complexes found in those genera most nearly related to *Crepis,* namely, *Taraxacum, Hieracium, Chondrilla, Youngia,* and *Ixeris,* differ from it chiefly in the number of sexual species, and therefore in the size and degree of diversity in the complex. The latter three are probably smaller, and the former two are certainly larger than *Crepis.* The two complexes of *Hieracium,* one in the subgenus *Archieracium* and one in subg. *Pilosella,* have been insufficiently explored as to their sexual species, but the range of variation in both of them is very great. In *Taraxacum,* no less than twelve species are known to be diploid and sexual and to have close relatives among the apomicts. But cytological explorations over large parts of the range of this complex are either lacking or are very superficial, and many more diploids undoubtedly exist. *Taraxacum,* therefore, is two or three times as extensive as *Crepis* in the range of its morphological variability, as well as in its geographical distribution. Gustafsson (1947c) has considered that *Hieracium* and *Taraxacum* belong to a different type of agamic complex from that of *Crepis* because the sexual relatives of the apomicts are considered (p. 254) to be "depauperate and relict." This distinction seems hardly justified, since five out of seven of the sexual species of *Crepis* are of this nature and would not have been found if a thorough search had not been made for them. No such search has been undertaken in the case of *Hieracium* and *Taraxacum.*

The agamic complex of *Antennaria* is nearly as extensive as that of *Taraxacum,* since it is based on at least twelve to fifteen different sexual diploid species. Hybrids involving three of these diploids have been found in nature (Stebbins 1932a, 1935), and the one studied cytologically has fairly regular meiosis. The apomicts, which are mostly tetraploid and hexaploid, reproduce by diplospory and by autonomous parthenogenesis. The great majority are obligate apomicts, but the high frequency in certain areas of male plants in apparently apomictic forms of some of the species (*A. parlinii, A. fallax*), suggests that facultative apomicts occur in some regions occupied jointly by sexual diploids and their close relatives among the apomicts. In external morphology, many of the apomicts are intermediate between two or more sexual diploids, and they tend to obscure the original species distinctions, just as in *Crepis, Taraxacum,* and *Hieracium.* A further complication which may exist in this complex is the presence of polyploid sexual species. Bergman (1935c) found $2n = 42$ in *Antennaria carpathica,* which may be related to some of the apomicts of arctic North America. In *A. media,* which possesses both sexual and apomictic biotypes, the writer found $n = 28$ in staminate plants from an apparently sexual colony.

Parthenium differs from all the complexes mentioned above in its smaller size, since its total range is confined to a single desert area, and probably only two original sexual diploid species have entered into it. The basic haploid chromosome number, $x = 18$, is possessed by the relatively restricted sexual forms of both *P. argentatum* and *P. incanum,* and the polyploids include forms with somatic numbers as high as 144, although nearly all the apomicts with numbers higher than $2n = 90$ are dwarf aberrants, which are rare or absent in nature. The method of apomixis is diplospory followed by pseudogamous parthenogenesis, and nearly all the apomicts are facultative. The conspicuous divergence from *Crepis* in the variation pattern of this complex is that the apomicts are grouped around two modes representing the morphological characteristics of the two sexual diploids, and intermediates between them, although clearly identified by the brilliant and painstaking work of Rollins (1944, 1945a, 1946), are less common than types which appear to be autopolyploid. *Parthenium argentatum* and *P. incanum* are sympatric throughout most of their geographic range, but they occupy different habitats and nearly everywhere are distinct enough so that they can be recognized even by people without botanical training. The largest stands of apomicts intermediate between these two species are in parts of Texas in which neither of the parental species is found.

The facts presented above suggest that the complexes most similar to *Crepis* are those found in other genera of the family Compositae. The best-known agamic complexes in other plant families, namely, those of *Rubus, Potentilla,* and *Poa,* differ from *Crepis* in most or all of the four characteristics mentioned above. The best known of these is *Rubus,* due to the exhaustive study by Gustafsson (1942, 1943a). Although polyploidy is known in several of the twelve subgenera of this large genus, apomixis is found chiefly in one of the six sections of a single subgenus, sect. *Moriferi* of subg. *Eubatus.* This section has a disjunct distribution, with one series of species in Europe and western Asia and the other in eastern North America, from Canada to Guatemala, plus a single species in Japan. Apomixis is predominant in the European members of this section. The hybridization experiments of Brainerd and Peitersen (1919, Peitersen 1921) indicate that many of the North American species are entirely or chiefly sexual, but the cytological studies of Longley (1924) and Einset (1947b) suggest that apomictic forms also exist, and Einset (unpublished, cited by Gustafsson) has confirmed this supposition.

Gustafsson recognizes five primary diploid sexual species among the European Eubati, but shows that these will by no means account for the amount of variation represented by the polyploid apomicts. The range of variation in these forms goes beyond that of the sexual species in three directions. One series, represented chiefly by the group *Suberecti,* varies in the direction of the North American sexual diploid *R. allegheniensis,* and it could be explained on the assumption that a close relative of this species existed in Europe at some time in the past. The second series, represented by the *Glandulosi,* has its center of variation in the Caucasus Mountains, and further exploration in this area may reveal a sexual diploid with the characteristics necessary to explain the morphological characteristics of this group. The third series centers about the anomalous species *R. caesius,* an apomictic allopolyploid of which the diploid ancestors are completely unknown. *Rubus,* therefore, differs from *Crepis* in that several of the sexual diploid ancestors of its agamic complex appear to be extinct, but may have close relatives still existing in regions not occupied by the apomicts. *Rubus* differs also in the presence of polyploid sexual species, and in the fact that most of its apomicts are facultative, while even the obligate apomicts have such good pollen that they frequently serve as the parents of hybrids.

The wide Holarctic range and disjunct distribution of the

section to which the agamic complex belongs indicates a greater geological age than that of the North American section of *Crepis,* but nevertheless some of the sexual members of the *Rubus* complex have remained common, aggressive, and capable of contributing genes on a large scale to its apomictic members, although other sexual ancestors are apparently extinct. Due to the facultative apomixis, the origin of new forms through hybridization between apomicts is taking place on a much larger scale and over a much wider area than it is in *Crepis.* In fact, one whole group, the Corylifolii, appear to be derived from recent (post-human) hybridization between *Rubus caesius* and apomicts belonging to other sections. Thus, although the total range of morphological and ecological variability in the agamic complex of *Rubus* is about the same as that in *Crepis,* and although perhaps only half of the original sexual ancestors are now living within the range of the apomicts, the greater ability for hybridization, plus the vigor of the plants and their adaptability to areas disturbed by man, has kept the complex of *Rubus* in a more "youthful" self-perpetuating condition than is that of *Crepis.*

Neither of the two remaining well-known and extensive agamic complexes, those of *Potentilla* and *Poa,* has been given comprehensive treatment by a single author. The diploid species of the *Potentilla* complex may prove to be more numerous than those in any other, and the range of chromosome numbers, from $2n = 14$ to $2n = 109$ (the latter in the western American *P. gracilis*), is likewise the greatest known in any agamic complex. *Potentilla* is unique in the great development of apomixis through apospory and pseudogamy among two diploid species complexes, those of *P. arguta* (Popoff 1935) and *P. argentea* (A. and G. Müntzing 1941). Like the apomicts of *Rubus,* those of *Potentilla* are largely facultative, and hybridization between them is common in nature. Gustafsson suggests that one whole group of apomicts, the Collinae, like the Corylifoliae of *Rubus,* has arisen through recent hybridization between two other groups. Although sexual diploid species of *Potentilla* related to the apomicts are known, the precise relationships between them have not been worked out, so that the structure and history of this agamic complex is likewise obscure.

The genus *Poa* may contain one or several agamic complexes,

Apomixis is known or strongly suspected in seven sections of this genus, and since these sections are very difficult to separate from each other, and wide intersectional hybrids prove to be vigorous and productive of apomictic seed (Clausen, Keck, and Hiesey 1945b), all the apomicts of *Poa* may eventually prove to be interconnected and to form a single gigantic agamic complex. This would be undoubtedly the largest in the plant kingdom. Furthermore, two North American sections, the *Nevadenses* and the *Scabrellae,* resemble in some respects the neighboring genus *Puccinellia,* which also contains species that may be apomictic. It is not inconceivable, therefore, that this agamic complex is spread over two world-wide genera.

Poa, along with the apparently similar but little-known complex of *Alchemilla* (Gustafsson 1947c, p. 254), is distinctive in that only two sexual diploids are known which are likely ancestors for some of its apomicts. These are *Poa trivialis* (Kiellander 1942) and a sexual diploid of *P. alpina* reported by Christoff (1943). Furthermore, sexual polyploids form a conspicuous feature in two series, those of *P. alpina* and of *P. pratensis.* In both of them there is the additional unusual feature of aneuploidy. In particular, Müntzing (1933) has found somatic numbers of 22, 24, 25, 31, 33, and 38 in European strains of *P. alpina.* The type of apomixis present varies from one group to another. Several species are viviparous; the best-known method, found in *P. pratensis* and *P. alpina,* is apospory followed by pseudogamous parthenogenesis; and other mechanisms are likely to be found. Several species of western North America, such as *P. nervosa* and *P. epilis,* consist entirely of plants with abortive anthers and no pollen over large sections of their range (Keck, oral communication); these must be obligate apomicts with autonomous parthenogenesis. The range of chromosome numbers varies with the section. That in *P. alpina* is probably the lowest; that in the Scabrellae of western North America, with somatic numbers from $2n = 42$ to $2n = 104$, and a mode at 84 (Hartung 1946), is probably the highest. *Poa,* therefore, has more in common with *Potentilla* than with any other agamic complex, and agrees with it to a large extent in geographic distribution. Although additional diploid sexual species of *Poa* will almost certainly be discovered, most of those which contributed originally to the agamic complex are probably extinct. Furthermore, many polyploid sexual species of *Poa* are known, and these may have contributed very largely to the agamic complex. In fact, when this genus is better known, it may have to be regarded as a single huge polyploid complex, which is in part purely sexual, in part facultatively apomictic, and which contains in addition obligate apomicts. Like *Rubus* and *Potentilla,* the bulk of this agamic complex has retained its youthfulness and its ability to produce new forms through the activity of the facultative apomicts.

The agamic complexes just discussed are all classified by Gustafsson in the second of his two series of "apomictic and amphi-apomictic complexes." They are characterized by the statement that (Gustafsson 1947c, p. 235) "The apomicts belong to two or more different complexes which merge into one another," while in the first series (p. 222) "the apomicts are included in a single complex." The distinction between two merging complexes and a single one would seem to be rather difficult to make in many instances. The groups which Gustafsson places in his first series seem to the present writer rather heterogeneous, and many of them are relatively little known. Some, like *Ranunculus ficaria, Allium carinatum-oleraceum, Saxifraga stellaris-foliolosa,* and *Hypericum perforatum,* may be relatively small and simple agamic complexes similar to the larger ones described above. Others, like *Rubus idaeus, Poa bulbosa,* and *P. compressa,* may be portions of the main agamic complex found in each of these genera, and may appear to be isolated only because their relationships have not been well studied. Still others, like *Deschampsia caespitosa* and *D. alpina, Nigritella nigra, Festuca ovina, Stellaria crassipes, Polygonum viviparum, Cardamine bulbifera, Saxifraga cernua,* and *Gagea spathacea,* may be apomictically or asexually reproducing units of otherwise sexual polyploid complexes. Finally, there is the example of *Houttuynia cordata,* which so far as known is an isolated apomictic species without living sexual relatives, and may be the last relict of an ancient agamic complex (Babcock and Stebbins 1938).

CAUSES OF VARIATION IN AGAMIC COMPLEXES

The information reviewed in the preceding section makes possible an explanation of an apparent paradox mentioned by several authors, namely, that partial or complete abandonment of sexual reproduction results in a great increase in polymorphy of the group in which this phenomenon occurs. The causes of this polymorphy may be summarized as follows: first, hybridization and allopolyploidy between the original sexual ancestors of the agamic complex; second, hybridization between facultative apomicts, with the resulting segregation, or between apomicts and sexual species; third, chromosomal and genic changes within the apomictic clones themselves.

Although the causal relationships of hybridization, polyploidy, and apomixis are still to a certain extent matters for debate, there is no denying the fact that in the great majority, if not all, of the known agamic complexes extensive hybridization has occurred in the past between the ancestral sexual species. The products of this initial hybridization were undoubtedly highly sterile, except for those which happened to acquire a combination of genes favoring apomixis. Nevertheless, since the sexual species ancestral to most agamic complexes are rather closely related to each other, their F_1 hybrids might be expected to set occasional seeds, particularly when fertilized by pollen from one or another of their parents, and so to produce a strongly segregating series of backcross types, many of which could be vigorous and well adapted to some natural habitat. Thus, the complex would be likely to possess a high degree of polymorphism, particularly among the partially sterile hybrid derivatives. In such types gene combinations producing apomixis, whether acquired by mutation or by hybridization with apomictic forms, would have a particularly high selective value. Therefore, in any group which possesses the genetic potentialities for apomixis, this process would tend to perpetuate types which because of their sterility would not persist in a sexual group. It would crystallize an initial polymorphism which existed before the onset of apomixis and which, but for the arrival of this process, would have disappeared.

The second process, hybridization between facultative apomicts and sexual forms, or between different facultative apomicts, has been demonstrated experimentally as a cause of variability, particularly in *Rubus, Potentilla,* and *Poa,* in which nearly all of the apomicts are facultative. Gustafsson (1947b, pp. 141–146) has summarized a series of hybridization experiments in these genera, all of which have had essentially similar results. Hybridization between two apomictic forms usually yields sexual F_1 plants and a great range of segregant types in the F_2 generation. Many of these are sterile, inviable, or both, but others are vigorous, and in a few of the individuals in the F_2 or later generations apomixis may be restored. Such plants, if they find an ecological niche to which they are adapted, may be the progenitors of new apomictic clones. Facultative apomixis may therefore give a pattern of variation and a type of evolution which is in some ways similar to that found in self-fertilized organisms. The constancy of such apomictic clones, like that of pure lines, is periodically interrupted by bursts of new variability and selection, caused by hybridization. The major difference between the patterns created by these two types of reproduction is that in predominantly self-fertilized species the modes of variation represented by genetically isolated species are usually maintained, while in agamic complexes the reproductive ability of sexually sterile hybrid derivatives tends to obscure or to obliterate the original species boundaries.

Variability resulting from hybridization is still possible in some obligate apomicts. In the genus *Hieracium*, the classic hybridization experiments of Mendel, as well as the more recent ones by Ostenfeld (1910) and by Christoff (1942), have shown that fertilization of a sexual species by pollen from an apomict yields a great variety of segregant types in the F_1 generation, because of the heterozygosity of the parents. Some of these are wholly or partly apomictic and may give rise immediately to new clones, while others may segregate further and produce still more variant apomicts in later generations. It is obvious, however, that when obligate apomixis has set in, new variability through hybridization and segregation is confined to the regions in which sexual species occur and to those apomicts which still have functional pollen. For this reason, we should expect to find a conspicuous difference between the variation patterns of agamic complexes having facultative as compared to those with obligate apomixis. In the former, variability would be rather evenly spread throughout the geographic range of the complex, while in the latter it would be highly concentrated in those regions containing sexual species. The former situation appears to be true in *Rubus, Potentilla,* and *Poa,* while the latter is the case in *Crepis, Antennaria,* and probably *Taraxacum.*

The total amount of polymorphism would be expected to be greater in complexes containing facultative apomicts as compared to those with obligate apomicts, but this does not appear to be the case. The probable explanation of this situation is that obligate apomicts do not arise directly from sexual forms, but from facultative apomicts. This was postulated by Babcock and Stebbins (1938, Stebbins 1941a) on the basis of their study of *Crepis*. But Gustafsson (1935a) believed that the obligate apomicts arise directly in *Taraxacum* and *Hieracium*. He based this belief on the observation that facultative apomixis is usually associated with apospory, and obligate apomixis with diplospory. He maintained that "an entirely new type of division had to be introduced in the peculiar kind of parthenogenesis occurring in *Antennaria* and *Hieracium*" and that this type of division cannot occur in the same plant with the formation of reduced megaspores through normal meiosis. However, precisely this situation has been found in *Parthenium argentatum* (Esau 1946), fulfilling the prediction of Stebbins (1941a, p. 531). There is good reason, therefore, for believing that much of the polymorphism now found in *Antennaria, Taraxacum, Hieracium, Calamagrostis,* and other complexes with obligate apomixis was produced in an earlier stage of their evolution, when facultative apomixis was still widespread in them.

Various types of chromosomal and genic changes can occur within the apomicts. The first is that postulated by Darlington (1937) as a result of meiotic pairing and crossing over in the megaspore mother cells, with subsequent restoration of the diploid chromosome number in the egg through the formation of restitution nuclei or by some other means. Crane and Thomas (1939) postulate this mechanism for the origin of certain variants in *Rubus,* but do not present cytological evidence. Rollins (1945b) has suggested a similar explanation for variation in the apomictic progeny of *Parthenium*. Gustafsson (1943, 1947c) called the phenomenon autosegregation and discussed several cytological mechanisms, none of which is known to occur in these apomicts. Second, chromosomal aberrations may occur, particularly in the divisions leading to the diplosporous embryo sacs, which will cause the egg to have one or more chromosomes less than the normal number. The best example of this is the series of monosomics found by Sörenson and Gudjónsson (1946) in *Taraxacum.* In *Taraxacum,* as in most other genera, however, these chromosomal aberrants are less viable than euploid plants, and for this reason do not become established in nature. Only in *Poa* and in *Potentilla* are aneuploid apomicts abundant. Finally, apomictic clones may produce new variants by means of somatic mutations. These have been described as occasional occurrences in *Rubus* (Gustafsson 1943, p. 78), but they are better known in *Citrus* (Frost 1926). Because of the genetic heterozygosity of most apomicts, such mutations will usually affect the appearance of the phenotype, and if they occur in the tissue of the inflorescence or flower, they will produce mutant offspring.

SPECIES CONCEPTS IN AGAMIC COMPLEXES

The species concept developed in Chapter VI, which resembles the concepts maintained by the majority of those who desire a truly biological concept of species, centers about the possibility for exchange of genes between members of the same species and the separation of different species by barriers to the exchange of genes. As Dobzhansky (1941) and Babcock and Stebbins (1938) have pointed out, such a species concept cannot be applied to agamic complexes. Free interchange of genes between apomicts is prevented by the very nature of their types of reproduction, while the origin of many apomictic clones is from genotypes which have combined the genes of previously isolated sexual species, and which without apomixis would not be able to persist because of their sexual sterility. It is not strange, therefore, that systematists have not been able to agree on the boundaries of species in such genera as *Taraxacum, Hieracium, Antennaria, Rubus, Potentilla, Poa,* and *Calamagrostis.* In attempting to set up species like those found in sexual groups, they are looking for entities which in the biological sense are not there. Nor are those "splitters" who make a separate species out of every apomictic clone of *Taraxacum, Hieracium,* or *Rubus* likely to provide any better concepts of the variation patterns in the genera concerned. Criticisms of this method have been given by Müntzing, Tedin, and Turesson (1931), by Fernald (1933), by Turrill (1938c), by Babcock and Stebbins (1938), and by Stebbins (1941a). One criticism is that the number of apomicts in any well-developed complex is so large that recognizing them as separate species makes the comprehension of the group as a whole difficult or impossible; "one cannot see the forest for the trees." But more important is the fact that in many complexes the great majority of the apomicts, and in nearly all complexes at least some of them, are only partly or facultatively apomictic. From time to time they reproduce sexually, and on such occasions a whole series of new clones, or "species," may arise in the offspring of a single individual. Such forms may have sexual relatives in which a comparable amount of genetic variation occurs regularly in each generation. A *reductio ad absurdum* would be reached by the treatment as separate species of the apomicts in an entity like *Crepis acuminata*. Here the apomicts, indistinguishable in external morphology from their sexual ancestors, differ from each other in precisely those characteristics which segregate among the sister individuals of any sexual progeny (see diagram, Fig. 38).

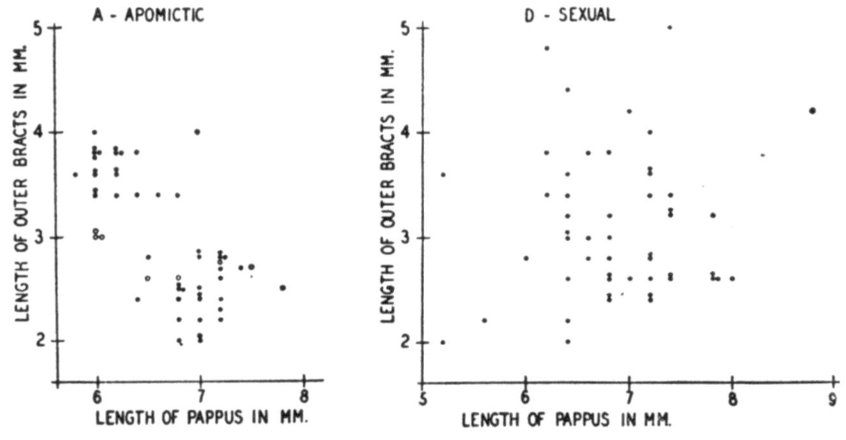

Fig. 38. Diagrams showing the range of variation in respect to length of outer involucral bracts and of pappus on the mature fruits within a colony of apomictic and of sexual *Crepis acuminata*. From Babcock and Stebbins 1938.

As Babcock and Stebbins (1938) have pointed out, the needs of classification and the clearest conception of the structure, interrelationships, and phylogeny of the agamic complex are obtained by drawing arbitrary species boundaries which are related to the original boundaries of the sexual species ancestral to the complex. The number of species recognized should not be smaller than the number of sexual ancestors which are known to exist or can be reasonably inferred to have existed in the past. In most complexes, furthermore, the number of species should be larger, since groups of apomicts combining the characteristics of sexual species which are widely different from each other should be recognized as agamospecies (Turesson 1929). The boundaries between these species must be drawn where most convenient, usually on the basis of one or a few easily recognized "key characters" which are known to separate the ancestral sexual species.

If, as in *Rubus* and in *Poa*, the sexual ancestors of the complex are mostly extinct, this method is difficult or impossible to apply. Here the method, suggested by various European systematists and followed by Gustafsson (1943, 1947c, p. 245), of recognizing as "circle microspecies" groups of apomicts morphologically similar to each other, with each species centered about one or a few particularly widespread and common apomicts, is perhaps the only feasible one. Consciously or unconsciously, it has been regularly used in practice. Nevertheless, the use of such a method must be recognized simply as a means of bringing some type of order to a situation which from the biological point of view is incapable of resolution. Disputes between systematists not fully conscious of the biological situation in these groups are particularly academic and futile.

AGAMIC COMPLEXES AND PLANT GEOGRAPHY

There are some features of agamic complexes which make them even better tools for analyses of plant distribution than are sexual polyploid complexes (Gustafsson 1935b, 1947c; Turrill 1938c,d; Babcock and Stebbins 1938; Stebbins 1941a). In the first place, the apomicts must have arisen from sexual ancestors and cannot have diverged very far from them in morphological characteristics. The identification of these ancestors should in most instances be possible by careful morphological and cytogenetical studies. If the apomicts are of allopolyploid origin, the same inferences may be drawn about past distributions of the sexual diploid members of the complex as is possible with sexual polyploids (see page 350). Secondly, the relative constancy of apomicts through hundreds or even thousands of generations makes them valuable indicators of ecological conditions, both in their present habitats and in past migration routes.

In the case of agamic complexes like that of *Crepis,* in which all of the sexual ancestors still exist and are concentrated in one or a few geographic areas, these regions must be regarded as primary centers of differentiation, from which migrants have traveled to the peripheral portions of the range of the complex. Obviously, one method of ascertaining centers of migration for entire floras would be to analyze the distribution of the sexual members of all the agamic complexes found in them. This might be particularly successful in floras like those of the arctic regions, which contain many agamic complexes, and in which present distribution patterns are the result of relatively recent migrations. Two suggestions are already possible in this connection.

In *Taraxacum,* the most numerous apomicts of the arctic regions, belonging to the section Ceratophora, have their closest relatives among the sexual species in those from central Asia, particularly the Thian Shan Mountains, studied by Poddubnaja-Arnoldi and Dianova (1934). This central Asiatic *refugium* is therefore the most likely center from which the arctic *Taraxaca* radiated in postglacial times. In *Antennaria,* on the other hand, the most common apomicts in the arctic regions belong to the *A. alpina* group, of which the sexual members are confined to the New World. The probable sexual species (judging by the abundance of male plants) which are morphologically most similar to *A. alpina* and its relatives are *A. monocephala* and *A. alaskana* (Malte 1934). Comparison by the present writer of specimens of these three makes rather likely the hypothesis that *A. alpina* and the arctic American "species" related to it are allopolyploid derivatives of these two sexual species. Since *A. alaskana* is confined to Alaska, and *A. monocephala* is more abundant there than anywhere else, the apomicts of the *A. alpina* complex are best explained as radiants from the Beringian center of Hultén (1937).

As described above, the agamic complex of *Rubus* is one in which certain ancestral diploid sexual species are no longer living in the region occupied by most of the apomicts, but occur in North America. On these grounds, Gustafsson has made the plausible inference that at some time in the past, probably during the latter part of the Tertiary period, some species of *Rubus,* morphologically and ecologically similar to the modern North American *R. allegheniensis* or *R. argutus*, existed in Europe, so as to give rise to the *Suberecti* and related sections. This evidence is in line with much of that from paleontology which indicates that the Tertiary floras of the Old and the New Worlds had considerably more in common than do the modern ones, and it runs parallel to the evidence from *Oryzopsis* allopolyploids (see page 353) that Eurasian species of this genus were once represented in North America.

The constancy of apomicts, and the fact that the ancestry of many of them may be determined, makes them particularly good indicators of past migration routes of floras. Gustafsson (1935b, 1947c) used apomicts of *Taraxacum, Hieracium,* and *Poa* to help solve certain problems of local distribution in Scandinavia, while Babcock and Stebbins (1938, pp. 39–44) have used the apomicts of *Crepis* in a similar manner in connection with the flora of the western United States. If the detailed study and classification of the endless "microspecies," or apomictic clones of the agamic complexes, has any general scientific value, it is in this direction. A broader problem of this nature, toward the solution of which apomicts may be valuable, is that of the disjunct ranges of many species in the temperate zones of the northern and southern hemispheres — the "bipolar" distributions discussed by DuRietz (1940). The fact that apomicts of both *Taraxacum* and *Antennaria* occur in temperate and subantarctic South America, while their possible sexual ancestors occur only in the temperate and subarctic regions of the Northern Hemisphere, indicates that these apomicts must somehow have migrated across the tropics. Because of the relatively narrow limits of ecological tolerance possessed by most apomictic clones, we must suppose that they followed a route which had ecological conditions essentially similar to those found in their present habitat, unless they were dispersed southward by long-distance migration. Some idea of their ability to cross the tropics along the crests of mountain ranges could be obtained by suitable experiments on growing them in such regions, from which they are at present completely absent. By the use of apomicts, therefore, at least one phase of this fascinating and puzzling problem of plant geography is open to experimental attack.

THE EVOLUTIONARY SIGNIFICANCE OF AGAMIC COMPLEXES

Apomixis has three principal effects on the groups in which it occurs. In the first place, it makes possible the survival of many genotypes which are vigorous and well adapted to their surroundings, but which because of their sexual sterility would be unable to survive were it not for this process. In a group of strictly diploid sexual organisms, new combinations of genes for adaptation to new habitats must be formed from a series of separate "gene funds," each of which is largely limited by the isolating barriers which bound the different species. Allopolyploidy permits the "pooling" of the genetic resources of two or more species to an extent limited by the factors which render allopolyploids fertile and constant, and to that extent a polyploid complex is better equipped than a group of comparable diploid species for the colonization of newly available habitats. With apomixis, the possibilities for pooling the genetic resources of several distinct species are still further increased, so that agamic complexes are even better equipped than are sexual polyploid complexes for the rapid evolution of new genotypes adapted to new habitats suddenly made available. They represent evolutionary opportunism carried to its limit. It is no accident, therefore, that apomicts of such genera as *Taraxacum, Antennaria, Hieracium, Rubus,* and *Crataegus* have "weedy" tendencies, and that agamic complexes are prominent in the arctic and subarctic floras which have recently been subjected to alternations of glacial and temperate climates.

Thus, the concept of Darlington (1939), that apomixis is an "escape from sterility," can be taken only in the relatively limited sense that it permits the survival of well-adapted, but sexually sterile, genotypes. Gustafsson (1947b, p. 173) has pointed out that many apomicts are still able to produce viable and fertile sexual offspring, so that the selective value of apomixis cannot be confined to this function.

The second effect of apomixis is similar to that of self-fertilization; it permits the building up of large populations of genetically similar individuals for the rapid colonization of newly available habitats. Thus the "opportunism" of agamic complexes results both from the increase in the number of adaptive genotypes available and from the ease with which such genotypes may be perpetuated. The fact is doubtless significant that apomixis and self-fertilization appear as mutually exclusive genetic systems. So far as the present writer is aware, the sexual ancestors of all known agamic complexes are self-incompatible, dioecious, or possess other mechanisms which promote allogamy. Apomixis has not developed in self-fertilizing plants.

One probable reason for this situation is that the type of growth habit which favors apomixis is apparently the opposite of that which promotes self-fertilization. Data presented in Chapter V (see Table 4) show that self-fertilization is predominant in annuals and in short-lived perennials, and is rare or absent in perennials with rhizomes or other effective means of vegetative reproduction. On the other hand, Gustafsson (1948a) has emphasized the fact that the sexual ancestors of agamic complexes are nearly all perennials, and many of them possess accessory methods of vegetative reproduction, such as rhizomes or stolons. This situation is probably explained by the fact that the change from cross- to self-fertilization can be accomplished by a simple genetic mechanism which does not necessarily affect fertility, while the acquisition of an agamospermous method of apomixis is usually associated with sexual sterility, either through interspecific hybridization or through genetic factors causing a disturbance of meiosis. So, just as in the case of polyploidy, the evolutionary line which is acquiring apomixis must pass through a bottleneck of partial sterility, and efficient vegetative means of growth and reproduction preadapt plants to these conditions. It is probably for this reason that, although apomixis would be expected to promote weedy tendencies, apomicts are not conspicuous among the weedy flora of cultivated fields and roadsides. The plants of such habitats are mostly annuals, and acquire through self-fertilization the constancy which helps them in colonization.

The third effect of apomixis is the limitation of genetic variability imposed on those plants which have adopted this method of reproduction. Over a short period of time, agamic complexes are capable of a rapid burst of evolution in terms of the production and establishment of new genotypes. But once an agamic complex has come to consist chiefly of apomicts, its evolutionary future is decidedly limited. If obligate apomixis has become predominant, new evolutionary progress is bound up with the distribution and potentialities of the remaining sexual species. If these species become much restricted geographically and ecologically, the evolutionary plasticity of the complex is lost, and it can no longer evolve in response to changing environments. Babcock and Stebbins (1938, pp. 61–62) have pointed out that aggressive, weedy tendencies are evident in those apomicts of *Crepis* related to the widespread sexual *C. acuminata,* while the apomicts related to the highly restricted sexual species *C. pleurocarpa* and *C. monticola* are themselves relictual types, showing little or no tendency to spread. An agamic complex with predominant obligate apomixis is a closed system which can produce little or no variation beyond that already present in the sum total of its sexual ancestors. As long as recombinations of these ancestral characteristics will yield gene combinations which are adapted to some available environment, evolution can continue, but when climatic changes or the evolution of more efficient competitors has started to encroach on the environments to which these genetic recombinations are suited, the agamic complex will gradually decline and die out.

In complexes like those of *Rubus, Poa,* and *Potentilla,* which contain chiefly facultative apomicts, the potentiality for new variation is independent of the fate of the sexual species. Such complexes have a longer lease on life, acquired by a favorably adjusted balance between apomictic and sexual reproduction (Gustafsson 1947b, pp. 173–178). Judging from their present distribution, the complexes of *Rubus, Potentilla,* and *Poa* are relatively old, since they are well represented in the temperate regions of both hemispheres and appear to have acquired their initial distribution along with the mid-Tertiary Holarctic flora (see Chapter XIV). On the other hand, most of the complexes with obligate apomixis show their relative youth by their restriction to a single hemisphere or to one hemisphere plus arctic-alpine regions elsewhere on the earth. Furthermore, the apomicts of these complexes with facultative apomixis have for the most part retained their aggressive tendencies, while in most complexes with obligate apomixis there are some groups of apomicts which appear as "conservative," relictual types, such as the *Crepis* apomicts mentioned above and certain forms of *Antennaria* and *Taraxacum* in Newfoundland and eastern Quebec (Fernald 1933).

Nevertheless, even in complexes with facultative apomixis, there is little evidence that they can give rise to anything except new recombinations of the genes present in the original sexual ancestors. There is no evidence that apomicts have ever been able to evolve a new genus or even a subgenus. In this sense, all agamic complexes are closed systems and evolutionary "blind alleys." It is true, as Gustafsson (1947b, p. 178) has maintained, that both sexual and apomictic groups are born, live their span upon the earth, and sooner or later die out. But other things being equal, the life expectancy of agamic complexes is shorter than that of sexual groups. Furthermore, while sexual species may, during the course of their existence, give rise to entirely new types by means of progressive mutation and gene recombination, agamic complexes are destined to produce only new variations on an old theme.

APOMIXIS AND PLANT BREEDING

The fact that apomixis occurs in a number of economically valuable crop plants has rendered a more than academic interest to studies of this phenomenon. The most important apomictic crop plants are the various species of *Citrus,* but apomixis is known or suspected in other cultivated fruits, such as the mango (H. J. Webber 1931, Juliano and Cuevas 1932, Juliano 1934, 1937), the mangosteen (Sprecher 1919, Horn 1940), and blackberries (see above). It occurs in the rubber-bearing plant guayule (*Parthenium argentatum,* see above), as well as in species of *Taraxacum* related to another rubber plant, *T. kok-saghyz.* Among forage grasses, it is conspicuous in *Poa pratensis* and apparently also in *Paspalum* (Burton 1948), while it occurs in some groups of ornamental shrubs, particularly *Malus* (crabapples, Dermen 1936b) and *Eugenia* (Pijl 1934, Johnson 1936). It is obvious that breeding programs on such groups must be conducted differently from those on sexual plants, and plant breeders who are planning a program on some group unknown cytogenetically, particularly if it belongs to the Gramineae, the Rosaceae, or the Compositae, should include among their preliminary explorations studies designed to determine whether or not apomixis is present.

As is evident from the discussion in this chapter, the presence or absence of apomixis in a group cannot be determined simply by castration experiments or by a few casually conducted experiments of emasculation and cross-pollination to determine whether or not hybrids can be obtained. A large proportion of apomicts are pseudogamous and need pollination for the successful production of seed just as much as do sexual species. Furthermore, many apomicts, being facultative, can yield occasional hybrids when pollinated by a different species, so that the production of hybrids in small numbers is insufficient indication that no apomixis is present.

Positive evidence for the presence or absence of apomixis can be obtained only from laborious and time-consuming studies of megaspore, embryo sac, and embryo development. But properly conducted breeding tests on a sufficiently large scale should in most groups provide reasonably decisive indirect evidence. If emasculation is possible, this should be carried out on a large scale, and the stigmas of the emasculated flowers should be pollinated with either a foreign species pollen or, preferably, with pollen from a different genotype bearing a dominant marker gene, as was done by Burton (1948) in *Paspalum.* If cross-pollinations of this type repeatedly yield a consistent percentage of strictly maternal individuals, apomixis can be suspected. If the flowers and anthers of the species are so small that emasculation is impracticable, the pollen should at least be examined through the microscope to see whether it stains well, and stained dissections of the styles and the stigmas of flowers bagged at anthesis should be examined for growing pollen tubes. If this is found, and if the resulting progeny are uniform, self-fertilization rather than apomixis is to be suspected; since, as mentioned in the preceding section, these two processes do not usually occur in the same group.

In groups known to be apomictic, the degree of apomixis should be determined by appropriate progeny tests (Powers and Rollins 1945), since facultative apomicts can be treated entirely differently from obligate ones. Fortunately, all the economically important groups mentioned above are, so far as is known, facultatively apomictic, and interspecific hybrids have been obtained in many of them. If any economic group is found to contain mostly obligate apomicts, the genus should be explored for sexual species, and pollen from these should be employed for breeding whenever possible. Even in groups containing mostly facultative apomicts, the sexual species may contain a supply of genes for disease resistance and other valuable properties which can by no means be ignored. Obviously, a study of the method of inheritance of apomixis in such groups is one of the next essential steps, so that estimates can be made of the extent to which apomixis will be recovered in the progeny of hybrids.

Selection and progeny testing in apomictic groups must obviously be conducted on an entirely different basis from that in sexual ones. Tinney and Aamodt (1940) in *Poa pratensis* and Rollins, Catcheside, and Gerstel (1947) in *Parthenium argentatum* have made some suggestions as to how this can be done. In *Poa,* the genetic constancy of seed produced by apomixis makes possible the testing of some varieties without the time-consuming process of establishing numerous clonal divisions from a single plant. Selection of new strains could be made from the occasional sexual plants found in most strains. In *Parthenium,* variety testing could be carried out as in *Poa.* But here there is a particularly useful source of new genetic variability, namely, the 36-chromosome "haploids" which occasionally develop from reduced eggs of normally apomictic 72-chromosome plants. These haploids inherit from their mothers the tendency to produce chiefly unreduced gametes; therefore, when pollinated by 72-chromosome plants they produce some offspring with this number of chromosomes. Since genetic segregation has occurred both in the production of the egg which gave rise to the haploid plant and in the pollen grains which fertilize its eggs, considerable genetic variation can be expected among the progeny of haploids pollinated by diploids. But each individual of this variant progeny will breed true through apomixis. This ingenious scheme for obtaining both the variability needed for selection and the constancy needed for testing, increase, and commercial use of a variety may be applicable only to guayule and to a few other apomictic species with similar properties. Nevertheless, there is little doubt that intensive studies of the fundamental nature of apomixis in other groups of economically important plants will be a valuable and perhaps essential prelude to the breeding of improved varieties in them.

CHAPTER XI

Structural Hybridity and the Genetic System

In Chapter III the occurrence of structural changes in the chromosomes, chiefly inversions and interchanges, was described, and their significance was said to be connected mainly with their effect on the genetic system and on the production of isolating barriers between species. When they affect the genetic system, these structural changes act mainly as a means of holding together certain favorable gene combinations (Darlington 1937, 1939) and therefore of promoting immediate fitness at the expense of flexibility. They do this in various ways, depending on the type and number of structural changes which have accumulated in the species population.

As has been already stated by Darlington (1937), by Dobzhansky (1941, p. 126), and by others, plants showing at meiosis chromatin bridges and acentric fragments, which reveal the condition of heterozygosity for inversions, are well known not only among interspecific hybrids but also in typical individuals of a number of different species. In some of these, like *Paris quadrifolia* (Geitler 1937, 1938) and various species of *Paeonia* (Dark 1936, Stebbins 1938a), every individual studied was found to be heterozygous for one or more inversions, and these structural changes were distributed through every chromosome arm of the entire complement.

Since plants heterozygous even for a considerable number of inversions are usually fertile enough so that their capacity for sexual reproduction is not seriously hindered, inversions by themselves are rarely, if ever, responsible for the isolation barriers separating species. Their importance, therefore, is mainly their effect on the genetic system, which is due to the fact that chromosomes which result from crossing over in an inverted segment are usually inviable because of a deficiency for several genes

(Darlington 1937, Sturtevant and Beadle 1939). The genes in the inverted segment are effectively and permanently linked, and any combination of mutations which becomes fixed in a particular inverted segment will thereafter remain constant. If such a combination has a selective value in some habitat occupied by the species, the inversion will acquire a corresponding value as a preserver of the combination and will therefore aid in subdividing the species into races. Although the existence of inversions with such a role has not been established in any plant species, there is indirect evidence for it. Darlington and Gairdner (1937) found that in *Campanula persicifolia* crosses between plants from different geographic regions show in general more structural hybridity than wild plants from any one region. Stebbins (1938a) found a greater degree of structural hybridity for inversions in species of *Paeonia* which are subdivided into a number of geographical races than in those which are more nearly uniform. The more complex situation found in *Drosophila pseudoobscura* (Dobzhansky 1947a,b), in which the inversion heterozygote contains a gene combination superior to those of any structural homozygote, has not yet been detected in plants. The importance of this discovery in connection with heterozygosity for translocations is, however, considerable, and will be discussed below.

Inversions, therefore, appear from time to time in most species and exist in their populations as "floating" structural differences, neutral in effect. If they become associated by chance with gene combinations of selective value, they may become established as permanent acquisitions of the species population, either in the homozygous condition, in which they promote racial differentiation, or as heterozygotes. In either case they alter the genetic mechanism in such a way that it produces more constancy and fitness at the expense of flexibility.

Structural hybridity for translocations, or segmental interchanges, has played a more important role in plant evolution than has inversion heterozygosity, and in some groups it has come to dominate the genetic system. Occasional translocation heterozygotes, recognizable by the formation at meiosis of rings or chains of four chromosomes, are known in species of *Tradescantia, Allium, Tulipa, Salix, Paeonia, Matthiola, Pisum, Gaura, Clarkia, Godetia, Polemonium, Galeopsis, Datura, Notonia, Lactuca,* and many other genera. Extensive references to literature can be found in Darlington (1937) and Dobzhansky (1941). In most species these translocations are probably "floating" in the cross-fertilized population, as has been demonstrated for *Campanula persicifolia* (Darlington and Gairdner 1937). In this condition they have very little evolutionary significance, and any individual translocation type may exist only temporarily in the population. At a later stage of evolution, however, certain interchange types may spread so widely in the population that they can be recognized as chromosomal races or "prime types," as they are known in *Datura* (Blakeslee, Bergner, and Avery 1937, Bergner 1943).

THE GENETIC SYSTEMS OF *Oenothera*

The classic group of plants for studies of hybridity for segmental interchanges is the subgenus *Euoenothera* of *Oenothera*.[1] The difficulties of delimiting species in this group have long been recognized by systematists, and the anomalous cytological and genetic condition of the group was made known by a series of workers beginning soon after the completion of De Vries's well-known studies. Cleland (1936) has presented a review of the extensive literature on this group and a clear exposition of its cytogenetic peculiarities. In later publications (1940, 1944) he has surveyed the variation pattern in the subgenus as a whole, particularly in relation to the geographic distribution of the different cytogenetic types. The present discussion and interpretation is based partly on these three publications.

The most normal members of this subgenus are the group of *O. hookeri*, found in California and the adjacent states. These plants are large-flowered and open-pollinated, so that, although they are self-compatible and therefore may be inbred, they nevertheless are frequently cross-fertilized and have some degree of genetic heterozygosity (Cleland 1935). They are for the most part structurally homozygous in respect to segmental interchanges, but plants heterozygous for one or two interchanges are not uncommon. The situation here is probably not unlike that found in *Datura stramonium*, although the geographic distribution of particular segmental arrangements within this group has not been worked out.

[1] Munz (1949) has recently shown that the correct name for the subgenus referred to by most authors as *Onagra* is *Euoenothera*.

The next group, designated by Cleland the *irrigua* alliance, is centered in New Mexico, but forms with similar cytological behavior have been found in Utah, Colorado, Oklahoma, and Mexico. Most of these forms are considered by Munz (1949) to be varieties of *Oenothera hookeri*. The members of this alliance are structural heterozygotes which form at meiosis rings of intermediate size, mostly with four, six, or eight chromosomes. These rings, however, are not permanent, since progenies of these plants always contain some individuals with seven pairs. The frequency of structural heterozygosity in these forms is apparently due to the fact that, like the *hookeri* alliance, they are open-pollinated and frequently outcrossed, and several different structural arrangements occur in the group with about equal frequency.

There are two other isolated species of the subgenus *Euoenothera* which, like the *hookeri* alliance, are essentially normal cytogenetically. One is *O. argillicola,* a large-flowered, open-pollinated, seven-paired type endemic to the shale barrens of the Appalachian mountain system, and consequently rather local in distribution. The other is a seven-paired form of the *grandiflora* alliance, recently found in Alabama (Cleland, oral communication). The distribution and relationships of the latter type are as yet imperfectly known. *O. argillicola* differs strikingly in external morphology from the western structural homozygotes, particularly in its glabrous character, narrow leaves, and bent stems. Its chromosomal arrangement, however, differs by only two interchanges from the one most common in the *hookeri* alliance.

The remaining Oenotheras of the subgenus *Euoenothera* are of the anomalous cytogenetic type known as complex-heterozygotes. They regularly form rings rather than pairs of chromosomes at meiosis, the complete ring of 14 being by far the most common arrangement in naturally occurring races. They are usually self-pollinated, and generally seeds grown from any individual produce offspring which are exactly like each other and the maternal plant, both in external morphology and in the presence of chromosome-ring formation at meiosis. Phenotypically, they appear like the individuals of a pure line of a typical self-fertilized plant. Genotypically, however, they are entirely different. This fact becomes evident immediately when reciprocal crosses are made, either between two races of this type or between such a race and a structurally homozygous seven-paired Oenothera. In either case the reciprocal hybrids are likely to be entirely different from each other. If, for instance, a race such as "Iowa 1," belonging to the group usually recognized by systematists as *O. biennis,* is crossed with one of *O. hookeri,* most of the hybrids produced by the egg of "Iowa" and the pollen of *O. hookeri* are strongly maternal in appearance, while the reciprocal hybrids are very different and more like *O. hookeri.* Similar reciprocal differences and resemblances to the maternal parent would be found on crossing "Iowa 1" to a small-flowered, grayish-pubescent, ring-forming race such as *O. cockerellii,* from Colorado. The selfed progeny of these two sets of hybrids, however, would be very different from each other. Those from the hybrids with *O. hookeri* would usually segregate for both their morphological and cytological characteristics, while the F_1 hybrids between two ring-forming races would produce relatively constant progeny, and in many cases would breed absolutely true.

A great mass of evidence accumulated by a number of workers has established firmly the following explanation for this anomalous behavior. In the first place, the chromosome rings are nearly always oriented on the meiotic metaphase spindle in a particular way, so that alternate chromosomes go to the same pole (Fig. 39).

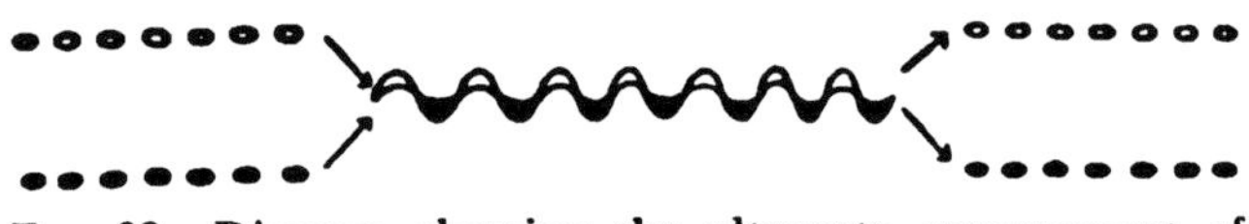

FIG. 39. Diagram showing the alternate arrangement of the chromosomes at first metaphase of meiosis in a ring-forming type of *Oenothera* and showing how this arrangement leads to the segregation to opposite poles of the spindle of an entire set of chromosomes of paternal origin (white) and of maternal origin (black). From Cleland 1935.

Because of this fact, only two types of gametes are regularly formed by any particular race. The seven chromosomes which enter any gamete, therefore, bear one of two particular complements of genes and a definite arrangement of the chromosome ends. They are termed *complexes,* or *Renner complexes,* since they were first recognized by Renner (1917, 1921). The complex, although it contains seven chromosomes, can be regarded genetically as a single linkage group. Each race of the ring-forming Oenotheras is termed a complex-heterozygote because it must contain two complexes which differ from each other in both the genic content of the chromosomes and the arrangement of their ends. This is because the complexes contain systems of *balanced lethal factors* (genes or chromosomal segments), which usually cause the death of zygotes containing two identical complexes. In most Oenotheras, however, such homozygous zygotes are rarely or never formed, because of a peculiar property of the lethal factors which causes one of the two complexes to be carried only through the pollen and to cause the death of the megaspore bearing it, while the other complex is lethal in the pollen, but viable in the megaspore and the egg. Races of this type, which is the most common one, have 50 percent of abortive pollen, but an almost perfect seed set, due to the fact that in nearly every ovule one of the megaspores bearing the viable complex, regardless of its position in the tetrad, develops at the expense of its sisters. This is the phenomenon known as the "Renner effect." In a smaller number of races, typified by *O. lamarckiana* and a few of its relatives, which are known only in gardens and as adventive weeds in Europe, both complexes are carried through the pollen as well as the egg, but the zygotes receiving similar complexes fail to develop. In races of this type the pollen is highly fertile, but one half or more of the seeds are abortive. The two situations are illustrated in the diagram, Fig. 40.

The variation pattern produced by such a genetic system is superficially similar to that found in ordinary self-fertilizing species, in that there is a large series of closely related biotypes, most of them represented in nature by thousands of identical or very similar individuals and breeding absolutely true under ordinary conditions. The course of evolution in such a group, as well as its evolutionary potentialities, is, however, entirely different. As was mentioned above, the result of occasional crossing between different biotypes of a self-fertilized species is a short burst of evolution, in which the strong segregation of genes in the F_2 and later generations presents a large number of new biotypes for the action of natural selection. Occasional crossing between complex-heterozygote biotypes of *Oenothera*, on the other hand, has a much more limited effect. As a rule, only two

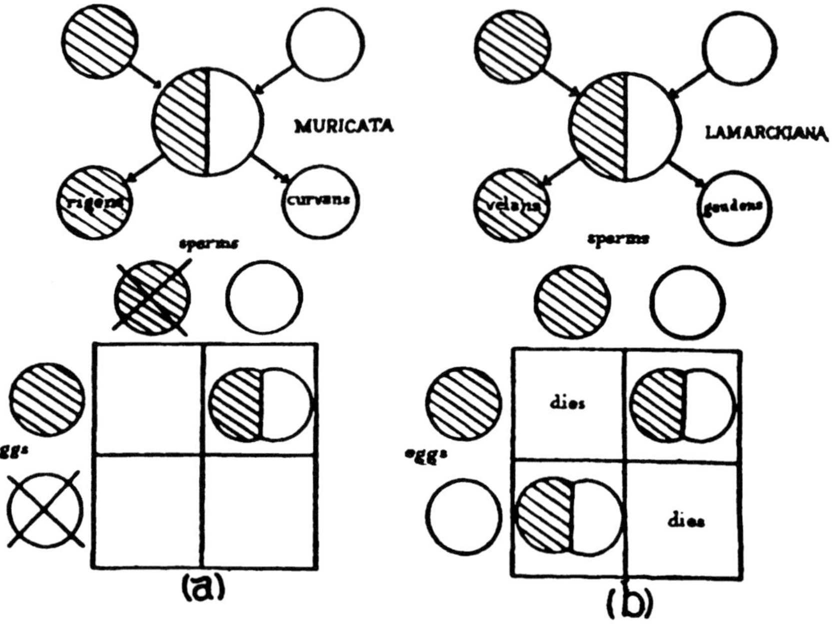

FIG. 40. Diagram showing the two types of balanced lethal mechanisms in *Oenothera*. At left, gametic lethals in *O. muricata*. The rigens complex is prevented by a lethal from functioning as sperm; *curvans* is prevented from functioning as egg. At right, zygotic lethals in *O. lamarckiana*. Further explanation in the text. From Cleland 1935.

new biotypes are produced, which may be either closely similar to or widely different from the parental ones. Some of these are capable of a limited amount of segregation, which usually involves rearrangement of the chromosomes or chromosomal segments in the ring, but this is much less than that found in biotypes of ordinary species which have a comparable degree of heterozygosity.

Another difference between the evolutionary behavior of complex-heterozygotes and that of ordinary self-fertilized biotypes consists in the genetic changes which may take place in them independently of crossing. In the latter, the only source of variability other than hybridization is through gene or point mutations. These occur very rarely, and their chance of establishment in the population is extremely small. Furthermore, most of them either are harmful to the organism or affect the genotype so slightly that they can have very little effect on evolution. In the complex-heterozygotes, on the other hand, mutations can be retained much longer, since they are maintained in the heterozygous condition. Also, there are two additional types of genetic change which can give the phenotypic effect of mutations. One of these is the occurrence of additional interchanges between individual chromosomes, either those belonging to the same complex or to different ones. A large proportion of such interchanges will produce unbalanced, nonfunctional complexes, but some of them may give rise to new, balanced complexes with different gene contents and pairing relationships from their progenitors.

Variation also occurs through crossing over in segments, known as differential segments, which are homologous, but are located on chromosomes of which the ends are not homologous to each other. As Darlington (1931, 1937, 1939) first pointed out, the effect of the series of interchanges which produces the ring formation in complex-heterozygotes is to produce two sets of chromosomes in which the proximal or median segments of those chromosomes whose ends are paired are not homologous to each other, and consequently cannot pair and cross over in normal fashion. These proximal segments are therefore known as *differential segments* and are capable of independent, divergent evolution through the accumulation of gene mutations. The systems of balanced lethal factors which maintain the heterozygous condition of these forms must reside in the differential segments. Any complex-heterozygote, in order to be viable, must have two of each genetic type of differential segment, but the location of these segments in relation to the pairing ends will be very different in the two complexes.

If, therefore, pairing, chiasma formation, and crossing over occur between two homologous differential segments, there results a type of figure-of-eight, in which two chromosomes are paired interstitially with each other and terminally with four other chromosome ends (see diagrams in Darlington 1937, pp. 345–356). The best evidence that this type of behavior can occur is the observance of occasional interstitial chiasmata and figures-of-eight at meiosis in *Oenothera* (Darlington 1931, Sweet 1938). Its result is in effect the crossing over of whole complexes. One half or more of the combinations of chromosomes which result from the segregation of the elements of the figure-of-eight will be nondisjunctional and therefore inviable, and two of them will represent the two original complexes (Emerson 1936, Sweet 1938). But for each such crossover, two new viable combinations of chromosomes or complexes will be formed, each of which will contain some of the chromosomal material from one and some from the other of the original two complexes. Gametes bearing one or the other of these two new complexes, when mated with unaltered gametes, will give zygotes homozygous for half of a complex and heterozygous for the other half. From these types, which are known as half mutants, completely homozygous, pair-forming types, known as full mutants, can be formed by segregation. According to the genetic evidence of Emerson (1936), crossing over in the differential segments of some types of complex-heterozygotes, such as the wild race of the *O. biennis* group known as *O. pratincola,* can yield a regular percentage of "mutants" which retain the balanced lethal condition.

That the origin of new complexes by this "mutational" process may be and perhaps usually is more complex than the two steps just outlined is evident from the studies of Catcheside (1940) and Cleland (1942) on derivatives of *O. lamarckiana.* Catcheside found that three different complexes which originated in experimental cultures of *O. lamarckiana* differ from either of the two complexes of that species by two or three interchanges. Cleland independently verified this fact for ^{h}*decipiens,*[2] which has the same arrangement as *rubricalyx,* analyzed by Catcheside. He found that a number of possible combinations of interchanges could have given ^{h}*decipiens,* but the most likely ones appear to be those which involve an interchange, first, between two chromosomes belonging to the same complex and, second, between two chromosomes belonging either to the same or to opposite complexes. If these interchanges are to be explained as a result of crossing over in differential segments, the assumption would have to be made that one or both of the complexes of *O. lamarckiana* contain duplications for a differential segment. Support for such an assumption is provided by Catcheside's discovery (1932) that pairing and chiasma formation occur frequently in certain chromosomal segments of the haploid *O. blandina,* a derivative of *O. lamarckiana.* One mechanism which could initiate such more complex series of interchanges is the anomalous type of crossing over and segrega-

[2] In *Oenothera* terminology, the superscript h before a complex indicates that it is an alethal component of a structurally homozygous race.

tion which could take place in trisomic derivatives, due to the masking by the extra chromosome of the effects of recessive lethal factors. Catcheside (1936) has described the numerous types of trisomics which can occur in a ring-forming complex-heterozygote, mostly as a result of nondisjunctional segregation of adjacent chromosomes in the ring, and he has discussed the anomalous types of chromosomal configurations which they will produce. De Vries (1923) has recorded that *O. blandina,* one of the mutant derivatives of *O. lamarckiana* analyzed by Catcheside (1940), arose in the fourth-generation progeny of a cross between two trisomic "mutant" types, *O. lata* × *semilata.* Gates and Nandi (1935) have reported that a wild race from Massachusetts, apparently related to *O. biennis* and like it forming a ring of 14, is characterized by the frequent segregation of trisomic derivatives.

Based on this discussion we can recognize in the complex-heterozygote Oenotheras two agents which tend to stabilize and maintain the constancy of the biotypes – self-pollination and the condition of balanced lethals – and three which may produce occasional abrupt changes or "mutations" and the origin of new, often constant biotypes. The latter agents are occasional hybridization between biotypes; the heterozygosity of the differential segments, which is responsible for the changes resulting from occasional crossing over in these segments; and the cytological condition which makes possible the occasional formation of gametes with abnormal chromosomal constitutions through nondisjunctional segregation of the chromosomes in the ring. These three types of change obviously involve nothing more than a much restricted amount of the recombination and segregation of genetic factors which takes place regularly in a cross-fertilized, structurally homozygous, but genetically heterozygous, organism. The genetic system of *Oenothera,* therefore, like that of other self-fertilized organisms, serves primarily to restrict genetic recombination and segregation. The essential difference between the genetic systems in these two types of self-fertilized organisms is that the temporary heterozygosity, following interracial hybridization, in normally homozygous types causes evolution to proceed in a series of short bursts; while the permanent structural and genic heterozygosity of the complex-heterozygotes tends to produce a relatively constant, though usually slow, rate of recombination and selection through periodic release of this heterozygosity. In addition, the complex-heterozygote races benefit from whatever advantages exist in the condition of heterozygosity *per se,* with its accompanying possibility for hybrid vigor. Finally, the complex-heterozygotes have an advantage over self-pollinated homozygotes in that recessive mutations can accumulate in them, perhaps to an even greater extent than in cross-fertilized organisms.

The aggressive, "weedy" character of the complex-heterozygote Oenotheras and their rapid spread in habitats disturbed by man is in itself evidence that their genetic system adapts them very well to the colonization of new habitats and to the rapid evolution and establishment of many new biotypes over a short period of time. The ultimate fate of these plants is, however, less secure. Darlington (1939, p. 92) has suggested that they represent an "evolutionary blind alley," because of the restriction of crossing over and gene recombination. It is certainly true that a population of complex-heterozygotes can never present to the environment at any one time more than a small fraction of the number of gene combinations which can exist in a normal cross-fertilized population of the same size. But, since progressive gene mutations can occur in the proximal, nonpairing, or differential segments of the chromosomes, and may occasionally be combined with similar mutations which have occurred in other races, progressive evolution is at least theoretically possible. Nevertheless, there is evidence that complex-heterozygosity has occurred repeatedly and independently in various groups of flowering plants, but there is no evidence that it has ever led to anything more than the multiplication of races or species within a subgenus. And Darlington has pointed out that in *Rhoeo discolor* we have a structural heterozygote which, so far as is known, is monotypic and very constant both genetically and cytologically. It may very well represent the last survivor of a group which a long time ago ventured along the road of complex-heterozygosity.

The problem of delimiting species in a group with the cytogenetic properties of *Oenothera,* subg. *Euoenothera,* is well-nigh insoluble (Cleland 1944). The *hookeri* alliance of California, which is mostly pair forming and without balanced lethals, represents a normal sexual species. Another such species is the highly localized endemic *O. argillicola,* found on shale-barren slopes of the Appalachian area. In their external morphology these two are entirely different and distinct (Cleland 1937, 1944), and in any ordinary genus their status as species would be unquestionable. When we turn to the structural heterozygotes, however, we find all transitions between the morphological characteristics of these two structurally homozygous species, and in addition a large number of forms with entirely new characteristics not known in any normal pair-forming species. And among all of the thousands of structurally heterozygous biotypes no sharply defined categories could be recognized above the individual biotype. These can be subdivided into several different groups, each having a recognizable set of morphological characteristics and a more or less definite geographical distribution (Cleland 1944). These groups, however, are by no means distinct enough to rank as species; they compare more nearly with the subspecies of normal sexual organisms.

If, therefore, the subgenus *Euoenothera* as a whole is considered, only two possible species concepts can be upheld. One is to consider each biotype a distinct species, as was done by De Vries and more recently by Gates and his co-workers (cf. Gates and Catcheside 1932). This is reasonably applicable in some parts of the range of the genus, such as eastern Canada, where each biotype is represented by many thousands of individuals and appears to be relatively distinct from other biotypes in a number of different characteristics. But in the central and southeastern United States it is more difficult to apply, since crossing appears to be more frequent, and the biotypes are closely similar to each other. In the *irrigua* alliance of the southwest the ring-forming structural hybrids, because of the absence of balanced lethals, are capable of relatively free genetic segregation, so that each biotype is represented by a relatively small number of individuals. The biotype concept of species can therefore be maintained for only a few ill-defined portions of the subgenus and therefore has neither scientific nor practical value.

The opposing alternative of putting the whole subgenus into one species is perhaps equally impractical. Morphologically, genetically, and cytologically, the amount of variation found in it is far greater than that within any normal species. Besides, there are included at least two clearly distinct, normally cross-fertilizing, structurally homozygous species, namely, *Oenothera hookeri* and *O. argillicola*. Furthermore, many of the structural hybrids are probably descended from hybrids between ancient, genetically distinct species (Cleland 1944). The most satisfactory procedure, therefore, is to recognize the subgenus *Euoenothera* as a series of genetic types in which, because of the peculiar genetic system prevailing, biological species comparable to those found in normal cross-fertilized organisms do not exist (Cleland 1944, Camp and Gilly 1943).

THE OCCURRENCE OF *Oenothera*-LIKE SYSTEMS IN OTHER GROUPS

This extreme development of genetic systems based on structural hybridity is not known in any other group of organisms. At least three other subgenera of *Oenothera; Hartmannia, Lavauxia,* and *Raimannia*[3] (Schwemmle 1938, Schwemmle and Zintl 1939, Haustein 1939a,b, Hecht 1941, Cleland 1944) have rings of 14 and balanced lethal mechanisms as highly developed as those of *Onagra*. Rings involving the entire chromosome complement are known in only five other species: *Gaura biennis* (Bhaduri 1941, 1942), *Paeonia brownii* (Stebbins and Ellerton 1939), *P. californica* (Walters 1942), *Rhoeo discolor* (Sax 1931a, Koller 1932), and *Hypericum punctatum* (Hoar 1931). The two short papers on *Gaura,* another genus of the Onagraceae, indicate that this genus may have the same series of genetic systems as *Oenothera,* subg. *Euoenothera*. The genetical properties of the rings in *Paeonia brownii* are not known, since this species is not well suited to genetical experiments. In northern California there are, however, pair-forming races of this species which apparently resemble the *hookeri* alliance of *Oenothera* in their genetic system, and intermediate rings have been found in northern California (Walters 1942 and unpublished). In the related *Paeonia californica,* sampling of populations throughout the range of the species has revealed only a small number of individuals with the complete ring of ten, and no entire populations with this configuration. In this species also, structural homozygotes are predominant in a considerable area near the center of the

[3] The species referred by Schwemmle to *Euoenothera* belong to the subgenus *Raimannia,* according to present treatments (Munz 1949).

range. Most of the populations, however, consist of individuals with rings of intermediate size, and are often very heterogeneous in this respect (Walters 1942). Furthermore, the plants appear to be predominantly cross-fertilized, since considerable individual differences are found in all populations. The genetic systems prevailing in *P. californica,* therefore, are in some respects different from any known in *Oenothera.*

Rhoeo discolor is known only as a structural heterozygote with a complete ring of 12 chromosomes. Only cultivated plants of this species have been studied, and these are remarkably constant when reproduced by seed. As a wild plant, *R. discolor* appears to be rare and geographically restricted to a small part of Central America. Since it represents a monotypic genus with no close relatives, Darlington's conclusion (1939), that it has suffered the inevitable fate of evolutionary stagnation and extinction which results from the adoption of its genetic system, is probably justified. About the fifth species, *Hypericum punctatum,* nothing is known except the cytology of certain plants collected in western Massachusetts. It should be a promising object for future study of this type of genetic system.

Permanent interchange heterozygotes with smaller rings are known in *Briza media* (Kattermann 1938b) and *Godetia whitneyi* (Hiorth 1942, Håkansson 1942). The latter species is of particular interest because it illustrates the first stages in the evolution of a genetic system involving interchange hybridity. The great majority of plants of *G. whitneyi* are structurally homozygous and have similar segmental arrangements. Hybrids between races derived from such remote localities as central California and British Columbia form seven pairs at meiosis, and the species is apparently not divided into chromosomal races, or "prime types," like those found in *Datura.* Nevertheless, in certain races with distinctive morphological characteristics, like one from the Willamette Valley, Oregon, a distinctive type of ring, but one without balanced lethals, recurs in many individuals. In a small area of northwestern California, in northern Mendocino County, and in southern Humboldt County, is found a complex of races of *G. whitneyi* which form rings of 4, 6, 8, 10, and 12 of their 14 chromosomes. In some of these, as well as in certain garden races, the existence of a balanced lethal mechanism has been established

(Håkansson 1942). The differentiation of these races has been aided by the fact that, in the region concerned, *G. whitneyi* exists in the form of isolated populations confined to the barer crests of the mountain ridges and separated from each other by broad, heavily forested valleys not available to the species.

ORIGIN OF COMPLEX-HETEROZYGOTE SYSTEMS

The additional data which have accumulated during the past fifteen years on both *Oenothera* and other groups containing structural heterozygotes enable us to evaluate with some degree of plausibility the various hypotheses which were suggested before this time as to how such systems can arise. There now seems to be little doubt that certain cytological conditions are prerequisite to their existence, since these conditions are shared by all groups which have this system, and in some instances are absent from related groups which do not. These are, as Darlington (1931, 1939) has suggested, the existence of "floating" interchanges, giving rise to occasional individuals with rings at meiosis; a relatively low basic chromosome number; chromosomes with mostly median centromeres; and a condition of the chromosomes which causes chiasma formation and crossing over to be much restricted in the proximal regions of the chromosomes, and largely confined to the regions near their ends. Cytological evidence for this situation was produced by Marquardt (1937) and Japha (1938) in *Oenothera*, by Coleman (1941) in *Rhoeo*, and by Stebbins and Ellerton (1939) in *Paeonia*. The only additional characteristic common to those groups in which the beginnings of balanced lethal formation and permanent rings have been noted is that they possess a certain degree of cross-pollination. The autogamy of the complex-heterozygote Oenotheras does not seem to be shared by any other group with this system, except perhaps for *Rhoeo*, and it is not found in any of the pair-forming species or in the *irrigua* alliance, which forms alethal rings.

This last fact suggests that the hypothesis favored by Darlington (1931), namely, that complete rings with balanced lethals arise from pair-forming types through "cumulative interchange with self-fertilization," is unlikely. Still more improbable is the hypothesis of Renner (1917) that they arise through interspecific hybridization. This is particularly clear in the case of *Godetia whitneyi*, since in this species the few wild races which have balanced lethals are found in the middle of the range of the species and in an area where no other species of the genus occurs. Furthermore, the only other species of *Godetia* which can hybridize with *G. whitneyi* and form bivalents in the F_1 hybrid, namely, *G. amoena*, has the same structural arrangement of the chromosomes as *G. whitneyi*. In the New World species of *Paeonia*, large rings and probable balanced lethals seem to have originated independently in *P. brownii* and *P. californica*.

The present facts, on the other hand, definitely favor the hypothesis of Catcheside, which he states as follows (1932, p. 107).

> The origin of the *Oenothera* mechanism proceeds originally from reciprocal translocations, subsequent crossing with defined segmental interchanges in the hybrid produced, and the establishment of the ring-complex mechanism in the hybrid. Later increases in the size of the ring occur (1) by exchanges between the ring and bivalents, (2) by crossing between forms bearing other translocations, (3) by a combination of the two processes.

In the same publication, he showed how crossing between two forms having the same arrangement of the chromosome ends, but both derived from an original type by translocations between the same two chromosomes occurring independently at different loci, could give rise to hybrids which usually form pairs at meiosis, but in which some of these pairs contain differential segments; that is, small homologous interstitial segments in otherwise nonhomologous chromosomes. Occasional crossing over in these differential segments can produce from such hybrids gametes containing deficiencies and duplications of chromosomal segments (Catcheside 1947b). Judging from the behavior of these and other experimentally produced deficiencies, as well as of monosomic and nullisomic plants (McClintock 1944, Clausen and Cameron 1944, Sears 1944b), some of these gametes would be lethal both as pollen and as eggs, some would be viable chiefly as pollen, and others, chiefly as eggs. Union between two gametes containing the same duplications and deficiencies would almost certainly produce inviable zygotes, but if two gametes containing complementary duplications and deficiencies should join, there would result a plant containing balanced lethals and forming rings at meiosis. Depending on the transmissibility of the deficiency-duplication systems such plants would have two complexes carried by either the eggs or the pollen grains, as in *O. lamarckiana*, or more probably would form one type of viable gamete in the egg and a different one in the sperm, as in most of the other complex-heterozygotes. Both of the types of balanced lethal systems found in *Oenothera*, therefore, probably have a parallel and independent origin; one was not derived from the other. The balanced lethal types which would arise by this method would differ from those most common in *Oenothera* only by the fact that the corresponding differential segments responsible for the lethal condition would lie in the same, not in opposite complexes. Increase in size of the ring could occur through translocations between its chromosomes and those of bivalents, which can take place most easily as a result of interlocking at meiotic prophase, a phenomenon which, according to Catcheside, is common in *Oenothera*. During this process, differential segments located in opposite complexes, which are characteristic of *Oenothera*, will automatically arise, through translocations involving such chromosomes.

The facts which favor this hypothesis are as follows. First, the groups of races which appear to be the immediate ancestors of the complex-heterozygotes in *Oenothera* (*irrigua* alliance), in *Godetia whitneyi* (racial complex of Humboldt County, California), and in *Paeonia californica* (several populations) are all open-pollinated and divided into partly isolated populations, some small, some large, which permit some divergence of racial types, with occasional crossing between them. Second, a large number of interchanges is present in these populations, so that temporary rings not containing balanced lethals are formed very often. With this very frequent occurrence of interchanges, the probability of the existence in the same population of two interchange types involving the same two chromosomes at different loci is high. Third, the permanent rings in *Godetia* and in *Paeonia* are entirely or predominantly relatively small ones, involving only a part of the chromosomal complement. It is possible that a thorough study of the races of *Oenothera* in Colorado, New Mexico, and Utah, where the alethal *irrigua* alliance and the typical complex-heterozygote *strigosa* alliance come into contact with each other, will reveal similar small rings containing balanced lethals. Fourth, the hypothesis of later hybridization between complex-heterozygote races as a source of the formation of new races and consequent increase in complexity of the group as a whole is supported by the data of Cleland (1944) on races of eastern North America belonging to the *biennis* series. In the egg-borne complex of one subgroup of this series, one particular chromosomal arrangement appeared in nine different races and was in each instance associated with a pollen-borne complex having a different arrangement. Such situations could most easily arise through interracial hybridization.

Three further characteristics of complex-heterozygotes need explanation. The first, which is characteristic of all the groups concerned, is the presence of complex-heterozygotes and permanent rings in some parts of the geographic range of a species or species complex, and its absence in others. The second, which is found in *Oenothera* and perhaps in *Paeonia brownii,* but not in *Godetia whitneyi* and *Paeonia californica,* is the predominant frequency of pairs or alethal rings, on the one hand, and permanent rings including the entire chromosomal complement, on the other, with a corresponding scarcity of types with permanent rings of intermediate size. The third, which is at present known only in *Oenothera,* is the predominance of autogamy among the complex-heterozygotes, thus greatly increasing their genotypic constancy.

The first characteristic is best explained through the action of natural selection, which, as Darlington (1939) pointed out, must act to secure the spread of complex-heterozygote races. The importance of structural heterozygosity in maintaining in the heterozygous condition certain gene combinations which have a high selective value has been already mentioned in preceding chapters. Some of the results of Hiorth and Håkansson on *Godetia whitneyi* strongly suggest that rings are being maintained in this species because of the selective value of genes which they contain. In a type called the Willamette race, distributed in the Willamette Valley of northwestern Oregon and characterized by its robust, vigorous growth, its weedy tendencies, and certain distinctive morphological features, one particular chromosomal configuration, consisting of a ring of three chromosomes and a univalent, appears repeatedly in individuals from different localities. Since it does not carry balanced lethals, it could survive and spread in natural populations only if the zygotes containing these chromosomes had an advantage under certain environmental conditions over those with more normal chromosomal complements. In *Oenothera,* the complex-heterozygote races are chiefly weedy plants of ruderal habitats, which have spread in the areas disturbed by man. The pair-forming races of California likewise occupy disturbed soil, but in the dry, open country where they are most frequent such sites have been continuously available even before the coming of man. They are therefore not occupying newly available habitats, as are the *biennis* and the *parviflora* alliances of the central and eastern states. It should be further noted that the races of the latter two alliances contain for the most part two complexes which bear genes responsible for rather different and contrasting morphological characteristics, and which may have arisen in geographic races or ecotypes adapted to different ecological conditions. It is thus possible that many of the complex-heterozygote races in *Oenothera* owe their success to the fact that they are preserving more or less intact the wide range of ecological tolerance and the adaptability to new conditions which is often found in F_1 and F_2 progeny of interracial hybrids (Clausen, Keck, and Hiesey 1947).

The scarcity of permanent rings containing only part of the chromosome complement of complex-heterozygotes was explained by Darlington (1931) as due to the reproductive inefficiency of such types. Segregation of chromosomes from small rings may be more irregular than from large ones, and if two rings, both containing balanced lethals, exist in the same organism, partial sterility of gametes is almost certain to result. Natural selection will therefore tend to favor genotypes with increasingly large rings, on account of their greater fertility. This hypothesis is not contradicted by the situations in *Godetia whitneyi* and in *Paeonia californica.* In *Godetia whitneyi,* balanced lethals are so rare that they appear to have been established only recently in the history of the species. One might suggest that with further lapse of time, and with selection favoring the complex-heterozygotes, the course of evolution in this species would follow that taken ages ago by *Oenothera. Paeonia californica,* on the other hand, is certainly an old species, but its manner of growth and life history are entirely different from those of *Oenothera* and *Godetia.* It is a long-lived perennial which reproduces rather infrequently from seed. Furthermore, recent observations of Dr. J. L. Walters (unpublished) indicate that in nature seed sterility often occurs because of failure of pollination. Follicles of wild plants are often completely devoid of seed, while other follicles on the same plant may contain an abundance of fully mature, apparently viable seeds. This great influence of environmental conditions on seed setting would cause the heritability of genetic factors affecting seed fertility to be relatively low, while selection pressure in favor of fertility must be much lower than it is in annuals or in biennials like *Oenothera* and *Godetia*. The explanation of the predominance of rings of intermediate size in *Paeonia californica*, as well as its tendency for nondisjunctional separation of the rings at meiosis and for the low chiasma frequency, which often causes the potential ring to exist as two or three chains at meiosis, can all be based on the low intensity and inefficiency of selection in the direction of high pollen and seed fertility.

The selective value of autogamy in *Oenothera* is probably the same as that in other weedy annuals and biennials. It assures a high seed production and a constancy of the offspring which fits them for rapid colonization of newly opened, temporarily available habitats. There is good reason for believing that autogamy arose in *Oenothera* after the complex-heterozygote condition had already been evolved. All the pair-forming races in the subg. *Onagra* are open-pollinated. Furthermore, the two types of genes which promote autogamy, those for small corollas and short styles, are mostly located on the distal pairing segments of the chromosomes, in contrast to the great majority of other genes differentiating the races, which are located in the differential segments (Emerson and Sturtevant 1932). If the genes promoting autogamy arose as mutations in complex-heterozygotes, mutations of this character which arose in the distal segments would have a much greater chance of being transferred from one race to another by hybridization, and hence of spreading through the geographic range of the complex, than similar mutations which might happen to take place in the differential segments.

The probable course of evolution in the subgenus *Onagra* of *Oenothera* and similar highly developed systems of complex-heterozygote races can be summarized as follows. They existed originally as a typical, open-pollinated diploid species, forming seven pairs of chromosomes at meiosis and characterized by median centromeres and the restriction of pairing and crossing over to the distal segments of the chromosomes. This species evolved first by the formation of ecotypes and geographical subspecies in the usual manner, but accompanied by frequent segmental interchange between nonhomologous chromosomes, perhaps caused principally by interlocking of bivalents at the meiotic prophase. In time, the gene complexes existing in some of these translocated segments acquired particular selective values and caused the translocations to spread through the populations, often as structural heterozygotes, forming impermanent rings of four and six chromosomes. When these translocations became sufficiently numerous in the populations, crossing would almost certainly occur between two forms possessing the same arrangement of chromosome ends, but derived from an original common type with a different arrangement by independent translocations at different loci of the chromosomes concerned. The products of such hybridization, being partly sterile, would be at a selective disadvantage in direct competition with their parents, because of their lowered fertility. But if they should occur in a new habitat to which they were better adapted than any other forms of *Oenothera,* and particularly if they displayed the hybrid vigor often found in interracial crosses, they would have a good chance of surviving and of producing successful progeny. As described above, a few of the gametes produced by such interracial hybrids would possess the rudiments of a balanced lethal system, and if such gametes also contained gene combinations with a high selective value, they would have an additional advantage through the possession of a mechanism for preserving intact these favorable combinations. Once the complex-heterozygotes arising from these gametes became established, evolution would be likely to proceed in four ways. First, additional interchanges and balanced lethal forms would become established in each race, thereby increasing the size of the ring to its maximum extent. Second, the origin and establishment of genes favoring autogamy would help to preserve the constancy of the superior races. Third, occasional hybridization between complex-heterozygotes and pair-forming or ring-forming strains without balanced lethals would give rise to new complex-heterozygotes with different morphological and physiological properties. And, finally, intercrossing between complex-heterozygotes, and the "mutational" processes described earlier, would increase the complexity of the group. Having reached its climax of variability, such a system of complex-heterozygote races might persist and go on evolving in its own peculiar way for many ages. But because of self-fertilization and the high development of linkage, it would have lost much of its flexibility and would not adapt itself easily to radically different environmental conditions. If, therefore, the conditions of the habitat should change abruptly and radically, such systems of complex-heterozygotes might be the first types to become decimated and eventually they would perish.

CHAPTER XII

Evolutionary Trends I: The Karyotype

THERE HAS BEEN a tendency among many systematists to regard differences between species and genera in the number, form, and size of their chromosomes as merely additional taxonomic characters. For those who are interested in "pigeon hole" classification and nothing else, this attitude may be justified. But systematists should realize that, considering the amount of time and labor needed to prepare and to observe them, chromosomes often provide inefficient and uncertain diagnostic characters. On the other hand, they often provide to the student of evolution and phylogeny valuable signposts indicating the nature of the evolutionary processes at work and the trends which evolution has taken.

The important fact to remember here is that chromosomes are not only structures which result as end products of a series of gene-controlled developmental processes; they are themselves the bearers of the genes or hereditary factors. This at once puts them into a different category from all other structures of the body. Furthermore, as White (1945, p. 152) has observed in this connection, chromosomes are not merely aggregates of discrete genic units. To a certain extent they are units in themselves. We should expect changes in the chromosomes to bear a more direct relationship to genetic-evolutionary processes than do any other types of changes.

Before studying visible chromosomal differences, however, we must remind ourselves of the fact that chromosomes which resemble each other in outward appearance are not necessarily alike in genic content or in hereditary potentiality. The changes in the chromosomes produced by gene mutations are by definition invisible even under the most powerful microscope. In some genera, like *Pinus* and *Quercus* (see Chapter II), differentiation of species appears to have been entirely by this process, since in their gross structure the chromosomes of all the species are so similar as to be indistinguishable. This is shown both by comparative karyology of somatic chromosomes and by the regular pairing of the chromosomes at meiosis in species hybrids. Furthermore, the chromosomes of different species may look exactly alike as to size and form, but may nevertheless possess many differences in gross structure, such as translocations and inversions, which become evident only when they pair with each other in species hybrids. This is notably true of the genus *Paeonia* (Stebbins 1938a).

On the other hand, the superficial appearance of the chromosomes may be completely altered in two entirely different ways, without comparable changes in the genotype. One way is by means of large unequal reciprocal translocations of chromosomal segments. Such changes have little or no effect on the external morphology or the physiological reactions of the plant. They have been induced artificially in *Crepis tectorum* (Gerassimova 1939) and in several other examples. Another is by the accumulation of chromosomes with little or no genetic activity, such as the "B-type" chromosomes found in certain strains of *Zea mays* (Longley 1927, Randolph 1941b, Darlington and Upcott 1941), and the fragment-type chromosomes sometimes found in species of *Tradescantia* (Darlington 1929, Whitaker 1936). We cannot, therefore, estimate either the amount or the directional trend of evolution by studying only the external, visible characteristics of the chromosomes. Nevertheless, comparison of the chromosomes of related species has in many groups disclosed certain regular differences, which are often correlated with trends of specialization in the external morphology of the plant. The full explanation of these correlations has not yet been obtained for any example, but they nevertheless deserve careful consideration, since they very likely will provide valuable guides toward the solution of fundamental problems concerning the relation between chromosomes, genes, and visible characters or character complexes.

THE CONCEPT OF THE KARYOTYPE

The pioneering work of the Russian School of cytologists,

headed by S. Navashin, established firmly the fact that most species of plants and animals possess a definite individuality in their somatic chromosomes, which is evident in their size, shape, position of primary constrictions or centromeres, and in such additional features as secondary constrictions and satellites. Furthermore, closely related species are usually similar in these respects, and distantly related ones are often recognizably different, at least in those species with large, easily studied chromosomes. These facts led to the early formulation of the concept of the *karyotype*. Delaunay (1926) formulated the concept of the karyotype as a group of species resembling each other in the number, size, and form of their chromosomes. In the liliaceous genera with which he worked (*Muscari, Bellevalia, Ornithogalum*), the karyotype seemed to correspond to the genus, but Levitzky (1924, 1931b) soon showed that this situation is not at all general, and that in many genera the evolution of the karyotype can be traced through a series of gradual changes in the external appearance of the chromosomes. He therefore redefined the term karyotype as the phenotypic appearance of the somatic chromosomes, in contrast to their genic contents. An earlier term, the idiogram, was used in a somewhat similar sense by S. Navashin, but the term karyotype is the one which has been established by general usage, whereas the term idiogram is now applied to the diagrammatic representation of the karyotype.

The principal ways in which karyotypes differ from each other are well described in the classic work of Levitzky (1931a,b), and additional information has been supplied by Heitz (1926, 1928, 1932) and others (cf. Darlington 1937). These distinguishing characteristics are as follows: first, basic chromosome number; second, form and relative size of different chromosomes of the same set; third, number and size of the satellites and secondary constrictions; fourth, absolute size of the chromosomes; and fifth, distribution of material with different staining properties, that is, "euchromatin" and "heterochromatin." These five types of difference will be discussed in turn.

In this discussion, frequent reference will be made to the primitive or advanced condition of the various genera and species in respect to various characteristics of external morphology. In most instances, the reasons for assuming that the characteristics in question are primitive or advanced are discussed by the authors whose papers are cited, and the present writer believes that their conclusions are based on satisfactory evidence. Other examples are judged on the basis of criteria familiar to and accepted by most systematists and morphologists. Some of the principal ones of these are that the annual habit is usually derived from the perennial one; that genera or species groups which form connecting links between other groups are likely to be primitive, unless they are of allopolyploid origin; and that such conditions as sympetaly, epigyny, zygomorphy, and dioecism are derived conditions. Some of these points are discussed in Chapter V, and particularly in Chapter XIII.

CHANGES IN BASIC NUMBER

The commonest type of change in chromosome number found in the higher plants is polyploidy. This type of change is an irreversible one, so that in any polyploid series the oldest, most primitive members are those with the lowest chromosome numbers. But in many genera and families of the higher plants we find an entirely different type of change in chromosome number, occurring most often on the diploid level, without multiplication of chromosome sets or even of whole chromosomes. In the best-known examples these changes in basic haploid number may be seen to involve an increase or decrease by one chromosome at a time. Furthermore, the trend may be in either direction, depending on the group concerned.

The principles governing this stepwise, aneuploid alteration of the basic chromosome number are now well established in cytological literature. They are summarized by Darlington (1937, Chap. XII) and by White (1945, Chaps. IV, VIII). The postulate of M. Navashin (1932), that centromeres (that is, "kinetochores" or "primary constrictions") cannot arise *de novo,* is now well supported by a body of observational and experimental evidence, as is also that of the permanence of chromosomal ends. Furthermore, chromosomes with two centromeres cannot function efficiently and are unknown in natural populations of plant species. Once these principles are understood, the impossibility of changes in basic chromosome number through simple transverse fragmentation and end-to-end fusion of single chromosomes becomes evident. Reduction of the basic number must involve loss of a centromere plus at least a small amount of adjacent chromosomal material, while increase must involve at first a duplicated chromosome, or centromere-bearing fragment.

Darlington (1937, pp. 559–560) has shown how conditions favoring the loss or gain of a chromosome can be produced by

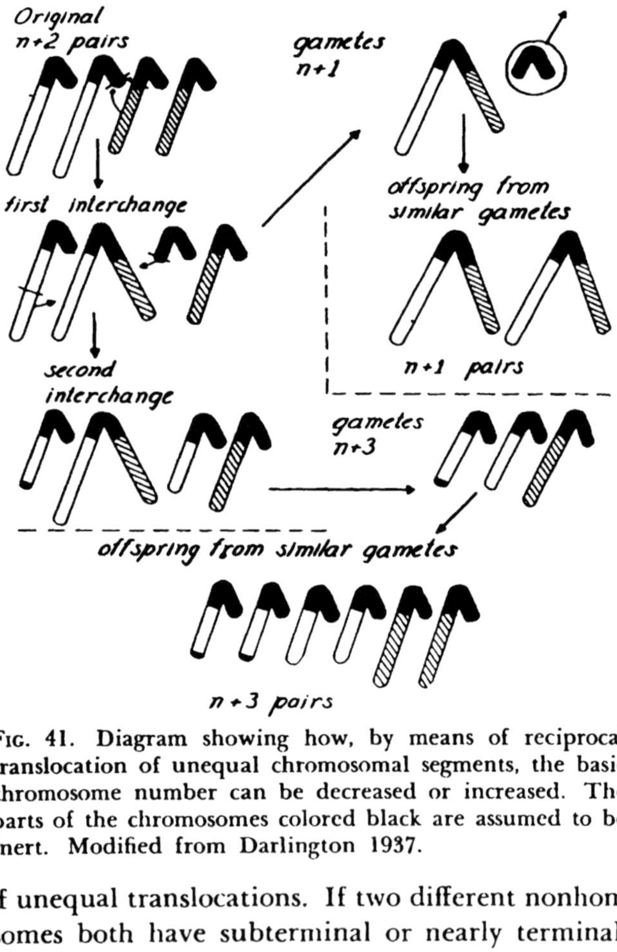

FIG. 41. Diagram showing how, by means of reciprocal translocation of unequal chromosomal segments, the basic chromosome number can be decreased or increased. The parts of the chromosomes colored black are assumed to be inert. Modified from Darlington 1937.

means of unequal translocations. If two different nonhomologous chromosomes both have subterminal or nearly terminal centromeres (the "acrocentric" chromosomes of White, 1945), then a segmental interchange involving the long arm of one chromosome and the short arm of the other will produce one long chromosome with a median centromere ("metacentric," in White's terminology) and one very short, fragment type of chromosome (see Fig. 41). If such an interchange type becomes homozygous, it will form at meiotic metaphase one exceptionally large bivalent and one very small one. In the latter bivalent, both chromosomes are likely to pass to the same pole of the spindle, as Gerassimova (1939) observed in a chromosomal race of this nature produced artificially in *Crepis tectorum.* This will yield, in addition to normal gametes, others with either two or none of the small chromosomes. The behavior of such gametes will depend on whether these small chromosomes are genetically active or consist entirely of inert material. The latter is likely to be the case in many organisms, since the portions of the chromosomes on either side of the centromere are often inert. If this is true, gametes lacking such small chromosomes will function as well as normal ones, and the fusion of two such gametes will give rise to a true-breeding strain with the basic number reduced by one (see Fig. 41). If, on the other hand, the small chromosomes are genetically active, reduction of the basic number is impossible by this method, but individuals with an increased number, possessing three or four of the small chromosomes, are likely to occur frequently (see Fig. 41). These will at first be impermanent tri- or tetrasomics, but they could be converted into races or species with a permanently increased basic number in two ways, either by divergent gene mutations in the duplicated small chromosomes or by reciprocal translocation between one of these and one of the large chromosomes of the set (see Fig. 41). Therefore, as Darlington has pointed out, the question of whether the basic number will be increased or decreased as a consequence of unequal translocations depends largely on whether the regions about the centromeres are active or inert.

This aneuploid alteration of the basic number has been carefully studied in *Crepis* and its relatives (Babcock and Cameron 1934, Babcock, Stebbins, and Jenkins 1937, 1942, Babcock and Jenkins 1943, Babcock 1942, 1947). Here the evidence is conclusive that the most primitive species have the highest basic numbers, and that the trend has been toward reduction. In the genus *Dubyaea,* which in both vegetative and floral characteristics is more primitive and generalized than *Crepis,* and which forms a connecting link between the large genera *Crepis, Lactuca, Prenanthes,* and *Hieracium* (Stebbins 1940), all the species counted have the haploid number $x = 8$, while the basic numbers of most of the genera related to *Crepis*, namely, *Youngia, Prenanthes, Hieracium, Lactuca, Sonchus, Launaea*, and *Taraxacum*, are either $x = 8$ or $x = 9$. In *Crepis* itself, the species which in one way or another approach one of these other genera have

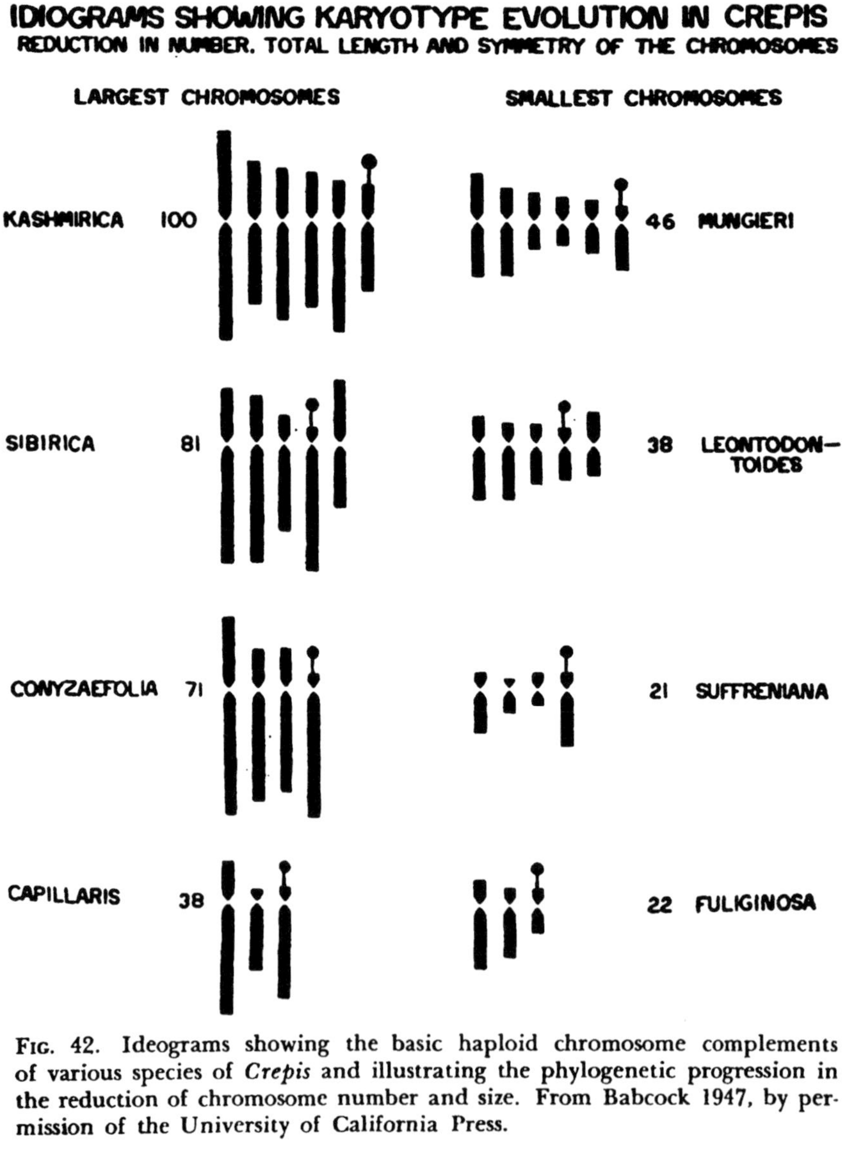

FIG. 42. Ideograms showing the basic haploid chromosome complements of various species of *Crepis* and illustrating the phylogenetic progression in the reduction of chromosome number and size. From Babcock 1947, by permission of the University of California Press.

$x = 7$, $x = 6$, or $x = 5$, while those with $x = 4$ and $x = 3$ are the most typical of the genus and the farthest removed from the other genera (see Fig. 42). In addition, the reduction in basic number has been accompanied by certain definite trends of specialization in external morphology. The most primitive species of *Crepis* are those with $x = 6$ (the seven-paired species are a specialized offshoot, perhaps forming a transition toward the related genus *Ixeris*). Their primitiveness consists in the perennial habit, the presence of shallow rooting rhizomes, entire or shallowly dissected leaves, relatively large involucres in a few-headed inflorescence, more or less imbricated, unspecialized involucral bracts, and unspecialized, unbeaked fruits or achenes. The species with $x = 5$, $x = 4$, and $x = 3$ possess to an increasing degree some of the following specializations: annual habit; deep taproots; deeply pinnatifid leaves; smaller and more numerous involucres; specialized involucral bracts in two series, the outer reduced and the inner variously thickened and otherwise modified; and beaked, sometimes strongly dimorphic, achenes. This correlation between reduction in basic number and increasing morphological specialization is not complete; some species with $x = 5$ (*C. foetida* et aff.) are highly specialized in nearly all of their characteristics, while at least one three-paired species, *C. capillaris*, is less specialized in most respects than are some species with $x = 4$. But when all the facts are considered, the series can be read in only one way. The three-paired species are related only to four-paired species of *Crepis;* all the latter are typical of the genus and show clear connections with the former and with the five-paired species, while it is only in those species with six and five pairs that we can see the evolutionary connection between *Crepis* and other genera. Both morphological and genetic evidence show that the lower numbers have arisen independently several times. The three species with $x = 3$, *C. capillaris, C. fuliginosa,* and *C. zacintha,* are all more closely related to various unconnected four-paired species than they are to each other. Similar independent connections can be traced between various four-paired species and related ones with $x = 5$.

Darlington's postulates concerning the mechanism by which basic chromosome numbers are reduced have now been fully confirmed for this material. Tobgy (1943) has shown that the three-paired *C. fuliginosa* was derived from the four-paired *C. neglecta* or its ancestor through a system of reciprocal translocations (Fig. 43). One of the four chromosomes of *C. neglecta* (designated as C) is heterochromatic, and probably inert genetically except for the distal part of one arm. In *C. fuliginosa*, the active material of

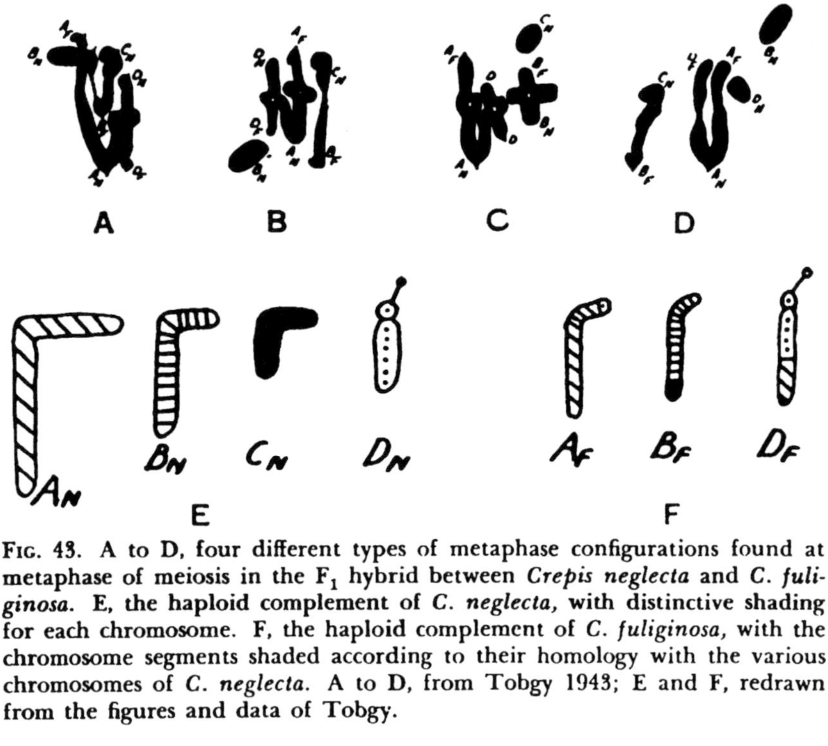

FIG. 43. A to D, four different types of metaphase configurations found at metaphase of meiosis in the F_1 hybrid between *Crepis neglecta* and *C. fuliginosa*. E, the haploid complement of *C. neglecta*, with distinctive shading for each chromosome. F, the haploid complement of *C. fuliginosa*, with the chromosome segments shaded according to their homology with the various chromosomes of *C. neglecta*. A to D, from Tobgy 1943; E and F, redrawn from the figures and data of Tobgy.

the *neglecta* C has been translocated onto the short arm of another chromosome, designated as A. This is clearly demonstrated by the pairing found in the F_1 hybrid between the two species. Tobgy believes, on the basis of indirect evidence, that the translocation was reciprocal, but that the segment of the ancestral A translocated onto the ancestral C of *fuliginosa* was so small that the latter chromosome consisted entirely of inert material, and so could be lost with impunity. Sherman (1946) has shown that the four-paired *C. kotschyana* was derived in a similar manner from a five-paired form related to *C. foetida*. From these examples we may make the general statement that if the region of a chromosome near the centromere is genetically inert, its active distal portions may occasionally be translocated to another, nonhomologous chromosome, its centromere may be lost, and reduction of the basic number will result. This type of reduction has taken place several times in the evolution of the genera of the tribe Cichorieae, family Compositae. In *Youngia*, a close relative of *Crepis*, most of the species have the basic haploid number $x = 8$, but in *Y. tenuifolia* the number has been reduced to $x = 5$ (Babcock and Stebbins 1937). In *Ixeris*, the progression has been from $x = 8$ (*I. chinensis*, *I. japonica*, *I. stolonifera*) to $x = 7$ (*I. dentata*) and $x = 5$ (*I. denticulata*, *I. lanceolata*, etc., Babcock, Stebbins, and Jenkins 1937).

In *Godetia whitneyi*, the combined efforts of Hiorth and Håkansson (1946a) have demonstrated the production of a fertile type with six pairs of chromosomes out of a species which normally has seven. Hiorth crossed two structurally heterozygous monosomic plants of this species, and among the progeny obtained a relatively fertile F_1 plant with six pairs and essentially regular meiosis. The fact that this result was obtained only when structural heterozygotes were used as parents supports strongly the other evidence indicating that reduction of the basic chromosome number usually, if not always, involves rearrangement of chromosomal segments.

Progressive increase in basic chromosome number has been suggested in a much smaller number of examples among plants, and no natural example of such increase has as yet been supported by experimental evidence. Darlington (1936c) has suggested that *Fritillaria pudica*, with $x = 13$, has arisen from species with $x = 12$, since the latter number is the most common one in the genus and the only one found in the related genera *Lilium* and *Tulipa*. Levan (1932, 1935) has concluded that the most primitive species of *Allium* are the North American ones with $x = 7$, while the Eurasian species with $x = 8$ are derived from them. Krause (1931) has shown that in the genus *Dorstenia* (Moraceae) the primitive number is $x = 14$, corresponding with that for the Urticales as a whole. From this, the basic number of 12 and probably 10 have evolved in the Old World species, while in the New World the more specialized species have progressively increasing basic numbers of $x = 15$ and $x = 16$.

The simplest process by which the basic number could be progressively increased would be by the duplication of one whole pair of chromosomes. Plants containing such extra chromosomes, known as tetrasomics, have been found as occasional aberrants in a number of species grown experimentally, such as *Datura* (Blakeslee and Belling 1924), *Nicotiana* (Goodspeed and Avery 1939), and Maize (McClintock 1929). They are usually unstable, and since they are not isolated genetically from their normal relatives, they would under natural conditions lose their identity through crossing with plants having the normal chromosome number, followed by selection of the more viable, genetically balanced normal disomic types. Somewhat similar types of plants with increased chromosome numbers are the alien addition races produced by Gerstel (1945a,b) in *Nicotiana.* He crossed tetraploid *N. tabacum* ($2n = 96$) with diploid *N. glutinosa* ($2n = 24$) and pollinated F_1 plants of this cross with diploid *N. tabacum.* A great variety of segregant types resulted from these pollinations, but among them were constant, true-breeding lines with 25 and 26 pairs of chromosomes. These contained the complete diploid set of *N. tabacum* plus one or two pairs from *N. glutinosa.* Crossing between these alien addition races and normal *N. tabacum* produced trisomic or tetrasomic types in the F_1 generation, as expected, and only a small proportion of 25- or 26-paired types was recovered in the F_2 generation. Gerstel therefore concluded that alien addition races could exist in nature only in plants with regular self-fertilization. Although similar types have been produced as a result of hybridization by Florell (1931) and by O'Mara (1940) in wheat, with the addition of a chromosome pair from rye, and by Beasley and Brown (1943) in cotton, no example is know to the writer of such types existing in nature.

Such evidence as is available suggests that in nature progressive increase of the basic chromosome number is usually brought about by a process involving duplication of a centromere plus a system of translocations. This may occur most easily through the medium of extra centric fragment chromosomes, as described above (p. 446, and Fig. 41). Unfortunately, no clear examples of this type of alteration are yet available.

Another type of aneuploidy which simulates a progressive increase in chromosome number can be derived by a combination of progressive decrease or increase in basic number followed or accompanied by amphidiploidy. If, for instance, a genus with an initial basic number of $x = 7$ produces by progressive reduction according to the *Crepis* scheme derivatives with $x = 6$ and $x = 5$, then various polyploid and amphidiploid combinations from these species can produce every haploid number from $n = 10$ upward. Polyploids of five-paired species will produce those with $n = 10$, while $n = 11$ can arise from amphidiploids of species with $x = 6$ and $x = 5$. In a similar manner may be derived $n = 12$ $(6 + 6$ or $7 + 5)$, $n = 13$ $(7 + 6)$, $n = 14$ $(7 + 7)$, $n = 15$ $(5 + 10)$, $n = 16$ $(6 + 10$ or $5 + 11)$, $n = 17$ $(7 + 10, 6 + 11$, or $5 + 12)$, and so on. It will be seen that the higher the number, the more the different ways in which it can be derived; hence, in such series the higher numbers should be more common and found in more different unrelated species than the lower ones.

Part of such a series has been produced experimentally in *Brassica* by Nagahuru U (1935) and by Frandsen (1943). The basic diploid numbers in this genus are $x = 8$, 9, and 10, which probably represent a phylogenetically ascending series (Manton 1932). The amphidiploid combinations possible are those with $n = 16$, 17, 18, 19, 20, 24, 25, 26, 27, 28, 29, 30, and so on, and of these the numbers $n = 17$, 18, 19, 27, and 29 are known either in natural or in artificially produced amphidiploids. In *Erophila* (or *Draba* subg. *Erophila*) Winge (1933, 1940) has synthesized by amphidiploidy higher numbers in an aneuploid series. In the genus *Stipa*, which has an extensive aneuploid series extending from $x = 12$ to $n = c.42$, Love (1946 and unpublished) has shown that among the four species, *S. leucotricha* $(n = 14)$, *S. lepida* $(n = 17)$, *S. pulchra* $(n = 32)$, and *S. cernua* $(n = 35)$, the chromosome behavior in hybrids indicates that the two latter species are allopolyploids containing genomes derived from one or both of the two former or from their relatives.

The most extensive aneuploid series in the plant kingdom is that in the genus *Carex*, in which haploid numbers ranging from $n = 6$ to $n = 56$ have been reported, and every number from 12 to 43 is represented by one or more species (Heilborn 1924, 1928, 1932, 1939, Tanaka 1937, 1939, Wahl 1940). Heilborn (1939) has considered that the most important processes in the origin of this series were structural changes of the chromosomes and autopolyploidy. The evidence produced by him in favor of structural changes consists of the demonstration that within the complements of several species there exist chromosomes of very different sizes, and that in some instances there are pairs of closely related species with haploid numbers differing by one, of which the species with the smaller number of chromosomes possesses a large pair, while that with the larger number lacks it. Wahl (1940) has produced even more convincing evidence of structural differences between species. In several interspecific hybrids, both artificial and natural (*C. pennsylvanica* × *umbellata, C. platyphylla* × *plantaginea, C. swanii* × *gracillima),* chains of three to six or eight chromosomes occur regularly at meiosis, although bivalents are the rule in the parental species. These interspecific hybrids, therefore, are structural heterozygotes for translocations. Since the chromosomes not forming these higher associations usually pair as bivalents, so that meiosis in the hybrids is more or less regular, this evidence also indicates a high degree of homology between the chromosomes of different species belonging to the same section.

This evidence weakens the assumption by Heilborn of autopolyploidy, which is based chiefly on the presence of associations of three to eight chromosomes in *C. glauca,* which he therefore interprets as an autopolyploid. For the evidence of Wahl suggests that *C. glauca* could just as well be a segmental amphidiploid derived from a hybrid between two species belonging to the same section. At present, therefore, the only safe conclusion is that in *Carex* structural changes and polyploidy or amphiploidy have been the major processes responsible for its aneuploid series, but that the type of polyploidy or amphiploidy involved is uncertain and will remain so until a systematic series of hybridizations is undertaken to unravel species relationships in at least one portion of this most interesting and highly complex genus. Wahl has provided indirect evidence that at least four original basic numbers existed in *Carex,* namely, $x = 5$, 6, 7, and 8, with 7 probably the most common. The similar extensive aneuploid series found in other Cyperaceae, as well as in the Juncaceae, have probably originated in a manner similar to that in *Carex.*

We can thus classify naturally occurring aneuploid series of chromosome numbers in sexually reproducing plants into three types, as follows: 1) descending basic (*Crepis* and other Cicho-

Table 8

Summary of the Types of Aneuploid Series in Higher Plants

Type and genus	Growth habit (P = perennial) (A = annual) (W = woody)	Range of numbers involved	Authority
Descending basic			
Cycadaceae	P	11 (13)–8	Sax and Beal 1934
Dorstenia	P	14–12	Krause 1931
Polygonaceae	P	11–7	Jaretzky 1928
Delphinium-Nigella	P–A	8–6	Levitzky 1931b
Arabis	P	8–6	Rollins 1941
Lesquerella-Physaria	P–A	8–4	Rollins 1939
Cruciferae, several groups	P, P–A, A	8–7 and 7–6	Manton 1932, Jaretzky 1932
Primula	P	12–9	Bruun 1932
Polemoniaceae	P	9–7	Flory 1937, Wherry 1939
Phacelia (3 series)	P–A (3)	11–7, 11–9, 9–5	Cave and Constance 1942, 1944, 1947
Verbena-Glandularia	P–A	7–5	Dermen 1936a, Schnack and Covas 1944, Covas and Schnack 1944
Nicotiana (?)	P–A	12–9	Goodspeed 1934, 1945
Plantago	P–A	6–4	McCullagh 1934
Hemizonia	A	7(?)–4	Johansen 1933
Calycadenia	A	7–4	Clausen, Keck, and Heusi 1934
Crepis	P–A	6–3	Babcock 1947; see text

TABLE 8 (*Continued*)

SUMMARY OF THE TYPES OF ANEUPLOID SERIES IN HIGHER PLANTS

Type and genus	Growth habit (P = perennial) (A = annual) (W = woody)	Range of numbers involved	Authority
DESCENDING BASIC			
Youngia	P	8–5	Babcock and Stebbins 1937
Leontodon	P	7–4	Bergman 1935b
Briza	P–A	7–5	Avdulov 1931
Hierochloe-Anthoxanthum	P–A	7–5	Avdulov 1931
Pennisetum	P–A	9–7	Avdulov 1931, Krishnaswamy 1940
Scilla-Bellevalia	P	8–4	Delaunay 1926
Ornithogalum	P	8–5	Delaunay 1926, Darlington and Janaki-Ammal 1945
Triteleia	P	7–5	Burbanck, 1941, 1944
Fritillaria	P	12–9	Darlington 1936c
ASCENDING BASIC			
Dorstenia	P	14–16	Krause 1931
Biscutella	P	8–9	Manton 1932
Phacelia	A	11–13	Cave and Constance 1942, 1944, 1947
Allium	P	7–9	Levan 1932, 1935
Triteleia-Dichelostemma	P	7–9	Burbanck 1941, 1944
Fritillaria	P	12–13	Darlington 1936c
INTERCHANGE-AMPHIDIPLOID			
Draba, subg. Erophila	A	7, 15–32, 47	Winge 1933, 1940
Brassica	A	8–11, 17–19	Manton 1932, Nagahuru U 1935

TABLE 8 (*Continued*)

SUMMARY OF THE TYPES OF ANEUPLOID SERIES IN HIGHER PLANTS

Type and genus	Growth habit (P = perennial) (A = annual) (W = woody)	Range of numbers involved	Authority
INTERCHANGE-AMPHIDIPLOID			
Sedum	P + A	4–34, 56, 64	Baldwin 1935, 1937, 1939, 1940, R. T. Clausen 1942, Hollingshead 1942
Euphorbia	P + A	6–15, 20, 28, 30, c.50, c.100	Perry 1943
Viola	P + A	6–13, 17–24, 27, 28, 36, c.40	Miyaji 1929, Clausen1929,1931b, Gershoy 1934
Salvia	P + A	7–19, 22, 27, 32	Scheel 1931, Yakovleva 1933, Benoist 1938
Nicotiana	P–A	9, 10, 12, 16, 18–22, 24, 32	Kostoff 1943, Goodspeed 1947
Veronica	P + A	7–9, 14–21, 24, 26, 28, 32, 34	Hofelich 1935, Darlington and Janaki-Ammal 1945
Stipa	P	12, 14, 16–24, 32–35, 41	Avdulov 1931, Stebbins and Love 1941, Love (unpubl.)
Carex	P	6, 9, 12–56	Heilborn 1924, etc.; see text
Scirpus	P	13, 18, 20–34, 38, 39, 52, 55	Tischler 1931
Eleocharis	P	5, 8–10, 15–19, 23, 28	Tischler 1931
Iris	P	8–24, 27, 28, 30, 35, 36, 41–44, 54	Simonet 1934, Anderson 1936b, Foster 1937
Crocus	P	3–16, 20, 23	Mather 1932, Darlington and Janaki-Ammal 1945

rieae), 2) ascending basic (*Fritillaria pudica* et aff., *Allium*, New World *Dorstenia*), and 3) interchange polyploid-amphiploid (*Brassica, Erophila, Carex, Stipa*). A fourth type of aneuploid series, containing unbalanced numbers which have resulted from polyploidy and apomixis, was mentioned in Chapter X.

Table 8 gives a summary of all of the examples of these three types of series known to the writer. This summary shows that the number of descending basic series (25) is far greater than that of either ascending basic (6) or interchange-amphidiploid (13). More than half (12) of the descending series culminate in annual species or occur in strictly annual groups, while only one of the ascending series is of this nature. This supports the hypothesis, suggested in Chapter V, that a low chromosome number has a selective advantage in a cross-fertilized annual species, since it increases the amount of linkage and therefore the degree of constancy of a population over short periods of time. In the interchange-amphidiploid series, most of the higher numbers are derived, but the lowest basic number is not necessarily the original one. In fact, indirect evidence in five of the genera listed (*Euphorbia, Salvia, Veronica, Eleocharis,* and *Crocus*) suggests that the lowest numbers found in them have been derived by stepwise reduction.

Although more complete evidence may change somewhat the relative frequency of the known examples of these three types of aneuploid series, nevertheless we can safely say that changes in basic diploid chromosome number proceed more frequently in the direction of decrease than of increase. In plants higher chromosome numbers are usually produced by means of either euploid polyploidy or aneuploid amphidiploidy, and therefore through duplication of entire chromosome sets. It must be noted again, however, that in a number of families (Pinaceae, Fagaceae, Asclepiadaceae, Caprifoliaceae, Rubiaceae), a high degree of morphological divergence and differentiation of species and genera has been reached without any changes in chromosome number.

CHANGES IN FORM AND RELATIVE SIZE OF THE CHROMOSOMES

In many genera and families of flowering plants, conspicuous differences in the appearance of the karyotype have been found in species having the same chromosome number. In some instances these differences follow definite trends, associated with trends of morphological specialization. The most significant of these correlations, in addition to the ones in *Crepis* and its relatives, is that recorded by Levitzky (1931b) in the family Ranunculaceae, tribe Helleboreae. In this tribe the genus most primitive in floral structures, *Helleborus,* has a karyotype in which the chromosomes differ little from each other in size and most of them are V-shaped or isobrachial, with median or submedian centromeres. Karyotypes of this nature are the most common ones in the higher plants as a whole and are the only ones found in most of the families and genera which are relatively homogeneous karyologically. Therefore, such karyotypes, as Levitzky has pointed out, can be considered as generalized types from which various specialized ones have been derived. Two types of specialization were traced out in the tribe Helleboreae. The first is the reduction in length of one of the chromosome arms, altering the V-shaped chromosomes first to J-shaped and then to headed types, with subterminal centromeres. Rod chromosomes with terminal or apparently terminal centromeres, the acrocentric chromosomes of White (1945), are the ultimate stage in this increasing asymmetry, but these are rare in plants, although they are rather common in animals. The second type of specialization consists of the reduction in size of some chromosomes in relation to others of the same set, so that the specialized karyotype contains chromosomes of very unequal sizes. Specialization along both of these lines in the Helleboreae reaches its climax in *Aconitum* and *Delphinium,* the two genera which have highly specialized zygomorphic flowers. In the latter genus the more primitive species, like *D. staphysagria,* have the largest number of V- or J-shaped chromosomes, while the reduced annual species of the subgenus Consolida (*D. ajacis, D. consolida*) have the most specialized karyotypes in both the form and the relative size of the component chromosomes. A karyotype consisting of chromosomes all essentially similar to each other in size and with median or submedian centromeres may be termed a symmetrical one. Asymmetrical karyotypes possess many chromosomes with subterminal centromeres, or great differences in size between the largest and the smallest chromosome, or both.

It should be noted here that these trends of specialization in the karyotype do not occur in any tribes of the Ranunculaceae other than the Helleboreae. In the Clematideae and the Anemoneae, both of which are highly specialized in their achenes, the karyotypes are essentially symmetrical (Gregory 1941, Meurman and Therman 1939). The apparently sporadic distribution of this and other trends of divergence in the karyotype is characteristic of the higher plants.

In the tribe Cichorieae, family Compositae, karyotype specialization follows essentially the same lines as in the Ranunculaceae (Babcock, Stebbins, and Jenkins 1937, Babcock and Jenkins 1943, Babcock 1947). Species of *Dubyaea* which are morphologically unspecialized in both vegetative and floral characteristics have symmetrical karyotypes. In *Crepis,* most of the chromosomes have submedian or subterminal centromeres, but the species with $n = 6$ and relatively primitive external morphology have on the whole more symmetrical karyotypes than do most of the species with $n = 4$ or $n = 3$.

In *Lactuca* the same tendency accompanies the increasing specialization of the species in external morphology, though to a lesser extent. But in *Hieracium, Taraxacum,* and *Youngia,* considerable specialization in external morphology has been associated with the retention of a karyotype of the symmetrical, unspecialized type similar to that of *Dubyaea.*

Examples of such trends in chromosome morphology are scattered through the higher plants, but are easily studied only if the chromosomes are reasonably large. In *Vicia* and in *Lathyrus,* the perennial species with $x = 7$ have mostly symmetrical karyotypes. But in the annual species, *Vicia sativa, V. angustifolia, V. faba,* and others, the reduction of the number to $n = 6$ and $n = 5$ is accompanied by the appearance of chromosomes with subterminal centromeres and considerable differences in size between the largest and the smallest chromosomes (Sveshnikova 1927, 1936).

In the Gramineae, tribe Hordeae, most of the species have symmetrical karyotypes, but the genus *Aegilops* is a notable exception. In this genus of reduced annual grasses increasing specialization of the fertile scales or lemmas, associated with a fragile rachis of the inflorescence and the development of a system of awns which is a great aid to the transportation of seeds, is associated with increasing specialization of the karyotype. Elsewhere in the tribe, the karyotypes of species ordinarily placed in different genera are indistinguishable from each other, but in *Aegilops* each section has a distinctive karyotype (Senjaninova-Korczagina 1932, Kihara 1940). On the other hand, the genus *Hordeum* has developed a number of annual species with a fragile rachis on the inflorescence and a system of awns which aid in seed dispersal, and are therefore functionally similar, though not homologous, to the awn systems of *Aegilops*. But these annual *Hordeum* species (*H. murinum, H. gussoneanum, H. marinum*) have symmetrical karyotypes essentially similar to those of their perennial relatives.

The greatest diversity of karyotypes to be found in any single family of plants is probably that in the Liliaceae. Here, the most asymmetric examples are found in plants which are definitely specialized morphologically. The most striking of these is in *Yucca* and in *Agave,* which have five pairs of relatively large chromosomes with mostly subterminal centromeres and 25 much smaller ones (McKelvey and Sax 1933, Whitaker 1934a). A similar karyotype is found in the eastern Asiatic genus *Hosta,* but Whitaker's suggestion that this genus and *Yucca* are descended from a common ancestor is unlikely on both morphological and distributional grounds. Another type of highly asymmetrical karyotype is common to three morphologically specialized genera of the tribe Aloineae — *Aloe, Gasteria,* and *Haworthia* (Sato 1937, Resende 1937). This consists of seven pairs of chromosomes, four long and three very short, all with subterminal centromeres. It is noteworthy that *Kniphofia,* a genus of the Aloineae with much less specialization in its vegetative habit, has a much more symmetrical karyotype. Its six pairs of chromosomes are about equal in size, and three have median to submedian centromeres (Moffett 1932). Two other genera of Liliaceae, *Nothoscordum* (Levan 1935) and *Miersia* (Cave and Bradley 1943), are among the few plant genera with chromosomes having apparently terminal centromeres. Both of them are rather specialized as to flowers, as well as vegetative parts.

Speaking generally, the statement can be made that plants with asymmetrical karyotypes are usually specialized morphologically, sometimes in vegetative parts, sometimes in flowers or fruits, and sometimes in both. On the other hand, while symmetrical karyotypes are most often found in plants which are morphologically more or less generalized, they also occur not uncommonly in relatively specialized species or genera. In other words, the evidence indicates that the karyotypes of the original, unspecialized progenitors of most families of plants were essentially symmetrical. Increased asymmetry of the karyotype, consisting in the evolution both of chromosomes with subterminal centromeres and of inequality in size between the different chromosomes of the same karyotypes, has been a frequent, but far from universal, type of change accompanying increased specialization in external morphology.

That this kind of change has taken place at various times throughout the history of the seed plants is suggested by the presence of a definitely asymmetrical karotype in the genera *Cycas* (Sax and Beal 1934) and *Ginkgo* (Sax and Sax 1933). Both of these genera, although in many characteristics among the most primitive of seed plants, are in other respects highly specialized. Fossil evidence, reviewed in Chapter XIV, indicates that both represent the end products of evolutionary trends which took place during the Mesozoic era or even in the latter part of the Paleozoic era, hundreds of millions of years ago. On the other hand, equally specialized genera of conifers, such as *Pinus* and *Taxus,* have essentially symmetrical karyotypes (Sax and Sax 1933).

As compared with plants, the karyotypes of most animals are far more asymmetrical. Rod chromosomes with apparently terminal centromeres are much more common and in many groups, like the reptiles and the birds, the difference in size between the largest and the smallest chromosome of the set is enormous (White 1945). The statement of White, that spontaneous breakage and translocation of chromosomes usually involves whole arms that break near the centromere, does not hold for plants. In them translocations usually involve breaks near the middle or distal parts of the chromosomes, as is evident from the presence of "interstitial segments" in the translocation heterozygotes of *Oenothera* (see Chapter XI), of figure-of-eight, "necktie," and similar configurations in *Pisum* (Sansome 1932) in *Datura* (Bergner, Satina, and Blakeslee 1933), in *Paeonia* (Stebbins and Ellerton 1939), and in other plants, and from chromosome pairing in interspecific hybrids involving translocations (Tobgy 1943, Sherman 1946). It is tempting to speculate on the reason why animals and plants, which are in many other respects so very similar in their hereditary mechanism, should differ in the morphology of their karyotypes. The fact that animals possess a vastly more differentiated and specialized body than that of plants may be significant in this respect.

The reason for these trends in karyotype evolution — both the progressive reduction of the basic chromosome number and the increasing asymmetry — is not easy to find, although a tentative hypothesis in respect to the former has already been advanced. Both types of changes result directly from gross structural alterations of the chromosomes, either translocations or inversions. As explained above, progressive reduction is produced by unequal translocations, but it cannot take place unless some of the chromosomes possess genetically inert material near their centromeres. Also, it occurs more easily if they have at least one arm that is relatively short and completely inert. Increasing asymmetry may result either from unequal translocations or from inversions involving the centromere. If such an inversion occurs as a result of a break in one arm near the centromere, and in the other near its distal end, the chromosome can be converted at one step from a V-shaped one with a median centromere to a "headed" type with a subterminal centromere.

The difficulty of explaining these regular trends in karyotype differentiation as a result of structural alterations alone is that, in spite of a large body of knowledge about such alterations produced artificially by X radiation, no mechanism has been discovered by which they could lead to the progressive alteration of the karyotype which has been observed. To be sure, viable translocation and inversion types induced by X rays do not result from breaks occurring at random. The most frequent regions of breakage are the heterochromatic ones (Bauer, Demerec, and Kaufmann 1938, Heitz 1940). Furthermore, short arms have a greater chance of receiving chromosomal material from translocations, while long ones tend to give it up (Levitzky and Sizova 1934). Hence, if all viable translocation types had an equally high survival value, the tendency would be for translocations to alter an asymmetrical karyotype in the direction of greater symmetry rather than in that of progressively increasing asymmetry. The occurrence of this latter tendency, therefore, would be most easily explained if in certain groups of organisms translocations which lead toward a more asymmetrical karyotype have a relatively high selective value. At present, however, evidence is not available for the formulation of such a hypothesis.

In regard to the third series of differences between karyotypes, those affecting the nucleoli and the satellites, a vast literature has accumulated, which is reviewed in detail by Gates (1942). The work of Heitz (1931), Navashin (1934), and McClintock (1934) established firmly the fact that the small beadlike appendages known as satellites or trabants, which in many species occur on the ends of one or a few pairs of chromosomes, are directly related to those portions of the chromosomes which form the nucleoli. The satellite is separated from the rest of the chromosome either by a slender thread or by a "secondary constriction," which corresponds with the nucleolar organizing region of the chromosomes. The size of the satellite depends on the distance of this constriction from the end of the chromosome. Usually it is nearly terminal and the satellite is small; but in some plants the presence of an interstitial nucleolar organizing region and secondary constriction produces a chromosome with a large "satellite" which may include as much as one fourth of the chromosome (Fig. 44). Usually the satellite is borne on the end of the short arm of a chromosome with a subterminal centromere, but V-type chromosomes bearing satellites are not uncommon. In most diploid species, only one pair of satellites is found, but species with two, three, or more satellited chromosome pairs are known. As mentioned earlier (page 362), this weakens greatly the value of the number of satellites and nucleoli as evidence for the existence of polyploidy. Furthermore, the genetic experiments of Navashin (1934) and of McClintock (1934) have shown that nucleolar organizers vary greatly in the strength of their activity. If by hybridization a chromosome with a strong nucleolar organizer is placed in the same cell with a weak one, the activity of the latter may not be expressed, and the plant may have one less than the expected number of nucleoli and satellites.

Very little is known about evolutionary changes in the satellites and nucleoli. The wide distribution of these structures through-

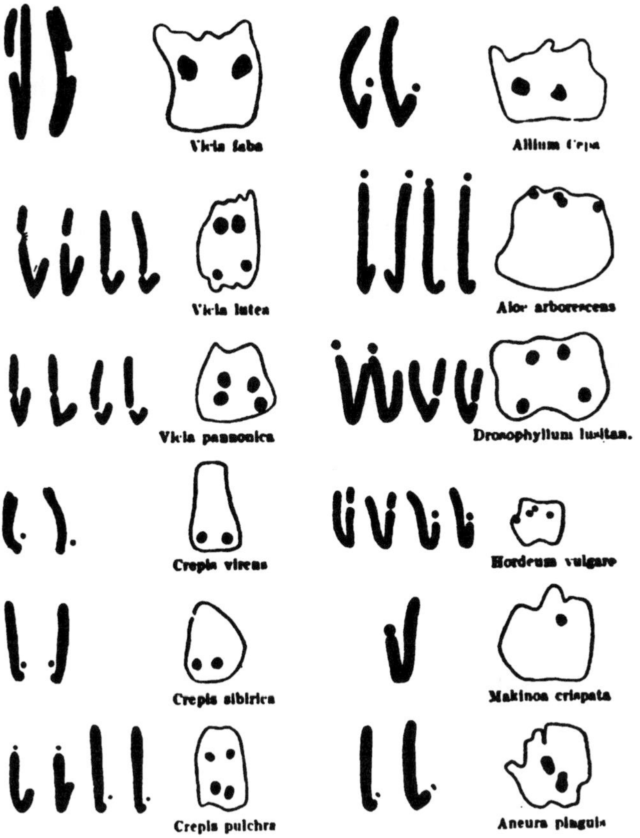

FIG. 44. Drawings showing various types of satellited chromosomes and their relationship to nucleoli. From Heitz 1931.

out the plant kingdom shows that they are a valuable, if not essential, part of the chromosomal complement. Furthermore, the distribution and relationships of species having different numbers and sizes of nucleoli and satellites suggests that a nucleolar organizer with its accompanying satellite may be either lost or gained during evolution, and that its size may either increase or decrease.

EVOLUTIONARY CHANGES IN CHROMOSOME SIZE

Differences in chromosome size are of two types. The first consists of differences in *relative* size between different chromosomes of the same set, as was discussed in a preceding section of this chapter. Such changes are produced by unequal translocations, which increase the size of some chromosomes at the expense of others, without changing the total amount of chromosomal material present. In contrast to these are differences in *absolute* size of all the chromosomes of a set. The fact that some organisms have very small chromosomes and others very large ones is well known to cytologists. Darlington (1937, p. 82) has estimated that in the protozoan *Aulacantha* the total bulk of the somatic chromosomes at metaphase is more than 10,000 times that in the fungus *Saprolegnia,* which has very small chromosomes and a low basic number. Even in comparisons between different genera of the same family, as *Drosera: Drosophyllum,* the ratio of bulk is 1:1,000.

These differences in absolute chromosome size are to some extent controlled by factors outside of the chromosomes themselves. Pierce (1937) found that lack of phosphorus in the nutrition of the plant causes considerable reduction in the size of the chromosomes of *Viola.* Navashin (1934) found that in certain hybrids between species of *Crepis* having different chromosome sizes, the differences in the sizes of the chromosomes in the hybrid nuclei were much less than would be expected from comparison between the cells of the parental species. On the other hand, Tobgy (1943) found that the size differences between the chromosomes of *Crepis neglecta* and *C. fuliginosa* were just as evident in the hybrid nuclei as in those of the parental species, and therefore were apparently controlled by the chromosomes themselves. The factors controlling absolute chromosome size need much further exploration before the full significance of the trends to be described below will be evident.

In contrast to alterations of the basic number and the symmetry of the karyotype, changes in absolute size of the chromosomes appear to have involved both phylogenetic increase and decrease with about equal frequency, and are probably reversible. Phylogenetic reduction in chromosome size was first described by Delaunay (1926), in the genus *Muscari* of the Liliaceae. In this genus the species which show greater morphological specialization, like *M. tenuiflorum, M. caucasicum,* and *M. monstrosum,* have a smaller absolute chromosome size than the relatively primitive species, like *M. longipes.* This is associated with a reduction in the amount of meristematic tissue in the root tips, and probably also in the shoots, since in *M. tenuiflorum* the rachis of the inflorescence is shorter and more slender, the flower pedicels are shorter, and the flowers are smaller than in *M. longipes.* In *Crepis,* a similar reduction in chromosome size, correlated with the appearance of the annual habit of growth and the reduction in size of all the parts of the involucre and the flower, was observed by Babcock and Cameron (1934). Other genera of the tribe Cichorieae in which phylogenetic reduction in chromosome size has taken place are *Youngia, Ixeris, Taraxacum,* and *Sonchus.* In these genera this reduction has not always been accompanied by appearance of the annual habit or decrease in size of the floral parts. The species of *Taraxacum* are nearly all long-lived perennials with large involucres and flowers, while some of the species of *Sonchus,* which nevertheless have relatively small chromosomes, are shrubby in habit and have large floral parts.

Other examples of phylogenetic reduction in chromosome size are scattered throughout the plant kingdom. In the leptosporangiate ferns the more primitive families Osmundaceae and Hymenophyllaceae have relatively large chromosomes; those in the Cyatheaceae and the Polypodiaceae are of intermediate size, while the smallest ones are in the vegetatively reduced and reproductively specialized heterosporous family Salviniaceae (de Litardière 1921). Among angiosperms, in addition to *Muscari* and the Cichorieae, phylogenetic reduction in chromosome size apparently occurs in the genus *Dianthus* of the Caryophyllaceae. Rohweder (1934) has reported that the reduced and specialized annual species, *D. armeria,* has smaller chromosomes than the perennials, while within the perennials themselves considerable differences in chromosome size exist. These are associated with resistance to cold, since the species with the largest chromosomes are found in high alpine areas. It is possible that the original species of *Dianthus* had chromosomes of an intermediate size, and that phylogenetic progression has been toward decrease in some lines and toward increase in others. The two families Juncaceae and Cyperaceae, both of which are almost certainly reduced, specialized derivatives of ancestors similar to the Liliaceae, have chromosomes considerably smaller in size than most species of the latter family.

The best example of phylogenetic increase in chromosome size is in the family Gramineae. Avdulov (1931) has pointed out that the most primitive grasses, of the tribes Bambuseae and his series Phragmitiformes, have relatively small chromosomes, as do also the more specialized grasses of tropical regions. But the grasses predominant in temperate regions, contained in his series Festucaeformes, usually have the basic number $x = 7$ and much larger chromosomes. Avdulov considers that the phylogenetic increase in size has occurred as an adaptation to the cool climate in which these grasses live, and he has suggested that it took place during the period of Pleistocene glaciation. The latter hypothesis is very unlikely in view of the wide distribution and extensive differentiation into tribes and genera found in the Festucaeformes, and in fact certain distribution patterns of species groups in this series suggest that many of its subgenera date back to the Tertiary period. But the correlation found in the Gramineae between large chromosome size and occupation of regions with a temperate climate may have some significance. Within certain groups of the Festucaeformes, on the other hand, the reverse tendency toward reduction in chromosome size may be observed. In the genus *Agropyron,* for instance, the reduced annual species *A. prostratum* has considerably smaller chromosomes than those of more typical species of the genus (Avdulov 1931, p. 260), while in another tribe the genus *Phalaris* has two specialized annual species, *P. canariensis* and *P. paradoxa,* which have smaller chromosomes than the perennial, more generalized species of *Phalaris.* The two genera *Poa* and *Puccinellia,* which have lemmas reduced in size and often with specialized, elongate cobwebby hairs, have chromosomes considerably smaller than those typical for the series Festucaeformes. This series of grass genera, therefore, provides the best evidence among plants for the reversibility of trends in absolute chromosome size.

Other examples of phylogenetic increase in chromosome size are scattered through the class of angiosperms. In the Polygonaceae, the group of *Rumex acetosa* and its relatives, which are specialized particularly in their dioecious condition, have larger chromosomes than any other species of the genus or family to which they belong (Jaretzky 1928). In the Cruciferae, of which most of the species have small chromosomes, two specialized groups of annual species in *Hesperis* and *Matthiola* have large chromosomes for the family (Manton 1932). In the Leguminosae, the genera *Vicia, Pisum,* and *Lathyrus,* all closely related to each other and specialized, tendril-bearing vines, have the largest chromosomes found in the family, and therefore certainly represent end points of a trend of increase in chromosome size (Senn 1938a). In the Onagraceae considerable differences in chromosome size are found within the genus *Godetia* (Håkansson 1941, 1943b). Most remarkable is the fact that two closely related species, *Godetia bottae* and *G. deflexa,* were found to be strikingly different in chromosome size, the chromosomes of *G. deflexa* being more than four times as large as those of *G. bottae.* In addition, two related species, *G. amoena* and *G. whitneyi,* have larger chromosomes than most of the other species of the genus. These species, although their phylogenetic position in *Godetia* is at present uncertain, are significant in being among the most northerly in distribution of the genus.

Another probable example of phylogenetic increase in chromosome size is in the Rubiaceae. The extensive survey of Fagerlind (1937) has revealed one genus, *Galium,* in which many of the species have larger chromosomes than those typical of the family. This is a specialized genus of herbaceous perennials and annuals, and it is significant in being one of the few genera of this predominantly tropical family which are distributed in the North Temperate Zone.

Turning to the Monocotyledons, two further groups deserve mention. In the family Commelinaceae the temperate species of *Tradescantia* are conspicuous in having larger chromosomes than any other members of the family, which are mostly tropical in distribution (Darlington 1929). The phylogenetic position of these species is uncertain, so that one cannot say whether this family is an example of phylogenetic reduction or of increase in chromosome size. But in this, as in several previous examples, the significant fact is that the species inhabiting the temperate zone have conspicuously larger chromosomes than their tropical relatives. Finally, in the tribe Paridae of the family Liliaceae, the genera *Trillium* and *Paris* have the largest chromosomes known in the plant kingdom (Huskins and Wilson 1938, Geitler 1937, 1938). These genera are certainly specialized in their vegetative characteristics, and the tribe as a whole is related to and perhaps descended from the tribe Uvularieae (Hutchinson 1934), the members of which have large chromosomes, but nevertheless smaller ones than those in *Trillium* and *Paris*.

The facts which stand out from this survey are, first, that phylogenetic reduction and phylogenetic increase in chromosome size are about equally common in the higher plants and that both of these processes are reversible. Second, in about one half of the groups reviewed there is no difference in geographic distribution between the species with large chromosomes and those with small chromosomes. But in the other half such a difference exists, and the species with larger chromosomes invariably occupy cooler climates than those with small ones.

Another correlation between chromosome size and a characteristic of external morphology was pointed out earlier by the writer (Stebbins 1938b). Woody angiosperms nearly all have small chromosomes, while in herbaceous members of this class all sizes of chromosomes exist. The explanation of this correlation may lie in two characteristics of these woody types. First, the wood of angiosperms contains fiber cells which are very small in their transverse dimensions and must originate from small-sized cambial initials. Second, in woody plants the slow rate of replacement of individuals in a population puts a high selective premium on a genetic system permitting the maximum amount of recombination, as explained in Chapter V. This would favor a relatively high chromosome number and a symmetrical karyotype, both of which are characteristic of most woody plants (Stebbins 1938b). If the absolute size of the chromosomes were large, such a karyotype would have a relatively large total volume, and because of the relative constancy of the nuclear-cytoplasmic ratio this would require that all of the meristematic cells of the plant be relatively large. This would be incompatible with the necessary small cambial initials mentioned above, so that in woody angiosperms we should expect genetic changes tending toward small absolute chromosome size to have a high selective value.

In support of this hypothesis is the fact that the principal woody gymnosperms, the Coniferales, which lack wood fibers and have cambial initials all about equal in size, have symmetrical karyotypes and a relatively high basic number (mostly $x = 11$,

12, or 13), but also have rather large chromosomes. Among the few woody dicotyledonous angiosperms with medium-sized rather than small chromosomes are members of the most primitive order, Ranales, namely, *Illicium* (Whitaker 1933) and *Michelia* (Sugiura 1936, p. 572) of the Magnoliaceae, *Schizandra* and *Kadsura* of the Schizandraceae (Whitaker 1933), and various genera of the Anonaceae (Asana and Adatia 1945, Bowden 1945). It is likely, therefore, that phylogenetic reduction of chromosome size took place early in the evolution of woody dicotyledons.

Throughout this discussion the question naturally arises as to whether these differences in chromosome size involve corresponding differences in the number or size of the genes themselves. To this question no direct answer can be given as yet. The only plant species thoroughly known genetically is *Zea mays,* which has medium-sized to small metaphase chromosomes. The next best-known species are *Lycopersicum esculentum, Antirrhinum majus,* and *Hordeum vulgare,* of which the two former species have small chromosomes and the latter, large chromosomes. There is no evidence for the presence of more gene loci in *Hordeum* than in the other species mentioned, although this might possibly become evident when the genetics of *H. vulgare* are better known. Four other familiar cytogenetic objects with large chromosomes which should be watched in this connection are *Secale cereale, Vicia faba, Lathyrus odoratus,* and *Pisum sativum.*

Indirect evidence on the nature of differences in absolute chromosome size is provided by Tobgy's analysis (1943) of the difference between *Crepis neglecta* and *C. fuliginosa.* The metaphase chromosomes of *C. fuliginosa* are all smaller than the corresponding ones of *C. neglecta,* and this difference is evident in the F_1 hybrid nuclei at both somatic and meiotic metaphase. On the other hand, observations of chromosome pairing in the prophase of meiosis (pachytene) revealed that the pairing threads at that stage are of equal length. The difference in size between the metaphase chromosomes of the two species results from the greater contraction and presumably the tighter coiling of the chromonema in *C. fuliginosa,* and from the smaller amount of heavily staining substance, presumably desoxyribose nucleic acid, which forms around this chromonema. Examination of chromosomes in segregating individuals of the F_2 generation showed that in spite of the fact that these chromosomes had acquired through crossing over in the bivalents of the F_1 plants chromosomal segments belonging to both parental species, each chromosome had dimensions characterizing it as typical of either *C. neglecta* or *C. fuliginosa.* Apparently, the dimensions of each chromosome are determined autonomously according to the genetic constitution of the region around its centromere.

A further difference between the nuclei of *C. neglecta* and *C. fuliginosa* is associated with the differences between them in chromosome size. The resting nuclei of *C. neglecta* contain many moderate-sized regions of heterochromatic material, that is, of substance which stains deeply after Feulgen treatment and is presumably desoxyribose nucleic acid. On the other hand, the nuclei of *C. fuliginosa* contain at the same stage a much smaller number of relatively large bodies of this substance. It is possible, therefore, that differences in absolute size of the chromosomes have nothing to do with the size of the chromonema or gene string, but are related to the coiling properties of this structure and to the amount and distribution of certain chemical substances in the chromosomes.

This possibility deserves further exploration in view of the fact that phylogenetic trends in absolute chromosome size, both toward decrease and toward increase, are often associated with a corresponding decrease or increase of the plant as a whole or of certain of its parts. Furthermore, if the relationship observed by Delaunay between reduction in chromosome size and decrease of the amount of meristematic tissue holds for other groups of plants, then there probably exists a definite influence of the amount of chromatin in the nucleus on the extent of meristematic activity of the tissue. In other words, it may become possible here to establish a definite relationship between the chemical constitution of the nucleus and certain highly significant morphological and physiological characteristics of the plant.

DIFFERENCES IN THE RESTING AND PROPHASE NUCLEUS

Until now we have been considering only differences and trends in the karyotype itself, that is, in the chromosomes as they appear at somatic metaphase. But in the preceding section it was shown that in one of these characteristics, absolute chromosome size, the differences observed in the metaphase karyotype are closely connected with and perhaps dependent upon differences in the physicochemical nature of the nucleus which are observable only in the resting and prophase stages. The subject of evolutionary trends in the karyotype is, therefore, not complete without a survey of the comparative karyology of the resting and prophase nucleus.

The background for such a survey is the pioneering and fundamental work of Heitz (1932), on which the following account is based. He recognized a number of different types of resting nucleus, depending on the distribution of the so-called heterochromatic substance, that is, the substance which stains darkly when nuclei are prepared according to the Feulgen method, and is presumably desoxyribose nucleic acid. In one type, exemplified by *Paeonia,* this substance is found almost throughout the nucleus, so that the chromosome threads can be stained at any stage of the resting or mitotic cycle. In addition, small centers which stain more strongly are found near the centromere. These are known as chromocenters. This type of nucleus is found in many groups of plants with large chromosomes (Ranunculaceae, Liliaceae, many Gramineae) and is probably the commonest type in such groups. A modification of it is found in *Vicia faba,* in which small chromocenters occur at definite positions on the distal parts of the chromosome arms, as well as at the centromere.

The second type of nucleus is termed by Heitz the "cap nucleus," and it is exemplified by *Hordeum vulgare.* In this, the chromosome substance at one side of the nucleus stains much more heavily than that at the other. The staining region can be identified at mitotic telophase as the one away from the equator and toward the pole of the spindle, so that the parts of the chromosomes which stain are large chromocenters situated about the centromere and the proximal portions of the arms. Cap nuclei are not common and are found usually in plants with relatively large chromosomes (*Hordeum, Collinsia, Scorzonera*).

The two other types of nucleus are those in which all the heterochromatic substance is aggregated in a more or less definite and relatively small number of chromocenters, the remainder of the resting nucleus being clear when stained by the Feulgen technique. These chromocenters may be located at various positions along the chromosome arms in addition to the region of the centromeres, or they may be confined to the latter regions, so that the number of chromocenters corresponds to the diploid number of chromosomes. Such large, localized chromocenters have been termed prochromosomes by many authors. Most resting nuclei, therefore, belong to one of four principal types, as follows.

Diffuse staining type. — Associated with large chromosomes; no conspicuous aggregation of heterochromatic material. Examples: *Paeonia, Bromus,* some species of *Allium.*

Cap nucleus. — Associated with large or medium-sized chromosomes. Heterochromatic material concentrated in the proximal regions of the chromosome arms. Examples: *Hordeum, Collinsia, Scorzonera.*

Multiple chromocenter nucleus. — Associated with medium-sized or small chromosomes. Heterochromatic substance contained in dark-staining chromocenters, which, at least in early prophase, are of a higher number than the diploid number of chromosomes and are found in the distal parts of the chromosome arms as well as near the centromeres. Examples: *Pellia* (Heitz 1928), *Crepis neglecta, C. fuliginosa.*

Prochromosome nucleus. — Associated chiefly with small chromosomes. Heterochromatic substance confined to a number of dark-staining bodies or chromocenters which, at least at early prophase, often equal the diploid chromosome number, each chromocenter consisting of the proximal portions of the two chromosome arms. Examples: *Thalictrum, Impatiens, Oenothera, Sorghum vulgare.*

Nuclei intermediate between some of these types are occasionally found. There is little doubt that the evolutionary trend from the diffuse to the multiple chromocenter type of nucleus and vice versa can proceed in either direction, since the differences between these two types are usually associated with differences in absolute chromosome size. It is likely, also, that the evolution from a cap nucleus to a prochromosome nucleus, as well as the reverse trend, can occasionally take place. On the other hand, the trend from the diffuse to the cap nucleus or from the multiple chromocenter to the prochromosome type may be irreversible. The few definitely established examples of cap nuclei are in relatively specialized species or genera, of which the more primitive relatives have the diffuse type. *Hordeum vulgare,* for instance, is one of the most specialized members of its genus. In most of the more primitive perennial species of *Hordeum,* as well as in most species of the more generalized and perhaps ancestral genus *Elymus,* the nuclei are of the diffuse type. The same may very well be true of prochromosome nuclei, which also seem to be found in relatively specialized groups. *Sorghum vulgare,* the best-known example in the Gramineae, is certainly a specialized end line in its tribe. The evidence suggests that in certain unrelated lines the tendency develops for the chromosomal regions with a high content of desoxyribose nucleic acid to become aggregated near the centromeres, but that the reverse tendency rarely, if ever, takes place. The significance of this tendency is at present obscure.

None of the examples presented in this chapter of evolutionary trends in the karyotype are consistent for all groups of plants. But every one of these trends has occurred repeatedly in several different unrelated groups of plants, and in many cases has persisted over the time needed to differentiate whole genera. Such long-continued trends cannot be explained by chance, and there appears to be nothing in the structure of the chromosomes themselves which would force them to continue evolving in the same direction. They can be explained best on the assumption that in certain types of organisms under some particular environmental conditions alteration of the chromosomes in the direction of one of these trends has a definite selective value. At present, the study of the comparative karyology is in its infancy, so that the nature of these selective values can only be suggested in the form of tentative hypotheses. But they may eventually provide valuable clues as to the relation between changes in the chromosomes and the evolution of the individual and the population in visible, external characteristics.

CHAPTER XIII

Evolutionary Trends II: External Morphology

THE PRESENT VOLUME is concerned with the dynamics of plant evolution in general, rather than with the evolutionary origin and phylogenetic history of any particular group or groups, however important they may be. Nevertheless, certain principles and trends of phylogeny are repeated so many times in various groups of plants that some consideration of them is essential to an understanding of the mechanism of evolution. Fortunately, most of these trends can be seen operating to a small extent even within groups of genetically related species, so that to this degree they are open to study by experimental methods.

Since the trends to be considered in this chapter are those characteristic of the vascular plants, the more primitive and generalized structural features of these plants must first be described. These are now well known in the fossil Psilophytales (Eames 1936, pp. 309–331), and the vegetative parts are essentially similar in the modern *Psilotum*. The aerial vegetative parts of these primitive rootless plants are simply branch systems which grow at the tips by means of apical meristems and fork repeatedly at regular intervals. This forking or branching is typically dichotomous, that is, the apical meristems of the two branches are of equal size and grow at the same rate. The parts bearing the reproductive structures are also branch systems, which differ principally in bearing sporangia at their apices. Typically, therefore, the primitive vascular plant is characterized by apical, indeterminate growth and the successive differentiation of similar organs, one at a time.

Very early in the evolution of the vascular plants there developed independently in a number of different lines an organization characterized by the presence of a main axis and a number of appendages. This is found in nearly all the living vascular plants. Evidence both from paleobotany and from comparative morphology strongly supports the telome theory, according to which the appendages of ferns and seed plants (Pteropsida), whether leaves, microsporophylls, or megasporophylls, are derived

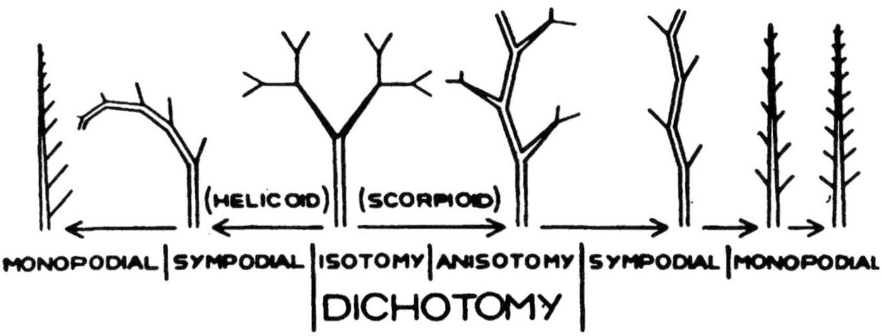

Fig. 45. The main principles of the overtopping process, showing two ways in which a monopodial branch system can evolve from a primitive type of dichotomous branching. From Lam 1948.

from branch systems whose ontogeny has been modified in successive phylogenetic stages by means of reduction of apical growth and fusion of branches (Fig. 46, cf. Zimmermann 1930, p. 67, Bower 1935, Chap. XXX, Eames 1936, p. 348). These three processes of reduction, fusion, and change in symmetry will be discussed in more detail below. In its most generalized form, the shoot-appendage organization of the plant body agrees with the dichotomous branch systems in that organs are differentiated apically and serially, one at a time, but differs in that the meristem undergoes two types of branchings. One of these divides the growing tissue into a portion with finite, determinate growth (appendage) and another with indeterminate growth (continuation of shoot); while the other type, the true branching of the shoot, produces two shoots, each with indeterminate growth (Fig. 45). As Foster and Barkley (1933), as well as Anderson (1937a) have pointed out, the most characteristic feature of plant growth is the constant repetition of similar appendages and similar branches. The spiral arrangement of appendages which is most common in vascular plants is presumably derived originally from a dichotomy in more than one plane (Zimmermann 1930, p. 76).

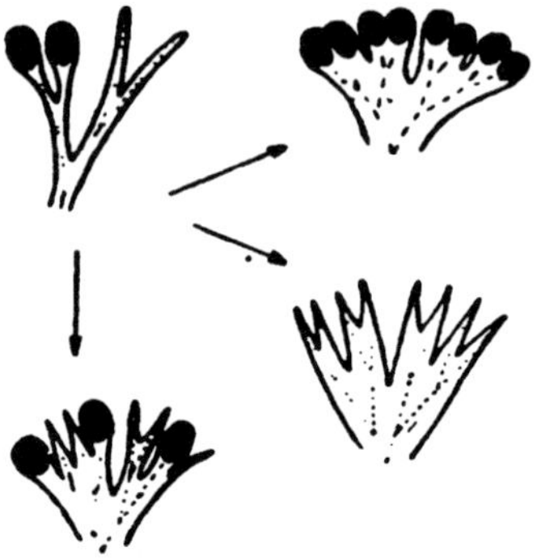

Fig. 46. The main principles of Zimmermann's telome theory, showing the evolution of primitive types of leaves and sporophylls from dichotomous branch systems. From Lam 1948.

SOME COMMON EVOLUTIONARY TRENDS

The great majority of the evolutionary trends toward increased specialization in the vascular plants can be explained as a result of three types of morphological trends, acting either separately or in conjunction with each other. These are *reduction, fusion,* and *change in symmetry.*

The most widespread of these trends has been reduction. There is no living species of vascular plant which has not been affected by this trend in one or another of its parts. As mentioned above, the leaves of all ferns and seed plants are reduced from branch systems. In many plants, moreover, reduction has affected the growth of the shoot, reducing in particular the period of growth between the successive differentiation of appendages or the length of the internodes. The limit of this type of reduction is the rosette habit, in which the shoot axis itself is hardly evident and the vegetative part of the plant consists almost entirely of appendages. This has been attained repeatedly in many different phylogenetic lines, such as most ferns, cycads, sedums, dandelions, aloes, yuccas, and a host of others. Another common type of reduction in the vegetative parts of the plants is the conversion of the shoot from an indeterminate one, which produces an indefinite number of appendages and branches, to a determinate one which, after producing a more or less definite number of leaves, either stops growing or starts to differentiate reproductive structures. The best examples of determinate vegetative shoots are the short shoots of many conifers, in particular the fascicles of needles characteristic of *Pinus*. Determinate vegetative-reproductive shoots are characteristic of many spring-blooming perennial flowering plants, particularly those that grow from rhizomes or corms, like *Trillium, Claytonia,* and *Anemone.* They also occur in many annuals, like wheat, oats, barley, and many other crop plants.

Although trends of reduction are common in the vegetative parts of the plant, they are even more characteristic of its reproductive structures. Both the cone or strobilus of gymnosperms and the flower of angiosperms are reduced from shoots or shoot systems. These systems have repeatedly suffered further reduction in both the number and the diversity of their parts and in the structural and anatomical complexity of individual organs. Details of these trends can be found in many books and papers on plant morphology and anatomy (Jeffrey 1916, Goebel 1928, Zimmermann 1930, Eames 1936, Florin 1944a,b). Present knowledge makes possible the generalization that in modern seed plants reproductive organs which are simple in structure and development are many times less likely to be primitively simple than they are to be reduced and simplified from organs which were both structurally and ontogenetically more complex in the phylogenetic ancestors of the modern form. With respect to the angiosperms, we can accept with some assurance the dicta of Bessey (1915), Hutchinson (1926), and others, that apetalous flowers are derived from polypetalous (or sometimes sympetalous) ones; that unisexual flowers are derived from bisexual (perfect) ones; that flowers with few stamens are derived from those with more numerous ones; and that gynaecia with solitary carpels, as well as carpels or ovaries with one ovule, are derived by reduction from more complex structures. The reduced or missing parts are often present as vestigial rudiments or are represented by vascular bundles which pass in the direction of their expected position (Eames 1931).

A noteworthy fact which has not previously been emphasized in discussions of phylogenetic reduction is that reduction in size and complexity of individual structures is often accompanied by an increase in the number of these structures produced, while reduction in number may be balanced by increase in size and complexity of individual structures. The best illustrations of these interrelationships are found in the ovules and carpels of the angiosperms. In many groups, such as the genus *Nicotiana* and the family Orchidaceae, the individual ovules are much reduced in size and in the development of their integuments, but their number per carpel is greatly increased over that found in the nearest relatives of these groups. In some genera with much reduced one-seeded carpels that form achenes, the number of carpels is much greater than that found in any of their relatives. This is true in *Fragaria* of the Rosaceae and in *Myosurus,* as well as in some species of *Ranunculus* and *Anemone* in the Ranunculaceae. On the other hand, in such genera as *Quercus, Juglans,* and *Prunus,* the single ovule which remains in the ovary has become much enlarged and elaborated in structure. In many families of angiosperms, such as Umbelliferae, Compositae, Gramineae, Cyperaceae and Juncaceae, all of the parts of the flower have become greatly reduced and simplified, but the number of flowers per reproductive shoot or inflorescence has been greatly increased.

These interrelationships between reduction and multiplication deserve attention because they suggest that phylogenetic trends of this nature may in many instances be produced by changes in the timing relationships of growth and differentiation phases in a structure with determinate growth. As stated in Chapter IV, Tedin (1925) found that in *Camelina sativa* the same gene which increases seed size also reduces the number of seeds per pod.

From the developmental point of view, reduction in size and complexity of a mature organ in successive generations of a phylogenetic line must result from the accumulation of mutations which reduce either the rate of growth of that organ or the time during which it grows. And as pointed out above, and illustrated in the example of *Camelina* (page 133), this reduction may be compensated by an increase in the number of parts or in other cases by a corresponding increase in ultimate size, and therefore in speed or length of growth of other parts of the plant. On this basis the phylogenetic trend of reduction may be regarded as one phase of a much more general phenomenon, namely, a change in relative growth rates, or allometry. The great importance of allometry for understanding evolutionary trends in animals has been emphasized by Huxley (1932, 1942, pp. 525–555). In plants, genetic differences in fruit shape between different lines of various species of Cucurbitaceae and some other families have been explained on this basis by Sinnott (1936, 1937, 1939) and his associates (Sinnott and Kaiser 1934, Houghtaling 1935). Riley (1942, 1943, 1944) has similarly explained differences between species of *Iris* in respect to their ovaries and other floral parts. Foster (1932a, Foster and Barkley 1933) has shown that the difference between leaves and bud scales depends largely on differences in growth rate between different parts of their primordia, while Boke (1940) has demonstrated that the difference between leaves, phyllodes, and transitional structures in *Acacia melanoxylon* depends on the relative growth rate of the adaxial part of the foliar or rachis primordium, as compared to the primordia of the leaflets. Nevertheless, so far as this writer is aware, no general evolutionary trends have yet been analyzed on this basis. The phenomena of reduction are so similar in the two kingdoms that their explanation is probably the same.

In particular, allometry explains two generally observed facts about reduced or vestigial structures. As Huxley (1932) has pointed out, we would expect on the basis of allometric growth that the first formed embryonic primordium of a reduced organ would be as large or nearly as large as the corresponding primordium of an organ capable of developing to normal size. If the absence or reduction of the organ were of value to the organism, then selection would establish mutations slowing the growth rate of this primordium up to the point needed to produce the necessary reduction in size at the time of development when the selective factors were operating. The direct action of selection, therefore, would cause an organ to become reduced and vestigial, but not to disappear entirely. On the other hand, if the relative growth rate of one organ in respect to another is a constant function, for instance, if one organ is growing half as fast as another, the relative adult size of the two organs will be determined by the time over which growth is operating, that is, absolute size of the organism as a whole. In other words, given a constant allometric growth rate, reduction of an organ can be produced secondarily by the action of genes increasing absolute size. If the allometric constant is high and the growth persists long enough, this reduction can result in virtual disappearance of the vestigial organ. On this basis we would expect that direct selection would tend to reduce an organ until it became vestigial, and further reduction, although of no selective value in itself, could come about secondarily through selectively induced changes in absolute size. We can in this way explain on the basis of allometry apparently nonadaptive differences in organs that are themselves of no value to the species. Furthermore, Huxley has pointed out that small differences in growth ratio, if they operate over a wide range of absolute size, can greatly affect the relative size of the organs at maturity. This explains further the variability of vestigial, apparently useless organs.

The second of the three widespread morphological trends, namely, fusion or union of parts, has been nearly as common and important in the evolution of the higher plants as reduction. The term fusion, though it is widely used, is misleading in this connection because of its ontogenetic connotation. The "fused" parts do not usually grow first as separate organs and become united later by growing together. On the other hand, the union is accomplished by a reduction in differentiation of the primordia of the individual organs and the growth of a new, relatively undifferentiated type of tissue. For instance, the growth of a sympetalous corolla with strongly developed lobes, like that of *Lysimachia* (Pfeffer 1872), *Rhododendron, Hemerocallis* (personal observation), or the tomato (Cooper 1927), is about as follows. The first parts to appear are the primordia of the apical lobes, which correspond to the separate petals of the ancestral form. Growth is at first most active in separate centers located in each of these primordia, but at some time in ontogeny there appears a new type of growth at the base of the corolla. This has been termed by Coulter, Barnes, and Cowles (1910, p. 254) zonal development, a term which to the present writer seems the most appropriate of those suggested. Goebel (1928, p. 463) has termed the same growth zone a "Ringwall" (circular rampart), while Thompson (1934) uses the term toral growth. Zonal development produces growth as rapidly below the sinuses of the corolla as below the lobes themselves. In all later stages, growth in size of the lobes depends on the activity of the growth centers in their primordia, while elongation of the tube depends on the rapidity of zonal development. Since the lobes have in most corollas a head start over the tube, they will at certain intermediate stages be relatively large and the tube correspondingly short. But in strongly sympetalous corollas, like that of *Cucurbita* (Goebel 1933, p. 1861), the timing of zonal development is shifted forward so that it coincides with the first initiation of the corolla lobe primordia, and consequently the latter are not distinct from each other at any stage in ontogeny. In various zygomorphic flowers, such as the corolla of various Labiateae and the genus *Veronica,* as well as the calyx of *Utricularia,* some of the perianth lobes may begin their existence as separate primordia, while other lobes of the same perianth may be fused from the start (Goebel 1933, pp. 1859–1860).

It is obvious, therefore, that the eventual degree of "fusion" of the petals will be determined by three factors: first, the relative time of initiation of zonal development as compared to differentiation of the lobe primordia; second, the ratio between the rates of growth of the lobe primordia and the region of zonal development; and, third, the absolute size of the corolla. Thus, while the process of "fusion" is begun by the initiation of a new type of growth center, the degree of union, like that of reduction, is determined chiefly by allometry. On the basis of the conclusions reached in the discussion of reduction, we may expect that the size of the corolla lobes in relation to the tube will be subject to nonadaptive differences which depend on the absolute size of the corolla, or of the entire flower, as well as on slight differences in the degree of allometry.

The description given above holds with slight modifications for most examples of the union of similar parts, such as synsepaly, the uniting of filaments or anthers (Goebel 1933, p. 1860); of carpels to form a syncarpous ovary (cf. the description of *Nigella* by Thompson 1934); of branch systems to form a leaf; of the edges of a leaf to form a sheath, as in many grasses, sedges, rushes, *Allium,* and so forth; or of the edges of a megasporophyll to form a closed carpel (Goebel 1933, p. 1901). On this basis, the contentions of Thompson (1934) and Grégoire (1938), that the closed carpel and the syncarpous ovary are organs *sui generis,* are to a certain extent justified from the ontogenetic viewpoint. The regions of growth which produce the greater part of these structures did not exist in the more primitive ancestral forms. But from the phylogenetic viewpoint, both separate carpels and the syncarpous ovary are derived from distinct megasporophylls by means of successive modifications of ontogeny, induced by genes affecting the distribution, timing, and rate of growth.

A second type of union, which has been particularly common in the flower of angiosperms, has a parallel, but somewhat different, explanation. This is adnation, or the union of appendages belonging to different series or whorls. The best-known examples are the adnation of stamens to corolla and of calyx or receptacle to the gynaecium to produce the perigynous and epigynous type of flower. For a complete review of the literature on the latter condition, see Douglas (1944). Adnation agrees with the union of similar parts in that the united portions are produced by the appearance of zonal development. Furthermore, the degree of adnation must likewise be affected mainly by allometric growth. But the zonal development responsible for adnation must originate in a somewhat different manner from that which brings about union of similar parts. There is apparently a condensation of the timing stages of differentiation, so that parts which should normally be differentiated one after the other become differentiated at the same time. In fact, the normal successive differentiation of sepals, petals, stamens, and carpels may actually be partly reversed, as in *Sisyrinchium, Helianthus, Galium,* and *Valeriana* (Hannah 1916). As Schaffner (1937) has expressed it, "There is, so to speak, a telescoping of the processes of cell development and cell differentiation. . . . " Schaffner has suggested that this tendency is associated with a determinative process, which is responsible for the shortening of the floral axis and for the reduction in number of its parts. This determinative tendency, plus two others, namely, a tendency toward lateral expansion and one toward intercalary growth between stamen and anther primordia, are believed by Schaffner to explain in their interactions all the fundamental modifications of the flower. The tendency toward lateral expansion is simply a type of zonal development involving the entire basal part of the floral primordium (Coulter 1885, Goebel 1933, pp. 1916–1920, Hannah 1916, Thompson 1933, 1934).

The third tendency, change of symmetry, is less common than reduction and fusion and can only occasionally be recognized as a long-continued trend. Probably the most significant example is that which, according to the telome theory, converted the primitive branch systems into leaves and sporophylls. Most of the other good examples are in the flowers of various angiosperms, converting them from regular types with radial symmetry to various zygomorphic, irregular types. All stages of this change can be observed in different living species of such families as the Leguminosae and the Scrophulariaceae. It is obvious that the change from radial to bilateral symmetry involves the development of unequal rates of growth, affecting either different directions of growth in the same organ or the relative growth of different organs. This third trend, like reduction and fusion, is also associated developmentally with allometry. There is every reason to believe, therefore, that in plants, as in animals, the key to the understanding of the development and genetic basis of the major evolutionary trends in external morphology lies in the study of genetic factors affecting the distribution, timing, and relative as well as absolute rates of growth.

The fact may seem paradoxical at first sight that increase in complexity of plants results mainly from two processes, reduction and fusion, which would appear to be degenerative. The explanation is, of course, that in a plant, which normally differentiates organs serially one after the other, the way of producing a single complex structure with the least modification of ontogeny is by compressing together and modifying several relatively simple organs. How this has occurred can be best illustrated by describing the known and probable homologies of the various parts of a fruit or "seed" of wild oats (*Avena fatua*), such as one would be likely to pick up in any dry hillside or field in California and many other parts of the world. This structure is definitely a single unit, which with its dorsal, bent awn and stiff, upward-projecting hairs is ideally adapted for transportation by animals and burying in the ground once it has landed, as the rapid spread and common occurrence of the species in favorable climates amply testify.

The derivation of the various parts of this structure is as follows. In the central position is the single ovule, containing the endosperm and the embryo. The latter two parts will be left out of consideration, since they belong to a genetically different individual from the rest of the structure and were differentiated at a totally different time. The ovule itself, however, is descended from a megasporangium which became surrounded by parts of the reproductive telome system, the latter having become by reduction and fusion the ovule integuments (de Haan 1920, Zimmermann 1930). The ovule is contained within an ovary, which in the oat is merely the thin, hairy outer wall of the caryopsis, but which according to all evidence from the comparative morphology of grasses is derived phylogenetically from a syncarpous ovary with three carpels. Each of these carpels is derived by phylogenetic reduction and fusion from a megasporophyll, which at a much earlier time in the ancestry of the grass was in turn derived in a similar manner from a branch system. The caryopsis of the oat, therefore, contains the homologues of four different branch systems, enormously modified by reduction and fusion. Next outside of the caryopsis, the oat fruit contains the shriveled but persistent filaments of the three stamens. Each of these is a modified microsporophyll and is therefore descended from another branch system. At the base of the caryopsis are found two tiny scales, known as lodicules. These are much reduced petals, which in turn are descended from foliaceous or bractlike appendages, and ultimately from branch systems. Then comes the thin inner husk, or palea. Its homology is somewhat obscure, but it very likely represents two fused sepals, which again are modified from leaf- or bractlike appendages. Finally, enclosing the whole structure, is the firm outer husk or lemma. This is again an appendage, which has acquired its distinctive appearance through allometric reduction and fusion. The main body of the lemma is probably homologous to a leaf sheath (Philipson 1934), while the awn represents a blade which has suffered reduction both in length and in width through negative allometry. The pointed, apical lobes of the lemma are homologous to the ligule, but are relatively enlarged by positive allometry. The dorsal position of the awn on the lemma is due to the fact that the part of that structure above the awn represents the "fused" basal portions of the lobes; that is, it was formed by an area of zonal development which originated between the awn and the lobes. Thus, the entire oat "fruit," or grain with its husks, contains, in addition to the embryo and the endosperm, all or part of eleven different appendages, which were originally modified from as many different branch systems. Although this is an extreme case, nearly all the organs of plants have to a certain extent had a similar history. The excellent paleontological evidence of Florin (1944a,b), for instance, would enable one to give a very similar picture of the derivation of the cone of a gymnosperm.

There is no doubt, however, that a fourth trend, namely, differentiation or elaboration, must be added to the three just discussed. But it is a remarkable fact that elaboration has played a relatively minor role in molding the external form of the plant or its organs. We see its effects in such remarkable leaves as those of the pitcher plants (*Sarracenia, Darlingtonia, Nepenthes*), the Venus' fly trap (*Dionaea*) and other insectivorous plants, and more commonly in various specialized parts of the flower, such as the lip of the orchids, the keel of the Leguminosae, the nectar-bearing spurs on the petals of the columbine (*Aquilegia*), the spurs, nectaries, and pollen clips of the milkweeds (*Asclepias*), and the elaborate nectaries on the petals of the mariposa tulip (*Calochortus*). But all these structures are extreme specializations which have appeared at a relatively late stage of plant evolution and have little to do with the basic evolutionary trends. Some of them, like the petal spurs of *Aquilegia*, are clearly the result of progressively increasing allometric growth (Anderson and Schafer 1931). The most fundamental role of differentiation or elaboration in plant evolution has been in the cellular or histological organization of certain structures, particularly the vascular system. The successive differentiation of various types of tracheids and sieve tubes, and later of vessels, fibers, and wood parenchyma of various types (Jeffrey 1916), is one of the most important trends of morphological plant evolution. We may mention in addition the differentiation and elaboration of epidermal hairs, stomata, and in families like the Gramineae, various other specialized types of epidermal cells (Prat 1932, 1936), of the phenomenal cells known as sclereids which occur in the leaves of certain genera (Foster 1944, 1945, 1946), and of the various fleshy and hard, sclerenchymatous tissues found in many fruits. All of the latter structures, however, are specializations which seem to appear rather suddenly and can only occasionally be followed as evolutionary trends through series of species or genera. Elaboration, therefore, in contrast to reduction and fusion, appears in plants most often as short-time tendencies in certain organs of particular, specialized families, genera, or species.

RECAPITULATION AND EMBRYONIC SIMILARITY

The evidence from embryology, which since the time of Darwin has been one of the most important lines of evidence, both for the occurrence of evolution and the course which it has taken, has recently been reviewed for animals by De Beer (1940). He has concluded that the so-called "bio-genetic law" as stated by Haeckel, namely, that ontogeny is a brief recapitulation of phylogeny, is more misleading than helpful, and should be rejected. On the other hand, he finds that the principle of Von Baer, which was also that accepted by Darwin, namely, that the embryos and young of related forms usually resemble each other more than do the adult organisms, is of wide application and has a firm basis in developmental morphology. Judging from the evidence presented below, the same is partly true of plants, though to a considerably lesser extent than it is in animals.

In discussing the ontogeny of plants, we must first remember that because of the presence of an apical meristem and the serial differentiation of organs, a plant has two types of ontogeny. One is that of the embryo and seedling or sporeling and the other is that of the various lateral appendages, starting from the time when they become differentiated from the apical meristem of the shoot. In both of these ontogenies embryonic similarity is evident, but its manifestation might be expected to be different in the two types.

In the sporelings and seedlings of many plants, the appendages first formed often have a more primitive character than those produced by older shoots. In many ferns, the first formed leaf of the sporophyte is often lobed dichotomously, so that it recalls a flattened dichotomous branch system (Fig. 47, Bower 1923, p. 85, Orth 1939). Among seed plants, examples of "recapitulation," that is, seedling similarity, are numerous. Most species of conifers of the family Cupressaceae, such as *Juniperus, Thuja,* and *Libocedrus,* have complex, scalelike leaves on their mature branches, but the earliest formed leaves on the seedlings are needles like those found generally in the Coniferales (Goebel 1928, pp. 494–498). Perhaps the most striking examples among angiosperms are in the species of *Acacia* which at maturity bear only phyllodes, consisting of the expanded, flattened homologues of the leaf petiole and rachis (Boke 1940). As is well known and is illustrated in most textbooks of elementary botany, the seed-

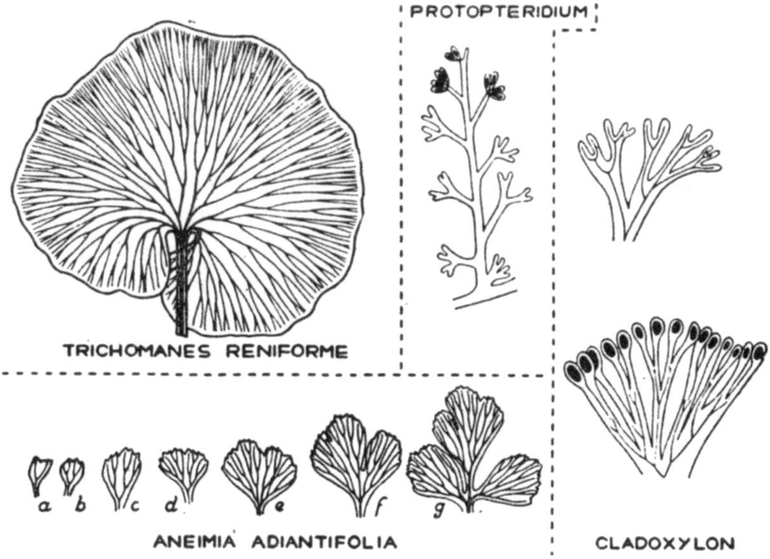

Fig. 47. Similarity in ontogeny and probable organogeny in ferns, showing evidence for the telome theory based on the principle of recapitulation, and from the fossil genera *Protopteridium* and *Cladoxylon*. Partly after Bower, from Lam 1948.

lings of many of these species possess normal, bipinnate leaves. Almost equally striking is the situation in *Eucalyptus*. In contrast to the mature twigs, which bear alternately arranged, pendent, bilaterally symmetrical leaves of a firm, sclerophyllous texture, the seedlings produce opposite, decussately arranged dorsiventral leaves which are broader, thinner, and more herbaceous in texture. Two further remarkable examples are found in species of sclerophyllous shrubs of coastal California. The bush poppy (*Dendromecon rigidum*) bears at maturity narrowly elliptic, entire leaves, quite different from those ordinarily found in the Papaveraceae. The seedlings of this species, however, as observed by Mirov and Kraebel (1939) and by the writer, have relatively broader leaves which are usually three-lobed at the apex, a common condition in seedlings of species of the family and one which in other genera is followed by the variously lobed, pinnatifid, or pinnate leaves characteristic of the family. The chamise bush (*Adenostoma fasciculatum*) bears at maturity clusters of hard, linear, needlelike leaves with revolute margins, quite unlike those of other genera of Rosaceae, to which the genus belongs. But young seedlings of *Adenostoma* are characterized by relatively large, delicate, finely pinnatifid and dissected leaves, resembling seedling as well as mature leaves of species belonging to the related genera *Alchemilla* and *Acaena.*

A notable fact about all these examples of great differences between seedling leaves and those borne by the mature plant is that the species concerned all have at maturity relatively simple, reduced leaves, and are adapted to semiarid climates, particularly to regions with great seasonal differences in precipitation. This suggests that the retention of more complex leaves in the seedling has a selective value in these species. Seed germination and the early growth of the seedlings would naturally occur during the wet season, and the seedlings would be likely to be shaded by their parents and by other plants. For this reason, they would be growing under mesophytic conditions, so that the development of a large leaf surface for promoting rapid growth would have a higher selective value than the conservation of water. But later, with the onset of the dry season, the exposure of the plants to sunlight, as well as with the greater problems of conduction which develop as the plant increases in height, the need for conservation of water would become paramount. Therefore, the highest premium would be placed upon mutations tending to modify the leaves of the adult plant, but not of the young seedling. Evidence for this explanation is provided by the fact that in both *Acacia* and *Eucalyptus,* the cultivated species in which the seedling type of leaves persist the longest (*Acacia melanoxylon* and *Eucalyptus globulus*) are native to regions which are moister than those inhabited by most of the species of these genera, in which the adult, reduced type of leaf appears at a relatively early age.

Nevertheless, many types of seedling resemblances cannot be explained on this basis. For instance, in all the species of oaks (*Quercus*) which at maturity have deeply lobed leaves — both the white oak (*Q. alba, Q. lobata, Q. garryana*) and the red or black oak type (*Q. borealis, Q. coccinea, Q. velutina, Q. kelloggii*) – the first leaves of the seedlings are shallowly lobed or merely toothed, as are both the seedling and the adult leaves of oaks like *Q. prinus* and *Q. agrifolia,* whose counterparts are definitely older in the fossil record than are the oaks with lobed leaves (Schwarz 1936b, Arnold 1947, p. 361). In the Gramineae, the typical leaves of mature twigs of the bamboos (tribe Bambuseae) have a complex structure consisting of a broad blade, a constricted, hardened petiole, and a thickened, tough sheath. But the leaves on the young primary shoots have strap- or wedge-shaped blades without petioles and thin sheaths with typical ligules, in all these respects resembling the leaves of ordinary grasses. These and numerous similar examples are difficult to explain on the basis of selection, except to the extent that the more complex type of leaf may have a selective advantage in the mature plant, but not in the seedling or young shoot, a suggestion offered by Massart (1894).

One of the most striking examples of seedling similarity, which is particularly difficult to explain on a selective basis, is that in some groups of Cactaceae. Certain genera of this family have forsaken the xerophytic habitat characteristic of most of its species and have become epiphytes in tropical forests. A notable example is the genus *Phyllocactus* (Massart 1894, Goebel 1928, pp. 511–512). The adult branches in this genus are flattened and nearly or completely spineless, in agreement with the photosynthetic organs of most other mesophytic plants. But the seedlings of *Phyllocactus* are columnar, four- to many-angled, and strongly spiny, as are both the seedlings and the mature plants of the related and more primitive genus *Cereus.*

A partial explanation of these phenomena in terms of developmental genetics is provided by the work of Swingle (1932), Frost (1938), and Hodgson and Cameron (1938) on the genus *Citrus.* These workers have shown that the appearance of the juvenile type of foliage depends not on the incidence of a new sexual generation, but upon the physiological conditions prevailing in the seed and the seedling. In *Citrus* two types of asexual propagation are possible. Mature twigs may be grafted or budded, while in the ovule the embryo often arises asexually as an adventitious bud from the nucellar or integumentary tissue. Trees grown from grafts or buds possess from the start foliage resembling that of mature branches, but all seedlings, both those from sexual reproduction and the asexual ones derived from adventitious embryos, are characterized by smaller leaves and a more spiny growth. In this they resemble the seedlings of a number of the more xerophytic genera of the Rutaceae. From this evidence we can surmise that the seedling similarity or pseudo-recapitulation found in many plants is due to the fact that the genetic factors which alter the apical meristem so that it produces appendages of a more specialized type often begin their action relatively late in development. Haldane (1932b) has suggested a similar explanation for the related phenomena in animals, and it was used for plants by Takhtadzhian (1943). They also suggest that if many gene mutations produce their effect by alteration of growth rates, a hypothesis which results directly from applying to developmental studies the principle of allometry, we should on this basis alone expect a certain degree of embryonic similarity.

The probable basis of seedling and embryonic similarity in plants may be summarized as follows. First, mutations which act relatively late in ontogeny are less likely to disorganize the whole process of development and so have a general deleterious effect than those which alter the early stages (Haldane 1932b, Goldschmidt 1940, pp. 389–390). The mutations established will therefore usually be those which affect development at the latest possible stage for the modification of the mature structure in the direction of a selectively advantageous shape. That part of the ontogeny which is less modified by the sum total of these late-acting genes will show embryonic similarity. Second, most mutations which alter the form of the mature organ change the rates of growth processes. Therefore, the longer the genetically different organs grow, the greater is the difference between them at maturity; and, conversely, the younger they are, the more alike they will be. Finally, the fact that seedlings often live under different environments from adult plants, and usually make different demands on their environment, may cause a particular type of structure to retain its selective value in the seedling after this advantage has already been lost in the adult plant.

THE PRINCIPLE OF IRREVERSIBILITY

A concept which figures in all considerations of phylogenetic trends is the principle or "law" of irreversibility, often referred to as Dollo's Law. This generalization states with varying degrees of definiteness that evolutionary trends cannot be reversed. The principle has no genetic basis, as Muller (1939) has pointed out, since reverse mutations occur in nearly all characters. Furthermore, mutations which reverse major phylogenetic trends, that is, atavistic mutations, are also known. Examples are the *radialis* mutation of *Antirrhinum majus,* which converts the corolla from the normal zygomorphic type into one with radial symmetry (Baur 1924, 1930, see Chapter III), and the mutation in *Crepis capillaris* which restores the receptacular paleae, a primitive feature of the family Compositae which is absent in all but one of the species of *Crepis* (Collins 1921, 1924).

As a rule, such atavistic mutations produce a disharmonious combination in association with the other characteristics of the organism, and so are promptly eliminated by selection. This is definitely true in the examples of *Antirrhinum.* But in *Crepis* there is one probable example of the relatively recent establishment in a gene complex restoring the receptacular paleae. The genus *Rodigia* was originally founded to include one species, *R. cummutata,* which is identical with species of *Crepis* except for the presence of paleae on the receptacle. But the experiments of Babcock and Cave (1938) showed that this "species" is interfertile with the common and widespread *Crepis foetida,* and they therefore reduced the "genus" *Rodigia* to the status of a subspecies under *C. foetida.* Furthermore, the presence or absence of paleae is conditioned by only two genetic factors. Now *Crepis foetida,* because of its annual habit, strongly thickened involucral bracts, and long-beaked, dimorphic achenes, is in one of the most advanced and specialized sections of the genus (Babcock 1947), while within *C. foetida,* subsp. *commutata* is more specialized and reduced in habit than subsp. *typica.* Hence, the possibility is very remote that subsp. *commutata* received its gene complex for the presence of paleae from some more primitive ancestors of *Crepis* which regularly had them, and we can assume safely that their existence in this subspecies is due to the establishment of two atavistic mutations.

Long-term reversals of trend in the structure of single organs also have almost certainly taken place in plant evolution. For instance, the most primitive angiosperms probably had palmately lobed leaves. This assumption is in accordance with the telome theory of the origin of the leaf (p. 477, cf. Zimmermann 1930, p. 343) and is suggested by the fact that the presence of three primary leaf traces is the primitive condition in the class (Sinnott 1914). In the genus *Paeonia,* all the evidence from the comparative morphology of the species indicates that a major trend has been from species with lobed leaflets to those with entire ones (Stebbins 1939). The same trend is suggested by the character of the seedlings in the genus *Dendromecon* of the Papaveraceae, as stated earlier in this chapter. But in the genus *Quercus,* the primitiveness of dentate leaves and the advanced character of lobed leaves is attested by both paleontological and developmental evidence. The evidence from paleontology, from comparative morphology and anatomy, and from development, all combine to suggest that evolutionary trends in the leaves of angiosperms may progress from the lobed to the entire condition and back again toward lobing. Similar reversals of trend have undoubtedly affected the size of parts, both vegetative and floral. Anatomical and developmental evidence indicates that the leaves of the original monocotyledons were relatively small, simple structures, which had suffered considerable reduction in the differentiation of this subclass from their less specialized ancestors (Arber 1924, Chap. V). In such groups as the asparagus tribe of the Liliaceae this reductional trend has continued. On the other hand, reversals in the direction of increasing leaf size have undoubtedly taken place rather frequently, as in bananas, palms, and aroids. In the case of the flower, the general trend has been toward reduction in size, particularly in stamens and carpels, which are almost certainly smaller than were the micro- and megasporophylls of the ancestors of the angiosperms. But this trend seems to have been reversed not infrequently. Probable examples are *Lilium, Tulipa, Godetia,* and many genera of the Cucurbitaceae. An example of reversals of trend within a group of closely related, specialized genera is provided by the tribe Cichorieae of the family Compositae. As Babcock (1942, 1947) has shown for *Crepis* and the writer for the more primitive genus *Dubyaea* and its relatives

(Stebbins 1940b), the general trend in this tribe has been toward reduction in size of involucres, flowers, and achenes. However, the largest floral parts known in the tribe are found, not in a primitive genus, but in *Tragopogon,* which in its vegetative characters, involucres, and achenes is one of the most specialized genera of the tribe. It is related to and probably descended from the larger genus *Scorzonera,* which also has large heads, though not so large as those of *Tragopogon.* In *Scorzonera* the species least specialized in vegetative habit, involucres, and achenes have the smallest heads, indicating that in these two genera the evolutionary trend has been toward increase in head size, or the reverse of that in most genera of the tribe. A similar trend toward increase is found in the American genera *Lygodesmia* and *Agoseris.* It is a noteworthy fact that the large florets of these specialized genera have a simplified, reduced vascular anatomy similar to that in specialized types with small florets, and quite different from the complex vascular anatomy found in the large-headed species of *Dubyaea* (Stebbins 1940b).

Reversals in habitat preferences and mode of life have also occurred frequently. Like the higher animals, all the higher land plants are descended from aquatic ancestors. And as in animals, examples of reversions to aquatic life are well known. The Salviniaceae and the Marsileaceae among ferns and the Najadaceae (pondweeds), the Lemnaceae (duckweeds), and the Nympheaceae (water lilies) among seed plants are good examples. The trend from mesophytic to xerophytic adaptation has also been reversed in a number of different lines. One example is the genus *Tragopogon* of the Compositae, tribe Cichorieae, mentioned above. Its probable ancestors in the genus *Scorzonera* are nearly all more or less xerophytic, but *Tragopogon* itself lives in relatively moist climates. Another rather frequent reversal is in the type of pollination. It is most probable that the ancestors of the flowering plants were wind-pollinated, but the earliest angiosperms themselves, on the other hand, were in all likelihood pollinated by insects. Nevertheless, a large number of families of angiosperms — Fagaceae, Betulaceae, Chenopodiaceae, Gramineae, Cyperaceae, and many others — have reverted to wind pollination. But in at least one case this trend has been reversed again. It is most probable that the flowers of the Euphorbiaceae became apetalous and unisexual as an adaptation to wind pollination, and that the ancestors of this family were insect-pollinated relatives of the order Geraniales. But in the most advanced genus of its family, *Euphorbia*, the presence of nectaries and brightly colored bracts surrounding the still monoecious and apetalous flowers indicates that the species of this genus have reverted to insect pollination.

We thus have evidence of reversibility in single mutations or gene complexes, in the size and morphology of individual organs, and in the general habit and functions of the plant. There are, nevertheless, certain ways in which reversal of trends occurs little or not at all. In particular, lost parts are seldom regained. This is apparently not because reverse mutations restoring these parts cannot occur – the example of receptacular paleae in *Crepis* is evidence of that – but because mutations altering the growth rates and the final shapes of existing parts probably occur with greater frequency and in any case can become established more easily because they cause less disturbance of the growth pattern of the plant as a whole. Thus, when the primitive xerophytic pines became adapted to more mesic conditions, the greater leaf surface necessary for this adaptation was more easily accomplished by the establishment of mutations increasing the length of the leaves and the number of short shoots per branch than by the restoration of the long leafy shoots which probably existed in the remote ancestors of the pine. In the asparagus tribe of the Liliaceae, in which the leaves have been reduced to small, colorless scales, probably as an adaptation to aridity, the reversion to more mesic habitats could occur more easily through factors increasing the number of minute branchlets, or causing a broadening and flattening of individual branchlets, as in various species of *Asparagus*, than by factors causing the reappearance of chlorophyll-bearing parenchyma in the original leaves. In the example mentioned above of the reversion to a tendency toward increasing size in the florets of the genus *Tragopogon*, the vascular supply needed for these larger parts is obtained more easily through genetic factors increasing the size of the bundles already present than it would be through the restoration of bundles lost through reduction. Bailey (1944) has made similar observations regarding the recovery of xylem vessels, once these structures have been lost during the evolution of a group. Thus, that part of the "law" or principle of irreversibility which states that lost parts cannot be regained is the only one which has even partial validity in plants. In fact, most of the trends of plant evolution are best expressed by disregarding this law entirely and by substituting for it the fourth and fifth of the "cardinal principles of morphology" suggested by Ganong (1901). One of these, which is perhaps stated too strongly, is

> the principle of indeterminate anatomical plasticity, that is, in all anatomical characters (size, shape, number, position, color, cellular texture) plant-organs, or, if one pleases, plant-members, are not limited by anything in their morphological nature, but under proper influence, may be led to wax and wane indefinitely in any of these respects.

The other is the

> principle of *metamorphosis along the lines of least resistance. . . .* This means that when, through a change in some condition of the environment, the necessity arises for the performance of a new function, it will be assumed by that part which happens at the moment to be most available for that purpose, regardless of its morphological nature, either because that part happens to have already a structure most nearly answering to the demands of the new function, or because it happens to be set free from its former function by change of habit, or for some other non-morphological reason.

As a matter of fact, recent work on comparative morphology and phylogenetic trends has emphasized more and more the validity of the realistic point of view toward morphology advocated by Ganong and has suggested a broadening of this viewpoint to encompass all evidence on phylogeny. According to Ganong, this realistic point of view constitutes a synthesis of evidence both from comparative anatomy and from embryology, in which the changes in the mature structures are visualized as the outcome of alterations in ontogeny. Some recent authors, however, have merely substituted for the idealistic morphology, based entirely on comparative anatomy, an idealistic embryology. This is based on the assumption, first, that in all the higher plants serially homologous organs would have to be formed according to a fixed sequence and, second, that the ontogenetic development of homologous organs is always similar, so that the phylogenetic origin of a structure can always be reconstructed on the basis of its ontogeny. The fallacies inherent in these assumptions have been pointed out by Clapham (1934), by Kozo-Poljansky (1936), and by Arber (1937), in their criticisms of the work of Grégoire and Thompson. They fall down completely when tested by examining the ontogeny of any series of organs which are becoming progressively more different from each other, and they are not in accord with the realistic modern concepts of genetics. In particular, we should emphasize here the principle brought out in Chapter III, that genes affect primarily not the adult characters, but developmental processes. Hence, gene mutations, which are the building stones of phylogenetic changes, alter various developmental processes, some occurring early and some late in ontogeny. Similarity in ontogeny between organs which are descended one from the other will be found only if the evolution of the organs concerned has taken place by means of mutations which have affected chiefly the later stages of development.

The realistic point of view toward phylogeny may be summarized as follows. As stated by De Beer, phylogeny consists of a succession of modified ontogenies. The geneticist can add that these modifications have been induced by gene mutations altering the nature or timing of various growth processes. Furthermore, the overwhelming body of evidence from genetic analysis of hybrids between species and subspecies indicates that for the most part alterations have been produced through a long succession of small mutational steps, rather than through the sudden action of a few mutations, each with a radical effect on development. Hence, Ganong's principle of continuity of origin is upheld. This means that we must also uphold Ganong's seventh principle, that "in the progressive development of metamorphoses, difference of degree passes over gradually into difference of kind." As Ganong has pointed out, and Goebel (1933, p. 1862) has also made clear, the full acceptance of this principle renders futile any arguments about such points as whether the inferior ovary is axial or appendicular in nature. It is both in the phylogenetic sense, since both the floral axis and the bases of the various appendages were greatly modified during its formation; and it is neither, in the sense that no part of the ontogeny, histology, or anatomy of any inferior ovary resembles closely that of any particular structure or series of structures in a plant with a superior ovary. Finally, we must state that such regularities of phylogenetic progression which may be observed are based, not on any predetermined laws or principles, but on regularities in the selection and establishment of certain types of mutations and in the rejection of others. These regularities, and therefore the phylogenetic principles, can have numerous exceptions and so must be applied cautiously and critically, in correlation with all other available evidence on the group of organisms concerned. The great frequency of reduction sequences, in which mutations curtailing growth after the formation of vascular tissue seem to be established more often than those which reduce vasculation, gives value to evidence on phylogeny from comparative anatomy, which makes use of vestigial vascular bundles. Similarly, the greater alteration of the later processes of ontogeny renders qualified validity to evidence from comparative embryology, as based on similarities both in seedlings and in the primordia of the foliar and reproductive appendages of the mature plant. But, since these anatomical and embryological regularities are based on the establishment, in most instances by natural selection, of certain types of mutations, they can be greatly modified or even obliterated by the selective action of unusual environmental factors, as well as by the occasional random establishment of mutations without selective value. Each proposed phylogenetic sequence must be approached with all these considerations in mind, and the most probable course of evolution must be determined by weighing all the lines of evidence against each other. As Bailey (1949) has pointed out, this is the approach which should characterize the "new morphology."

ORTHOGENESIS, SPECIALIZATION, AND THE DIFFERENTIATION OF PLANT FAMILIES

The causal explanation of any evolutionary trend must have two parts. The first part is the developmental explanation. We must explain how the altered genotype acts to produce the visible changes seen in the phenotype. This can be done largely in terms of time action of genes, growth substances, allometric growth, and the influence of the environment, as indicated earlier in this chapter. But more fundamental from the standpoint of evolution in general is the second part, namely, an explanation of how the mutations originate which produce the altered genotype and, more particularly, how they become established in natural populations. The origin and establishment of mutations was discussed in Chapters III and IV, in which the points emphasized were the random nature of individual mutations, the fact that most mutations which become established in populations are those which have a relatively slight effect on the phenotype, and the fact that most of the qualities which give a selective advantage to individuals or populations are produced by combinations of different, genetically independent, morphological and physiological characteristics. In addition, the process of establishment itself was described as one which cannot be accomplished by chance alone, and for which some positive force is necessary in order to change the frequency of genes in populations. When these points are kept in mind, the conclusion becomes inevitable that all long-continued evolutionary trends must be governed by some guiding force.

Only two such guiding forces have ever been postulated, and it is unlikely that any others could be imagined. One of these is natural selection, and the other is some unexplained force which, presumably by causing the more frequent or predominant occurrence of mutations which are genetically unconnected, but have a similar morphological and physiological effect on the phenotype, directs or canalizes the course of evolution. The latter force is specifically designated or at least implied in explanations of evolutionary trends on the basis of orthogenesis.

The explanation of evolutionary trends on this strict orthogenetic basis has been done chiefly by animal paleontologists. The fallacies in their reasoning have already been pointed out by Goldschmidt (1940, pp. 321–322), by Mayr (1942, pp. 291–294), and particularly by Simpson (1944, Chap. V). The only serious attempt known to the writer to interpret evolutionary trends in plants on the basis of orthogenesis is the work of Schaffner (1929, 1930, 1932). In the examples which he cites, as in the zoological ones reviewed by Simpson, explanations on the basis of selection are in no case impossible or even improbable.

In the first place, the "series" described by Schaffner are all based on different living forms, which may have been derived from a common ancestor, but are certainly not derived one from another. In fact, Florin (1944b) has shown by fossil evidence that some of the "series" described by Schaffner (1932) in the conifers are actually different end points of a number of adaptively radiating lines, of which the common ancestor was much more primitive than, and in some ways intermediate between, the modern extremes. Furthermore, the "series" of Schaffner are not cited for the organism as a whole, but for certain structures. In some of the "series" cited, like the fruits of various species of *Stipa* (Schaffner 1930), only certain features, like awn length, are progressive. The other characteristics of the species concerned vary in many different directions. Finally, nearly all of the structures described are highly adaptive, and it is only through a peculiar type of logic, the fallacy of which was pointed out in Chapter IV, that this adaptiveness is dismissed by Schaffner as not significant.

The conclusion is inevitable, therefore, that all long-continued evolutionary trends in plants, as well as in animals, are guided by natural selection. This means that the differences between families and orders of flowering plants must originally have had a selective basis, at least in part. Nevertheless, an explanation of the nature of this basis is admittedly one of the most difficult problems in plant evolution. The difficulty lies in the fact that not only are plants now living which appear to represent the earliest, most primitive stages of these evolutionary trends, as well as their highly specialized end products, but in addition there is surprisingly little difference between some primitive types and many advanced types, either in their abundance of individuals or in geographic distribution and the habitats they occupy. The differentiation of orders and families of flowering plants through the action of natural selection under present conditions is well-nigh impossible to imagine.

The solution to this difficulty probably lies in the following facts. First, all the trends leading to the differentiation of families of flowering plants probably took place simultaneously and at a relatively early stage in angiosperm evolution. For instance, both distributional and paleontological evidence indicates that the Compositae, the most highly specialized family of dicotyledons, already existed in the latter part of the Cretaceous period (Stebbins 1941c), and distributional evidence indicates a similar age for the most advanced families of monocotyledons, the

Orchidaceae and the Gramineae. It is likely, therefore, that the major part of angiosperm evolution, involving the principal trends in the modification of the flower, took place during the Mesozoic era. The environmental conditions prevailing during that time were quite different from those now operating. In the first place, during at least half of this era angiosperms were in the minority, at least in those regions, now rather numerous, from which fossil records have been obtained. They were competing less with each other than with various types of gymnosperms, many of which were less efficient in their reproductive mechanisms. Second, the climate prevailing throughout the earth was at that time more uniform than at present, so that selective pressures favoring the differentiation of the vegetative parts of the plant in adaptation to new climatic and edaphic conditions were relatively low. Finally, the various types of flower-pollinating insects, as well as seed- and fruit-eating birds, were also probably undergoing relatively rapid evolutionary differentiation at that time. One would expect, therefore, that gene combinations adapting plants in new ways to pollination or seed dispersal by insects and birds would have had a particularly high selective value during the Jurassic and the Cretaceous periods.

The next set of facts to be remembered is that the flower is not just a collection of structures which have convenient diagnostic characters, but a harmonious unit which performs more or less efficiently the two vital functions of cross-fertilization and the maturation and dissemination of seed. Therefore, an alteration of one of its parts will immediately change the selective value of modifications in all the others, as well as the value of such general characteristics as the size, number, and arrangement of the flowers produced.

Furthermore, the nature of the functions performed by the flower is obviously related to the economy of the plant as a whole. It was pointed out in Chapter V that the value of self-pollination as against cross-pollination is different in plants with different types of life cycle and with populations of different size and structure. The nature of the pollinating agents available will likewise differ, depending on whether the plant is terrestrial or aquatic, mesophytic or xerophytic, temperate or tropical, and so forth. Also, the selective value of various types of seed dispersal will vary similarly with the structure and habitat of the plant as a whole.

To illustrate this situation, the writer has made a tabulation of the families of flowering plants according to their primitiveness or advancement in respect to the eight diagnostic characteristics used most commonly and prominently in keys to plant families. These are: apetaly vs. presence of sepals and petals; choripetaly vs. sympetaly; actinomorphic (regular) vs. zygomorphic (irregular) flowers; many vs. few stamens; apocarpy vs. syncarpy; many vs. few ovules; axial vs. parietal placentation; and superior vs. inferior ovary. One can see that a family, or part of one, can be classified as primitive in all eight characteristics (corolla present, choripetalous, regular; stamens numerous; apocarpous; ovules several to many; placentation axial; ovary superior), as advanced in all eight, or as advanced in any one of a number of different combinations of two to seven characteristics.

The details of this study will be published elsewhere, but its principal results can be summarized as follows. In the first place, the total number of combinations realized is only 86, or 34 percent of the 256 possible, and 37 of these are represented by only one or two groups. Of the 438 groups (families or part families) which are classified, 200 are contained in only 12 combinations, or less than 5 percent of the total number possible. On the other hand, certain combinations are represented by a large number of groups, the largest numbers of groups per combination being 36, 24, and 20. Thus, the eight characters studied are far from being combined at random in the different families and genera of angiosperms.

The following evidence indicates that this nonrandom distribution of character combinations is connected with adaptation, so that the combinations realized by very few or no genera can be considered as adaptive "valleys" and those characteristic of many genera and families as adaptive "peaks." In the first place, most of these "valleys" represent combinations of which the inadaptive character can be clearly recognized. Many of the combinations not represented are impossible or nearly so from the structural point of view. These are all of the combinations which include epigyny or parietal placentation without union of carpels. Another series consists of those combinations which include sympetaly or zygomorphy together with a large number of stamens.

The poorly adaptive nature of most combinations of this type probably results from the fact that sympetalous and zygomorphic corollas are usually highly adapted for insect pollination and function best in this respect if the stamens are definite in both number and position. Another combination, absence of petals with zygomorphy, is rare because the former characteristic is usually associated with wind pollination, the latter with pollination by insects. The majority of the "valleys" can be explained on the basis of these five impossible or poorly adaptive combinations.

Among those character combinations which represent "peaks," some are found in a relatively large number of unrelated groups, none of which contains many genera and species. For instance, the combination of polypetaly, regular corollas, few stamens, hypogyny or perigyny, syncarpy, few ovules, and axial placentation is found in 34 different groups of various relationships. The largest of these are the Anacardiaceae, the Rutaceae (except the Aurantoideae), and the Celastraceae, but some Liliaceae and Commelinaceae are also found here. Combinations of characters such as this may be considered high peaks that are relatively easy to climb, that is, they may be reached by relatively little modification from a number of different floral types with more primitive combinations. Other combinations are represented by only a few groups, but one of these constitutes all or the major part of one of the very large plant families, and therefore is represented by a larger number of genera than all but a few of the other peaks. For instance, the combination of polypetaly, zygomorphic corollas, stamens of a definite number, and apocarpous, hypogynous gynaecium with several ovules is the exclusive property of the Leguminosae (*sens. lat.*) and therefore is represented by 400 genera. Similarly, the combinations found in the large families Orchidaceae, Gramineae, Cruciferae, Malvaceae, Scrophulariaceae, Rubiaceae subfamily Coffeoideae, and Compositae are shared by relatively few other groups. These can be considered as peaks which are difficult to climb, but have plenty of room on top; that is, they permit a wide range of variation in the adaptive types of floral structure based on the combination of fundamental characters which forms the general plan.

The next fact which has become evident is that certain combinations of two or three characters are particularly prevalent.

For instance, apetaly is usually accompanied by reduction in ovule number. Furthermore, certain tendencies in the flower are correlated with characteristics of the inflorescence. Zygomorphic flowers are correlated with racemose, spicate, or axillary inflorescences, while actimomorphy is most commonly associated with cymose, corymbose, umbellate, paniculate, or capitate inflorescences. Reduction in number of ovules or seeds per flower is usually correlated with increase in the number of flowers per inflorescence. Many other similar correlations can be cited. All have their exceptions, but the correlations are close enough so that their origin by chance is highly improbable.

In addition, most of the different families occupying certain of the higher peaks are similar in general habit and in habitat preferences. For instance, the high peak mentioned above as occupied by the Anacardiaceae and other families contains mostly woody families of primarily tropical and subtropical distribution. Another high peak, characterized by sympetalous, actinomorphic corollas, stamens definite in number, and hypogynous, syncarpous ovaries with many ovules and axial placentation, is occupied chiefly by mesophytic herbs, of both temperate and tropical distribution (Polemoniaceae, Solanaceae, Plantaginaceae, some Bromeliaceae, some Liliaceae, and so on).

Finally, the characters chosen, though diagnostic of families more often than any other characters, break down in many cases, so that different parts of the same family have different combinations of the eight character pairs and therefore are placed in different groups. The significance of this in connection with the nature of families and other higher categories will be discussed later in this chapter. The noteworthy fact relevant to the present discussion is that groups consisting of part of a family are in a considerably higher percentage in the combinations with a low degree of specialization than they are in those which are more highly specialized. In other words, evolutionary experiments with single ones of these fundamental characters on the level of the genus or the species are more likely to be successful when the flower as a whole is more generalized than when it has reached a high degree of specialization. Primitive groups have more diversity in the "fundamental" characteristics than advanced ones.

When all these facts and interrelationships are considered, the hypothesis becomes most probable that the differentiation of families and genera of flowering plants is an example of adaptive radiation, which nevertheless differs quantitatively from the ones well known in animals. These quantitative differences are as follows. First, the number of possible adaptive combinations, which solve adequately the problems of cross-pollination and seed dispersal, is relatively large, so that many more different lines of radiation exist in the higher plants than in the higher vertebrate animals. Second, the relative simplicity of plant structure restricts the distance to which specialization can be carried, so that each line of adaptive radiation must be relatively short. At the same time, this simplicity makes possible an exceptionally large amount of parallel variation, so that morphological similarity is much less indicative of phylogenetic relationship in plants than it is in animals. Finally, this adaptive radiation affects only the reproductive characteristics of the plant. Since the importance to the plant of an efficient reproductive mechanism varies greatly with its habitat and vegetative life cycle, the selective pressure in favor of specialization along some line of adaptive radiation is much greater in some groups of plants than in others. This makes possible the existence side by side of relatively generalized and highly specialized types to a much greater degree than in most groups of animals.

As emphasized in Chapter IV, recent studies of natural selection have shown that the existence of a selective basis for an evolutionary change is relatively easy to demonstrate, while the actual selective agent responsible is much more difficult to identify. This generalization holds with particular force in respect to the differentiation of the families of flowering plants, since this differentiation took place many millions of years ago, when both the climate and the biota of the world were entirely different from what they are now. We can, therefore, do little more than speculate on the probable nature of these selective agents. The three particular agents which probably played the largest role were insect pollinators, the wind as a pollinating agent, and the needs of the seedling, as they affected ovule and seed size.

The fact is well known that many flowers are adapted in size, color, and form to the needs of insect pollination. Nevertheless, these adaptations have appeared independently within many different families, and there is little evidence that they affected materially the differentiation of the families themselves. Zygomorphy is perhaps the principal tendency which is based on adaptation to insect pollination. It is rare or absent in wind-pollinated flowers, and among entomophilous types it is closely correlated with the racemose or spicate character of the inflorescence, which causes the flower to be borne horizontally rather than vertically. When the horizontal position is adopted, differentiation of the corolla into an upper and lower lip, as well as the grouping of the stamens on the roof or the floor of the corolla, greatly increases the efficiency of insect pollination. Wind pollination, on the other hand, often brings about a reduction or disappearance of the perianth, as well as the loss of stamens and the evolution of unisexual (monoecious or dioecious) flowers.

The selective significance of seed size and its relation to alterations in the structure of the gynaecium were fully discussed in Chapter IV. On the basis of that discussion, we should expect to find a relatively high proportion of small-seeded types, which very often have capsules with parietal, axial, or free central placentation, among mesophytic herbs, particularly those of meadows and among plants like the Ericaceae and the Orchidaceae, of which the roots possess symbiosis with mycorrhizal fungi. On the other hand, large-seeded types, with reduction of the number of ovules per gynaecium, might be expected most frequently among herbs, shrubs, and trees of forest areas and among xerophytes. These conditions were observed to be true to a considerable extent in the survey of families and genera of angiosperms on which the present discussion is based.

It is possible, however, that one general selective agent has been operating in different ways in the differentiation of many if not most of the families of angiosperms. This is the value to the plant of increasing rapidity in the cycle of flowering and fruiting, and of economy in the use of available food for the manufacture of reproductive structures, particularly seeds. This value would be especially great in plants of desert areas and of the cool temperate zone, in which the growing season has been much reduced in length. Mutations tending to speed up the development of flowers would almost automatically cause a reduction in the size and number of their parts. Furthermore, the generalized growth region responsible for zonal development, and therefore for "fusion" and "adnation," might be expected to originate and develop more rapidly than the series of separate growth centers in each primordium.

If such a generalized selective agent were operating, we should expect to find a great amount of parallel evolution in the fundamental organization of the flower, and this appears to have taken place. Divergence, on the other hand, could be produced by special environmental agents acting concurrently with this general tendency, as well as by isolation and chance reassortment of different combinations of genetic factors promoting increased rapidity in development.

We may now construct a tentative picture of how the families and orders of flowering plants became differentiated. Let us imagine the existence at an early geological period, perhaps the Jurassic, of an original stock of angiosperms, all of which are relatively similar to each other and possess the generalized characteristics associated with the order Ranales. They are scattered throughout the earth, but being still relatively inefficient competitors against long-established, abundant gymnosperms, like Bennettitales, cycads, conifers, Ginkgoales, and Gnetales, they are not present, except locally, in large numbers of either individuals or species. Spatial and ecological isolating mechanisms are therefore rather well developed. Under these conditions, mutations toward one or more of the various fundamental specializations, such as zygomorphy, syncarpy, or reduction of ovules or stamens, may have arisen in one or more localities and may have become established, either through accidental association with some highly adaptive vegetative characteristic or because of some local adaptive value in connection with the mode of life of the species in which the specialization arose. Even when such a mutation or series of mutations became fully established, the population bearing it would by no means have been a newly evolved family or even a new genus, any more than is a saxifrage with a zygomorphic corolla, a *Hypericum* with a one-celled ovary and parietal rather than axial placentation, or an *Oxalis* with the petals united at the base. But the establishment in a population of one such morphological change, or even of the beginnings of one, would increase the selective value of mutations in the direction of any one of several other specializations, so that the population might soon come to possess a combination of characteristics quite different from the original, undifferentiated populations. Once this was achieved, the individuals with an intermediate condition would tend to be eliminated by competition, and the population would be separated by a morphological gap from its altered relatives. Now, the high adaptive value of this new combination would enable the population bearing it to spread rapidly, particularly in areas opened up by climatic or edaphic changes. It would therefore become geographically diverse rather quickly. At the same time, it would meet primitive, unchanged relatives of its ancestral population. If these relatives were separated from the changed population by genetic isolation barriers, they would either be eliminated by the latter or they would stop its advance, depending on the relative efficiency of the two species in both vegetative and reproductive characteristics. But if reproductive isolating mechanisms were absent or weakly developed, hybridization would occur, and from segregation of these hybrids there would arise new types, possessing the selectively advantageous combination of new floral characteristics associated with factors for vegetative characteristics, and perhaps also for size and arrangement of flowers, quite different from those present in the first altered population. This might increase the selective value of still further specializations in floral structure, and thus continue the adaptive trend, in addition to increasing diversity in other respects. By repetition of this process, a series of populations or species could arise, all possessing the same selectively advantageous and diagnostic modifications of the flower which arose monophyletically, but different from each other in various characteristics acquired both through independent variation under isolation and through hybridization. Such a series would constitute a new genus. Its further expansion and diversification, along with more complete extinction of its ancestors and more primitive relatives, would elevate it to the status of a family. If this hypothesis is a fair approximation of what actually happened in the early history of angiosperm evolution, we can assume that the earliest appearance, on the species level, of individual characteristics now considered to be diagnostic of families and orders was largely associated with chance, but that the molding of the higher categories themselves through the assembling of successful combinations of characteristics was guided mainly by natural selection. In the adaptive radiation of both plants and animals, the higher the level of specialization reached, the less is the possibility of chance variation in fundamental characteristics, and the more powerful is the guiding influence of natural selection in a particular direction.

The hypothesis presented, therefore, suggests that natural selection has played a large role in directing evolutionary specialization during the differentiation of the families of flowering plants, but that this role has been much obscured by the effects of chance in the first establishment of individual characters, by parallel evolution, and by hybridization. The first stages of adaptive radiation were mega-evolution in the sense of Simpson (1944, p. 123), and probably took place rather rapidly.

The same course of evolution was probably followed by other highly differentiated groups of plants. Heitz (1944) has suggested that the great diversity of structure in the peristome of the capsule of mosses, which has provided the best characters for dividing the class into orders, families, and genera, is mainly the result of random diversification in evolution. He bases his conclusions on the fact that the different types of peristome, as well as capsules devoid of any peristome, are found in different species living apparently the same type of life in the same habitat. The condition is precisely what one would find if he studied the distribution of a single important diagnostic character in angiosperms. But, as Heitz recognizes, the peristome certainly has a function, namely, to sift the spores and prevent their too rapid dissemination. The selective value of this function would depend greatly upon such matters as the size and shape of the capsule, the length and flexibility of its stalk, the precise habitat and length of fruiting season of the maternal gametophyte, and many other factors. Furthermore, the mosses are an ancient group, and some families may have originated under rather different conditions from those in which most of their representatives are found today. One would expect, therefore, that the guiding action of selection would have been obscured in mosses just as it has been in flowering plants and would become more evident only after a detailed ecological as well as systematic study of the group.

THE NATURE OF THE HIGHER CATEGORIES

No discussion of evolutionary trends is complete without a consideration of the nature of genera, families, and still larger subdivisions, all of which are, at least ideally, the nomenclatorial framework supporting our concepts of lines of evolution. The opinions of botanists as to the nature of these categories have differed as widely as their opinions about species. But in contrast to the recent history of the species problem, which has been one of increasing agreement as our knowledge of the biology of species formation has increased, opinions about the higher categories have if anything become more diverse, and the tendency has increased to regard genera, families, and orders as partly or wholly subjective groupings for convenience in classification.

Yet there is no doubt that many genera and families are as real as the species which compose them. Some genera, like *Pinus, Picea, Abies, Salix, Populus, Betula,* and *Alnus,* have been far more stable and uniformly delimited in the history of classification than their component species. If that group is the most natural which is the most easily recognized and has been the most consistently delimited by one careful observer after another, then the conclusion is inevitable that in some families of plants (Pinaceae, Salicaceae, Betulaceae) the genus is more natural than the species, while in others (Leguminosae, Cruciferae, Asclepiadaceae, Orchidaceae) the species is more natural than the genus.

The difficulty of recognizing genera and other higher categories as natural units lies partly in the fact that they cannot be defined on an experimental basis, as can most species (see Chapter VI). As a rule, members of different genera cannot be hybridized, and if they can, the resulting F_1 hybrid is completely sterile, barring the rare occurrence of allopolyploidy. If two species supposed to belong to different genera do form partly fertile hybrids, as with maize × teosinte (*Zea* × *Tripsacum*), the two "genera" are probably artificial groups which should be merged.

Anderson (1937a) has suggested a fundamental difference between generic and specific characters. He found that the species of such genera as *Aquilegia* and *Narcissus* differ from each other chiefly in "change of emphasis," that is, in the shapes and relative proportions of their different parts. When genera were compared, however, their differences appeared to represent "actual differences in the pattern [of variation] itself." For instance, species of *Narcissus* differ from each other largely in the size of the flower and the proportion of its parts, but the chief difference between *Narcissus* and the related genus *Cooperia* is that the former possesses and the latter lacks a crown or corona at the apex of its corolla tube.

Epling (1938, 1939) attempted to apply this criterion to generic distinctions in the large family Labiatae, but he found that obvious differences in the pattern were not usually evident. On the other hand, species groups in this family are usually separated by "an accumulation of changes in emphasis," affecting many different parts of the flower. They are thus quantitatively, but not qualitatively, different from the differences between related species. The same may be said about the differences between species groups or sections of *Crepis,* as recognized by Babcock (1947).

As a matter of fact, most systematists recognize genera on the basis of morphological similarities between groups of species and of actual or presumed gaps in the variation patterns separating the different genera. In general, these gaps may be presumed to have resulted from the extinction of ancestral connecting links between the various groups of species, so that the greater the amount of extinction a family or order has suffered in its evolutionary history, the more easily recognizable are its genera. If a particular type of structural pattern has a specific selective advantage, various modifications of this pattern in its more perfect form have a better chance of survival than conditions intermediate between this and some other adaptive combination of structures. When this is true, generic boundaries coincide with changes in pattern, as in *Narcissus.* But in many instances, the species groups do not possess structures with distinctive adaptational features, and the gaps have resulted at least partly from accidental extinction of species. This is true in the case of many groups of genera in large families such as the Gramineae and the Compositae. In them there is no easy way of defining genera, and many of these groupings must be frankly subjective. Although they are based on discontinuities between groups of species, the particular discontinuities which are recognized as of generic value, as compared to those which have been used to separate sections or subgenera, are based largely on tradition and convenience. Mayr (1942, Chap. X) has discussed the great divergences of opinion among ornithologists as to the delimitation of genera and the difficulty of reaching an agreement about situations in which the probable phylogeny of the group is generally accepted. Similar disagreements have arisen among botanists, particularly in regard to such genera as *Saxifraga* and *Astragalus,* and there has been the same difficulty in finding an objective basis on which to found a generally acceptable opinion. In general, the genera of the higher plants are much larger than those of vertebrate animals. This may be partly the result of tradition, but it is to a greater extent the result of such processes as hybridization and allopolyploidy, which have woven together networks of species. In genera such as *Stipa* and *Carex,* clusters of twenty or thirty species exist which cannot be subdivided on any basis except for artificial key characters. Splitting of genera in such groups would be more confusing than helpful.

In the opinion of the writer, there are no characteristics of genera, families, or any other higher categories of plants which would suggest that the mechanism of their evolution was anything more than a continuation of the processes which give rise to subspecies and species. Mutation, recombination, and selection, as well as isolation and extinction, are responsible for groups and categories of organisms at all levels. As we go higher up the scale, reproductive isolation becomes less important as the basis for the delimitation of categories, and extinction more so, but this is a quantitative rather than a qualitative change. Once species have appeared, the origin of genera and families, as well as orders, classes, and phyla, is largely a matter of time and further genetic plus environmental change.

CHAPTER XIV

Fossils, Modern Distribution Patterns and Rates of Evolution

THE ULTIMATE problem for students of the dynamics of evolution is to determine how fast evolution progresses and under what influences its speed can be increased or decreased. A complete solution of this problem should make possible the control of evolutionary rates by man and consequently, at least in certain organisms, the artificial acceleration of evolutionary progress to the point where it can be directly observed and analyzed during the span of a human life.

Unfortunately, complete direct evidence relating to this problem is impossible to obtain. In order to have a direct bearing on the causes and dynamics of evolution, these rates must be expressed in terms of numbers of genic changes per unit of chronological time or per generation, and of the corresponding rate at which isolating barriers develop. All these processes occur so slowly under natural conditions that time must be reckoned in geological terms of tens or hundreds or thousands, or even millions of years, so that ideally the perfect evidence for the solution of this problem would be obtained from coordinated paleontological, genetic, and systematic studies on the same group of organisms. Unfortunately, no group of animals or plants is yet known on which all these types of evidence can be obtained. We are thus forced to analyze as best we can the evidence from the fossil record of organisms which have living counterparts and that from historical records of changes in the composition of populations of organisms amenable to genetic studies, and to correlate these two lines of evidence by means of the most plausible inferences we can make. One line of evidence which can help to bridge the gap between genetics and paleontology is that from the geographic distribution of living organisms which are closely enough related so that they can be intercrossed and their genetic relationships determined.

THE NATURE AND VALUE OF PALEOBOTANICAL EVIDENCE

There is no question that the evidence from fossil animals, particularly vertebrates and mollusks, is much more complete and definite than any now available from plants. Simpson (1944) has presented a convincing analysis of evolutionary rates based on the paleozoological evidence, and in general the paleobotanical evidence can do little more than support the principles which he has deduced. The one advantage of fossil plants over fossils of land animals is that, when sufficiently similar to modern types, they indicate much more accurately the nature of the climate at the time and site of deposition. For a student of the dynamics of evolution, therefore, the plant fossils which have the greatest value are those most nearly related to modern species.

As Simpson (1944, Chap. III) has emphasized, the gaps in the fossil record are as important as the fossils themselves. The remains available are not only a minute sample of the plants which existed on the earth; they are far from a random sample. Nearly all of the sites favorable for deposition were moist lowlands; plants adapted to upland areas and to dry climates, although they must have been as abundant in some past epochs as they are now, have only rarely been preserved. Furthermore, only certain types of plants, particularly the large, woody species, have been preserved with any degree of frequency. Fossil evidence in general provides us with a conception of the dominant species of certain sites during past geological ages, but we can expect other habitats and types of plants to be absent from the fossil record.

Another weakness of the fossil record which must be considered in its positive as well as in its negative aspect is the fragmentary nature of nearly all plant fossils. The diffuse structure of the plant body does not lend itself to preservation as a single unit. By far the most common fossils are detached leaves, tree trunks or bits of wood, and spores or pollen, while next in order of frequency are isolated seeds. Fruits and flowers are still less common, and in the case of the angiosperms the flowers are rarely well enough preserved to be identifiable. Now, the botanist judges the systematic position of a species on the basis of the whole plant: arrangement of leaves, character of inflorescence, nature of the floral parts, as well as that of the fruits and seeds. It is a well-known fact that the commonly preserved fossils, particularly the leaf impressions of the flowering plants, are the least diagnostic of all plant parts. Leaves of *Cercidiphyllum* (order Ranales) have been mistaken for *Grewia* (Malvales) or *Populus* (Salicales); those of *Arbutus* (Ericales), for *Ficus* (Urticales); and so on. Even a gymnosperm (*Gnetum*) has leaves which to the average observer are strikingly similar to those of some flowering plants (Thymeleaceae).

Some authors have suggested that these difficulties are so great as to be unsurmountable and that the paleobotanist can never be certain of the identity of the remains which he describes, except in the case of a few angiosperm fruits of a distinctive character and of those parts of gymnosperms and spore-bearing plants which can be sectioned and studied under the microscope. This is, however, an exaggeration. Seward (1931) and Cain (1944, Chap. IV) have given careful, impartial estimates of the accuracy with which fossil leaves can be identified. The method of identification is, of course, simple comparison between the fossil and the leaves of living species, but various approaches which have been adopted by modern paleobotanists have greatly increased its accuracy. In the first place, every observable detail is considered: shape, texture, character of margin and apex; position, distribution, and endings of primary veins; character of secondary venation; and, when possible, the microscopic structure of the epidermis (Bandulska 1924, Edwards 1935, Florin 1931). Secondly, decisions are not made on the basis of one or a few leaves of either the fossil or its possible living relatives, but on the range of variability found among large numbers of leaves of both fossil and living forms. Finally, the identifications are based on the assumption that in the past, as at present, plant associations formed harmonious communities, all of the members of which were adapted to similar environmental conditions. Going on the assumption that in the past, as at present, camphor trees did not grow in or near forests of redwood (*Sequoia*), Chaney (1925a) became doubtful of the reference to the genus *Cinnamomum* of certain leaves found in the mid-Tertiary flora of western America, and he started to compare them with the leaves of various species which now grow in or near forests of *Sequoia*. As a result, the leaves of one of these species of "*Cinnamomum*" were found to be a close match for those of the modern *Philadelphus lewisii* (Hydrangeaceae). This species is common in the forests bordering those of the modern *Sequoia sempervirens,* and their similar position in the Tertiary period is entirely to be expected.

The correctness of many of the identifications of angiosperms based on leaf impressions has been verified by the discovery of unmistakable fruits of the postulated species in the same deposits. One of the most striking examples was the discovery by Brown of seeds of the genus *Cercidiphyllum,* of which the only living species is confined to Japan and China, in company with leaves previously assigned by Chaney to that genus and now known to be widespread and abundant in mid-Tertiary deposits of western North America. Other genera which are known from well-preserved leaves and fruits or seeds occurring in the same deposit are *Populus, Alnus, Quercus, Lithocarpus, Juglans, Carya, Engelhardtia, Platycarya, Chaetoptelea, Liquidambar, Platanus, Cercis, Acer, Cedrela, Koelreuteria, Firmania, Gordonia, Nyssa,* and *Terminalia.* In the famous London Clay Flora of Eocene age, Reid and Chandler (1933) described a whole series of fruits and seeds, many of which were certainly identified as belonging to modern genera. Nearly all the modern counterparts of these early Tertiary species of northwestern Europe are found in the tropics of the Indo-Malayan region, a conclusion which would have been predicted on the basis of other Eocene floras of this region which contained only leaf impressions. There is little reason to doubt, therefore, that the majority of the angiosperm remains which have been assigned by progressive, modern paleobotanists to particular modern genera actually belonged to the genera with which they have been identified. In the case of fossils dating from the latter part of the Tertiary period, the comparison with modern species is usually reliable (cf. Chaney, Condit, and Axelrod 1944, Hu and Chaney 1940).

The positive conclusion which follows directly from the nature of the paleobotanical record of the angiosperms is that forms similar to modern species and genera are recognized with relative ease, while radically different types, including those which might have formed connecting links between existing genera or families, could not be recognized or assigned to their correct phylogenetic position even if they were found. Evolutionary conservatism and stability are much easier to demonstrate by means of fossil evidence than is rapid progress or the differentiation of the modern families and orders.

With this in mind, we can make a brief summary of the fossil record of the modern seed plants and of the forms known to be directly ancestral to them. Among the gymnosperms, the most primitive living order, the Cycadales, are known from scattered remains as old as the beginning of the Mesozoic era (Schuster 1931, Florin 1933), as well as from leaf impressions of Cretaceous and early Tertiary age (Seward 1931, Chaney and Sanborn 1933). The reproductive structures found by Harris (1941) in the lower Jurassic rocks of Yorkshire, England, are very similar to those of modern cycads. This author has concluded that by the beginning of the Jurassic period the Cycadales had evolved to their full extent in respect to the differentiation of genera. The conspicuous, amply preserved, and highly evolved cycadlike plants of the latter part of the Mesozoic era, particularly the Bennetitales, are a line of parallel evolution which is only distantly related to the true cycads and became extinct without giving rise to any modern groups of plants.

The Ginkgoales, represented by the single modern species *Ginkgo biloba,* are known in the fossil record chiefly from leaves, so that their course of evolution is difficult to trace. Forms similar to the modern species are known from the Jurassic period (Shaparenko 1936, Florin 1936). Since this time *Ginkgo* appears to have remained essentially static from the evolutionary point of view, having gradually reduced its area of distribution, until now it is almost extinct except under cultivation.

The most flourishing of the modern orders of gymnosperms, the Coniferales, has also the most completely known fossil history. The thorough and discerning studies of Florin (1944a,b) have produced convincing evidence that, starting with the ancestral order Cordaitales in the upper Carboniferous period, the evolution of the conifers progressed steadily through the latter part of the Paleozoic era and the first half of the Mesozoic era, until forms essentially similar to modern genera appeared in the Jurassic period. Other modern genera of this order are recorded from the Cretaceous period, so that by the end of the Mesozoic era the coniferous flora of the world was not materially different from that of modern times. Fitting this record with the most widely accepted time scale of the geological periods, we reach the conclusion that for about 100 to 150 million years after the appearance in the fossil record of their earliest recognizable ancestors, the progress of conifer evolution was relatively steady and continuous. There were, however, fluctuations in rate, and a good many side lines were thrown off, which progressed for some time and then became extinct. For the next 40 to 80 million years progress was considerably slower, with many lines ceasing to evolve any new types. During the past 60 to 70 million years, since the beginning of the Tertiary period the conifers have remained remarkably stable. They have produced a good many additional species, but these appear to have been merely variations on a series of complex architectural patterns which were built up as genera during the first 150 to 200 million years of conifer evolution.

The fossil record of the final order of gymnosperms, the Gnetales, is scanty and inconclusive. Nevertheless, analysis of the pattern of distribution of modern representatives of the order, based on principles to be discussed in the next section, suggests that they, too, have evolved very little since the end of the Cretaceous period, 70 million years ago.

In the gymnosperms as a whole, therefore, progressive evolution was confined largely to the end of the Paleozoic era and the major part of the Mesozoic era. Since that time they have remained relatively static, except for the adaptation of old morphological types to new climatic conditions. Nevertheless, they obviously have not become "senescent" or ready to die out. Many species of several genera are still widespread and variable, particularly in *Pinus, Abies, Picea, Juniperus,* and *Podocarpus.* Some of these actually have "weedy" tendencies in land disturbed by man. The gymnosperms, therefore, like many groups of angiosperms, show us that neither chronological age nor evolutionary stability necessarily produces the characteristics attributed by some authors to "senescence."

For the great group of flowering plants, or angiosperms, no series of fossil ancestors exists like those known for the conifers. The Jurassic order Caytoniales (Thomas 1925, Seward 1931, pp. 366–367) and some other Mesozoic groups related to the Pteridosperms or Cycadofilicales may be near to the ancestral angiosperm stock, but the interrelationships of all of these groups are highly problematical. Except for a few isolated earlier records (Arnold 1947, pp. 336–337), angiosperms enter the fossil record in the lower part of the Cretaceous period, and at that time they already included many surprisingly modern types. Many botanists have argued from this fact that the angiosperms must be vastly older than these records of them, and that they originated in the Triassic or even in the Permian period. But evidence produced by Simpson (1944) and additional facts reviewed in this chapter speak strongly in favor of the assumption that the same evolutionary line can progress very rapidly during some periods of its history and very slowly or not at all during others. If we accept this postulate, we can imagine that the origin of the angiosperms was, geologically speaking, not much before their earliest appearance in the fossil record, but that just before this time they were undergoing a period of particularly rapid evolution. The length of the period of angiosperm evolution prior to their appearance in the fossil record is entirely an open question.

Although the leaves and fruits of some Cretaceous angiosperms are similar enough to modern ones to suggest that they belonged to existing genera, many of them are of doubtful affinity and have been assigned by different authors to various unrelated modern genera (Dorf 1942). Toward the end of the Cretaceous period, however, and particularly during the interval between this period and the Eocene epoch of the Tertiary period, the number of modern types increased rapidly. In the Wilcox flora of the southeastern United States (Berry 1930), the Goshen and Chalk Bluffs floras of the Pacific coast of North America (Chaney and Sanborn 1933, MacGinitie 1941), and several contemporary Eocene floras of Eurasia (Kryshtofovich 1929, 1930, Chaney 1940a,b), the majority of the species belonged to or closely approximated modern genera. These conclusions, based largely on comparisons of leaf impressions, are well supported by the large series of fruits and seeds in the flora of the lower Eocene London Clay, studied intensively by Reid and Chandler (1933). They found that nearly all of the species could be referred to living families, and that most of them were close to or identical with modern genera, although they did not feel justified in using modern generic names, because of the uncertainties of identification.

The floras of the middle and latter part of the Tertiary period contain only modern genera, and an increasing number of fossils from deposits of these ages are indistinguishable from living forms. In the Miocene Tehachapi flora of Southern California, which is an association largely of sclerophyllous shrubs indicating a semi-arid climate, most of the species are likewise closely similar to modern ones (Axelrod 1939). By the middle of the Pliocene epoch, as shown by remains from several different localities (Chaney, Condit, and Axelrod 1944), the woody species of the California flora were nearly all similar to modern ones. In fact, a comparison of the extensive table of species published by Chaney, Condit, and Axelrod (1944) with Jepson's manual of the California flora (Jepson 1925) has revealed the fact that all the species of no less than twenty-eight of the woody genera of this flora are represented by identical or nearly identical precursors in the floras of the Miocene and the Pliocene epochs. In many other genera, the species not represented are at present rare and occur in sites unfavorable for preservation. In fact, the abundant fossil evidence suggests that during the past five million years new species of woody plants have been added to the California flora only in such large and complex genera as *Eriogonum, Ceanothus, Arctostaphylos,* and various Compositae.

Although there is no other region in the world in which the fossil floras of the Tertiary period have been so carefully correlated with each other and with modern floras as they have in the western United States, all the available evidence indicates that the rate of evolution among woody plants since the middle of this period has everywhere been as slow or slower than it has in the western United States. The Miocene Shanwang flora of northern China consists almost entirely of modern species (Hu and Chaney 1940), as do the early Pliocene floras of Japan (Miki 1941), while the essentially modern character of various mid-Tertiary floras from Siberia is reviewed by Kryshtofovich (1929, 1930, 1935). The Miocene and Pliocene floras of Europe are in some respects very different from the modern ones (Seward 1931, pp. 450–453), but this is a result of the great climatic changes which took place on that continent during the glacial epoch of the Pleistocene, causing the extinction of large numbers of species. It is not certain what proportion of the relatively small number of woody species in the modern European flora are Tertiary types which survived the ice age in various refugial areas, and how many of them have been newly evolved during that epoch, but the safest inference is that the number of such new species is relatively small.

The reader may justly ask at this point, "How can we be sure that these fossil forms known to have borne leaves or sometimes seeds and fruits similar to those of modern species were actually the genetic equivalents of the modern species which they resemble?" The answer to this question is, of course, that we cannot actually be sure. No conclusions in the field of paleontology can have the degree of certainty and predictability which may be reached in experimental sciences such as physiology or genetics. About all of the events of the remote past we must be content with the inferences which have the greatest degree of probability, based on the widest variety of evidence which can be obtained.

In the present instance, three different lines of evidence point to genetic continuity between most of the woody species of the middle and later part of the Tertiary period and their modern descendants. In the first place, the comparison is usually based on the range of variability within a large series of leaves of the fossil form and its modern equivalent, rather than on a few isolated leaves of each. Secondly, in the more recent and better-known assemblages of fossils, the association of species is similar to that in modern floras.

Finally, the evidence from hybridization experiments between modern subspecies and species of woody plants, as discussed in Chapter VI, indicates that forms which differ greatly in external morphology are often essentially alike in their chromosomes and are not separated from each other by any barriers of hybrid sterility. Species which are closely similar in external morphology, but cannot be crossed or form sterile hybrids, are unknown in woody plants, except for those genera in which polyploidy is present. Hence, the most likely inference on the basis of all available evidence is that most of the woody species of today have existed for five million years or more, and that the evolution of the genetic isolating mechanisms separating them took place largely during the early and middle parts of the Tertiary period.

The most conspicuous feature of the early Tertiary floras is that all of those known consist of types adapted to climates much

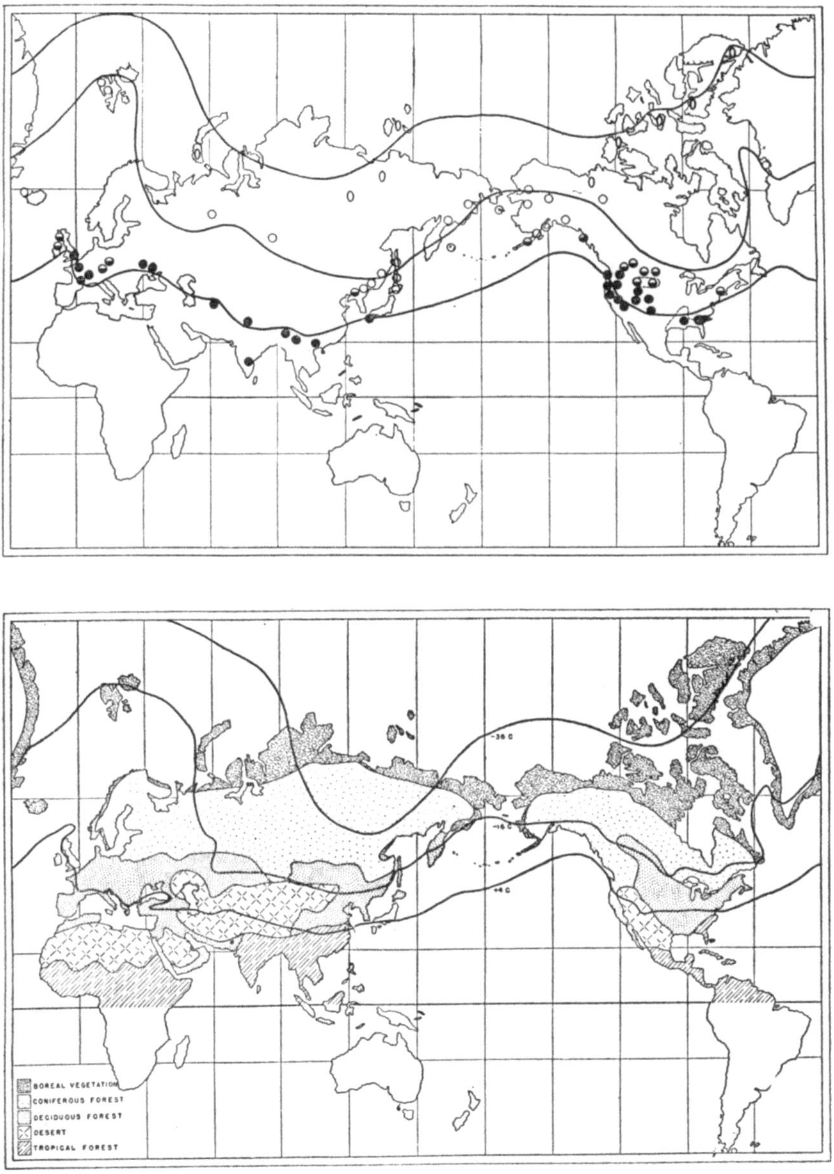

FIG. 48. Top, distribution in the Northern Hemisphere of Eocene fossil floras of tropical character (solid circles), of warm temperate character (open circles), and of cool temperate character (open ellipses), and the Eocene "isoflors" constructed from these distributions. Below, distribution in the Northern Hemisphere of modern vegetation and January isotherms. From Chaney 1940.

milder than that which now prevails at the sites of deposition. Reid and Chandler (1933) found that the large assemblage of fruits and seeds in the early Eocene London Clay flora resembles most closely the modern flora of the tropical Indo-Malayan region, with the *Nipa* palm as the dominant species. The numerous Eocene floras of the United States, such as the Wilcox (Berry 1930), the Goshen (Chaney and Sanborn 1933), and the Chalk Bluffs (MacGinitie 1941), likewise have tropical or subtropical affinities, particularly with the modern floras of Mexico, Central America, and the West Indies (Chaney 1947). Certain subtropical species, such as palms and cycads, extended even to southern Alaska (Hollick 1936), although most of the Alaskan floras of this epoch were temperate, containing *Metasequoia* in great abundance (Chaney 1948), accompanied by deciduous angiosperms like *Acer, Betula, Fagus, Liquidambar, Platanus, Populus,* and *Ulmus.* Similar temperate floras are known from Eocene deposits of central and eastern Asia, Greenland, and Spitzbergen, while the floras of arctic Siberia, Novaya Zemlya, Ellesmere, and Grinnell Land, the latter at 82° North Latitude, were cool temperate, containing chiefly *Picea, Salix, Populus,* and *Betula* (Chaney 1940a,b). The Southern Hemisphere is less well known paleobotanically, but subtropical early Tertiary floras are known from relatively high latitudes both in South America and in Australasia (Ettingshausen 1887a,b, Berry 1922, Florin 1940), while the Tertiary floras of the far southern island of Kerguelen, as well as of Graham Land, near Antarctica, are at least temperate in character (Dusén 1908, Hill 1929, Seward 1934). The evidence for this relatively warm, mild climate in the Eocene epoch is not confined to plant fossils, but is supported by the vertebrate and even the marine invertebrate fauna (Smith 1919). When jungles containing palms, figs, and lianas grew along the shores of California, Oregon, and Washington, the warm waters of the adjoining oceans were populated with corals and large clams (*Venericardia planicosta*) closely related to species now living in the Gulf of Panama. Evidence for similar warm climates near the poles in earlier geological periods is abundant and is well reviewed both by Seward (1931) and by Arnold (1947).

Emphasis must be placed upon the evidence that the Eocene climate, although certainly milder than that of the present, was nevertheless zonal, with warmer temperatures near the equator and cooler ones near the poles. Furthermore, Chaney (1940a,b) has shown that the position of Eocene floras was about the same distance north of their present counterparts at all longitudes (Fig. 48). The available data suggest a similar southward displacement of the floras of the Southern Hemisphere. This evidence is best explained on the assumption that the amelioration of the climate was world-wide and that the continents occupied approximately the same positions that they do now. The various climatic changes which may have been responsible for these alterations in the environmental conditions of the earth's biota have been discussed by Brooks (1926) and Simpson (1940), while Mason (1947) has raised the problem of the availability of sufficient light for the growth of shrubs or trees at high latitudes. The explanation of the climatic and edaphic conditions which permitted the existence of the Eocene floras and faunas in their present position undoubtedly raises many difficult and as yet unsolved problems. But, as Reid and Chandler (1933), as well as Chaney (1940b), have pointed out, problems of equal or even greater magnitude arise when attempts are made to explain all the available evidence by assuming extensive shifts in the position of the continents or poles, at least during the time since the advent of the flowering plants. There is not time in this chapter to discuss the much-debated hypotheses concerning the migration of continents; Just (1947) has presented a fine review of this subject, with full literature citations.

From the beginning to the end of the Tertiary period, the habitats available to subtropical and tropical species were contracting, while temperate and arctic species were finding larger and larger areas open to them. Furthermore, the latter part of this period, particularly the Pliocene epoch, saw the rise of great mountain chains in many parts of the world, including the Alps, Himalaya, the western Cordillera of North America, and the Andes. The appearance of semiarid steppes and deserts in the rain shadows of these mountain systems restricted greatly the areas occupied by mesophytic plants of both tropical and temperate climates, while the opportunities for expansion afforded to xerophytes, as well as to alpine types adapted to life on the crests of the mountains, were greatly increased. This progressive cooling and diversification of the Tertiary climate, with consequent segregation and diversification of floras, is well described by Chaney (1936, 1947), by Cain (1944, Chap. IX), and by Axelrod (1941, 1948). The latter author has pointed out that the origin of most of the modern ecotypes and ecospecies of woody plants must have taken place during the Pliocene and the Pleistocene epochs.

The fossil record of the woody angiosperms may therefore be summarized as follows. They first appear in great abundance in the early part of the Cretaceous, about 100 million years ago, with many families and genera already represented, some of them modern. For the next 40 to 50 million years, until the beginning or middle of the Eocene epoch, the differentiation of modern genera occurred at a gradually decreasing rate. During this time, the climates of the world were milder than they are at present, although there was alternation of cooler and warmer periods. The culmination of this period of warm climate came in the Eocene epoch, when plants now typical of subtropical floras reached latitudes of 55 to 60 degrees north, while woody plants now characteristic of cool temperate regions grew within ten degrees of the north pole, and forests covered at least part of the Antarctic continent. Since the Eocene epoch, in accompaniment with the cooling and diversification of the climate, the evolution of woody plants has consisted chiefly of the differentiation of new species and ecotypes, and at least in temperate regions few if any new genera have appeared. Most modern species of woody plants date back at least to the middle of the Pliocene epoch, about five million years ago, while several (*Liriodendron tulipifera, Populus trichocarpa, Castanopsis sempervirens, Quercus engelmannii, tomentella, chrysolepis,* and *palmeri, Celtis reticulata, Umbellularia californica, Lyonothamnus floribundus, Rhus integrifolia, Fremontia californica, Arbutus menziesii,* and so forth) have been practically unchanged since the beginning to the middle of the Miocene epoch, 20 to 30 million years ago, and some species (*Chaetoptelea mexicana, Cercidiphyllum japonicum*) have existed in essentially their present form since the Eocene epoch, 40 to 50 million years ago.

The evidence from fossils of herbaceous angiosperms, although it consists of only a few isolated examples, nevertheless suggests that evolution in herbs has not always progressed at the same rate as it has in the woody plants associated with them. The best preserved fossils of herbaceous species are usually fruits and seeds, and so do not give a record which is strictly comparable to the bulk of the record from woody angiosperms. The three best examples known to the writer are, fortunately, representatives of entirely different types of plants.

The first is the series of seeds assembled by Chandler (1923) of the aquatic genus *Stratiotes.* The oldest seeds in this series, of Eocene age, are small, broad, and heavily sculptured, while those of younger strata are progressively longer, narrower, and smoother, until in the uppermost Pliocene there occur seeds which are indistinguishable from the modern European *S. aloides.* The significance of this extensive evolution is difficult to estimate. *Stratiotes* is today a monotypic genus of the small, aquatic, or subaquatic family Hydrocharitaceae, in which it occupies a rather isolated position. Most members of this family multiply vegetatively to such an extent that plants with seeds are rarely collected, and the seeds of most species are not available for study, description, or comparison with those of *Stratiotes.* There is no way of deciding, therefore, whether the recorded changes in the seeds are on the species level, or whether the Eocene seeds belonged to plants of a different genus. *Stratiotes,* therefore, represents an isolated case of undoubted progressive evolution of which the causes and significance are obscure.

The second and most significant series is that of the grass seeds of the tribe Stipeae described by Elias (1942). This shows an undoubted progression from the beginning of the record in the lower Miocene epoch to its end in the middle of the Pliocene, a period of about 18 million years (Fig. 49). The fruits found in all the earlier deposits are different from any known in modern species of Stipeae, but many of those from the mid-Pliocene beds resemble modern species of the genus *Piptochaetium,* subg. *Podopogon,* nearly all of which are now confined to South America. Since the modern species of Stipeae of the North American Great Plains are fewer in number than the known Pliocene fossils, their history since the Pliocene has been mainly one of extinction. The four remaining species related to the fossil group are all polyploid with respect to the South American species of *Piptochaetium,* and presumably also with respect to the extinct

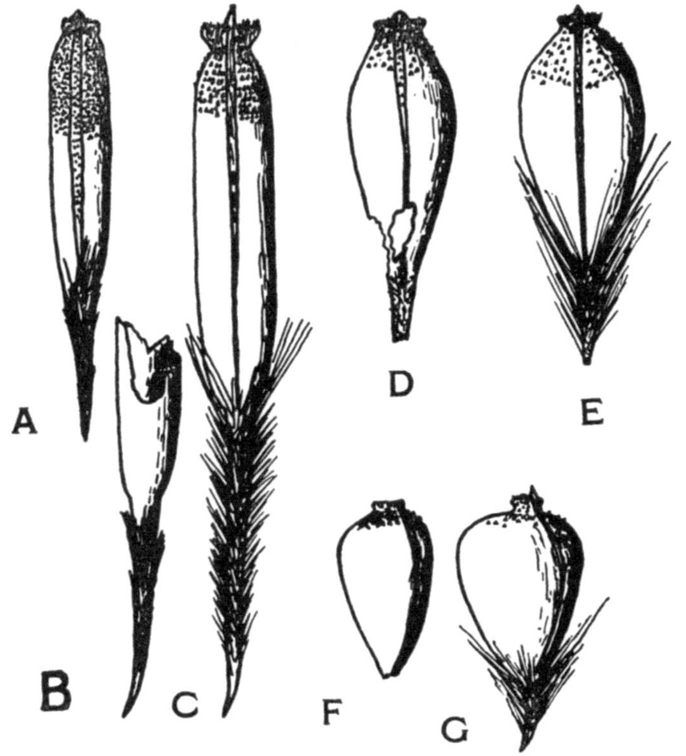

FIG. 49. Fruits of fossil *Stipidium* and *Berriochloa* and of living *Piptochaetium*. A and B, *Stipidium commune* Elias, redrawn from Elias 1942; C, *Piptochaetium napostaense,* drawn from herbarium specimen from Argentina; D, *Stipidium intermedium,* redrawn from Elias 1942; E, *Piptochaetium ruprechtianum,* drawn from herbarium specimen from Argentina; F, *Berriochloa amphoralis,* redrawn from Elias 1942; G, *Piptochaetium bicolor,* drawn from herbarium specimen from Argentina. From Stebbins 1947b.

North American ones, and they are probably allopolyploids. The period of extinction of Stipeae species, therefore, was apparently accompanied by hybridization and polyploidy. Moreover, Elias has noted that the only seeds of woody plants found in these deposits are identical with those of the modern *Celtis reticulata* and show no change from the oldest to the youngest deposits. In the Great Plains, therefore, the herbaceous species seem to have evolved rather rapidly during the latter part of the Tertiary period, while the woody ones, like the trees and shrubs of the forest belt, may have remained more constant.

The third group of fossils of herbaceous angiosperms is the extensive series of seeds known from the Pliocene deposits of northwestern Europe (Reid and Reid 1915, Reid 1920a,b, Mädler 1939). As was shown by Reid and Reid, many of these can be compared directly with the seeds of modern species now living in the mountains of China, and they thus represent species which have become extinct in part of their range, but remain unchanged in another part. Other seeds were identified as belonging to a modern genus, but they did not seem to represent any living species, while a small percentage were unidentifiable as to genus or even to family. It is, of course, possible that these seeds belong to some as yet unrecognized modern species, and there is no way of saying how many of them, instead of becoming extinct, gave rise to divergent modern European species of their genera. Nevertheless, the proportion of unidentifiable seeds to those resembling modern species should provide a rough estimate of the amount of evolution which has taken place in western Europe since the Pliocene. In particular, the fact that these floras contain seeds of herbaceous and woody species in about equal proportions should provide a basis for comparing the rate of evolution in the two types. The results of such a comparison, compiled from Reid and Reid's (1915) data on the early Pliocene Reuverian flora, are given in Table 9.

From these figures it seems likely that the herbs associated with

TABLE 9

COMPARISON BETWEEN RATE OF EXTINCTION IN WOODY AND HERBACEOUS SPECIES OF THE EARLY PLIOCENE REUVERIAN FLORA

	Woody		Herbaceous	
	Number of species	Percent	Number of species	Percent
Fossil seeds referred to modern species, either European or exotic	25	27	31	25
Fossils referred to modern genus, but not species	56	59	70	57
Fossils not identified	13	14	22	18
Total	94		123	

the forest existing in western Europe in early Pliocene times have not evolved any more rapidly than the woody species of this flora.

Thus, the scanty fossil evidence available suggests that under some conditions the herbaceous elements of the flora may have evolved at about the same rate as the woody species, but that at other times their evolution was considerably faster. Many more examples must be available before this evidence can be of major service in determining the rates of evolution of herbaceous groups.

MODERN PATTERNS OF DISTRIBUTION AND THEIR INTERPRETATION

Evidence on rates of evolution may be obtained from one direction by means of fossils, which tell more or less directly how fast certain particular groups of animals and plants have evolved under environmental conditions which may be elucidated, at least in part. And from another direction, the evidence from genetics, by explaining the mechanism and the dynamics of evolution, makes possible the formulation of hypotheses concerning the rates of evolution in entirely different organisms. Barring the unlikely discovery of an organism which is well represented in the fossil record and at the same time is favorable genetic material, the gap between these lines of evidence can be filled only to a limited extent. But the best means of filling this gap are provided by an entirely different discipline, namely, the study of the contemporary distribution patterns of modern animals and plants of all types. These patterns, like the external appearance and genetic constitution of the organisms themselves, are the end result of the interaction of various evolutionary processes and of changes in the earth's surface and climate over long periods of time. If, therefore, two or more unrelated groups of organisms have identical or similar modern patterns of distribution, one can reasonably infer that their evolutionary histories have been similar, at least in certain respects, and during the more recent periods of geological time. Furthermore, if a large enough number of such distribution patterns is known, comparisons can be made between those exhibited by organisms well represented in the fossil record, but unknown genetically, and those of other organisms of which the genetics and cytology are known, but which do not occur as fossils. But in all such studies of distribution, evidence from every possible direction must be considered, and the most probable hypothesis must be developed separately for each group of organisms, after the different lines of evidence have been compared. There is probably no field of biology in which broad generalizations are more dangerous than in plant or animal geography.

Two generalizations which must be greatly qualified or entirely discarded on the basis of modern knowledge are, first, that the age of a group of organisms may be determined by counting the number of members of that group now living and comparing its size with that of related groups and, second, that the age of a species or of any other systematic category is directly related to the size of its distributional area. These generalizations are the basis of the Age and Area hypothesis advanced by Willis (1922, 1940). The numerous criticisms of this hypothesis have been discussed by Wulff (1943) and by Cain (1944), and are ably summed up in the words of the latter author (1944, p. 230):

> According to Willis' age-and-area hypothesis, most endemic species are considered to be youthful. It is a truism of biology that populations tend to expand their areas in ever-increasing concentric circles, other things being equal. This ideal is seldom realized, for other things are seldom equal. Nevertheless, it is not possible to accept the fact, for an endemic species of narrow range, either that it is young or that it is old from a knowledge of its area alone. It is necessary to inquire into its other characteristics.

Babcock (1947) has inquired very thoroughly into other characteristics of the genus *Crepis*. As a result, he has been able to classify the species of this genus in respect to age and degree of advancement with greater probable accuracy than has been possible in any other genus not well represented in the fossil record. He has found (p. 129) that among the narrowly restricted endemic species of these genus there are two primitive, three intermediate, and five advanced types. Obviously, neither the age and area nor the relict hypothesis are valid as generalizations to explain the narrow endemics of *Crepis*. This is just what would be expected on the basis of theories which take into account the genetic nature of species populations (Cain 1940, 1944, Chap. XVI, Stebbins 1942b, Mason 1946a,b).

The same consideration holds in respect to the generalization which assumes that the age of a group is proportional to its size. It is probably true in general that any successful group of organisms is small at the beginning and gradually increases its size through the evolution of new species. But fossil evidence indicates that this increase may be slow or rapid, and may persist until the group has become very large, or it may stop and be followed by a decrease before many species have evolved (Simpson 1944). For these reasons, the elaborate calculations of Small (1937a,b, 1945, 1946, Small and Johnston 1937) are based on a fallacy. Furthermore, the interpretation by Willis and Small of the regular "hollow curves" which are obtained when the size of the groups is plotted against the frequency of groups of a particular size is also fallacious, as Wright (1941b) has pointed out. In fact, such hollow curves can be obtained by plotting the frequency of almost any series of categories of different sizes, such as the college graduates of any class with a certain income or the surnames in a telephone book.

Still another series of generalizations which are subject to numerous qualifications are those which seek to establish rules for determining the place of origin of a particular group. The best known of these is that the place of origin of a group is the region in which the largest number of representatives of that group exists at present. As applied to the origin of genera and families in terms of the distribution of living species, this generalization forms part of the Age and Area theory of Willis. But the gene center theory of Vavilov (1926), which states that the place of origin of a species of cultivated plant is that which contains at present the largest number of genetic varieties of that plant, is essentially similar. The center of origin theory as applied to genera and families has been criticized by Fernald (1926), Wulff (1943), Cain (1944), and many others, while the gene center theory has been similarly criticized by Schiemann (1939) and Turesson (1932). These criticisms have all been essentially similar. If the group in question is a young one, and if the selective forces of the environment, including competition with other organisms, have been operating in about the same manner throughout its evolutionary history, then the center of diversity and the primary center of origin are likely to be the same. But if the group is old, and particularly if it has formerly existed in regions where it is no longer found, or if it has survived great alterations in the environment, secondary centers of diversity are likely to have arisen in areas which more recently became favorable to the members of the group, and it may even have died out in the place where it originated.

Matthew (1939) has shown that for orders, families, and genera of mammals the fossil evidence is absolutely contradictory to the center of diversity–center of origin hypothesis. The horses, for instance, originated in North America, which now contains no indigenous species of this family. Its present centers of diversity, southwestern Asia and central Africa, were not occupied by the ancestors of its modern representatives until late Pliocene or Pleistocene time. For the order of marsupials, Australia is a center of diversity as prominent and striking as any which can be found for a group of animals or plants. But the evolution which produced this diversity is relatively recent and has obviously resulted from lack of competition with other mammals. The original marsupials probably entered Australia in the latter part of the Cretaceous period. With this example in mind, botanists should be very hesitant to assume that such genera as *Eucalyptus* and *Acacia,* which have equally prominent centers of diversity in Australia, originated on that continent. In fact, Diels (1934) has postulated that these and many others of the subtropical groups endemic to or strongly developed in Australia are of Malaysian origin. His hypotheses on the origin of the Australian flora are entirely in accord with the fossil evidence, both old and new.

On the basis of his evidence from mammals, Matthew produced another generalization about centers of origin, namely, that the center of origin is the region in which the most advanced species are found, while the most primitive species may be expected near the periphery of the range of the group. This generalization is, however, as dangerous as the previous ones. It is based on the following assumption (Matthew 1939, pp. 31–32).

Whatever be the causes of evolution, we must expect them to act with maximum force in some one region; and so long as the evolution is progressing steadily in one direction, we should expect them to continue to act with maximum force in that region. This point will be the center of dispersal of the race. At any period, the most advanced and progressive species of the race will be those inhabiting that region; the most primitive and unprogressive species will be those remote from this center.

If a group has lived through a period of great changes in climate, the selective forces guiding evolution will have changed their intensity and geographic position, and the above assumptions are obviously invalid. This has been true of all genera of plants and animals of the North Temperate Zone, since they all antedate the mountain-building period of the later Tertiary and also the glaciations of the Pleistocene. Direct evidence bearing on Matthew's hypothesis of the peripheral distribution of primitive types can be obtained from the cytology and distribution of polyploid complexes. Their diploid members must be older than the polyploids, although they are not more primitive in the sense that they are less specialized in structure. Nevertheless, on the basis of Matthew's assumption, their greater age should have permitted the diploids to migrate farther toward the periphery of the range of the group, while competition with more aggressive polyploids should tend to eliminate them near its center. Actually, as stated in Chapter IX, a survey of a large number of polyploid complexes has shown that the number of those in which the diploids are peripheral is about equal to those in which the diploids are centrally located and the polyploids are peripheral.

Matthew's hypothesis makes certain assumptions in addition to those which he states. The oldest members of a group will occupy a peripheral position only if their ability for migration and establishment is equal to that of the younger ones. But in many groups of plants, the principal trend of evolution is toward the development of more efficient means of migration and establishment. In many families and genera, the more specialized members differ from the primitive ones in possessing smaller and more numerous seeds and also in the higher development of specialized methods of seed dispersal, such as the pappus of the Compositae, the awns of the Gramineae, and the baccate fruits of many groups. Such species may be expected to overtake their less efficient ancestors in colonizing the globe. Furthermore, the trend in many groups is from long-lived perennials, which as members of climax formations establish themselves very slowly in a new region, to short-lived annuals, which become established quickly and easily. Finally, new polyploid forms, particularly amphiploids, may be able to establish themselves more easily than their diploid ancestors in new, peripheral areas, because of their greater supply of potential new gene combinations. Matthew's hypothesis may be expected to hold to a different degree in various groups, depending on certain specific properties of their course of evolution. It can thus be weighed in each example against the other hypotheses, according to the available evidence.

Still another type of generalization which may be as misleading as it is helpful consists of those hypotheses which attempt to establish a single region for the origin and differentiation of angiosperm groups and a single basis for their migration from one continent to another. Thus, Wulff (1943, Chap. V) speaks of hypotheses of a polar or a tropical origin of angiosperm groups in general, as if these were mutually exclusive, and he similarly indicates that intercontinental migrations must have been either entirely by land bridges across the present oceans, by migrations southward and northward from the two poles, or that they must all be explained on the basis of the previous union of the continents and the hypothesis of continental drift. Similarly, Camp (1947), by assembling a great number of distributional maps, has attempted to show that the differentiation of angiosperm families took place almost entirely in the Southern Hemisphere. But certain facts of distribution of both living and fossil organisms suggest that families have been differentiated on most if not all of the major land masses, and that migrations of plants have taken place in many different directions and via a number of different intercontinental connections, both past and present, as well as across the "stepping stones" afforded by groups of neighboring islands.

In the first place, certain specialized families of flowering plants are either endemic to or occur so predominantly on one continent that they must be assumed to have originated there. Thus, one can hardly doubt that the Bromeliaceae originated in South America, the Polemoniaceae and the Hydrophyllaceae in North America, the Valerianaceae and the Dipsacaceae in Eurasia, the Bruniaceae in Africa, and the Goodeniaceae and the Candolleaceae in Australia. Cain (1944, p. 245) has presented data from Irmscher to show that such endemic families are not predominant on any one continental mass. By analogy, one can suggest that the most likely situation is that the origin of the more widespread families has likewise been shared by various continental masses, with no one continent contributing an overwhelming proportion of them. In regard to the origin of the class of angiosperms as a whole, the present condition of our knowledge has been aptly expressed by Chaney (1947, p. 141) as follows.

> The ultimate origin of angiosperms is not clearly indicated by the Mesozoic record; much collecting by paleobotanists, followed by broad investigations both by paleobotanists and botanists, must precede our understanding of the nature and relationships of the earliest angiosperms, and our designation of the area or areas from which they have spread out to colonize the earth.

In regard to migration routes, Matthew (1939) has assembled a great body of fossil evidence to show that all migrations of land mammals from the Old World to the New, and vice versa, have been by Holarctic continental connections. But this does not mean that mammalian faunas have all been differentiated near the North Pole; on the contrary, the great centers of mammalian differentiation have been at middle latitudes in the centers of the large land masses of Eurasia, North America, and South America, with lesser centers of differentiation in Africa and Australia. The Holarctic higher latitudes have served as routes of migration, not as centers of differentiation. For the most part, these intercontinental connections have had a temperate climate, but, as previously mentioned, evidence from plant fossils indicates a subtropical climate during the Eocene epoch and perhaps at other earlier periods over at least a large part of the southern fringe of the Asiatic-American land bridge, in the site of the present Aleutian Islands and Alaska Peninsula.

On the other hand, Florin (1940, 1944b) has shown from fossil evidence that the conifers of the Southern Hemisphere have migrated freely from Australasia to South America and vice versa, while since the Jurassic or the Cretaceous period these conifers have remained entirely distinct and separated from those of the Northern Hemisphere. The apparent discrepancy between these two lines of evidence can be resolved on the basis of the different potentialities for migration possessed by these very different types of organisms. Mammals possess very limited capacities for crossing large bodies of water. This is a matter of observation and is also attested by their absence from most oceanic islands. On the other hand, the seeds of plants may occasionally be transported

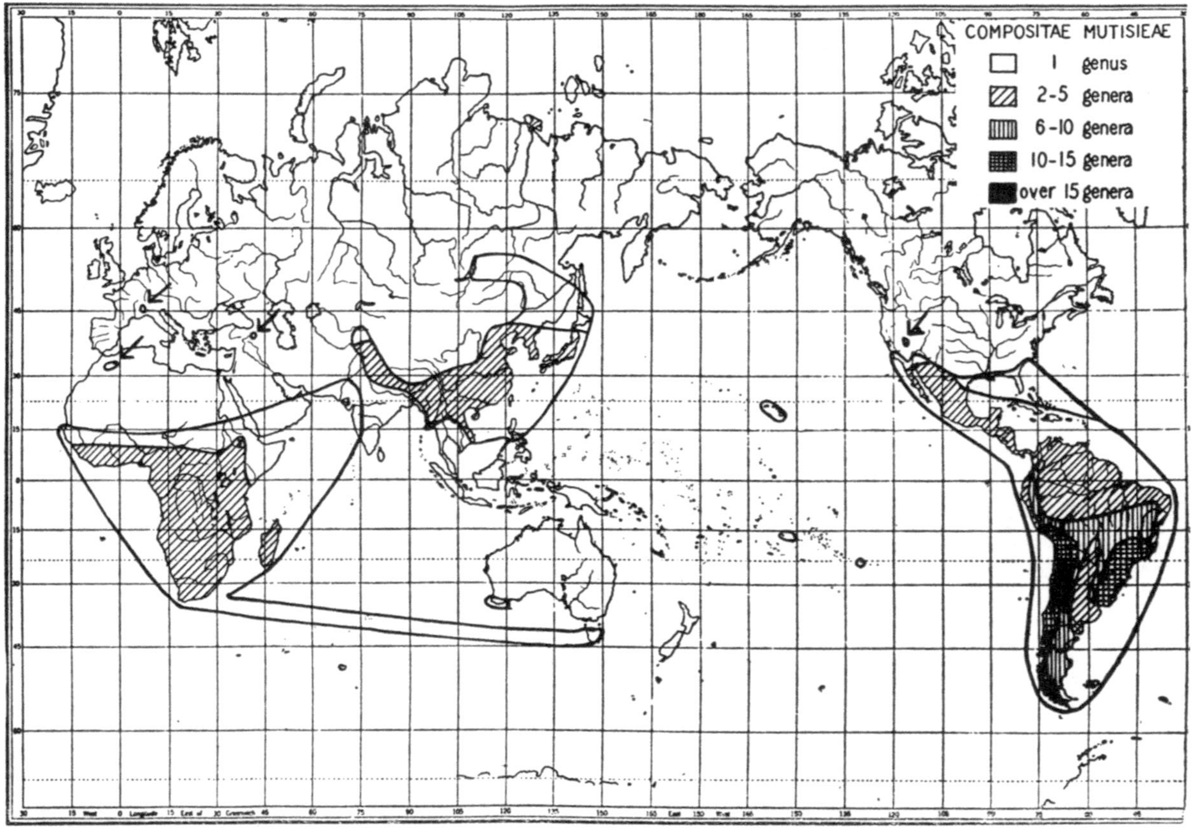

FIG. 50. Map showing the distribution of the tribe *Mutisieae* of the family Compositae, a group at present of austral distribution, but possessing relict genera in the Northern Hemisphere (shown by the arrows), which suggest a northern origin for the group. From Stebbins 1941c. Base map, Eubank's outline map of the world, by permission of Standard Process and Engraving Co., Berkeley, Calif.

over many hundreds of miles of ocean and may establish themselves on remote oceanic islands like Hawaii, Juan Fernandez, St. Helena, and the Canary Islands. We may therefore postulate that the antarctic connection between Australasia and South America existed for plants, but not for vertebrates. It probably consisted of an enlarged antarctic continent, with a temperate climate, which even now is near enough to South America so that seeds could be borne from one land mass to the other without difficulty, plus a series of islands partly connecting Antarctica with Australia and New Zealand (Hill 1929).

On the basis of this evidence, we should expect to find at least two different origins for those plant groups which are now predominant in the Southern Hemisphere. Some of them, like the southern mammals and reptiles, may be relict survivors of migrations from the north or groups derived secondarily from such relicts. Such an interpretation was suggested by the writer (Stebbins 1941c) for the primarily South American tribe Mutisieae of the family Compositae, on the basis of the virtual absence of this tribe from Australasia and the occurrence of isolated relict genera both in North America and in Eurasia (Fig. 50). As mentioned in Chapter IX, the fossil evidence suggests a similar northern origin for the numerous temperate South American species of the grass genus *Piptochaetium*. On the other hand, there is now little doubt that the Podocarpaceae, the Araucariaceae, and other southern groups of conifers originated in the south, or at least underwent their greatest differentiation on the southern continents. The same is most probably true of *Astelia, Luzuriaga, Acaena, Eucryphia,* and many other genera of flowering plants listed and mapped by Hill (1929), Skottsberg (1936), Camp (1947), and others (Fig. 51). DuRietz (1940), in his extended discussion of bipolar distributions, has clearly recognized these two types of origin.

DISJUNCT DISTRIBUTIONS AND THEIR SIGNIFICANCE

The data from plant geography which may be compared most directly with the data from paleobotany, on the one hand, and those from genetics and cytology, on the other, and thus provide indirect evidence for estimating evolutionary rates, are those data obtained from studies of disjunct distributions of species, genera,

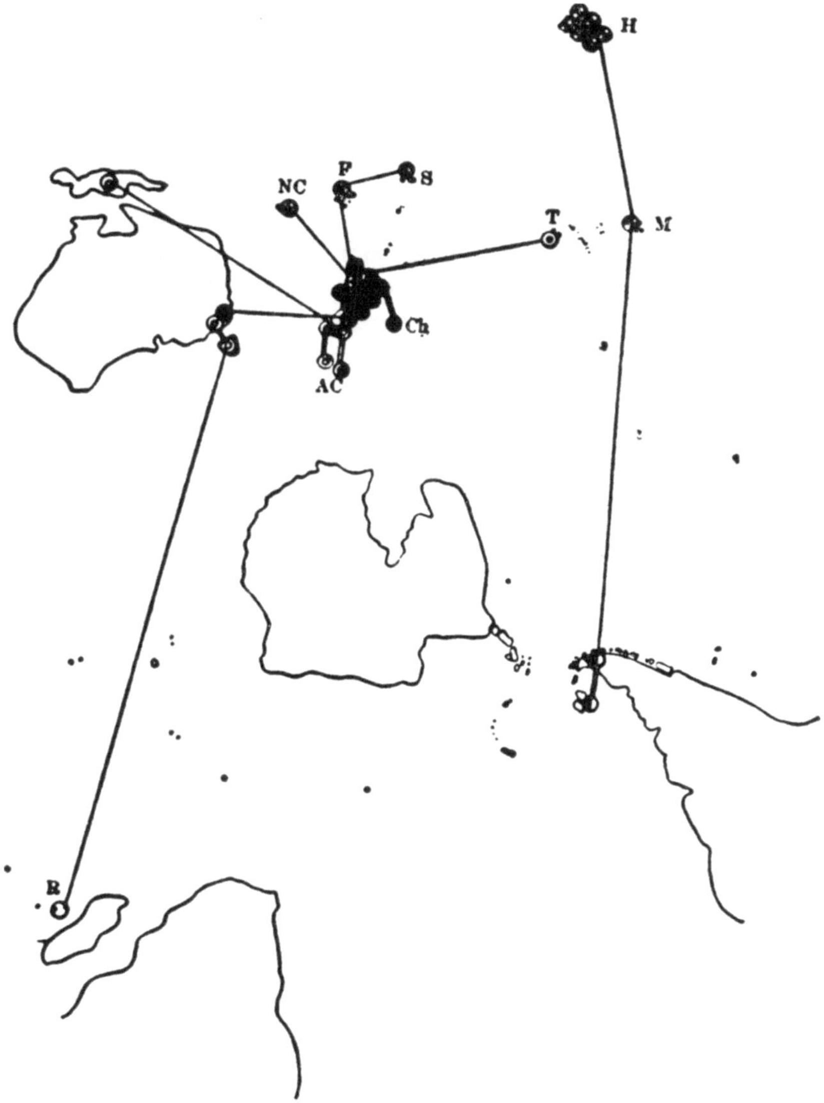

FIG. 51. Map showing the distribution of *Astelia* and *Collospermum,* two typical Antarctic genera. Antarctica is in the center of map, Australia at upper left, South America at lower right. From Skottsberg 1936. By permission of the University of California Press.

AC — Auckland and Cambell Islands
Ch — Chatham (Warikauri) Island
F — Fiji Islands
H — Hawaiian Islands
M — Marquesas Islands
NC — New Caledonia
R — Reunion
S — Samoa
T — Society Islands

and families of organisms. The significance of such distribution patterns was clearly recognized by the early evolutionists of the nineteenth century, such as Darwin, Wallace, Hooker, and Gray, and many of the most important patterns were already described by these authors. A great succession of plant geographers has amplified and partly clarified our knowledge of these patterns, while more recently several paleobotanists, like Seward, Reid and Chandler, Kryshtofovich, Florin, Berry, Chaney, and Axelrod, have related these patterns to the evidence from the fossil record, and in general have supported the major hypotheses developed by plant geographers. Finally, cytological and genetic studies are beginning to provide an evaluation of the amount of genetic difference between the isolated populations of some disjunct groups, and therefore of the amount of evolutionary divergence which has taken place since they became separated from each other.

Disjunct distributions may involve taxonomic categories of all degrees of size. Single species which occur in two widely separated areas are *Cypripedium arietinum, Symplocarpus foetidus, Brachyelytrum erectum,* and *Polygonum arifolium* (Fig. 52), all of which

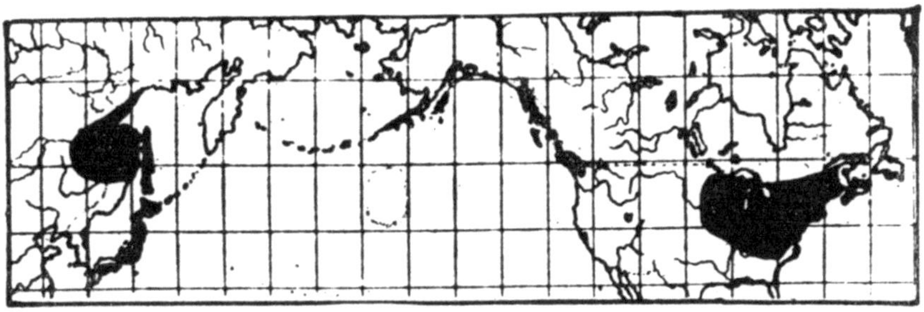

FIG. 52. Distribution of *Symplocarpus foetidus* in Asia and North America. From Fernald 1929.

have one area of distribution in eastern North America, and the other in eastern Asia (Fernald 1929). The distance between the separate areas of these species, which is nearly half the circumference of the earth, is probably the greatest which can be found. All degrees of separation can be found from this maximum down to the isolation of the separate populations of a species on different islands, mountain tops, or other areas which are only a few miles apart. If populations occupying separate areas are clearly descended from an immediate common ancestor, but are distinct enough in external morphology so that the systematist can always tell them apart, they are known as *vicarious species.* The different types of vicariads have been discussed and classified by Cain (1944,

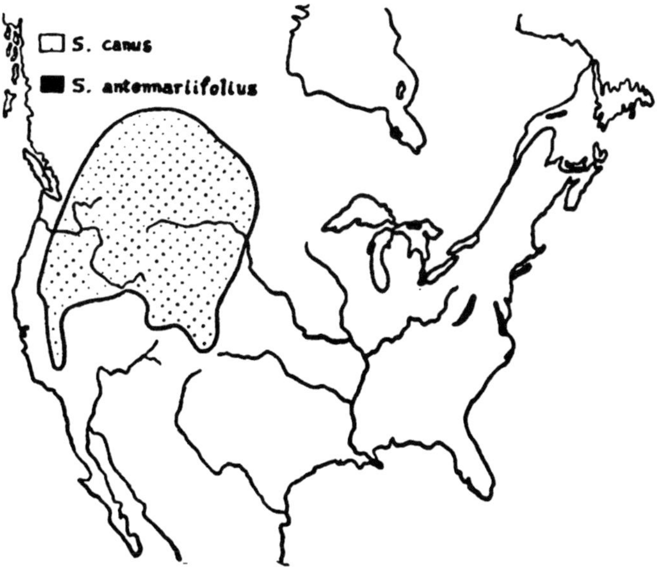

FIG. 53. Distribution of two closely related vicarious species, *Senecio canus* of the western Cordillera and *S. antennariifolius* of the Appalachian shale barrens. From Stebbins 1942b.

Chap. XVIII). Some of them are the sole representatives of their genus in each area, as in the genera *Diphylleia, Podophyllum, Cercis,* and *Platanus* of Eurasia and North America (Fernald 1931). In other instances, such as *Senecio canus* of the western United States and *S. antennariifolius* of the Appalachian shale barrens of the eastern states, the genera concerned are represented by a large number of species in each area, but the two vicarious species are more closely related to each other than either is to its associates in the same area (Fig. 53). Disjunct distributions for entire genera are well known. Among the widest disjunctions is that of *Hypochaeris,* with one group of species in Eurasia and another in temperate South America (Stebbins 1941c). Disjunct families or subfamilies are also well known; numerous examples are given by Fernald (1931), Cain (1944), and Camp (1947).

For the purpose of linking paleontological with cytogenetic

evidence on rates of evolution the significant disjunct patterns are obviously those in which the populations in the separate areas are related closely enough to each other so that they can be intercrossed, that is, disjunct species and vicarious species of the same genus. The significant data on such species are only in part those obtained by the systematist on the basis of external morphology. A much more accurate picture of the amount of evolutionary divergence between the vicarious forms can obviously be obtained by growing them under uniform conditions, by studying the number and morphology of their chromosomes, and especially by crossing them and determining the chromosome behavior and degree of fertility of their hybrids.

As has been brought out by Wulff (1943) and by Cain (1944), three explanations may be offered for any example of vicarious areas of distribution. The first assumes that the forms originated independently in the areas which they now occupy and that their present similarity is due to parallel or convergent evolution. The information which we now have on the genetic nature of the morphological differences between subspecies and species, as outlined in earlier chapters of this book, makes this hypothesis so improbable as to be untenable for all examples of single species with disjunct distribution patterns. Species differences are based largely on systems of multiple factors, which are built up by the occurrence and establishment of large numbers of genetically independent mutations. Hence, the probability that two isolated populations will evolve in exactly the same way in all of their characteristics is astronomically low, and the convergence in every respect of previously dissimilar organisms is even less probable. In regard to morphologically different, but apparently related, species, hybridization experiments can provide decisive evidence to show whether their similarity is due to true relationship or to convergent evolution. If the vicarious forms are more easily intercrossed and produce more fertile hybrids with each other than either of them does with any of its associates in its own area, the hypothesis of a common origin is by far the most probable one. But in many instances such elaborate experiments are not necessary. If the forms concerned possess no other relatives in one or both of their vicarious areas of distribution, their independent origin is extremely improbable. Wulff (1943, pp. 55–56) points out that isolated species of a number of arctic or boreal genera, such as *Primula, Draba, Saxifraga, Gentiana, Carex,* and *Phleum,* occur in the southern tip of South America, separated by many thousands of miles from their numerous northern relatives. These species could not have originated in their present habitat, since no possible ancestors exist there. They or their ancestors must have migrated southward from the north temperate regions.

Assuming that the two separated or vicarious populations have had a common origin or are immediately descended from a single common ancestor, two extreme hypotheses are often suggested to explain their present disjunction. Either they have always been separated, and migration took place by long-distance dispersal of seeds across the intervening territory, or the present disjunct areas are relicts of a former continuous distribution of the group. These two hypotheses are not sharply distinct or mutually exclusive. In many instances, two widely separated vicarious areas may never have been completely joined, but may have been closer to each other or partly connected by a series of intermediate, but still disjunct, areas. Furthermore, the hypothesis of a past continuous distribution need not imply that the former range of the group included both of the present areas plus the entire distance between them. They may have radiated from a third region and have become disjunct before they reached their present areas. Figure 54 illustrates five different methods of origin of vicarious areas: long-distance dispersal, past greater proximity, past continuity, "stepping stones," and divergent migration from a third area. Various combinations of these hypotheses may, of course, be imagined.

Evidence exists that each of these explanations is true for different ones of the modern vicarious areas. Long-distance dispersal has most probably been the method of colonization of remote oceanic islands, like Juan Fernandez, St. Helena, and the Canary Islands. Geological evidence is strongly against the hypothesis that these islands ever formed part of a major continent, and the absence or scarcity on them of mammals, fresh-water fishes, and other forms of life which could not possibly cross salt water is likewise significant. In recent years Skottsberg (1925, 1938, 1939) has championed the hypothesis of continuous distribution over past land bridges to explain the origin of the flora of all the Pacific

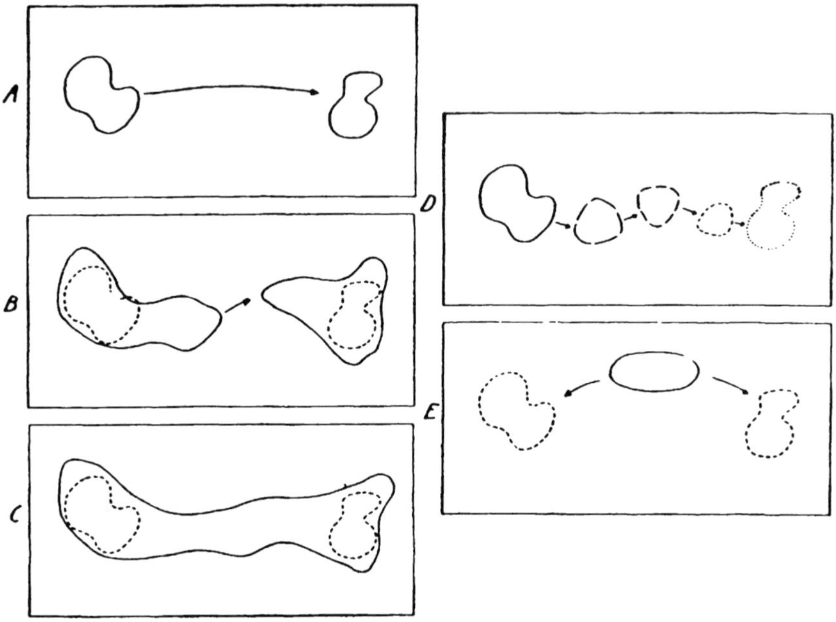

FIG. 54. Diagrams showing five different ways in which an instance of a disjunct distribution can arise. A, direct dispersal over a long distance; B, dispersal over a shorter distance through wider distribution and greater proximity of the two areas in the past; C, wider distribution with complete continuity between the two areas in the past; D, dispersal over a series of short distances by means of "stepping stones"; E, migration from a former area outside of the two present areas. Original.

islands. The chief arguments in favor of this have been the difficulty involved both in the transport of seeds over such long distances and in the establishment of the plants once these have arrived. But Mayr (1939, 1943) and Zimmerman (1942, 1948), basing their opinions, respectively, on the distribution of land birds and of insects, see no insuperable difficulties in the transoceanic migrations of these animals to all the islands which they now occupy. Zimmerman points out that such migrations need not have been over the entire distance now separating an island from the nearest land mass with similar ecological conditions. At least in the Pacific, numerous islands undoubtedly existed which are now represented only by low atolls, while other islands were formerly much larger than they are at present. Both authors agree that given sufficient time, extremely improbable events could occur. Their reasoning can be transferred to the colonization of mountain peaks by alpine plants and to other similar types of dispersal over land, as Mason (1946b) has pointed out.

The "stepping stone" method of long-distance dispersal is the most likely one to explain the antarctic migration and radiation of plants. Guppy (1906, cited by Cain 1944) has suggested this as the way in which arctic and cool temperate plants migrated from the north temperate to the antarctic regions, using as temporary refuges the higher mountain peaks of the tropics. This hypothesis has much to recommend it. A notable fact is that, with the exception of the subtropical and the desert floras, the vicarious elements which have a bipolar disjunction in North and South America (DuRietz 1940) are herbs with efficient methods for the transport and establishment of propagules. The floras of California and central Chile have many genera in common, such as *Chorizanthe, Acaena, Godetia, Phacelia, Mimulus, Blennosperma, Madia,* and *Agoseris,* and in most of these genera closely related vicarious species exist on the two continents. But the trees and shrubs of these two regions, although they possess very similar ecological adaptations, have entirely different relationships. Whereas the forests and the scrub areas of California are dominated by *Pinus, Quercus, Sequoia, Pseudotsuga, Arctostaphylos, Ceanothus,* and *Adenostoma,* those of Chile are characterized by *Araucaria, Podocarpus, Nothofagus, Fitzroya, Persea, Quillaja, Kageneckia, Adesmia,* and other types totally unfamiliar to northern botanists. The Magellanic flora, likewise, although it contains many species of arctic derivation, is entirely antarctic so far as its dominant elements are concerned. This condition is explained more easily on the assumption that the plants which crossed the equator possessed superior means of transport and establishment, rather than on the assumption that there was ever a continuous pathway of migration across the tropics for all types of temperate plants, as DuRietz (1940) has postulated.

The explanation of divergent migration is the most likely one for the disjunct distribution of many arctic and boreal species, as discussed by Hultén (1937). These have radiated outward from refuges which they occupied during the periods of glaciation, and in many instances they now occur in widely separated areas. On the other hand, there are undoubtedly many other vicarious and disjunct species of these regions which had continuous distributions before the last advance of the ice and have persisted only in unglaciated refugial habitats. The nunatak hypothesis of Fernald (1925) and the postglacial migration hypothesis of Wynne-Edwards (1937) in regard to the origin of the numerous isolated populations of far western species and local endemics centering about the Gulf of St. Lawrence may not be entirely contradictory. Some of these species may be preglacial relicts and others, postglacial immigrants.

The two hypotheses of complete and of partial continuity of distribution in the past are very difficult, if not impossible, to differentiate from each other. For both of them we have ample supporting evidence from fossils in the case of certain modern disjunctions. The most important of these are the tropical and subtropical discontinuities and the temperate discontinuities between Eurasia, particularly eastern Asia, and either the eastern or the western part of North America. The greater extension northward of subtropical floras in early Tertiary times has already been discussed. There is at present no evidence that these floras extended continuously across the Bering land bridge from Asia to America, but the Asiatic and American segments of these subtropical floras undoubtedly were closer to each other in the Eocene epoch than they are at present. They must have been at least close enough so that migration of subtropical species across this bridge was possible. The Asiatic-American disjuncts of temperate climates almost certainly ranged continuously from one continent to the other during the early part of the Tertiary period. Fossils of woody genera, such as *Carpinus, Castanea, Fagus, Juglans, Quercus,* and *Ulmus,* are known from a great number of localities in both the Old and the New World, in addition to Alaska (Chaney 1940a,b, 1947). During the Miocene epoch this bridge probably became closed to mesophytic trees and forest-loving types, except for boreal groups like *Picea, Salix, Alnus, Betula,* and *Acer.* The data of Smith (1919) indicate that the marine climate at that time was as cold as it is now, and the late Miocene floras of the western United States also indicate a climate as cold as the present one, as well as the beginning of arid conditions in the Great Plains area, effectively isolating the eastern American forests from the western ones.

The establishment of these past intercontinental connections between floras and the approximate determination of the time when they existed gives us, finally, an opportunity for estimating the rate of evolution, or the degree of evolutionary stability, of certain plant groups which have discontinuous distributions in these areas. We can assume with some degree of certainty that vicarious species with disjunct distributions in the tropics of the Old World and the New World have not been able to exchange genes since the middle of the Eocene epoch, 40 to 50 million years ago, while the numerous species with disjunct distributions in the temperate forests of Asia and America have not exchanged genes for 15 to 20 million years.

The subtropical and the tropical disjuncts consist mostly of genera with one area of distribution in the Old World and one in the New World. These are very irregularly distributed through the plant kingdom, as judged from a survey of distribution patterns as they are given in Engler and Prantl's *Die natürlichen Pflanzenfamilien.* Among the 84 genera of the Gesneriaceae, no examples are given. The genera of Palmaceae contain only a doubtful example in *Elaeis,* as well as *Cocos nucifera,* which is probably an example of recent long-distance transport by water or man. There are likewise very few examples of pantropical distributions among the 450 genera of Orchidaceae, the 100 genera of Araceae, and the 217 genera of Asclepiadaceae. But of the 40 genera of Convolvulaceae, 10 are pantropical, and in the Begoniaceae the only large genus, *Begonia,* is strongly represented in both the Old World and the New. Among the tropical genera of Gramineae and Cyperaceae, not only are there many which occur in both hemispheres, but in addition some of these contain closely related vicarious species. In the grasses, for instance, *Erianthus maximus* of Polynesia has its closest relative in *E. trinii,* of Colombia, Brazil, and Paraguay (Grassl 1946). In the related genus *Imperata, I. cylindrica* of subtropical regions in the Old World is closely related to *I. hookeri* of the southern United States. In the genus *Pennisetum, P. trachyphyllum* of Africa is vicarious to *P. latifolium, P. bambusaeforme,* and *P. tristachyum* of tropical America. These genera are not primitive, but highly advanced, specialized members of their family. The vicarious pantropical distribution of certain species groups indicates, nevertheless, that they were already well differentiated in the Eocene epoch, and suggests that much of the differentiation of genera in the Gramineae took place before this time. The same may be said of the Cyperaceae and the Convolvulaceae. Evolution in certain genera of these families during the last 50 million years has consisted only in the differentiation of species essentially similar to those already existing, and some of the modern species may be only slightly different from their Eocene ancestors. On the other hand, the fossil record of the Stipeae, mentioned earlier in this chapter, suggests that other genera of grasses, particularly those of temperate climates, were evolving actively during the Tertiary period.

Although the presence of vicarious pantropical patterns of distribution indicates great age and evolutionary stability for the group concerned, the absence of such patterns does not necessarily mean that evolution has been rapid. *Chaetoptelea mexicana*, which is the almost unchanged descendant of the Eocene *C. pseudofulva* (MacGinitie 1941), has a single area of distribution in Mexico, and the restricted relict distribution of many other early Tertiary and Cretaceous types forms, not a disjunct pattern, but a single area. We must not therefore assume that the genera of palms, aroids, orchids, asclepiads, and Gesneriaceae are nearly all recent as compared to *Erianthus, Pennisetum, Oryza,* and other tropical genera of grasses. Nevertheless, there is some reason for believing that generic differentiation has been considerably more active during the Cenozoic era in these families, than it has in the grasses, sedges, and Convolvulaceae.

Turning to the Asiatic-American disjunctions of the temperate zone, the North Pacific disjunctions of Wulff (1943), we find that nearly all of the genera of woody plants in these regions are involved. As has already been mentioned, ample fossil evidence has shown that these genera were well differentiated and practically in their modern form by the Eocene epoch. An equally large list of herbaceous genera could be cited, as was first noted by Asa Gray and has been repeatedly observed by other botanists (Cain 1944, pp. 251–257). Nearly all the herbaceous genera of the deciduous forests of the North Temperate Zone were probably well developed by the Eocene epoch 50 million years ago, and many of them may be much older (Fernald 1931). Furthermore, some species of herbaceous plants have disjunct distributions in eastern North America and eastern Asia (see Figure 52).

All available evidence suggests that these species have evolved little or not at all during the past 15 to 20 million years. In the case of two annual species, *Polygonum arifolium* and *P. sagittatum,* this means that they have gone through more than 15 million generations without evolutionary divergence.

The evidence from distribution thus supports that from fossils in indicating great evolutionary stability in the floras of the temperate forest belts, and in at least some elements of the mesophytic floras of the tropics. Both of these lines of evidence, however, are of such a nature that they make hypotheses of evolutionary stability much easier to support than those which favor evolutionary progress. For the plant groups which they represent, they indicate stability, but they do not preclude the possibility that much more active evolution has been going on contemporaneously in other groups.

DISTRIBUTIONAL PATTERNS SUGGESTING RAPID EVOLUTION

Patterns suggesting rapid evolution are those which consist of many closely related species occurring in adjacent localities within the same general area, particularly if that area is known to be a recently disturbed one. This postulate was suggested some time ago by Sinnott (1916) and has since been repeated by a number of authors. Thus, Marie-Victorin (1929, 1938) has called attention to the apparent sudden burst of speciation of the genera *Crataegus* and *Oenothera* in eastern Canada since the advent of the white man and the clearing of the forests, and to the similar phenomena in *Rosa, Hieracium,* and *Rubus* in Europe, as well as *Acacia* in Africa, *Sorbus* in the Orient, and *Hebe* (*Veronica*) in New Zealand. Most of these can be dismissed as special cases. As was mentioned in earlier chapters, the apomicts of the European *Hieracium* and *Rubus* are not equivalent genetically to species in sexual groups, while the peculiar "microspecies" of *Oenothera* in eastern North America and of the *canina* roses of Europe are the results of abnormal types of genetic segregation and are more comparable to the individual genotypes or, at most, to the ecotypes of cross-fertilizing sexual organisms than to their species. *Crataegus* and *Sorbus* are likewise probably agamic complexes, and it is likely that the new "species" of these genera which have populated cleared areas in China and America are mostly, if not entirely, apomicts of autopolyploid or hybrid derivation rather than true species. *Hebe* is known to form hybrid swarms (Allan 1940), so that the "species" of this genus may be largely unstable segregates from hybridization, like the "species" of *Iris* from the Mississippi Delta which have been described (cf. page 283).

Nevertheless, the conditions postulated by Marie-Victorin for the rapid evolution of new species, namely, the opening up of new environments to which particular organisms are especially well suited, are undoubtedly valid, and given sufficient time, genetically isolated sexual species can be expected to evolve in them in considerable numbers. Examples should be sought in the weedy floras of those regions which have for the longest time been subject to man's influence, particularly China, India, and the eastern Mediterranean region.

In the genus *Crepis* (Babcock 1947), active evolution of new, highly specialized species has undoubtedly been going on throughout the million years of the Pleistocene and recent epochs. The distribution patterns of the most advanced sections of this genus

FIG. 55. Distribution of the ten species comprising one of the phylogenetically most advanced and recent sections of the genus *Crepis*. From Babcock 1947; based on Goode base map No. 124, by permission of the University of Chicago Press.

show typical clusters of adjacent, closely related species (Fig. 55). In the case of two of the species included in Figure 55, *C. neglecta* and *C. fuliginosa,* the genetic basis of their differentiation is well understood (Tobgy 1943), and the time of this differentiation was probably the early part of the Pleistocene epoch. In a neighboring section, the polymorphic *C. vesicaria* contains eight subspecies, some of which, such as subspp. *myriocephala* and *taraxacifolia,* are beginning to develop sterility barriers between them and appear to be on the way to becoming species (Babcock 1947, pp. 825–863).

Another example of species clusters in a newly differentiated environment is presented by the tribe Madinae of the family Compositae in central California, already discussed in Chapter II (cf. Clausen, Keck, and Hiesey 1941, 1945a). In this tribe there are many examples of closely related species occupying localities which are near each other but nevertheless geographically and ecologically distinct and separated by imperfectly developed barriers of partial hybrid sterility. These are discussed more fully in another publication (Stebbins 1949a). Examples are also cited of probable rapid evolution in perennial herbs and even in woody groups, like *Ceanothus* and *Arctostaphylos.* These, however, are shrubs of pioneer formations; the dominant trees of the Californian savanna area are species of *Quercus* and *Pinus,* which have changed little or not at all since the Miocene epoch (see page 521).

THE POSSIBLE BASIS FOR DIFFERENTIAL EVOLUTIONARY RATES

The presence of widely different rates of evolution, both between different unrelated phyla and between related members of the same order, as well as within a single evolutionary line at different times of its history, has been firmly established for animals by Simpson (1944), on the basis of his analysis of the fossil evidence. He has found that most of the rates found within any order or class fall within a normal curve of distribution as to frequency, and he has designated these rates as horotelic. The normal rate, however, is not measured according to the geological time scale, but in reference to the group of animals under consideration. Thus, the horotelic rates for carnivorous mammals are about ten times as fast as those for pelecypod mollusks.

In addition to these normal rates, Simpson shows that in several groups of animals there have existed in the past exceptionally slow and probably also exceptionally fast rates of evolution. The slow rates, termed *bradytelic,* are exemplified by *Lingula,* the well-known modern brachiopod genus which has been virtually unchanged throughout the known fossil record; *Ostrea* (the oyster) and several other genera of mollusks, which have existed since the Carboniferous; and *Sphenodon,* the crocodiles, and the opossum, of which the modern representatives are very similar to their relatives which lived, respectively, in the Triassic, the early Cretaceous, and the late Cretaceous periods. Among plants, bradytelic genera are *Thyrsopteris,* a fern now living on Juan Fernandez Island which closely resembles fossils of the Jurassic period (Seward 1931); *Ginkgo, Sequoia,* and probably the living cycads and Gnetales. There is good reason to believe that many genera of thallophytes are as old and conservative as those of mollusks, but these plants are not preserved in the fossil record.

Exceptionally fast rates of evolution, designated by Simpson as *tachytelic,* cannot be directly observed in the fossil record. In even the best series of fossil deposits, events which happened in closer succession than once in a hundred thousand to once in a million years are impossible to observe because not enough fossils are available. Furthermore, on the basis of Wright's conclusion that evolution proceeds the most rapidly in populations of intermediate size which are broken up into partially isolated subunits (see Chapter IV), we should expect members of tachytelic evolutionary lines to be less abundant than those of bradytelic ones, and therefore less likely to be preserved in the fossil record. Simpson has further pointed out that rapid evolution would be expected in organisms imperfectly adapted to their environment or in those adapted to transitory environments, and that such organisms would certainly be uncommon and rarely preserved. For these reasons, Simpson has concluded that the apparent saltations or jumps in the evolutionary progress of many lines, which would include the "revolutions" in the plant world discussed by Sahni (1937), are due to short periods of tachytely rather than to the occurrence of single large mutations. This conclusion is fully in accord with the evidence obtained from living organisms, as summarized in this book. A noteworthy fact is that these bursts of rapid evolution usually occur in the beginning of the development of a line, at a time when several different adaptive trends are radiating from an unspecialized common ancestor. This agrees with the interpretation of these bursts on the basis of the occurrence and establishment by selection of many small mutations, since a generalized group with many specialized habitats or new ways of life open to it would be expected to evolve rapidly to occupy these untenanted "adaptive peaks."

It must be emphasized that rates termed tachytelic by paleontologists would not necessarily be rapid in genetic terms. For instance, the horotelic rate of evolution in a typical line, the horses, consists of the evolution of a new species about once every 2,500,000 years, or about every 250,000 generations. The production of a new species in a thousand generations would certainly be considered tachytely. The maximum rate at which new species of animals or plants could be produced by means of the occurrence and establishment by selection of mutations with relatively slight effects is not known, but a little more knowledge of the species-formation processes may make an estimate possible in the not too distant future. At present, the writer ventures to state that given selective forces acting at their maximum intensity, a normal rate of mutation, and the possibility of occasional hybridization between types widely different in morphology and their adaptive norms, a new species, adapted to a different environment from its immediate ancestor and isolated by a barrier of hybrid sterility, could evolve in fifty to a hundred generations. If this is true, there is ample possibility for the occurrence by means of the processes outlined in this book of all of the evolutionary changes which paleontologists have termed sudden, explosive, revolutionary, or tachytelic.

In discussing the causes of bradytely and tachytely, Simpson has concluded that neither the internal, genetic properties nor the influence of the environment can by themselves determine these exceptional rates. Their causes are to be sought in the nature of the relationship between the organism and its environment. This conclusion is amply borne out by the evidence from plants. Among the particular factors responsible for bradytely, Simpson has listed population size and degree of phylogenetic specialization. Bradytelic groups usually have large populations, and are specialized, not primitive, when they become bradytelic.

The importance of population size is borne out by the large proportion of slowly evolving types found among forest trees. But in plants, as well as in animals, exceptions exist in the form of ancient, relict types with restricted distributions. These presumably had large populations when they first became bradytelic, as is known to be true of *Gingko* and *Sequoia,* and their restriction is due to the fact that, as Simpson has suggested, once they are bradytelic, it is very difficult for them to become tachytelic or even horotelic again.

Simpson's next conclusion, that groups are not primitive, but specialized and advanced, when they become bradytelic, is likewise true for plants, as pointed out elsewhere by the present author (Stebbins 1949a). Furthermore, evidence presented earlier in this chapter indicates that another characteristic postulated by Simpson for bradytelic groups is found in woody angiosperms and gymnosperms; they probably evolved considerably more rapidly in past geological ages.

There is, moreover, no reason for assuming that these types show a greatly retarded or static evolution because of an innate quality of "senescence" due to their age. Cain (1944), in a thorough discussion of the examples of so-called senescence, has shown that the morphological constancy and inability for aggressive migration which are sometimes attributed to such a quality can be found in populations which are relatively young in terms of the geological time scale, while other, much older species populations show signs of aggressiveness and invading ability which are ascribed to youthfulness by the advocates of the concept of senescence. As both Cain and the writer (Stebbins 1942b) have pointed out, the characteristics sometimes attributed to senescence are most probably due to a depletion of the genetic variability of the species population, a narrowing of the range of ecological tolerance of its individual biotypes, a restriction of the habitats to which these biotypes are adapted, the possession of poor means of seed dispersal and establishment of seedlings, or to any combination of these four factors.

There is good reason, therefore, for botanists to follow Simpson, as well as Rubtzov (1945) and Schmalhausen (1946), in seeking for the principal factors governing rates of evolution by studying the adaptation of organisms to their environment (Simpson 1944, p. 140). Here also the situation in most woody plants is similar to that described for bradytelic animals. Their specialized characteristics are highly adaptive, as is evident from the fact that among trees of the temperate zone are found some of the most striking examples of parallel and convergent evolution. Such families as the Fagaceae and the Juglandaceae, or the Salicaceae and the Betulaceae, have developed very similar adaptations for pollination and seed dispersal, although the bulk of the evidence suggests that they have come from very different origins. Furthermore, these adaptations are of a broad, general type, as is evident from the very wide geographic and climatic range of most of the genera in these families, and in fact of many individual species. The trees of the temperate zone appear to be less sensitive to local differences in soil and other edaphic factors than are the herbs associated with them. Their distributions are governed more by climatic factors, and because of their wide range of climatic adaptability, these woody species could become adjusted to the major alterations which took place during various geological epochs with less change in their genetic composition than could many of the more particularly adapted herbs. The following remark of Simpson can be applied to them; they "are so well adapted to a particular, continuously available environment that almost any mutation occurring in them must be disadvantageous."

The herbs of temperate forests for the most part possess specializations of their vegetative organs which adapt them very well for growth in their chosen habitat, but limit very much their capability for adaptation to new conditions if the forest should be destroyed. Many of them, therefore, are becoming very rare or actually extinct in modern times, as the great temperate forests are being hewn down.

The epiphytes and the parasitic flowering plants of the tropical rain forests likewise are highly specialized for existence in a constant environment, so that each genus or family of such plants tends to possess a relatively constant, standardized type of organization of its vegetative parts. For instance, the epiphytic orchids, even those belonging to different genera or tribes, are remarkably similar in habit, as well as in the characteristics of their fruits and seeds. Their great diversity and extensive evolution is in the structure of the flower, which obviously represents a series of adaptations to cross-pollination by insects. This is associated with the fact that the upper stories of the tropical rain forest are very rich in flower-pollinating insects, which may still be undergoing considerable evolution. So while the secular environment of the rain forest epiphytes is remarkably constant, their biological environment has been and may still be undergoing considerable changes, which continually call forth new adaptive evolution in their flowers. The evolutionary significance of these plants, already recognized by Darwin, will be very great once their ecology and cytogenetic characteristics are better known. They deserve careful study by modern methods.

ENVIRONMENTAL CONDITIONS PROMOTING RAPID EVOLUTION

From the discussion in this and the preceding chapters, the conclusion is reached that a variable environment strongly promotes rapid evolution and may in fact be essential for speeding up evolutionary change. Both Simpson (1944) and Rubtzov (1945) have reviewed the geological and paleontological evidence indicating that the great mountain-building periods in the earth's history have seen the rise or the extinction of most of the larger groups of animals and plants. Simpson, however, has pointed out that the biological changes accompanying these geological revolutions have been not so much the origin or extinction of whole phyla or classes as the expansion and diversification of classes and the origin of new orders and families. The origin of archegoniate land plants may have coincided with the Caledonian and the Taconic revolutions of the Silurian period, while the diversification of vascular plants took place under the conditions of aridity and oscillation of land and sea which characterized the Devonian period. On the other hand, the mild, equable conditions of the subsequent Carboniferous period saw, not the origin of new groups, but the expansion and perfection of those already existing. The great, world-wide mountain-building revolution which ended the Paleozoic era caused the extinction of the giant Lepidodendrids and Calamites, as well as many archaic groups of ferns and seed plants, and the rise of modern orders of ferns and gymnosperms, such as the Cycadales and the Coniferales. In the early part of the Mesozoic era, which was much more quiescent geologically, the flora, so far as is known, was also relatively static.

The next change in the plant world, the replacement of cycadophytes and other archaic gymnosperms by the already highly developed angiosperms, took place in the lower part of the Cretaceous period, when relatively warm, moist conditions were being replaced by a cooler, but sunnier, climate.

The last two great geological revolutions of the earth's history, namely, the one which ended the Cretaceous period and that which culminated in the Pleistocene glaciation, were accompanied by major changes in the angiosperm flora. The first brought about the differentiation of many, and perhaps most, of the modern genera of woody plants, as well as many herbaceous ones; while during the Pliocene and the Pleistocene epochs there occurred the great expansion and diversification of modern genera of herbs and the breaking up of the more generalized floras into the complex pattern of plant associations which exists today. At present, the mountain-building and glacial revolution of the Pleistocene is being closely followed by a man-made "revolution," in which many preexisting habitats are being destroyed, while new, man-made habitats are being opened up to the weedy plants able to colonize them. At the same time, plant species have been transported to all corners of the earth with unprecedented speed. This "revolution," which is still in progress, began far too recently for us to observe more than the very beginnings of its effects on plant evolution. Nevertheless, there is every reason to believe that the dominant plants of the future will be those best able to colonize the habitats created by man, and that studies of these plants will give us an insight into the future course of evolution in the plant world.

BIOLOGICAL CONDITIONS PROMOTING RAPID EVOLUTION

The summary just given illustrates the fact that varying environments have caused the expansion and diversification of some groups, the decimation and extinction of others, and have left still other groups relatively unchanged. Although environmental change seems to be essential for rapid evolution, the ability of a group to respond to such change by evolving rapidly depends on certain internal, biological characteristics of that group. The most important of these are doubtless the possession of a high degree of genetic heterozygosity and a favorable population structure, as emphasized in Chapter IV and elsewhere in this book. Two other characteristics are, however, equally important.

The first of these is the possession of morphological or physiological characteristics which are potentially preadaptive in the direction of the environmental change. Thus, if the climate is becoming drier, species which have relatively small, thick leaves and hard seeds capable of surviving long periods of drought have an advantage over thin-leaved, small-seeded types. For instance, fossil evidence indicates that during the Miocene and the Pliocene epochs, when the climate of the North American Great Plains was becoming progressively drier, the essentially mesophytic genus *Celtis* remained static, but the tribe Stipeae of the grass family, the species of which have seeds and leaves admirably adapted to dry conditions, evolved rapidly (Elias 1942). In California during the Pliocene and the Pleistocene epochs the genus *Quercus,* as described earlier in this chapter, evolved little or not at all, while *Arctostaphylos* and *Ceanothus,* both of which have generations not much shorter than those of the oaks, evolved much more rapidly. This difference might be due to differences in the adaptive character of the seeds of the two groups. The acorns of oaks are short-lived, need considerable moisture for germination, and so can initiate a new generation in arid regions only within a relatively limited range of climatic and edaphic conditions. On the other hand, the harder, more drought-resistant seeds of *Arctostaphylos* and *Ceanothus* could produce seedlings under a greater variety of external conditions and so permit the establishment of a larger number of genetic variants. In this way the populations of species of the latter genera might be given greater opportunities for exploring new adaptive peaks.

The final biological characteristic making for rapid evolution is the possession by the species of adaptive mechanisms which are not too general, so that they must be modified in response to the changing conditions. Schmalhausen (1946, cf. Stebbins 1949a) has shown that in animals the character of adaptation depends on the position of the organism in the hierarchy of nutrition. Animals in the lowest position are those which have no defense against aggressors except for rapidity of reproduction. These are indifferent to the particular type of aggressor, and hence will respond little or not at all to changes in their biological environment.

Plankton organisms, the best examples of this type, are well known for their slow rates of evolution. The next lowest position is occupied by all organisms with purely passive forms of defense against their enemies. This category includes all sedentary organisms and therefore practically the entire plant kingdom. According to Schmalhausen, this explains the fact that in general plants have evolved more slowly than animals. Most of the larger animals are contained in the two higher categories of the hierarchy of nutrition, namely, those which actively elude their predators or which themselves prey on other organisms, and thus compete with each other for food.

The writer has pointed out elsewhere (Stebbins 1949a) that an amplification of Schmalhausen's scheme will serve to classify plants and parts of plants according to the rapidity of evolution required by their particular type of adaptation. This is centered about differences in the ecology of their reproduction. The three levels recognized in this hierarchy of reproduction are as follows. The lowest is that containing those organisms which rely solely on the large number of gametes and zygotes they produce. A large number of aquatic plants, particularly among the algae, belong here, and these have tended to retain simple reproductive structures and reproductive cycles. Among land plants, this level is occupied by many of the bryophytes and by homosporous pteridophytes, particularly those with large, unspecialized sporangia which rely entirely on the wind for the dissemination of their spores. Here we find such bradytelic genera as *Psilotum, Botrychium, Marattia,* and *Lycopodium.* Among seed plants, the conifers, which produce large numbers of pollen grains and seeds adapted for wind dispersal, have been the most successful of the bradytelic lines. In the angiosperms, the relationship between large numbers of pollen grains or seeds produced and slowness of evolutionary change has produced not only slower evolution of the pollen- or seed-producing organs in certain families of genera, but also relative retardation of the evolutionary change of some organs as compared to others of the same plant. The stamens have in general shown less evolutionary change than other parts of the flower, and they have been particularly constant in wind-pollinated groups like the grasses and the sedges, which produce large amounts of pollen. Furthermore, many-seeded capsules show much less evolutionary differentiation in the groups in which they occur than do achenes, nuts, or other few-seeded types of fruits.

The middle position in the hierarchy of reproduction is occupied by many plants, ranging from the algae to the flowering plants, which produce large, heavily coated, highly resistant resting spores or seeds. Generic and family differences can often be found in such structures, and then tend to show more diversity than the light, wind-borne spores and seeds of the plants in the lower group. The highest position in this hierarchy is occupied by those angiosperms which are cross-pollinated by insects, which have fruits adapted to dispersal by animals, or which rely on animals for both pollination and seed dispersal. Since there are many more diverse kinds of animals in any locality than there are different climatic or edaphic conditions, many more adaptive gene combinations are possible in plants which rely on animals for their vital reproductive functions. This explains to a large extent the evolutionary diversity and plasticity of such families as Leguminosae, Malvaceae, Labiatae, Compositae, Gramineae, and Orchidaceae. In addition, it explains the fact that this diversity has involved principally those organs involved in the dispersal by animals. Thus, the Orchidaceae and the Asclepiadaceae have highly specialized and complex floral structures, which show an amazing diversity in their specializations for insect pollination, while the seed capsules, which bear enormous numbers of wind-borne seeds, are remarkably similar throughout whole genera and tribes. On the other hand, the wind-pollinated grasses have relatively uniform structures of the flower itself, including the palea, lodicules, anthers, and stigmas. But their great diversity lies in the scales and rachises of their inflorescence, which are structures connected chiefly with seed dispersal, as described in Chapter IV. In grasses, this function is carried out to a large extent by animals.

THE BASIS OF EVOLUTIONARY RATES: SUMMARY AND CONCLUSIONS

The discussion presented in this chapter may be summarized in the following hypothesis, which is essentially that of Simpson (1944) and Rubtzov (1945, cf. also Stebbins 1949a). Rates of evolutionary change are very diverse, not only between different groups of organisms living in different environments or even in the same environment but also within the same line at different periods in its evolutionary history and between different parts of the same organism. They may range all the way from the extreme rapidity which is considered sudden or explosive in terms of the geological time scale down to rates so slow that the line is essentially static. They are determined primarily by the relationship between the evolving population and its environment and secondarily by forces inherent in the population itself. If the environment is constant in respect to all the forces affecting the adaptive character of the population, evolution will slow down until it stops at a level which represents the attainment in the population as a whole of the most adaptive set of gene combinations possible. This evolutionary stability is an equilibrium, maintained by the selective elimination of new mutations, which are constantly occurring.

If the environment changes, the population will either become reduced in size and eventually extinct or it will evolve in response to this change. The principal factors governing the rate of this change are as follows. First, the amount of genetic variability in the population, in terms of heterozygosity of individuals, and genetic differences between actually or potentially interbreeding races. Second, the structure of the population, whether large or small and whether continuous and panmictic or divided into partly isolated subpopulations (see Chapter IV). Third, the nature of the adaptation of the population to its environment. This includes the degree to which potentially preadaptive gene combinations are possible, as well as the position of the organism in the hierarchies of nutrition and reproduction. Fourth, the intrinsic mutation rate.

The various points of this hypothesis can be tested on a variety of organisms, both plant and animal, by observation, experimentation, and application of data already available. If they are correct, then an intimate study of a group of organisms with these points in mind should enable us to determine their evolutionary possibilities and to accelerate greatly their rate of evolution, provided that the group can be handled experimentally. The control by man of organic evolution is now an attainable goal.

Works Cited

Aaronsohn, A. 1910. Agricultural explorations in Palestine. U.S. Dept. Agr. Bur. Pl. Ind., Bull. No. 180. 64 pp.

Aase, H. C. 1935. Cytology of cereals. Bot. Rev., 1: 467–496.

— 1946. Cytology of cereals. II. Bot. Rev., 12: 255–334.

Abrams, L. 1905. The theory of isolation as applied to plants. Science, n.s. 22: 836–838.

Airy-Shaw, H. K. 1947. The botanical name of the tetraploid watercress. Kew Bull., 1947: 39–46.

Allan, H. H. 1940. Natural hybridization in relation to taxonomy. In J. Huxley, ed.; The New Systematics: 515–528.

Allard, H. A. 1932. A progeny study of the so-called oak species *Quercus saulii,* with notes on other probable hybrids found in or near the District of Columbia. Bull. Torrey Bot. Club, 59: 267–277.

— 1942. The hybrid oak, *Quercus rudkini,* at Arlington, Virginia. Rhodora, 44: 262–266.

Ames, O. 1937, Pollination of orchids through pseudocopulation. Bot. Mus. Leaflets, Harvard Univ., 5(1): 1–30.

Anderson, E. 1929. Variation in *Aster anomalus.* Ann. Missouri Bot. Gard., 16: 129–144.

— 1931. Internal factors affecting discontinuity between species. Amer. Nat., 65: 144–148.

— 1934. Origin of the angiosperms. Nature, 133: 462.

— 1936a. An experimental study of hybridization in the genus *Apocynum.* Ann. Missouri Bot. Gard., 23: 159–167.

— 1936b. The species problem in *Iris.* Ann. Missouri Bot. Gard., 23: 457–509.

— 1936c. Hybridization in American Tradescantias. Ann. Missouri Bot. Gard., 23: 511–525.

— 1937a. Supra-specific variation in nature and in classification, from the viewpoint of botany. Amer. Nat., 71: 223–235.

— 1937b. Cytology in its relation to taxonomy. Bot. Rev., 3: 335–363.

— 1939. Recombination in species crosses. Genetics, 24: 668–698.

— 1941. The technique and use of mass collections in plant taxonomy. Ann. Missouri Bot. Gard., 28: 287–292.

— 1943. Mass collections. Chronica Botanica, 7: 378–380.

— 1946. Maize in Mexico, a preliminary survey. Ann. Missouri Bot. Gard., 33: 147–247.

— 1948. Hybridization of the habitat. Evolution, 2: 1–9.

— 1949. Introgressive hybridization. New York, Wiley and Sons. 109 pp.

Anderson, E., and E. C. Abbe. 1934. A quantitative comparison of specific and generic differences in the Betulaceae. Jour. Arnold Arboretum, 15: 43–49.

Anderson, E., and L. N. Abbe. 1933. A comparative anatomical study of a mutant *Aquilegia*. Amer. Nat., 47: 380–384.

Anderson, E., and D. DeWinton. 1935. The genetics of *Primula sinensis*. IV. Indications as to the ontogenetic relationship of leaf and inflorescence. Ann. Bot., 49: 671–688.

Anderson, E., and L. Hubricht. 1938a. Hybridization in *Tradescantia*. III. The evidence for introgressive hybridization. Amer. Jour. Bot., 25: 396–402.

—— 1938b. American sugar maples. I. Phylogenetic relationships as deduced from a study of leaf variation. Bot. Gaz., 100: 312–323.

Anderson, E., and R. P. Ownbey. 1939. The genetic coefficients of specific difference. Ann. Missouri Bot. Gard., 26: 325–346.

Anderson, E., and K. Sax. 1936. A cytological monograph of the American species of *Tradescantia*. Bot. Gaz., 97: 433–476.

Anderson, E., and B. Schafer. 1931. Species hybrids in *Aquilegia*. Ann. Bot., 45: 639–646.

—— 1933. Vicinism in *Aquilegia vulgaris*. Amer. Nat., 67: 1–3.

Anderson, E., and W. B. Turrill. 1935. Biometrical studies on herbarium material. Nature, 136: 986–987.

Anderson, E., and T. W. Whitaker. 1934. Speciation in *Uvularia*. Jour. Arnold Arboretum, 15: 28–42.

Anderson, E. G. 1935. Chromosomal interchanges in maize. Genetics, 20: 70–83.

Andersson-Kottö, I. 1932. Observations on the inheritance of apospory and alternation of generations. Svensk. Bot. Tidskr., 26: 99–106.

Arber, A. 1925. Monocotyledons. A morphological study. Cambridge, Cambridge University Press. 258 pp.

—— 1937. The interpretation of the flower. Biol. Rev., 12: 157–184.

Armstrong, J. M. 1945. Investigations in *Triticum-Agropyron* hybridization. Empire Jour. Exp. Agr., 13(49): 41–53.

Arnold, C. A. 1947. An introduction to paleobotany. New York, McGraw-Hill. 433 pp.

Arzt, H. 1926. Serologische Untersuchungen über die Verwandschaftsverhältnisse der Gerste mit besonderer Berücksichtigung des Eiweissausgleichs innerhalb der präzipitierender Lösungen. Bot. Arch., 13: 117–148.

Asana, J. J., and R. D. Adatia. 1945. The chromosome numbers in the family Anonaceae. Current Science, 14: 74–75.

Ascherson, P., and P. Graebner. 1913. Synopsis der mitteleuropäischen Flora, Vol. 4. 885 pp.

Atchison, E. 1947. Studies in the Leguminosae. I. Chromosome numbers in *Erythrina* L. Amer. Jour. Bot., 34: 407–414.

Atwood, S. S. 1944. The behavior of oppositional alleles in polyploids of *Trifolium repens*. Proc. Nat. Acad. Sci., 30: 69–79.

Auerbach, C., J. M. Robson, and J. G. Carr. 1947. The chemical production of mutations. Science, 105: 243–247.

Avdulov, N. P. 1931. Karyo-systematische Untersuchung der Familie Gramineen. Bull. Appl. Bot., Suppl. 44. 428 pp.

Avery, P. 1938. Cytogenetic evidences of *Nicotiana* phylesis in the *alata* group. Univ. of Calif. Publ. Bot., 18: 153–194.

Axelrod, D. 1939. A Miocene flora from the western border of the Mohave Desert. Carnegie Inst. Washington, Publ. No. 516. 129 pp.

Axelrod, D. I. 1941. The concept of ecospecies in Tertiary paleobotany. Proc. Nat. Acad. Sci., 27: 545–551.

—— 1948. Climate and evolution in western North America during Middle Pliocene time. Evolution, 2: 127–144.

Babcock, E. B. 1942. Systematics, cytogenetics, and evolution in *Crepis*. Bot. Rev., 8: 139–190.

—— 1947. The Genus *Crepis*, I and II. Univ. of Calif. Publ. Bot., Vols. 21 and 22. 1,030 pp.

Babcock, E. B., and D. R. Cameron. 1934. Chromosomes and phylogeny in *Crepis*. II. The relationships of 108 species. Univ. of Calif. Publ. Agr. Sci., 6: 287–324.

Babcock, E. B., and M. S. Cave. 1938. A study of intra- and interspecific relations of *Crepis foetida* L. Zeitschr. Ind. Abst. u. Vererbungsl., 75: 124–160.

Babcock, E. B., and S. L. Emsweller. 1936. Meiosis in certain interspecific hybrids in *Crepis*, and its bearing on taxonomic relationship. Univ. of Calif. Publ. Agr. Sci., 6: 325–368.

Babcock, E. B., and J. A. Jenkins. 1943. Chromosomes and phylogeny in *Crepis*. III. The relationships of 113 species. Univ. of Calif. Publ. Bot., 18: 241–292.

Babcock, E. B., and G. L. Stebbins, Jr. 1937. The genus *Youngia*. Carnegie Inst. Washington, Publ. No. 484. 106 pp.

—— 1938. The American species of *Crepis*: their relationships and distribution as affected by polyploidy and apomixis. Carnegie Inst. Washington, Publ. No. 504. 200 pp.

Babcock, E. B., G. L. Stebbins, Jr., and J. A. Jenkins. 1937. Chromosomes and phylogeny in some genera of the Crepidinae. Cytologia, Fujii Jubil. Vol.: 188–210.

—— 1942. Genetic evolutionary processes in *Crepis*. Amer. Nat., 76: 337–363.

Bailey, I. W. 1944. The development of vessels in angiosperms and its significance in morphological research. Amer. Jour. Bot., 31: 421–428.

—— 1949. Origin of the angiosperms: need for a broadened outlook. Jour. Arnold Arboretum, 30: 64–70.

Bailey, I. W., and C. G. Nast. 1943. The comparative morphology of the Winteraceae. II. Carpels. Jour. Arnold Arboretum, 24: 472–481.

—— 1945. The comparative morphology of the Winteraceae. VII.

Summary and conclusions. Jour. Arnold Arboretum, 26: 37–47.
—— 1948. Morphology and relationships of *Illicium, Schisandra,* and *Kadsura.* I. Stem and leaf. Jour. Arnold Arboretum, 29: 77–89.
Baker, H. G. 1948. Stages in invasion and replacement demonstrated by species of *Melandrium.* Jour. Ecology, 36: 96–119.
Baldwin, J. T., Jr. 1935. Somatic chromosome numbers in the genus *Sedum.* Bot. Gaz., 96: 558–564.
—— 1937. The cytotaxonomy of the Telephium section of *Sedum.* Amer. Jour. Bot., 24: 126–132.
—— 1938. *Kalanchoe*: The genus and its chromosomes. Amer. Jour. Bot., 25: 572–579.
—— 1939. Certain cytophyletic relations of Crassulaceae. Chronica Botanica, 5: 415–417.
—— 1940. Cytophyletic analysis of certain annual and biennial Crassulaceae. Madroño, 5: 184–192.
—— 1941. *Galax*: The genus and its chromosomes. Jour. Hered., 32: 249–254.
—— 1942a. Polyploidy in *Sedum ternatum* Michx. II. Cytogeography. Amer. Jour. Bot., 29: 283–286.
—— 1942b. Cytological basis for specific segregation in the *Sedum nevii* complex. Rhodora, 44: 11–14.
—— 1943. Polyploidy in *Sedum pulchellum.* I. Cytogeography. Bull. Torrey Bot. Club, 70: 26–33.
—— 1947. *Hevea*: A first interpretation. Jour. Hered., 38: 54–64.
Bandulska, H. 1924. On the cuticles of some recent and fossil Fagaceae. Jour. Linn. Soc. Bot., 46: 427–441.
Barber, H. N. 1941. Evolution in the genus *Paeonia.* Nature, 148 (3747): 227–228.
Barthelmess, A. 1941. Mutationsversuche mit einem Laubmoos, Physcomitrium piriforme. II. Morphologische und physiologische Analyse der univalenten und bivalenten Protonemen einiger Mutanten. Zeitschr. Ind. Abst. u. Vererbungsl., 79: 153–170.
Bartsch, P. 1920. Experiments in the breeding of Cerions. Dept. Marine Biol., Carnegie Inst. Washington, Publ. No. 14. 54 pp.
Battaglia, E. 1945. Sulla terminologia dei processi apomittici. Nuov. Giorn. Bot. Ital., n.s. 52: 42–57.
—— 1946a. Fenomeni citologici nuovi nella embriogenesi ("Semigamia") e nella microsporogenesi ("Doppio nucleo di restituzione") di *Rudbeckia laciniata* L. Nuovo Giorn. Bot. Ital., n.s. 52: 34–38.
—— 1946b. Ricerche cariologiche e embriologiche sul genere *Rudbeckia* (Asteraceae). Nuovo Giorn. Bot. Ital., n.s. 53: 1–69.
Bauer, H., M. Demerec, and B. P. Kaufmann. 1938. X-ray induced chromosomal alterations in *Drosophila melanogaster.* Genetics, 23: 610–630.
Baur, E. 1924. Untersuchungen über das Wesen, die Entstehung, und die Vererbung von Rassenunterscheiden bei *Antirrhinum majus.* Bibliotheca Genetica, 4: 1–170.

— 1930. Einführung in die Vererbungslehre. 7–11 Auflage, Berlin. 478 pp.

— 1932. Artumgrenzung und Artbildung in der Gattung *Antirrhinum,* Sektion Antirrhinastrum. Zeitschr. Ind. Abst. u. Vererbungsl., 63: 256–302.

Beadle, G. W. 1930. Genetical and cytological studies of Mendelian asynapsis in *Zea mays.* Cornell Univ. Agr. Exp. Sta., 129: 1–23.

— 1945. Biochemical genetics. Chemical Reviews, 37: 15–96.

— 1946. The gene. Proc. Amer. Phil. Soc., 90: 422–431.

Beadle, G. W., and V. Coonradt. 1944. Heterocaryosis in *Neurospora crassa.* Genetics, 29: 291–298.

Beasley, J. O. 1940a. The origin of American tetraploid *Gossypium* species. Amer. Nat., 74: 285–286.

— 1940b. The production of polyploids in *Gossypium.* Jour. Hered., 31: 39–48.

— 1942. Meiotic chromosome behavior in species hybrids, haploids and induced polyploids of *Gossypium.* Genetics, 27: 25–54.

Beasley, J. O., and M. S. Brown. 1943. The production of plants having an extra pair of chromosomes from species hybrids of cotton. Records Genet. Soc. Amer., No. 12: 43.

Becker, G. 1931. Experimentelle Analyse der Genom- und Plasmonwirkung bei Moosen. III. Osmotischer Wert heteroploider Pflanzen. Zeitschr. Ind. Abst. u. Vererbungsl., 60: 17–38.

Beddows, A. R. 1931. Seed setting and flowering in various grasses. Bull. Welsh Plant Breed. Sta., Ser. H, No. 12: 5–99.

Beetle, D. E. 1944. A monograph of the North American species of *Fritillaria.* Madroño, 7: 133–159.

Benoist, E. 1938. Recherches caryologiques sur les *Salvia.* Rev. Cyt. et Cytophys. Veg., 2: 392–413.

Berg, R. L. 1944. Relation between the degree of frequency of mutations in the heterozygous state and their accumulation in natural populations of *Drosophila melanogaster.* Bull. Acad. Sci. USSR Cl. Sci., Math et Nat., Ser. Biol., 1944: 121–128. (In Russian, with English summary.)

Bergman, B. 1935a. Zytologische Studien über sexuelles und asexuelles *Hieracium umbellatum.* Hereditas, 20: 47–64.

— 1935b. Zytologische Studien über die Fortpflanzung bei den Gattungen *Leontodon* und *Picris.* Svensk. Bot. Tidskr., 29: 155–301.

— 1935c. Zur Kenntnis der Zytologie der skandinavischen *Antennaria*-Arten. Hereditas, 20: 214–226.

Bergner, A. D. 1943. Chromosomal interchange among six species of *Datura* in nature. Amer. Jour. Bot., 30: 431–440.

Bergner, A. D., S. Satina, and A. F. Blakeslee. 1933. Prime types in *Datura.* Proc. Nat. Acad. Sci., 19: 103–115.

Berry, E. W. 1922. The flora of the Concepcion-Arauco-Coal measures of Chile. John Hopkins Studies in Geology, 4: 73–142.

—— 1930. Revision of the Eocene Wilcox flora of the southeastern states. U. S. Geol. Survey Prof. Paper, 156: 1–196.

Bessey, C. E. 1915. The phylogeny and taxonomy of angiosperms. Ann. Missouri Bot. Gard., 2: 109–164.

Beyerle, R. 1932. Untersuchungen über die Regeneration von Farnprimarblattern. Planta, 16: 633–665.

Bhaduri, P. N. 1941. Cytological studies in the genus *Gaura*. Ann. Bot., n.s. 5: 1–14.

—— 1942. Further cytogenetical investigations in the genus *Gaura*. Ann. Bot., n.s. 6: 229–244.

Blackburn, K. B., and J. W. H. Harrison. 1921. The status of the British rose forms as determined by their cytological behaviour. Ann. Bot., 35: 159–188.

Blair, A. P. 1941. Variation, isolating mechanisms, and hybridization in certain toads. Genetics, 26: 398–417.

Blake, M. A. 1939. Some results of crosses of early ripening varieties of peaches. Proc. Amer. Soc. Hort. Sci., 37: 232–241.

Blakeslee, A. F. 1929. Cryptic types in *Datura*. Jour. Hered., 20: 177–190.

——1939. The present and potential service of chemistry to plant breeding. Amer. Jour. Bot., 26: 163–172.

—— 1945. Removing some of the barriers to crossability in plants. Proc. Amer. Phil. Soc., 89: 561–574.

Blakeslee, A. F., and A. G. Avery. 1937. Methods of inducing chromosome doubling in plants. Jour. Hered., 28: 393–411.

Blakeslee, A. F., and J. Belling. 1924. Chromosomal mutations in the Jimson weed, *Datura stramonium*. Jour. Hered., 15: 195–206.

Blakeslee, A. F., D. Bergner, and A. G. Avery. 1937. Geographical distribution of prime types in *Datura*. Cytologia, Fujii Jubil. Vol.: 1070–1093.

Blakeslee, A. F., and S. Satina. 1944. New hybrids from incompatible crosses in *Datura* through culture of excised embryos on malt media. Science, 99: 331–334.

Böcher, T. W. 1936. Cytological studies on *Campanula rotundifolia*. Hereditas, 22: 269–277.

—— 1943. Studies on variation and biology in *Plantago lanceolata* L. Dansk. Bot. Ark., 11(3): 1–18.

—— 1944. The leaf size of *Veronica officinalis* in relation to geographic and environmental factors. Dansk. Bot. Ark., 11: 1–20.

Boke, N. H. 1940. Histogenesis and morphology of the phyllode in certain species of *Acacia*. Amer. Jour. Bot., 27: 73–90.

Bowden, W. M. 1940. Diploidy, polyploidy, and winter hardiness relationships in the flowering plants. Amer. Jour. Bot., 27: 357–371.

—— 1945. A list of chromosome numbers in higher plants. I. Acanthaceae to Myrtaceae. Amer. Jour. Bot., 32: 81–92.

Bower, F. O. 1923. The ferns, I. Cambridge, Cambridge University Press. 359 pp.

—— 1935. Primitive land plants. London, MacMillan. 658 pp.

Boyes, J. W., and W. P. Thompson. 1937. The development of the endosperm and embryo in reciprocal interspecific crosses in cereals. Jour. Genet., 34: 203–227.

Boyle, W. S. 1945. A cytotaxonomic study of the North American species of *Melica*. Madroño, 8: 1–26.

Brainerd, E. 1924. Some natural violet hybrids of North America. Vermont Agr. Exp. Sta., Bull. No. 239. 205 pp.

Brainerd, E., and A. K. Peitersen. 1920. Blackberries of New England – their classification. Vermont Agr. Exp. Sta., Bull. No. 217. 84 pp.

Braun, W. 1946a. Dissociation in *Brucella abortus*: A demonstration of the role of inherent and environmental factors in bacterial variation. Jour. Bact., 51: 327–349.

—— 1946b. The effect of serum upon dissociation in *Brucella abortus*: A demonstration of the role of selective environments on bacterial variation. Jour. Bact., 52: 243–249.

Bremer, G. 1928. De Cytologie van het suikerriet. 4. Een cytologisch Onderzoek der Bastarden tusschen *Saccharum officinarum* en *Saccharum spontaneum*. Med. Proefst. v. Java Jahrg., 1928: 565–696.

Brieger, F. G. 1944a. Considerações sôbre o mecanismo da evolução. An. Esc. Sup. Agr. Luiz de Queiroz, Univ. São Paulo, 1944: 177–211.

—— 1944b. Estudos experimentais sôbre a origem do milho. An. Esc. Sup. Agr. Luiz de Queiroz, Univ. São Paulo, 1944: 226–276.

Brierley, W. B. 1929. Variation in fungi and bacteria. Proc. Int. Cong. Plant Sci., Ithaca, 2: 1629–1654.

Briggs, F. N. 1938. The use of the backcross method in crop improvement. Amer. Nat., 72: 285–292.

Brink, R. A., D. C. Cooper, and L. E. Ausherman. 1944. A hybrid between *Hordeum jubatum* and *Secale cereale* reared from an artificially cultivated embryo. Jour. Hered., 35: 67–75.

Brooks, C. E. P. 1926. Climate through the ages. London, Ernest Benn Ltd. 439 pp.

Brown, S. W. 1943. The origin and nature of variability in the Pacific Coast blackberries (*Rubus ursinus* Cham. and Schlecht and *R. lemurum* sp. nov.). Amer. Jour. Bot., 30: 686–697.

Brücher, H. 1943. Die reziprok verschiedenen Art- und Rassenbastarde von *Epilobium* und ihre Ursachen. III. Plasmon- und Genomwirkung bei *Epilobium adenocaulon* Kreuzungen. Jahrb. Wiss. Bot., 91: 331–351.

Bruun, H. G. 1932. Cytological studies in *Primula*. Symb. Bot. Upsalienses, 1: 1–239.

Buchholz, J. T. 1939. The embryogeny of *Sequoia sempervirens* with a comparison of the Sequoias. Amer. Jour. Bot., 26: 248–257.

Buchholz, J. T., L. F. Williams, and A. F. Blakeslee. 1935. Pollen-tube growth of ten species of *Datura* in interspecific pollinations. Proc. Nat. Acad. Sci., 21: 651–656.

Buller, A. H. R. 1941. The diploid cell and the diploidization process in the higher fungi. Bot. Rev., 7: 335–431.

Burbanck, M. P. 1941. Cytological and taxonomic studies in the genus *Brodiaea*. I. Bot. Gaz., 103: 247–265.

— 1944. Cytological and taxonomic studies in the genus *Brodiaea*. II. Bot. Gaz., 105: 339–345.

Burger, H. 1941. Fichten und Föhren verschiedener Herkunft auf verschiedenen Kulturorten. Mitt. Schweiz. Anstalt f. forstliche Versuchswesen, 22: 10–60.

Burton, G. W. 1948. The method of reproduction in common Bahia grass, *Paspalum notatum*. Jour. Amer. Soc. Agron., 40: 443–452.

Cadman, C. H. 1943. Nature of tetraploidy in cultivated European potatoes. Nature, 152: 103–104.

Cain, S. A. 1940. Some observations on the concept of species senescence. Ecology, 21: 213–215.

— 1944. Foundations of plant geography. New York, Harper. 556 pp.

Calder, J. W. 1937. A cytological study of some New Zealand species and varieties of *Danthonia*. Jour. Linn. Soc. London, 51: 1–9.

Callan, H. G. 1941. The cytology of *Gaulthettya wisleyensis* (Marchant) Rehder; a new mode of species formation. Ann. Bot., 5: 579–585.

Camp, W. H. 1942a. Ecological problems and species concepts in *Crataegus*. Ecology, 23: 368–369.

— 1942b. The *Crataegus* problem. Castanea, 7: 51–55.

— 1944. A preliminary account of the biosystematy of *Oxycoccus*. Bull. Torrey Bot. Club, 71: 426–437.

— 1945. The North American blueberries, with notes on other groups of Vacciniaceae. Brittonia, 5: 203–275.

— 1947. Distribution patterns in modern plants and problems of ancient dispersals. Ecological Monographs, 17: 159–183.

Camp, W. H., and C. L. Gilly. 1943. The structure and origin of species. Brittonia, 4: 323–385.

Caspari, E. 1948. Cytoplasmic inheritance. Advances in Genetics, 2: 1–66.

Catcheside, D. G. 1932. The chromosomes of a new haploid *Oenothera*. Cytologia, 4: 68–113.

— 1936. Origin, nature, and breeding behavior of *Oenothera lamarckiana* trisomics. Jour. Genet., 33: 1–23.

— 1937. Secondary pairing in *Brassica oleracea*. Cytologia, Fujii Jubil. Vol.: 366–378.

— 1939. A position effect in *Oenothera*. Jour. Genet., 38: 345–352.

— 1940. Structural analysis of *Oenothera* complexes. Proc. Roy. Soc. London, Ser. B, 128: 509–535.

—1947a. The P-locus position effect in *Oenothera*. Jour. Genet., 48: 31–42.

—— 1947b. A duplication and a deficiency in *Oenothera*. Jour. Genet., 48: 99–110.

Cave, M. S., and M. V. Bradley. 1943. Alteration of chromosome number in *Miersia chilensis*. Amer. Jour. Bot., 30: 142–149.

Cave, M. S., and L. Constance. 1942. Chromosome numbers in the Hydrophyllaceae. Univ. of Calif. Publ. Bot., 18: 205–216.

—— 1944. Chromosome numbers in the Hydrophyllaceae: II. Univ. of Calif. Publ. Bot., 18: 293–298.

—— 1947. Chromosome numbers in the Hydrophyllaceae: III. Univ. of Calif. Publ. Bot., 18: 449–465.

Chandler, M. E. J. 1923. The geological history of the genus *Stratiotes*; an account of the evolutionary changes which have occurred in the genus during Tertiary and Quaternary times. Quart. Jour. Geol. Soc. London, 79: 117–138.

Chaney, R. W. 1924. Quantitative studies of the Bridge Creek flora. Amer. Jour. Sci., 8: 127–144.

—— 1925a. A comparative study of the Bridge Creek flora and the modern redwood forest. Carnegie Inst. Washington, Publ. No. 349: 1–22.

—— 1925b. The Mascall flora – its distribution and climatic relation. Carnegie Inst. Washington, Publ. No. 349: 23–48.

—— 1936. The succession and distribution of Cenozoic floras around the northern Pacific basin. Essays in Geobotany in Honor of W. A. Setchell: 55–85.

—— 1940a. Tertiary forests and continental history. Bull. Geol. Soc. Amer., 51: 469–488.

—— 1940b. Bearing of forests on the theory of continental drift. Scient. Monthly, 51: 489–499.

—— 1947. Tertiary centers and migration routes. Ecological Monographs, 17: 139–148.

—— 1948. The bearing of the living *Metasequoia* on problems of Tertiary paleobotany. Proc. Nat. Acad. Sci., 34: 503–515.

Chaney, R. W., C. Condit, and D. I. Axelrod. 1944. Pliocene floras of California and Oregon. Carnegie Inst. Washington, Publ. No. 553. 407 pp.

Chaney, R. W., and E. I. Sanborn. 1933. The Goshen flora of west central Oregon. Carnegie Inst. Washington, Publ. No. 439: 1–103.

Cheeseman, E. S., and L. N. H. Larter. 1935. Genetical and cytological studies of *Musa*. III. Chromosome numbers in the Musaceae. Jour. Genet., 30: 31–52.

Chen, T. T. 1940a. Polyploidy and its origin in *Paramecium*. Jour. Hered., 31: 175–184.

—— 1940b. A further study on polyploidy in *Paramecium*. Chromosomes and mating types in *Paramecium bursaria*. Jour. Hered., 31(5): 249–251.

Chester, K. S. 1937. A critique of plant serology, I–III. Quart. Rev. Biol., 12: 19–46, 165–190, 294–321.

Christie, W., and H. H. Gran. 1926. Die Einwirkung verschiedener Klimaverhältnisse auf reine Linien von Hafer and Gerste. Hereditas, 8: 207–228.

Christoff, M. 1942. Die genetische Grundlage der apomiktischen Fortpflanzung bei *Hieracium aurantiacium* L. Zeitschr. Ind. Abst. u. Vererbungsl., 80: 103–125.

Christoff, M., and M. A. Christoff. 1948. Meiosis in the somatic tissue responsible for the reduction of chromosome number in the progeny of *Hieracium hoppeanum* Schult. Genetics, 33: 36–42.

Christoff, M., and G. Papasova. 1943. Die genetischen Grundlagen der apomiktischen Fortpflanzung in der Gattung *Potentilla*. Zeitschr. Ind. Abst. u. Vererbungsl., 81: 1–27.

Christoff, M. A. 1943. Polyploidie, apomiktische Samenbildung und Polyembryonie bei *Poa*-Arten. (In Bulgarian, with German summary.) Jahrb. Univ. Heilig. Kliment von Ochrid, Sofia, Land. u. Forstwirt., 21: 221–236.

Clapham, A. R. 1934. Advancing sterility in plants. Nature, 133: 704–705.

Clark, F. J. 1940. Cytogenetic studies of divergent meiotic spindle formation in *Zea mays*. Amer. Jour. Bot., 27: 547–559.

Clausen, J. 1921. Studies on the collective species *Viola tricolor* L. Bot. Tidsskr., 37: 205–221.

—— 1922. Studies on the collective species *Viola tricolor* L. II. Bot. Tidsskr., 37: 363–416.

—— 1926. Genetical and cytological investigations on *Viola tricolor* L. and *V. arvensis* Murr. Hereditas, 8: 1–156.

—— 1929. Chromosome number and relationship in some North American species of *Viola*. Ann. Bot., 43: 741–764.

—— 1931a. Genetic studies in *Polemonium*. III. Preliminary account on the cytology of species and specific hybrids. Hereditas, 15: 62–66.

—— 1931b. Cytogenetic and taxonomic investigations on Melanium violets. Hereditas, 15: 219–308.

—— 1933. Cytological evidence for the hybrid origin of *Pentstemon neotericus* Keck. Hereditas, 18: 65–76.

—— 1949. Genetics of climatic races of *Potentilla glandulosa*. Proc. 8th Int. Congr. Genetics: 162–172.

Clausen, J., D. D. Keck, and W. M. Heusi. 1934. Experimental taxonomy. Carnegie Inst. Washington, Yearbook, 1934: 173–177.

Clausen, J., D. D. Keck, and W. M. Hiesey. 1939. The concept of species based on experiment. Amer. Jour. Bot., 26: 103–106.

—— 1940. Experimental studies on the nature of species. I. The effect of varied environments on western North American plants. Carnegie Inst. Washington, Publ. No. 520. 452 pp.

—— 1941. Experimental taxonomy. Carnegie Inst. Washington, Yearbook, No. 40: 160–170.

—— 1945a. Experimental studies on the nature of species. II. Plant evolution through amphiploidy and autoploidy, with examples from the Madiinae. Carnegie Inst. Washington, Publ. No. 564. 174 pp.

— 1945b. Experimental taxonomy. Carnegie Inst. Washington, Yearbook, No. 44: 71–83.

— 1947. Heredity of geographically and ecologically isolated races. Amer. Nat., 81: 114–133.

— 1948. Experimental studies on the nature of species. III. Environmental responses of climatic races of *Achillea*. Carnegie Inst. Washington, Publ. No. 581. 129 pp.

Clausen, R. E. 1941. Polyploidy in *Nicotiana*. Amer. Nat., 75: 291–306.

Clausen, R. E., and D. R. Cameron. 1944. Inheritance in *Nicotiana tabacum* XVIII. Monosomic analysis. Genetics, 29: 447–477.

Clausen, R. E., and T. H. Goodspeed. 1925. Interspecific hybridization in *Nicotiana*. II. A tetraploid *glutinosa-Tabacum* hybrid, an experimental verification of Winge's hypothesis. Genetics, 10: 279–284.

Clausen, R. T. 1942. Studies in the Crassulaceae. III. *Sedum*, subgenus *Gormania*, section *Eugormania*. Bull. Torrey Bot. Club, 69: 27–40.

Clausen, R. T., and C. H. Uhl. 1943. Revision of *Sedum cockerellii* and related species. Brittonia, 5: 33–46.

— 1944. The taxonomy and cytology of the subgenus *Gormania* of *Sedum*. Madroño, 7: 161–180.

Cleland, R. E. 1935. Cytotaxonomic studies on certain Oenotheras from California. Proc. Amer. Phil. Soc., 75: 339–429.

— 1936. Some aspects of the cytogenetics of *Oenothera*. Bot. Rev., 2: 316–348.

— 1937. Species relationships in *Onagra*. Proc. Amer. Phil. Soc., 77: 477–542.

— 1940. Analysis of wild American races of *Oenothera* (Onagra). Genetics, 25: 636–644.

— 1942. The origin of *ʰdecipiens* from the complexes of *Oenothera lamarckiana* and its bearing upon the phylogenetic significance of similarities in gene arrangement. Genetics, 27: 55–83.

— 1944. The problem of species in *Oenothera*. Amer. Nat., 78: 5–28.

Coker, W. C., and H. R. Totten. 1934. Trees of the southeastern states. Chapel Hill, Univ. North Carolina Press. 399 pp.

Coleman, L. C. 1941. The relation of chromocenters to differential segments in *Rhoeo discolor* Hance. Amer. Jour. Bot., 28: 742–748.

Collins, J. L. 1921. Reversion in composites. Jour. Hered., 12: 129–133.

— 1924. Inheritance in *Crepis capillaris* (L.) Wallr. III. Nineteen morphological and three physiological characters. Univ. of Calif. Publ. Agr. Sci., 2: 249–296.

Cooper, D. C. 1927. Anatomy and development of the tomato flower. Bot. Gaz., 83: 399–411.

Copeland, H. F. 1938. The kingdoms of organisms. Quart. Rev. Biol., 13: 383–420.

Correns, C., and F. von Wettstein. 1937. Nicht mendelnde Vererbung. Handb. d. Vererbungswiss. Berlin: IIH. 159 pp.

Coulter, J. M. 1885. On the appearance of the relation of ovary and perianth in the development of dicotyledons. Bot. Gaz., 10: 360–363.

Coulter, J. M., C. R. Barnes, and H. C. Cowles. 1910. A textbook of botany. I. Morphology. Chicago, Univ. of Chicago Press. 484 pp.

Covas, G., and B. Schnack. 1944. Tres nuevas especies de *Glandularia* de la flora Argentina. Rev. Argent. Agron., 11: 89–97.

—— 1945. El valor taxonomico de la relación "longitud del pistilo: Volumen del grano de polen." Darwiniana, 7: 80–90.

Crane, M. B. 1940. The origin and behavior of cultivated plants. In J. Huxley, ed., The New Systematics: 529–547.

Crane, M. B., and C. D. Darlington. 1927. The origin of new forms in *Rubus* I. Genetica, 9: 241–277.

Crane, M. B., and D. Lewis. 1942. Genetical studies in pears. III. Incompatibility and sterility. Jour. Genet., 43: 31–43.

Crane, M. B., and P. T. Thomas. 1939. Segregation in asexual (apomictic) offspring in *Rubus*. Nature, 143: 684.

Cross, J. C. 1938. Chromosomes of the genus *Peromyscus* (deer mouse). Cytologia, 8: 408–419.

Cuénot, L. 1911. La genèse des espèces animales. Paris, Félix Alcan. 496 pp.

Danser, B. H. 1929. Über die Begriffe Komparium, Kommiskuum und Konvivium und über die Entstehungsweise der Konvivien. Genetica, 11: 399–450.

Dansereau, P. M. 1939. Monographie du genre *Cistus* L. Thesis No. 1003, Fac. des Sciences, Univ. of Geneva. 90 pp.

—— 1941a. Études sur les hybrides de Cistes. IV. Corrélation des caractères du *C. salviifolius* L. Canad. Jour. Res., 19C: 27–39.

—— 1941b. Études sur les hybrides de Cistes. VI. Introgression dans la section *Ladanum*. Canad. Jour. Res., 19C: 59–67.

—— 1943. Études sur les hybrides de Cistes. V. Le comportement du × *Cistus florentinus* Lam. Candollea, 10: 9–22.

Dansereau, P. M., and Y. Desmarais. 1947. Introgression in sugar maples. II. Amer. Midl. Nat., 37: 146–161.

Dansereau, P. M., and André Lafond. 1941. Introgression des caractères de l'*Acer saccharophorum* K. Koch et de l'*Acer nigrum* Michx. Contr. Inst. Bot., Univ. Montréal, No. 37. 31 pp.

Dark, S. O. S. 1936. Meiosis in diploid and tetraploid *Paeonia* species. Jour. Genet., 32: 353–372.

Darlington, C. D. 1928. Studies in *Prunus*, I and II. Jour. Genet., 19: 213–256.

— 1929. Chromosomal behavior and structural hybridity in the Tradescantiae. Jour. Genet., 21: 207–286.
— 1931. The cytological theory of inheritance in *Oenothera.* Jour. Genet., 24: 405–474.
— 1936a. The limitation of crossing over in *Oenothera.* Jour. Genet., 32: 343–351.
— 1936b. Crossing over and its mechanical relationships in *Chorthippus* and *Stauroderus.* Jour. Genet., 33: 465–500.
— 1936c. The external mechanics of the chromosomes. I–V. Proc. Roy. Soc. B, 121: 264–319.
— 1937. Recent advances in cytology. 2d ed. Blakiston, Philadelphia. 671 pp.
— 1939. The evolution of genetic systems. Cambridge, Cambridge Univ. Press. 149 pp.
— 1940. Taxonomic species and genetic systems. In J. Huxley, ed., The New Systematics: 137–160.
Darlington, C. D., and A. E. Gairdner. 1937. The variation system in *Campanula persicifolia.* Jour. Genet., 35: 97–128.
Darlington, C. D., and E. K. Janaki Ammal. 1945. Chromosome atlas of cultivated plants. London, Allen and Unwin. 397 pp.
Darlington, C. D., and A. A. Moffett. 1930. Primary and secondary chromosome balance in Pyrus. Jour. Genet., 22: 129–151.
Darlington, C. D., and M. B. Upcott. 1941. The activity of inert chromosomes in *Zea mays.* Jour. Genet., 41: 275–296.
Darrow, G. M., and W. H. Camp. 1945. *Vaccinium* hybrids and the development of new horticultural material. Bull. Torrey Bot. Club, 72: 1–21.
Darrow, G. M., W. H. Camp, H. E. Fischer, and H. Dermen. 1944. Chromosome numbers in *Vaccinium* and related groups. Bull. Torrey Bot. Club, 71: 498–506.
Darwin, C. 1862. The various contrivances by which orchids are fertilized by insects. 2d ed. 1877. 300 pp.
Davidson, J. F. 1947. The polygonal graph for simultaneous portrayal of several variables in population analyses. Madroño, 9: 105–110.
Davis, W. T. 1892. Interesting oaks recently discovered on Staten Island. Bull. Torrey Bot. Club, 19: 301–303.
Dawson, C. D. R. 1941. Tetrasomic inheritance in *Lotus corniculatus* L. Jour. Genet., 42: 49–72.
De Beer, G. 1940. Embryos and ancestors. Oxford, Clarendon Press. 108 pp.
Delaunay, L. 1926. Phylogenetische Chromosomenverkürzung. Zeitschr. Zellf. u. mikr. Anat., 4: 338–364.
Demerec, M. 1945a. Genetic aspects of changes in *Staphylococcus aureus* producing strains resistant to various concentrations of penicillin. Ann. Missouri Bot. Gard., 32: 131–138.

—— 1945b. Production of *Staphylococcus* strains resistant to various concentrations of penicillin. Proc. Nat. Acad. Sci., 31: 16–24.

Dermen, H. 1932. Chromosome numbers in the genus *Tilia*. Jour. Arnold Arboretum, 13: 49–51.

—— 1936a. Cytological study and hybridization in two sections of *Verbena*. Cytologia, 7: 160–175.

—— 1936b. Aposporic parthenogenesis in a triploid apple, *Malus hupehensis*. Jour. Arnold Arboretum, 17: 90–105.

Diels, L. 1934. Die flora Australiens und Wegeners Verschiebungs-Theorie. Sitzungsb. Preuss. Akad. Wiss., Berlin: 533–545.

Dienes, L. 1947. Complex reproductive processes in bacteria. Cold Spring Harbor Symp. Quant. Biol., 11: 51–59.

Digby, L. 1912. The cytology of *Primula kewensis* and of other related *Primula* hybrids. Ann. Bot., 26: 357–388.

Dobzhansky, Th. 1924. Die geographische und individuelle Variabilität von *Harmonia axyridis*. Biol. Zentralbl., 44: 401–421.

—— 1933. On the sterility of interracial hybrids in *Drosophila pseudoobscura*. Proc. Nat. Acad. Sci., 11: 950–953.

—— 1940. Speciation as a stage in evolutionary divergence. Amer. Nat., 74: 312–321.

—— 1941. Genetics and the origin of species. Revised ed. New York, Columbia University Press. 446 pp.

—— 1947a. Genetics of natural populations. XIV. A response of certain gene arrangements in the third chromosome of *Drosophila pseudoobscura* to natural selection. Genetics, 32: 142–160.

—— 1947b. Adaptive changes induced by natural selection in wild populations of *Drosophila*. Evolution, 1: 1–16.

Dobzhansky, Th., and C. Epling. 1944. Contributions to the genetics, taxonomy, and ecology of *Drosophila pseudoobscura* and its relatives. Carnegie Inst. Washington, Publ. No. 554. 183 pp.

Dobzhansky, Th., and B. Spassky. 1947. Evolutionary changes in laboratory cultures of *Drosophila pseudoobscura*. Evolution, 1: 191–216.

Dodds, K. S., and N. W. Simmonds. 1948. Genetical and cytological studies of *Musa*. IX. The origin of an edible diploid and the significance of interspecific hybridization in the banana complex. Jour. Genet., 48: 285–296.

Dodge, B. O. 1942. Heterokaryotic vigor in *Neurospora*. Bull. Torrey Bot. Club, 69: 75–91.

Dorf, E. 1942. Upper Cretaceous floras of the Rocky Mountain region. Carnegie Inst. Washington, Publ. No. 508: 1–168.

Douglas, G. E. 1944. The inferior ovary. Bot. Rev., 10: 125–186.

Drygalski, U. V. 1935. Über die Entstehung einer tetraploiden, genetisch ungleichmässigen F_2 aus Kreuzung *Saxifraga adscendens* L. × *Saxifraga tridactylites* L. Zeitschr. Ind. Abst. u. Vererbungsl., 69: 278–300.

Dubinin, N. P., and G. G. Tiniakov. 1945. Seasonal cycles and the concentration of inversions in populations of *Drosophila funebris.* Amer. Nat., 79: 570–572.

—— 1946. Natural selection and chromosomal variability in populations of *Drosophila funebris.* Jour. Hered., 37: 39–44.

Duffield, J. W. 1940. Chromosome counts in *Quercus.* Amer. Jour. Bot., 27: 787–788.

DuRietz, G. E. 1940. Problems of bipolar plant distribution. Acta Phytogeog. Suecica, 13: 215–282.

Dusén, P. 1908. Über die Tertiare Flora der Seymour-Insel. Wiss. Ergeb. Schwed Südpolar-Exped., 3 (3): 1–27.

Dustin, A. P., L. J. Havas, and F. Lits. 1937. Action de la colchicine sur les divisions cellulaires chez les vegetaux. C. R. Assoc. Anatomistes: 1–5. Marseilles.

Eames, A. J. 1931. The vascular anatomy of the flower, with refutation of the theory of carpel polymorphism. Amer. Jour. Bot., 18: 141–188.

—— 1936. Morphology of vascular plants, lower groups. New York, McGraw-Hill. 433 pp.

East, E. M. 1935a. Genetic reactions in *Nicotiana.* II. Phenotypic reaction patterns. Genetics, 20: 414–422.

—— 1935b. Genetic reactions in *Nicotiana.* III. Dominance. Genetics, 20: 443–451.

Edwards, W. N. 1935. The systematic value of cuticular characters in recent and fossil angiosperms. Biol. Rev., 10: 442–459.

Eghis, S. A. 1940. On the problem of the origin of *Nicotiana rustica.* Compt. Rend. (Doklady) Acad. Sci. URSS, 26: 952–956.

Eigsti, O. F. 1942. A cytological investigation of *Polygonatum* using the colchicine pollen tube technique. Amer. Jour. Bot., 29: 626–636.

—— 1947. Colchicine bibliography. Lloydia, 10: 65–114.

Einset, J. 1944. Cytological basis for sterility in induced autotetraploid lettuce (*Lactuca sativa* L.). Amer. Jour. Bot., 31: 336–342.

—— 1947a. Aneuploidy in relation to partial sterility in autotetraploid lettuce (*Lactuca sativa* L.). Amer. Jour. Bot., 34: 99–105.

—— 1947b. Chromosome studies in *Rubus.* Gentes Herbarum, 7: 181–192.

Elias, M. K. 1942. Tertiary prairie grasses and other herbs from the high plains. Spec. Papers Geol. Soc. Amer., No. 41. 176 pp.

Ellerton, S. 1939. The origin and geographical distribution of *Triticum sphaerococcum* Perc. and its cytological behavior in crosses with *T. vulgare* Vill. Jour. Genet., 38: 307–324.

Elton, C. S. 1930. Animal ecology and evolution. Oxford, Clarendon Press. 96 pp.

Emerson, S. 1929. The reduction division in a haploid *Oenothera.* La Cellule, 39: 159–166.

—— 1936. A genetic and cytological analysis of *Oenothera pratincola* and one of its revolute-leaved mutations. Jour. Genet., 32: 315–342.

—— 1947. Growth responses of a sulfonamide-requiring mutant strain of *Neurospora*. Jour. Bacteriol., 54: 195–207.

Emerson, S., and J. E. Cushing. 1946. Altered sulfonamide antagonism in *Neurospora*. Federation Proceedings, 3: 379–389.

Emerson, S., and A. H. Sturtevant. 1932. The linkage relations of certain genes in *Oenothera*. Genetics, 17: 393–412.

Emsweller, S. L., and H. A. Jones. 1935. Meiosis in *Allium fistulosum, Allium cepa,* and their hybrid. Hilgardia, 9: 277–288.

—— 1938. Crossing over, fragmentation, and formation of new chromosomes in an *Allium* species hybrid. Bot. Gaz., 99: 729–772.

—— 1945. Further studies on the chiasmata of the *Allium cepa* × *A. fistulosum* hybrid and its derivatives. Amer. Jour. Bot., 32: 370–379.

Emsweller, S. L., and M. L. Ruttle. 1941. Induced polyploidy in floriculture. Amer. Nat., 75: 310–326.

Engler, A. 1913. Einfluss der Provenienz des Samens auf die Eigenschaften der forstlichen Holzgewächse. Mitt. Schweiz. Centralanst. f. forstliche Versuchswesen, 10: 191–386.

Epling, Carl. 1938. Scilla, Charybdis, and Darwin. Amer. Nat., 72: 547–561.

—— 1939. An approach to classification. Scient. Monthly, 49: 1–8.

—— 1943. Taxonomy and genonomy. Science, 98: 515–516.

—— 1944. The living mosaic. Research Lecture, Univ. of Calif. Press. 26 pp.

—— 1947a. Actual and potential gene flow in natural populations. Amer. Nat., 81: 104–113.

—— 1947b. Natural hybridization of *Salvia apiana* and *S. mellifera.* Evolution, 1: 69–78.

Epling, Carl, and Th. Dobzhansky. 1942. Genetics of natural populations. VI. Microgeographic races in *Linanthus parryae*. Genetics, 27: 317–332.

Epling, Carl, and H. Lewis. 1946. Fertility and natural hybridization in *Delphinium* and its bearing upon gene exchange and the origin of diploid species. (Abstract.) Amer. Jour. Bot., 33: 20s.

Erickson, R. O. 1943. Population size and geographical distribution of *Clematis fremontii* var. *riehlii*. Ann. Missouri Bot. Gard., 30: 63–68.

—— 1945. The *Clematis fremontii* var. *riehlii* population in the Ozarks. Ann. Missouri Bot. Gard., 32: 413–460.

Erlanson, E. W. 1934. Experimental data for a revision of the North American wild roses. Bot. Gaz., 96: 197–259.

Ernst, A. 1918. Bastardierung als Ursache der Apogamie im Pflanzenreich. Jena, G. Fischer. 665 pp.

Esau, K. 1944. Apomixis in guayule. Proc. Nat. Acad. Sci., 30: 352–355.

— 1946. Morphology of reproduction in guayule and certain other species of *Parthenium*. Hilgardia, 17: 61–111.

Ettingshausen, C. von. 1887a. Beiträge zur Kenntniss der Tertiärflora Australiens, zweite Folge. Denks. Ak. Wiss. Wien, 53: 81–142.

— 1887b. Beiträge zur Kenntniss der fossilen flora Neuseelands. Denks. Ak. Wiss. Wien, 53: 143–192.

Ewan, J. 1945. A synopsis of the North American species of *Delphinium*. Univ. of Colorado Studies, Ser. D., 2: 55–244.

Fabergé, A. C. 1936. The physiological consequences of polyploidy. I. Growth and size in the tomato. Jour. Genet., 33: 367–382.

— 1943. Genetics of the *scapiflora* section of *Papaver*. II. The alpine poppy. Jour. Genet., 45: 139–170.

— 1944. Genetics of the *scapiflora* section of *Papaver*. III. Interspecific hybrids and genetic homology. Jour. Genet., 46: 125–149.

Faegri, K. 1937. Some fundamental problems of taxonomy and phylogenetics. Bot. Rev., 3: 400–423.

Fagerlind, F. 1937. Embryologische, zytologische, und bestäubungsexperimentelle Studien in der Familie Rubiaceae nebst Bemerkungen über einige Polyploiditäts-probleme. Acta Horti Bergiani, 11: 195–470.

— 1940a. Die terminologie der apomixis-prozesse. Hereditas, 26: 1–22.

— 1940b. Sind die *canina*-Rosen agamospermische Bastarde? Svensk. Bot. Tidskr., 34: 334–354.

— 1944a. Der Zusammengang zwischen Perennität, Apomixis, und Polyploidie. Hereditas, 30: 179–200.

— 1944b. Is my terminology of the apomictic phenomena of 1940 incorrect and inappropriate? Hereditas, 30: 590–596.

— 1945. Die Bastarde der *canina*-Rosen, ihre Syndese-und Formbildungsverhältnisse. Acta Horti Bergiani, 14: 9–37.

— 1946. Sporogenesis, Embryosackentwicklung, und pseudogame Samenbildung bei *Rudbeckia laciniata* L. Acta Horti Bergiani, 14: 39–90.

Fankhauser, G. 1945a. Maintenance of normal structure in heteroploid salamander larvae through compensation of changes in cell size by adjustment of cell number and cell shape. Jour. Exper. Zool., 100: 445–455.

— 1945b. The effects of changes in chromosome number on amphibian development. Quart. Rev. Biol., 20: 20–78.

Fassett, N. C. 1941. Mass collections: *Rubus odoratus* and *R. parviflorus*. Ann. Missouri Bot. Gard., 28: 299–374.

— 1942. Mass collections: *Diervilla lonicera*. Bull. Torrey Bot. Club, 69: 317–322.

— 1943. The validity of *Juniperus virginiana* var. *crebra*. Amer. Jour. Bot., 30: 469–477.

—— 1944a. *Juniperus virginiana, J. horizontalis,* and *J. scopulorum.* I. The specific characters. Bull. Torrey Bot. Club, 71: 410–418.
—— 1944b. *Juniperus virginiana, J. horizontalis,* and *J. scopulorum.* II. Hybrid swarms of *J. virginiana* and *J. scopulorum.* Bull. Torrey Bot. Club, 71: 475–483.
—— 1945a. *Juniperus virginiana, J. horizontalis,* and *J. scopulorum.* III. Possible hybridization of *J. horizontalis* and *J. scopulorum.* Bull. Torrey Bot. Club, 72: 42–46.
—— 1945b. *Juniperus virginiana, J. horizontalis,* and *J. scopulorum.* IV. Hybrid swarms of *J. virginiana* and *J. horizontalis.* Bull. Torrey Bot. Club, 72: 379–384.
—— 1945c. *Juniperus virginiana, J. horizontalis,* and *J. scopulorum.* V. Taxonomic treatment. Bull. Torrey Bot. Club, 72: 480–482.
Fedorova, N. J. 1946. Crossability and phylogenetic relations in the main European species of *Fragaria.* Compt. Rend. (Doklady) Acad. Sci. URSS, 52: 545–547.
Fernald, M. L. 1925. The persistence of plants in unglaciated areas of Boreal America. Mem. Amer. Acad. Sci., 15: 239–342.
—— 1926. The antiquity and dispersal of vascular plants. Quart. Rev. Biol., 1: 212–245.
—— 1929. Some relationships of the flora of the northern hemisphere. Proc. Int. Congr. Plant Sci., Ithaca, 2: 1487–1507.
—— 1931. Specific segregations and identities in some floras of eastern North America and the Old World. Rhodora, 33: 25–63.
—— 1933. Recent discoveries in the Newfoundland flora. Rhodora, 35: 327–347; 369–386.
—— 1934. Some beginnings of specific differentiation in plants. Science, 79: 573–578.
Fischer, A., and F. Schwanitz. 1936. Die Bedeutung der Polyploidie für die ökologische Anpassung und die Pflanzenzüchtung. Der Züchter, 8: 225–231.
Fisher, R. A. 1930. The genetical theory of natural selection. Oxford, Clarendon Press. 272 pp.
—— 1936. The use of multiple measurements in taxonomic problems. Ann. Eugenics, 7 (2): 179–188.
Florell, V. H. 1931. A cytologic study of wheat × rye hybrids and backcrosses. Jour. Agr. Res., 42: 341–362.
Florin, R. 1932. Die Chromosomenzahlen bei *Welwitschia* und einigen *Ephedra*-Arten. Svensk Bot. Tidskr., 26: 205–214.
—— 1933. Studien über die Cycadales des Mesozoikums. K. Svenska Vetensk. Akad. Handl. (3)12, No. 5: 1–134.
—— 1936. Die fossilen Ginkgophyten von Franz-Joseph-Land nebst Erörterungen über vermeintliche Cordaitales mesozoischen Alters. Paleontographica, 81(B): 71–173; *ibid.*, 82(B): 1–72.
—— 1940. The Tertiary fossil conifers of South Chile and their phytogeographical significance, with a review of the fossil conifers of

southern lands. K. Svenska Vetensk. Akad. Handl. Ser. 3, 19(2): 1–107.

—— 1944a. Die Koniferen des Oberkarbons und des unteren Perms. Vol. 6. Paleontographica, 85(B): 365–456.

—— 1944b. Die Koniferen des Oberkarbons und des unteren Perms. Vol. 7. Paleontographica, 85(B): 457–654.

Flory, W. S. 1936. Chromosome numbers and phylogeny in the gymnosperms. Jour. Arnold Arboretum, 17: 83–89.

—— 1937. Chromosome numbers in the Polemoniaceae. Cytologia, Fujii Jubil. Vol.: 171–180.

Flovik, K. 1938. Cytological studies of arctic grasses. Hereditas, 24: 265–376.

—— 1940. Chromosome numbers and polyploidy within the flora of Spitzbergen. Hereditas, 26: 430–440.

Foster, A. S. 1932a. Investigations on the morphology and comparative history of development of foliar organs. III. Cataphyll and foliage-leaf ontogeny in the black hickory (*Carya buckleyi* var. *arkansana*). Amer. Jour. Bot., 19: 75–99.

—— 1932b. Investigations on the morphology and comparative history of development of foliar organs. IV. The prophyll of *Carya buckleyi* var. *arkansana*. Amer. Jour. Bot., 19: 710–728.

—— 1935. A histogenetic study of foliar determination in *Carya buckleyi* var. *arkansana*. Amer. Jour. Bot., 22: 88–131.

—— 1936. Leaf differentiation in angiosperms. Bot. Rev., 2: 349–372.

—— 1944. Structure and development of sclereids in the petiole of *Camellia japonica* L. Bull. Torrey Bot. Club, 71: 302–326.

—— 1945. The foliar sclereids of *Trochodendron aralioides* Sieb. et Zucc. Jour. Arnold Arboretum, 26: 155–162.

—— 1946. Comparative morphology of the foliar sclereids in the genus *Mouriria* Aubl. Jour. Arnold Arboretum, 27: 253–271.

Foster, A. S., and F. A. Barkley. 1933. Organization and development of foliar organs in *Paeonia officinalis*. Amer. Jour. Bot., 20: 365–385.

Foster, R. C. 1937. A cytotaxonomic survey of the North American species of *Iris*. Contr. Gray Herb., 119: 1–82.

Frandsen, K. J. 1943. The experimental formation of *Brassica juncea* Czern. et Coss. Dansk. Botanisk Arkiv, 11 (4): 1–17.

Frankel, O. H. 1947. The theory of plant breeding for yield. Heredity, 1: 109–120.

Fritsch, F. E. 1916. The algal ancestry of the higher plants. New Phytologist, 5: 233–250.

—— 1935. The structure and reproduction of the Algae. Vol. 1. London and New York, Macmillan. 789 pp.

Frost, H. B. 1926. Polyembryony, heterozygosis, and chimeras in *Citrus*. Hilgardia, 1: 365–402.

—— 1938. Nucellar embryony and juvenile characters in clonal varieties of *Citrus*. Jour. Hered., 29: 423–432.

Gairdner, A. E., and C. D. Darlington. 1931. Ring formation in diploid and polyploid *Campanula persicifolia*. Genetica, 13: 113–150.
Gajewski, W. 1946. Cytogenetic investigations on *Anemone* L. I. *Anemone janczewskii*, a new amphidiploid species of hybrid origin. Acta Soc. Bot. Polon., 17: 129–194.
Ganong, W. F. 1901. The cardinal principles of morphology. Bot. Gaz., 31: 426–434.
Garber, E. D. 1947. The pachytene chromosomes of *Sorghum intrans*. Jour. Hered., 38: 251–252.
Gard, M. 1912. Recherches sur les hybrides artificiels de Cistes obtenus par Ed. Bornet. II. Les espèces et les hybrides binaires. Beih. Bot. Centralbl., 29 (2): 306–394.
Gardner, E. J. 1946. Sexual plants with high chromosome number from an individual plant selection in a natural population of guayule and mariola. Genetics, 31: 117–124.
Gates, R. R. 1909. The stature and chromosomes of *Oenothera gigas*, De Vries. Arch. f. Zellforsch., 3: 525–552.
—— 1942. Nucleoli and related nuclear structures. Bot. Rev., 8: 337–409.
Gates, R. R., and D. G. Catcheside. 1932. Gamolysis of various new *Oenotheras*. Jour. Genet., 26: 143–178.
Gates, R. R., and H. K. Nandi. 1935. The cytology of trisomic mutations in a wild species of *Oenothera*. Phil. Trans. Roy. Soc. London Ser. B., No. 524: 227–254.
Geitler, L. 1937. Cytogenetische Untersuchungen an natürlichen Populationen von *Paris quadrifolia*. Zeitschr. Ind. Abst. u. Vererbungsl., 73: 182–197.
—— 1938. Weitere cytogenetische Untersuchungen an natürlichen Populationen von *Paris quadrifolia*. Zeitschr. Ind. Abst. u. Vererbungsl., 75: 161–190.
Gentcheff, G., and A. Gustafsson. 1940. Parthenogenesis and pseudogamy in *Potentilla*. Bot. Not. (Lund), 1940: 109–132.
Gerassimova, H. 1939. Chromosome alterations as a factor of divergence of forms. I. New experimentally produced strains of *C. tectorum* which are physiologically isolated from the original forms owing to reciprocal translocation. Compt. Rend. Acad. Sci. URSS, 25: 148–154.
Gershenson, S. 1945a. Evolutionary studies on the distribution and dynamics of melanism in the hamster (*Cricetus cricetus* L.). I. Distribution of black hamsters in the Ukrainian and Bashkirian Soviet Socialist Republics. Genetics, 30: 207–232.
—— 1945b. Evolutionary studies on the distribution and dynamics of melanism in the hamster (*Cricetus cricetus* L.). II. Seasonal and annual changes in the frequency of black hamsters. Genetics, 30: 233–251.
Gershoy, A. 1928. Studies in North American violets. I. General considerations. Vermont Agr. Expt. Sta., Bull. No. 279. 18 pp.

—— 1932. Descriptive notes for *Viola* exhibit. The *Nominium* and *Chamaemelanium* sections. Sixth Int. Congr. Genetics, Ithaca, N.Y., Publ. Vermont Agr. Expt. Sta. 27 pp.

—— 1934. Studies in North American violets. III. Chromosome numbers and species characteristics. Bull. Vermont Agr. Expt. Sta., 367: 1–91.

Gerstel, D. U. 1943. Inheritance in *Nicotiana tabacum*. XVII. Cytogenetical analysis of *glutinosa*-type resistance to mosaic disease. Genetics, 28: 533–536.

—— 1945a. Inheritance in *Nicotiana tabacum*. XIX. Identification of the *tabacum* chromosome replaced by one from *N. glutinosa* in mosaic-resistant Holmes Samsoun tobacco. Genetics, 30: 448–454.

—— 1945b. Inheritance in *Nicotiana tabacum*. XX. The addition of *Nicotiana glutinosa* chromosomes to tobacco. Jour. Hered., 36: 197–206.

Ghimpu, V. 1929. Sur les chromosomes de quelques chênes. Rev. Bot. Appl. et Agric. Tropic., 9: 175–178.

—— 1930. Recherches cytologiques sur les genres *Hordeum, Acacia, Medicago, Vitis,* et *Quercus*. Archives d'Anatomie Microscopique, 26: 135–234.

Giles, N. 1940. Spontaneous chromosome aberrations in *Tradescantia*. Genetics, 25: 69–87.

—— 1941. Chromosome behavior at meiosis in triploid *Tradescantia* hybrids. Bull. Torrey Bot. Club, 68: 207–221.

Giles, N. H. 1942. Autopolyploidy and geographical distribution in *Cuthbertia graminea* Small. Amer. Jour. Bot., 29: 637–645.

Goebel, K. 1928. Organographie der Pflanzen, dritte Auflage, I. Jena, G. Fischer. 1,861 pp.

—— 1933. Organographie der Pflanzen, dritte Auflage, III. Jena, G. Fischer, pp. 1379–2078.

Goldschmidt, R. 1933. Some aspects of evolution. Science, n.s., 78: 539–547.

—— 1940. The material basis of evolution. New Haven, Yale Univ. Press. 436 pp.

Goodale, H. D. 1942. Further progress with artificial selection. Amer. Nat., 76: 515–519.

Goodspeed, T. H. 1934. *Nicotiana* phylesis in the light of chromosome number, morphology, and behavior. Univ. of Calif. Publ. Bot., 17 (13): 369–398.

—— 1944. *Nicotiana arentsii*. A new, naturally occurring amphidiploid species. Proc. Calif. Acad. Sci., 25(12): 291–306.

—— 1945. Cytotaxonomy of *Nicotiana*. Bot. Rev., 11: 533–592.

—— 1947. On the evolution of the genus *Nicotiana*. Proc. Nat. Acad. Sci., 33: 158–171.

Goodspeed, T. H., and P. Avery. 1939. Trisomic and other types in *Nicotiana sylvestris*. Jour. Genet., 38: 381–458.

Goodspeed, T. H., and R. E. Clausen. 1928. Interspecific hybridization in Nicotiana. VIII. The *sylvestris-tomentosa-tabacum* triangle and its bearing on the origin of *Tabacum*. Univ. of Calif. Publ. Bot., 11: 245–256.

Goodwin, R. H. 1937a. The cytogenetics of two species of *Solidago* and its bearing on their polymorphy in nature. Amer. Jour. Bot., 24: 425–432.

— 1937b. Notes on the distribution and hybrid origin of × *Solidago asperula*. Rhodora, 38: 22–28.

— 1941. The selective effect of climate on the flowering behavior of *Solidago sempervirens* L. Proc. Rochester Acad. Sci., 8: 22–27.

— 1944. The inheritance of flowering time in a short-day species, *Solidago sempervirens* L. Genetics, 29: 503–519.

Gould, F. W. 1945. Notes on the genus *Elymus*. Madroño, 8: 42–48.

Grassl, C. O. 1946. *Saccharum robustum* and other wild relatives of "noble" sugar canes. Jour. Arnold Arboretum, 27: 234–252.

Greenleaf, W. 1941. Sterile and fertile amphidiploids: their possible relation to the origin of *Nicotiana tabacum*. Genetics, 26: 301–324.

— 1942. Genic sterility in *tabacum*-like amphidiploids of *Nicotiana*. Jour. Genet., 43: 69–96.

Grégoire, V. 1938. La morphogénèse et l'autonomie morphologique de l'appareil floral. La Cellule, 47: 287–452.

Gregor, J. W. 1931. Experimental delimitation of species. New Phytologist, 30: 204–217.

— 1938a. Experimental taxonomy. II. Initial population differentiation in *Plantago maritima* L. of Britain. New Phytologist, 37: 15–49.

— 1938b. Reflections concerning new crop varieties. Herbage Reviews, 6: 234–239.

— 1939. Experimental taxonomy. IV. Population differentiation in North American and European sea plantains allied to *Plantago maritima* L. New Phytologist, 38: 293–322.

— 1942. The units of experimental taxonomy. Chronica Botanica, 7: 193–196.

— 1944. The ecotype. Biol. Rev., 19: 20–30.

— 1946. Ecotypic differentiation. New Phytologist, 45: 254–270.

Gregor, J. W., V. McM. Davey, and J. M. S. Lang. 1936. Experimental taxonomy. I. Experimental garden technique in relation to the recognition of small taxonomic units. New Phytologist, 35: 323–350.

Gregor, J. W., and F. W. Sansome. 1930. Experiments on the genetics of wild populations. II. *Phleum pratense* and the hybrid *P. pratense* × *P. alpinum* L. Jour. Genet., 22: 373–387.

Gregory, W. C. 1941. Phylogenetic and cytological studies in the Ranunculaceae Juss. Trans. Amer. Phil. Soc., n.s. 31: 443–520.

Griggs, R. F. 1937. Hybridity as a factor in evolution. Jour. Washington Acad. Sci., 27: 329–343.

Gustafsson, A. 1931a. Sind die *canina*-Rosen apomiktisch? Bot. Not. (Lund), 1931: 21–30.

—— 1931b. Weitere Kastrierungsversuche in der Gattung *Rosa*. Bot. Not. (Lund), 1931: 350–354.

—— 1932. Zytologische und experimentelle Studien in der Gattung *Taraxacum*. Hereditas, 16: 41–62.

—— 1935a. Studies on the mechanism of parthenogenesis. Hereditas, 21: 1–112.

—— 1935b. The importance of apomicts for plant geography. Bot. Not. (5): 325–330.

—— 1941a. Mutation experiments in barley. Hereditas, 27: 225–242.

—— 1941b. Preliminary yield experiments with ten induced mutations in barley. Hereditas, 27: 337–359.

—— 1942. The origin and properties of the European blackberry flora. Hereditas, 28: 249–277.

—— 1943. The genesis of the European blackberry flora. Lund. Kgl. Fysiogr. Sällsk. Handlingar, n.s. 54, No. 6. 200 pp.

—— 1944. The constitution of the *Rosa canina* complex. Hereditas, 30: 405–428.

—— 1946. Apomixis in the higher plants. I. The mechanism of apomixis. Lunds Univ. Arsskr. N. F. Avd. 2, 42 (3): 1–66.

—— 1947a. Mutations in agricultural plants. Hereditas, 33: 1–100.

—— 1947b. Apomixis in higher plants. II. The causal aspect of apomixis. Lunds Univ. Arsskr. N. F. Avd. 2, 43 (2): 71–178.

—— 1947c. Apomixis in higher plants. III. Biotype and species formation. Lunds Univ. Arsskr., 44 (2): 183–370.

—— 1948. Polyploidy, life-form, and vegetative reproduction. Hereditas, 34: 1–22.

Gustafsson, A., and A. Håkansson. 1942. Meiosis in some *Rosa*-hybrids. Bot. Not.: 331–343.

Haan, H. R. M. de. 1920. Contribution to the knowledge of the morphological value and the phylogeny of the ovule and its integuments. Rec. Trav. Bot. Neerl., 17: 217–324.

Hagedoorn, A. C., and A. L. Hagedoorn. 1921. The relative value of the processes causing evolution. The Hague, Martinus Nijhoff. 294 pp.

Hagerup, O. 1927. *Empetrum hermaphroditicum* (Lge.) Hagerup, a new tetraploid, bisexual species. Dansk Bot. Arkiv, 5: 1–17.

—— 1932. Über Polyploidie in Beziehung zu Klima, Ökologie, und Phylogenie. Hereditas, 16: 19–40.

—— 1933. Studies on polyploid ecotypes in *Vaccinium uliginosum* L. Hereditas, 18: 122–128.

—— 1940. Studies on the significance of polyploidy. IV. *Oxycoccus*. Hereditas, 26: 399–410.

—— 1944. On fertilization, polyploidy, and haploidy in *Orchis maculatus* L. (*sens. lat.*). Dansk Bot. Arkiv, 11 (5): 1–25.

Håkansson, A. 1940. Die Meiosis bei haploiden Pflanzen von *Godetia whitneyi*. Hereditas, 26: 411–429.

— 1941. Zur Zytologie von *Godetia* – Arten und Bastarden. Hereditas, 27: 319–336.

— 1942. Zytologische Studien an Rassen und Rassenbastarden von *Godetia whitneyi* und verwandten Arten. Lunds Univ. Arrskr. (Andra Med.) N. F. 2, 38(5): 1–69.

— 1943a. Die Entwicklung des Embryosacks und die Befruchtung bei *Poa alpina*. Hereditas, 29: 25–61.

— 1943b. Meiosis in a hybrid with one set of large and one set of small chromosomes. Hereditas, 29: 461–474.

— 1944a. Erganzende Beiträge zur Embryologie von *Poa alpina*. Bot. Not.: 299–311.

— 1944b. Studies on a peculiar chromosome condition in *Godetia whitneyi*. Hereditas, 30: 597–612.

— 1946a. Meiosis in hybrid nullisomics and certain other forms of *Godetia whitneyi*. Hereditas, 32: 495–513.

— 1947. Contributions to a cytological analysis of the species differences of *Godetia amoena* and *G. whitneyi*. Hereditas, 33: 235–260.

Haldane, J. B. S. 1930. Theoretical genetics of autopolyploids. Jour. Genet., 32: 359–372.

— 1932a. The causes of evolution. New York and London, Harper. 234 pp.

— 1932b. The time of action of genes, and its bearing on some evolutionary problems. Amer. Nat., 66: 5–24.

Hannah, M. 1916. A comparative account of epigyny in certain monocotyledons and dicotyledons. Trans. Amer. Micros. Soc., 35: 207–220.

Hansen, H. N. 1938. The dual phenomenon in imperfect fungi. Mycologia, 30: 442–455.

Hansen, H. N., and R. E. Smith. 1932. The mechanism of variation in imperfect fungi. *Botrytis cinerea*. Phytopathology, 22: 953–964.

— 1935. The origin of new types of imperfect fungi from interspecific co-cultures. Zentralbl. f. Bakt., Parasitenk. u. Infektionskrankheiten II Abt., 92: 272–279.

Harlan, H. V., and M. L. Martini. 1938. The effect of natural selection on a mixture of barley varieties. Jour. Agr. Res., 57: 189–199.

Harlan, J. R. 1945a. Cleistogamy and chasmogamy in *Bromus carinatus*. Hook. and Arn. Amer. Jour. Bot., 32: 66–72.

— 1945b. Natural breeding structure in the *Bromus carinatus* complex as determined by population analyses. Amer. Jour. Bot., 32: 142–148.

Harland, S. C. 1934. The genetical conception of the species. Tropical Agriculture, 11: 51–53.

— 1936. The genetical conception of the species. Biol. Rev., 11: 83–112.

—— 1937. The genetics of cotton. XVII. Increased mutability of a gene in *G. purpurascens* as a consequence of hybridization with *G. hirsutum*. Jour. Genet., 34: 153–168.

Harris, T. M. 1941. Cones of extinct Cycadales from the Jurassic rocks of Yorkshire. Phil. Trans. Roy. Soc. London, Ser. B., 231: 75–98.

Hartung, M. 1946. Chromosome numbers in *Poa, Agropyron* and *Elymus*. Amer. Jour. Bot., 33: 516–531.

Haustein, E. 1939a. Die Chromosomenanordnung bei einigen Bastarden der *Oenothera argentinea*. Zeitschr. Ind. Abst. u. Vererbungsl., 76: 411–421.

—— 1939b. Die Analyse der *Oenothera brachycephala*. Zeitschr. Ind. Abst. u. Vererbungsl., 76: 487–511.

Hayes, H. K., and R. J. Garber. 1927. Breeding crop plants. 2d ed. New York, McGraw-Hill. 438 pp.

Hayes, H. K., and F. R. Immer. 1942. Methods of plant breeding. New York and London, McGraw-Hill. 432 pp.

Hecht, A. 1941. Cytogenetic studies of *Oenothera*, subgenus *Raimannia*. (Abstract.) Amer. Jour. Bot., 28: 3s.

Hegi, G. 1906. Illustrierte Flora von Mittel-europa. Vol. IV. Munich, J. F. Lehmanns.

Heilborn, O. 1924. Chromosome numbers and dimensions, species-formation and phylogeny in the genus *Carex*. Hereditas, 5: 129–216.

—— 1928. Chromosome studies in Cyperaceae. Hereditas, 11: 182–191.

—— 1932. Aneuploidy and polyploidy in *Carex*. Svensk Bot. Tidskr., 26: 137–146.

—— 1936. The mechanics of so-called secondary association between chromosomes. Hereditas, 22: 167–188.

—— 1939. Chromosome studies in Cyperaceae. III-IV. Hereditas, 25: 224–240.

Heilbronn, A. 1932. Polyploidie und Generationswechsel. Ber. Deut. Bot. Ges., 50: 289–300.

Heiser, C. B., Jr. 1947. Hybridization between the sunflower species *Helianthus annuus* and *H. petiolaris*. Evolution, I: 249–262.

Heiser, C. B., Jr., and T. W. Whitaker. 1948. Chromosome number and growth habit in California weeds. Amer. Jour. Bot., 35: 179–186.

Heitz, E. 1931. Die Ursache der gesetzmässigen Zahl, Lage, Form, und Grösse pflanzlicher Nucleolen. Planta, 12: 774–844.

—— 1932. Die Herkunft der Chromocentren. Planta, 18: 571–636.

—— 1940. Durch Röntgenstrahlen ausgelöste Mutationen bei *Pellia neesiana*. Verhandl. Schweizer. Naturf. Ges., 1940: 170–171.

—— 1942. Über die Beziehung zwischen Polyploidie und Gemischtgeschlechtlichkeit bei Moosen. Arch. Julius Klaus Stift. f. Vererbungsforsch., etc., 17: 444–448.

—— 1944. Über einige Frage der Artbildung. Arch. Julius Klaus-Stift. f. Vererbungsf., etc., 19: 510–528.

Henry, A. 1910. On elm seedlings showing Mendelian results. Jour. Linn. Soc., 39: 290–300.

Heribert-Nilsson, N. 1918. Experimentelle Untersuchungen über Variabilität, Spaltung, Artbildung, und Evolution der Gattung *Salix*. Lunds. Univ. Arsskr. N.F., Avd., 2; 14, No. 28. 145 pp.

Hesse, R. 1938. Vergleichende Untersuchungen an diploiden und tetraploiden Petunien. Zeitschr. Ind. Abst. u. Vererbungsl., 75: 1–23.

Hiesey, W. M. 1940. Environmental influence and transplant experiments. Bot. Rev., 6: 181–203.

Hiesey, W. M., J. Clausen, and D. D. Keck. 1942. Relations between climate and intraspecific variability in plants. Amer. Nat., 76: 5–22.

Hill, A. W. 1929. Antarctica and problems of distribution. Proc. Int. Congr. Plant Sci., Ithaca, N.Y.: 1477–1486.

Hiorth, G. 1934. Genetische Versuche mit *Collinsia*. IV. Die Analyse eines nahezu sterilen Artbastardes. Pt. 1. Die diploiden Bastarde zwischen *Collinsia bicolor* und *C. bartsiaefolia*. Zeitschr. Ind. Abst. u. Vererbungsl., 66: 106–157.

—— 1941. Zur Genetik und Systematik der Gattung *Godetia*. Zeitschr. Ind. Abst. u. Vererbungsl., 79(2): 199–219.

—— 1942. Zur Genetik und Systematik der *amoena*-Gruppe der Gattung *Godetia*. Zeitschr. Ind. Abst. u. Vererbungsl., 80: 289–349.

Hitchcock, A. S., and A. Chase. 1931. Grass. In old and new plant lore. Smithsonian Science Series, 11: 201–250.

Hoar, C. S. 1931. Meiosis in *Hypericum punctatum* Lam. Bot. Gaz., 92: 396–406.

Hodgson, R. W., and S. H. Cameron. 1938. Effects of reproduction by nucellar embryony on clonal characteristics of *Citrus*. Jour. Hered., 29: 417–419.

Hoeg, E. 1929. Om Mellenformerne mellem *Quercus robur* L. og *Quercus sessiliflora* Martyn. Bot. Tidsskr., 40: 411–427.

Hoel, P. G. 1947. Introduction to mathematical statistics. New York, Wiley. See pp. 121–126.

Hofelich, A. 1935. Die Sektion Alsinebe Griseb. der Gattung *Veronica* in ihren chromosomalen Grundlagen. Jahrb. Wiss. Bot., 81: 541–572.

Hollick, A. 1936. Tertiary floras of Alaska. U. S. Geol. Survey, Prof. Paper, 182: 1–185.

Hollingshead, L. 1930a. A lethal factor in *Crepis* effective only in an interspecific hybrid. Genetics, 15: 114–140.

—— 1930b. Cytological investigations of hybrids and hybrid derivatives of *Crepis capillaris* and *Crepis tectorum*. Univ. of Calif. Publ. Agr. Sci., 6: 55–94.

—— 1942. Chromosome studies in *Sedum,* subgenus, *Gormania,* section *Eugormania*. Bull. Torrey Bot. Club, 69: 27–40.

Holmes, F. O. 1938. Inheritance of resistance to tobacco-mosaic disease in tobacco. Phytopathology, 28: 553–561.

Horn, C. L. 1940. Existence of only one variety of cultivated mangosteen explained by asexually formed "seed." Science, 92: 237–238.

Houghtaling, H. B. 1935. A developmental analysis of size and shape in tomato fruits. Bull. Torrey Bot. Club, 62: 243–252.

Howard, H. W. 1942. Self-incompatibility in polyploid forms of *Brassica* and *Raphanus*. Nature, 149: 302.

Howard, H. W., and I. Manton. 1940. Allopolyploid nature of the wild tetraploid watercress. Nature, 146: 303–304.

—— 1946. Autopolyploid and allopolyploid watercress (*Nasturtium*) with the description of a new species. Ann. Bot., n.s. 10: 1–13.

Hu, H. H., and R. W. Chaney. 1940. A Miocene flora from Shantung province, China. Carnegie Inst. Washington, Publ. No. 507. 147 pp.

Hubbs, C. L., and L. C. Hubbs. 1932. The increased growth, predominant maleness, and apparent infertility of hybrid sunfishes. Papers Michigan Acad. Sci., Arts and Letters, 17: 613–641.

Hubbs, C. L., and K. Kuronuma. 1942. Hybridization in nature between two genera of flounders in Japan. Papers Michigan Acad. Sci., Arts and Letters, 27: 267–306.

Hubbs, C. L., and R. R. Miller. 1943. Mass hybridization between two genera of Cyprinid fishes in the Mohave Desert, California. Papers Michigan Acad. Sci., Arts and Letters, 28: 343–378.

Hubbs, C. L., B. W. Walker, and R. E. Johnson. 1943. Hybridization in nature between species of American cyprinodont fishes. Contr. Lab. Vert. Biol. Univ. of Michigan, No. 23: 1–21.

Hultén, E. 1937. Outline of the history of arctic and boreal biota during the Quaternary period. Stockholm. 168 pp.

Hurst, C. C. 1925. Chromosomes and characters in *Rosa* and their significance in the origin of species. Experiments in Genetics, 38: 534–550. Cambridge Univ. Press.

—— 1928. Differential polyploidy in the genus *Rosa*. Zeitschr. Ind. Abst. u. Vererbungsl., Suppl.: 866–906.

—— 1932. The mechanism of creative evolution. New York, Macmillan. 365 pp.

Huskins, C. L. 1931. The origin of *Spartina townsendii*. Genetica, 12: 531–538.

—— 1941. Polyploidy and mutations. Amer. Nat., 75: 329–344.

—— 1946. Fatuoid, speltoid, and related mutations of oats and wheat. Bot. Rev., 12: 457–514.

Huskins, C. L., and G. B. Wilson. 1938. Probable causes of the changes in direction of the major spiral in *Trillium erectum* L. Ann. Bot., n.s. 2: 281–292.

Husted, L. 1936. Cytological studies on the peanut, *Arachis*. II. Chromosome number, morphology, and behavior, and their application to the origin of the cultivated forms. Cytologia, 7: 396–423.

Hutchinson, A. H. 1936. The polygonal presentation of polyphase phenomena. Trans. Roy. Soc. Canada, 3d Ser., Sec. 5, 30: 19–26.

Hutchinson, J. 1926. The families of flowering plants. I. Dicotyledons. London, Macmillan. 328 pp.
— 1934. The families of flowering plants. II. Monocotyledons. London, Macmillan. 243 pp.
Hutchinson, J. B. 1940. The application of genetics to plant breeding. Jour. Genet., 40: 271–282.
Hutchinson, J. B., and S. G. Stephens. 1947. The evolution of *Gossypium*. London and New York, Oxford Univ. Press. 160 pp.
Huxley, J. S. 1932. Problems of relative growth. London, Methuen. 276 pp.
— 1938. Clines: an auxiliary taxonomic principle. Nature, 142: 219.
— 1939. Clines: an auxiliary method in taxonomy. Bidr. tot de Dierkunde, 27: 491–520.
— (editor). 1940. The new systematics. Oxford, Clarendon Press. 334 pp.
— 1942. Evolution: the modern synthesis. New York, Harper. 645 pp.

Irwin, M. R. 1938. Immunogenetic studies of species relationships in the Columbidae. Jour. Genet., 35: 351–373.
Ishikawa, M. 1911. Cytologische Studien von Dahlien. Bot. Mag. Tokyo, 25: 1–8.
Ivanovskaja, E. V. 1939. A haploid plant of *Solanum tuberosum* L. Compt. Rend. Acad. Sci. URSS, 24: 517–520.

Jackson, H. S. 1931. Present evolutionary tendencies and the origin of life cycles in the Uredinales. Mem. Torrey Bot. Club, 18: 1–108.
Jacob, K. T. 1940. Chromosome numbers and the relationship between satellites and nucleoli in *Cassia* and certain other Leguminosae. Ann. Bot., n.s. 4: 201–226.
Japha, B. 1939. Die Meiosis von *Oenothera*. II. Zeitschr. f. Bot., 34: 321–369.
Jaretzky, R. 1928. Histologische und karyologische Studien an Polygonaceen. Jahrb. Wiss. Bot., 69: 357–490.
— 1930. Zur Zytologie der Fagales. Planta, 10: 120–137.
— 1932. Beziehungen zwischen Chromosomenzahl und Systematik bei den Cruciferen. Jahrb. Wiss. Bot., 76: 485–527.
Jeffrey, E. C. 1915. Some fundamental morphological objections to the mutation theory of De Vries. Amer. Nat., 49: 5–21.
— 1916. The anatomy of woody plants. Chicago, University of Chicago Press. 478 pp.
Jenkin, T. J. 1931. The method and technique of selection breeding and strain-building in grasses. Bull. Bur. Plant Genet. Aberystwyth, No. 3: 5–34.
— 1933. Interspecific and intergeneric hybrids in herbage grasses, initial crosses. Jour. Genet., 28: 205–264.

—— 1936. Natural selection in relation to grasses. Proc. Roy. Soc. London, Ser. B., 121: 52–56.

Jenkins, J. A. 1939. The cytogenetic relationships of four species of *Crepis*. Univ. of Calif. Publ. Agr. Sci., 6: 369–400.

Jepson, W. L. 1925. A manual of the flowering plants of California. Berkeley, Univ. of California. 1,238 pp.

—— 1939. A flora of California, Vol. 3, Pt. 1. Associated Students Store, Univ. of California.

Johannsen, W. 1926. Elemente der exakten Erblichkeitslehre. 3d ed. Jena, G. Fischer. 735 pp.

Johansen, D. A. 1933. Cytology of the tribe Madineae, family Compositae. Bot. Gaz., 95: 177–208.

Johnson, A. M. 1936. Polyembryony in *Eugenia hookeri*. Amer. Jour. Bot., 23: 83–85.

Johnson, B. L. 1945. Cytotaxonomic studies in *Oryzopsis*. Bot. Gaz., 107: 1–32.

Johnson, L. P. V. 1939. A descriptive list of natural and artificial interspecific hybrids in North American forest tree genera. Canad. Jour. Res., C, 17: 411–444.

—— 1942. Studies on the relation of growth rate to wood quality in *Populus* hybrids. Canad. Jour. Res., C, 20: 28–40.

—— 1947. A note on inheritance in F_1 and F_2 hybrids of *Populus alba* L. × *P. grandidentata* Michx. Canad. Jour. Res., C, 24: 313–317.

Jones, H. A., and A. E. Clarke. 1942. A natural amphidiploid from an onion species hybrid. Jour. Hered., 33: 25–32.

Joranson, P. N. 1944. The cytogenetics of hybrids, autotetraploids, and allotetraploids in the grass genus *Melica* L. Thesis (Ph.D.), Univ. of Calif.

Jordan, D. S. 1905. The origin of species through isolation. Science, n.s. 22: 545–562.

Jorgenson, C. A. 1928. The experimental formation of heteroploid plants in the genus *Solanum*. Jour. Genet., 19: 133–271.

Juel, O. 1918. Beiträge zur Blütenanatomie und Systematik der Rosaceen. K. Svensk Vet-Ab. Handl., 58, No. 5. 80 pp.

Juliano, J. B. 1934. Origin of embryos in the strawberry mango. Philippine Jour. Sci., 54: 553–556.

—— 1937. Embryos of Carabao mango (*Mangifera indica* L.). Philippine Agr., 25: 749–760.

Juliano, J. B., and N. L. Cuevas. 1932. Floral morphology of the mango (*Mangifera indica* L.) with special reference to the pico-variety from the Philippines. Philippine Agr., 21: 449–472.

Just, T. 1947. Geology and plant distribution. Ecological Monographs, 17: 127–137.

Karpechenko, G. D. 1927. Polyploid hybrids of *Raphanus sativus* L.

× *Brassica oleracea* L. Bull. Appl. Bot., Genet. Plant Breed., 17: 305–410.
— 1928. Polyploid hybrids of *Raphanus sativus* L. × *Brassica oleracea* L. Zeitschr. Ind. Abst. u. Vererbungsl., 39: 1–7.
Katayama, Y. 1935. Further investigations on synthesized octoploid *Aegilotriticum*. Jour. Coll. Agr. Tokyo Imp. Univ., 13: 397–414.
Kattermann, G. 1938a. Über konstante, halmbehaarte Stämme aus Weizenroggen bastardierung mit $2n = 42$ Chromosomen. Zeitschr. Ind. Abst. u. Vererbungsl., 74: 354–375.
— 1938b. Zur Kenntnis der strukturellen Hybriden von *Briza media* II and III. Planta, 27: 669–679.
Keck, D. D. 1935. Studies upon the taxonomy of the Madinae. Madroño, 3: 4–18.
— 1946. A revision of the *Artemisia vulgaris* complex in North America. Proc. Calif. Acad. Sci., 4th Ser., 25: 421–468.
Kemp, W. B. 1937. Natural selection within plant species as exemplified in a permanent pasture. Jour. Hered., 28: 329–333.
Kern, E. M., and C. Alper. 1945. Multidimensional graphical representation for analyzing variation in quantitative characters. Ann. Missouri Bot. Gard., 32: 279–286.
Kiellander, C. L. 1942. A subhaploid *Poa pratensis* L. and its progeny. Svensk Bot. Tidskr., 36: 200–220.
Kihara, H. 1924. Cytologische und genetische Studien bei wichtigen Getreiden Arten mit besonderer Rücksicht auf das Verhalten der Chromosomen und die Sterilität der Bastarden. Mem. Coll. Sci., Kyoto Imp. Univ., Ser. B., V, 1: 1–200.
— 1940. Verwandtschaft der *Aegilops*-Arten im Lichte der Genomanalyse. Ein Überlick. Der Züchter, 12: 49–62.
King, J. R., and R. Bamford. 1937. The chromosome number in *Ipomoea* and related genera. Jour. Hered., 28: 278–282.
Knowles, P. F. 1943. Improving an annual brome grass, *Bromus mollis* L., for range purposes. Jour. Amer. Soc. Agron., 35: 584–594.
Koller, P. C. 1932. Further studies in *Tradescantia virginiana* var. *humilis* and *Rhoeo discolor*. Jour. Genet., 26: 81–96.
— 1935. Cytological studies in *Crepis aurea* and *C. rubra*. Cytologia, 6: 281–288.
Koroleva, V. A. 1939. Interspecific hybridization in the genus *Taraxacum*. Compt. Rend. (Doklady) Acad. Sci. URSS, 24: 174–176.
Kostoff, D. 1930. Tumors and other malformations on certain *Nicotiana* hybrids. Zentrlbl. Bakt. Parasit. II, Abt. 81: 244–260.
— 1938a. Studies on polyploid plants. XVIII. Cytogenetic studies on *Nicotiana sylvestris* × *tomentosiformis* hybrids and amphidiploids and their bearing on the problem of the origin of *N. tabacum*. Compt. Rend. (Doklady) Acad. Sci. URSS, 18: 459–462.
— 1938b. Studies on polyploid plants. XXI. Cytogenetic behavior of the allopolyploid hybrids *Nicotiana glauca* Grah. × *Nicotiana*

langsdorffii Weinm and their evolutionary significance. Jour. Genet., 37: 129–209.
—— 1939. Autosyndesis and structural hybridity in F_1 hybrid *Helianthus tuberosus* L. × *Helianthus annuus* L. and their sequences. Genetics, 21: 285–300.
—— 1943. Cytogenetics of the Genus *Nicotiana*. State Printing House, Sofia, Bulgaria. 1,070 pp.
Kostoff, D., and E. Tiber. 1939. A tetraploid rubber plant, *Taraxacum kok-saghyz* obtained by colchicine treatment. Compt. Rend. Acad. Sci. URSS, 22: 119–120.
Kozo-Poljansky, B. 1936. On some "third" conceptions in floral morphology. New Phytologist, 36: 479–492.
Krause, O. 1931. Zytologische Studien bei den Urticales, unter besonderer Berücksichtigung der Gattung *Dorstenia*. Planta, 13: 29–84.
Krishnaswamy, N. 1940. Untersuchungen zur Cytologie und Systematik der Gramineen. Beih. Bot. Centralbl., Abt. A, 60: 1–56.
Kristofferson, K. B. 1926. Species crosses in *Malva*. Hereditas, 7: 233–354.
Krug, C. A., and A. J. T. Mendes. 1941. Observaçoes citológicas em Coffea. IV. Bragantia, 1: 467–482.
Kryshtofovich, A. N. 1929. Evolution of the Tertiary flora in Asia. New Phytologist, 28: 303–312.
—— 1930. Principal features of evolution of the flora of Asia in the Tertiary period. Proc. 4th Pacific Sci. Congr., Java, 3: 253–263.
—— 1935. A final link between the Tertiary of Asia and Europe. New Phytologist, 34: 339–344.
Kylin, H. 1938. Beziehungen zwischen Generationswechsel und Phylogenie. Arch. f. Protistenk., 90: 432–447.

Laibach, F. 1925. Das Taubwerden von Bastardsamen und die künstliche Aufzucht früh absterbender Bastardembryonen. Zeitschr. Bot., 17: 417–459.
Lamm, R. 1945. Cytogenetic studies in *Solanum* sect. *Tuberarium*. Hereditas, 31: 1–128.
Lammerts, W. E. 1942. Embryo culture an effective technique for shortening the breeding cycle of deciduous trees and increasing germination of hybrid seed. Amer. Jour. Bot., 29: 166–171.
Lamprecht, H. 1939. Über Blüten- und Komplex-Mutationen bei *Pisum*. Zeitschr. Ind. Abst. u. Vererbungsl., 77: 177–185.
—— 1941. Die Artgrenze zwischen *Phaseolus vulgaris* L. und *multiflorus* Lam. Hereditas, 27: 51–175.
—— 1944. Die genisch-plasmatisch Grundlage der Artbarriere. Agr. Hort. Genet., 11: 75–142.
—— 1945a. Intra- and interspecific genes. Agr. Hort. Genet., 3: 45–60.
—— 1945b. Durch Komplexmutation bedingte Sterilität und ihre Vererbung. Arch. Zurich, Julius Klaus-Stift, 20(Suppl.): 126–141.
Langlet, O. 1936. Studien über die physiologische Variabilität der

Kiefer und deren Zusammenhang mit dem Klima. Beiträge zur Kenntnis der Öcotypen von *Pinus silvestris* L. Meddel. f. Statens Skogsförsöksanstalt, 29: 219–470.

Larsen, P. 1943. The aspects of polyploidy in *Solanum*. II. Production of dry matter, rate of photosynthesis and respiration, and development of leaf area in some diploid, autotetraploid, and amphidiploid Solanums. Kg. Danske Vidensk. Selsk. Biol. Meddel., 18 (2): 1–52.

Laude, H. H., and A. F. Swanson. 1942. Natural selection in varietal mixtures of winter wheat. Jour. Amer. Soc. Agron., 34: 270–274.

Lawrence, W. E. 1945. Some ecotypic relations of *Deschampsia caespitosa*. Amer. Jour. Bot., 32: 298–314.

Lawrence, W. J. C. 1931. The secondary association of chromosomes. Cytologia, 2: 352–384.

Lawrence, W. J. C., and R. Scott-Moncrieff. 1935. The genetics and chemistry of flower color in *Dahlia*: a new theory of specific pigmentation. Jour. Genet., 30: 155–226.

Lawrence, W. J. C., R. Scott-Moncrieff, and V. C. Sturgess. 1939. Studies on *Streptocarpus*. I. Genetics and chemistry of flower color in the garden strains. Jour. Genet., 38: 299–306.

Lederburg, J. 1947. Gene recombinations and linked segregations in *Escherichia coli*. Genetics, 32: 505–525.

Lederburg, J., and E. L. Tatum. 1946. Gene recombination in *Escherichia coli*. Nature, 158: 558.

—— 1947. Novel genotypes in mixed cultures of biochemical mutants of bacteria. Cold Spring Harbor Symp. Quant. Biol., 11: 113–114.

Ledingham, G. F. 1940. Cytological and developmental studies of hybrids between *Medicago sativa* and a diploid form of *M. falcata*. Genetics, 25: 1–15.

Lehmann, E. 1931. Der Anteil von Kern und Plasma an den reziproken Verschiedenheiten von *Epilobium*-Bastarden. Zeitschr. f. Zuchtung, 17: 157–172.

—— 1936. Versuch zur Klärung der reziproken Verschiedenheiten von *Epilobium* Bastarden. I. Der Tatbestand und die Möglichkeit seiner Klärung durch differente Wuchsstoffbildung. Jahrb. Wiss. Bot., 82: 657–668.

—— 1939. Zur Genetik der Entwicklung in der Gattung *Epilobium*. Jahrb. Wiss. Bot., 87: 625–641.

—— 1942. Zur Genetik der Entwicklung in der Gattung *Epilobium*. IV. Das "Plasmon" in der Gattung *Epilobium* B. Die Aanalyse. Jahrb. Wiss. Bot., 90: 49–98.

—— 1944. Zur Genetik der Entwicklung in der Gattung Epilobium. V. Jahrb. Wiss. Bot., 91: 439–502.

Lerner, I. M., and E. R. Dempster. 1948. Some aspects of evolutionary theory in the light of recent work on animal breeding. Evolution, 2: 19–28.

Lesley, M. M., and J. W. Lesley. 1943. Hybrids of the Chilean tomato. Jour. Hered., 34: 199–205.

Levan, A. 1932. Cytological studies in *Allium*. I. Chromosome morphological contributions. Hereditas, 16: 257–294.

—— 1935. Cytological studies in *Allium*. VI. The chromosome morphology of some diploid species of *Allium*. Hereditas, 20: 289–330.

—— 1936a. Die Zytologie von *Allium cepa* × *fistulosum*. Hereditas, 21: 195–214.

—— 1936b. Zytologische Studien an *Allium schoenoprasum*. Hereditas, 22: 1–128.

—— 1937a. Polyploidy and self-fertility in *Allium*. Hereditas, 22: 278–280.

—— 1937b. Cytological studies in the *Allium paniculatum* group. Hereditas, 23: 317–370.

—— 1941a. Syncyte formation in the pollen mother-cells of haploid *Phleum pratense*. Hereditas, 27: 243–252.

—— 1941b. The cytology of the species hybrid *Allium cepa* × *fistulosum* and its polyploid derivative. Hereditas, 27: 253–272.

—— 1942a. Studies on the meiotic mechanism of haploid rye. Hereditas, 28: 177–211.

—— 1942b. Plant breeding by induction of polyploidy and some results in clover. Hereditas, 28: 245–246.

—— 1945. Aktuelle Probleme der Polyploidiezüchtung. Zurich, Arch. Julius Klaus-Stift., 20: 142–152.

Levan, A., and G. Ostergren. 1943. The mechanism of *c*-mitotic action. Hereditas, 29: 381–443.

Levene, H., and Th. Dobzhansky. 1945. Experiments on sexual isolation in Drosophila. V. The effect of varying proportions of *Drosophila pseudoobscura* and *Drosophila persimilis* on the frequency of insemination in mixed populations. Proc. Nat. Acad. Sci., 31: 274–281.

Levitzky, G. A. 1924. The material basis of heredity. Kiev, State Publication Office of the Ukraine. 166 pp. (In Russian.)

—— 1931a. The morphology of chromosomes. Bull. Appl. Bot. Genet. Plant Breed., 27: 19–174.

—— 1931b. The karyotype in systematics. Bull. Appl. Bot. Genet. Plant Breed., 27: 220–240.

—— 1940. A cytological study of the progeny of X-rayed *Crepis capillaris* Wallr. Cytologia, 11: 1–29.

Levitzky, G. A., and N. E. Kuzmina. 1927. Karyological investigation on the systematics and phylogenetics of the genus *Festuca* (Subgenus *Eu-Festuca*). Bull. Appl. Bot. Leningrad, 17(3): 3–36.

Levitzky, G., and M. Sizova. 1934. On regularities in chromosome transformations produced by X-rays. Compt. Rend. Acad. Sci. URSS, 86–87.

Lewis, D. 1943. Physiology of incompatibility in plants. III. Autopolyploids. Jour. Genet., 45: 171–185.

—— 1947. Competition and dominance of incompatibility alleles in diploid pollen. Heredity, 1: 85–108.

Lewis, D., and I. Modlibowska. 1942. Genetical studies in pears. IV. Pollen tube growth and incompatibility. Jour. Genet., 43: 211–222.

Lewis, H. 1947. Leaf variation in *Delphinium variegatum*. Bull. Torrey Bot. Club, 74: 57–59.

Lewis, H., and C. Epling. 1946. Formation of a diploid species of *Delphinium* by hybridization. (Abstract.) Amer. Jour. Bot., 33: 21s–22s.

Li, H. W., C. H. Li, and W. K. Pao. 1945. Cytological and genetical studies of the interspecific cross of the cultivated foxtail millet, *Setaria italica* (L.) Beauv., and the green foxtail millet, *S. viridis* L. Jour. Amer. Soc. Agron., 37: 32–54.

—— 1945. Desynapsis in the common wheat. Amer. Jour. Bot., 32: 92–101.

Li, H. W., and D. S. Tu. 1947. Studies on the chromosomal aberrations of the amphidiploid, *Triticum timopheevi* and *Aegilops bicornis*. Bot. Bull. Acad. Sinica, 1: 183–186.

Lilienfeld, F., and H. Kihara. 1934. Genomanalyse bei *Triticum* und *Aegilops*. V. *Triticum timopheevi* Zhuk. Cytologia, 6: 87–122.

Lindegren, C. C. 1934. The genetics of Neurospora. V. Self-sterile bisexual heterokaryons. Jour. Genet., 28: 425–435.

—— 1942. The use of the fungi in modern genetical analysis. Iowa State Coll. Jour. Sci., 16: 271–290.

Lindstrom, E. W. 1941. Genetic stability of haploid, diploid, and tetraploid genotypes in the tomato. Genetics, 26: 387–397.

Litardière, R. de. 1921. Recherches sur l'élément chromosomique dans la caryocinèse somatique des Filicinées. La Cellule, 31: 255–473.

—— 1939. Recherches sur les *Poa annua* subsp. *exilis* et subsp. *typica*. Relations taxonomiques, chorologiques et caryologiques. Rev. Cyt. et Cytophys. Veg., 3: 134–141.

Little, T. M. 1945. Gene segregation in autotetraploids. Bot. Rev., 11: 60–85.

Longley, A. E. 1924. Cytological studies in the genus *Rubus*. Amer. Jour. Bot., 11: 249–282.

—— 1927. Supernumerary chromosomes in *Zea Mays*. Jour. Agr. Res., 28: 673–682.

Lotsy, J. P. 1916. Evolution by means of hybridization. The Hague, M. Nijhoff. 166 pp.

—— 1932. On the species of the taxonomist in relation to evolution. Genetica, 13: 1–16.

Löve, A. 1944. Cytogenetic studies on *Rumex*, subg. *Acetosella*. Hereditas, 30: 1–136.

Löve, A., and D. Löve. 1943. The significance of differences in the distribution of diploids and polyploids. Hereditas, 29: 145–163.

Love, R. M. 1941. Chromosome behavior in F_1 wheat hybrids. I. Pentaploids. Canad. Jour. Res., C, 19: 351–369.

—— 1946. Interspecific and intergeneric hybridization in forage crop improvement. Jour. Amer. Soc. Agron., 39: 41–46.

Love, R. M., and C. A. Suneson. 1945. Cytogenetics of certain *Triticum-Agropyron* hybrids and their fertile derivatives. Amer. Jour. Bot., 32: 451–456.

Lush, J. P. 1946. Animal breeding plans. 2d ed. Ames, Iowa State College Press. 437 pp.

Lutz, A. M. 1907. A preliminary note on the chromosomes of *Oenothera lamarckiana* and one of its mutants, *O. gigas*. Science, n.s. 26: 151–152.

MacArthur, J. W., and L. P. Chiasson. 1947. Cytogenetic notes on tomato species and hybrids. Genetics, 32: 165–177.

McAtee, W. L. 1932. Effectiveness in nature of so-called protective adaptations in the animal kingdom, chiefly as illustrated by the food-habits of nearctic birds. Smithsonian Misc. Coll., Washington, 85, 7: 1–201.

McClintock, B. 1929. A cytological and genetical study of triploid maize. Genetics, 14: 180–222.

—— 1933. The association of nonhomologous parts of chromosomes in the mid-prophase of *Zea Mays*. Zeitschr. Zellf. Mikr. Anat., 19: 191–237.

—— 1934. The relation of a particular chromosomal element to the development of the nucleoli in *Zea Mays*. Zeitschr. Zellf. Mikr. Anat., 21: 294–328.

—— 1938. The production of homozygous tissues with mutant characteristics by means of the aberrant behavior of ring-shaped chromosomes. Genetics, 23: 315–376.

—— 1941. The association of mutants with homozygous deficiencies in *Zea mays*. Genetics, 26: 542–571.

—— 1942. Maize genetics. Carnegie Inst. Washington, Yearbook No. 41: 181–186.

—— 1944. The relation of homozygous deficiencies to mutations and allelic series in maize. Genetics, 29: 478–502.

McClintock, E., and C. Epling. 1946. A revision of *Teucrium* in the New World with observations on its variation, geographical distribution, and history. Brittonia, 5: 491–510.

McCullagh, D. 1934. Chromosome and chromosome morphology in Plantagiraceae. I. Genetica, 16: 1–44.

MacDougal, D. T. 1907. Hybridization of wild plants. Bot. Gaz., 43: 45–58.

McFadden, E. S., and E. R. Sears. 1946. The origin of *Triticum spelta* and its free-threshing hexaploid relatives. Jour. Hered., 37: 81–89; 107–116.

—— 1947. The genome approach in radical wheat breeding. Jour. Amer. Soc. Agron., 39: 1011–1026.

MacGinitie, H. D. 1941. A middle Eocene flora from the central

Sierra Nevada. Carnegie Inst. Washington, Publ. No. 534. 178 pp.
McKelvey, S. D., and K. Sax. 1933. Taxonomic and cytological relationships of *Yucca* and *Agave*. Jour. Arnold Arboretum, 14: 76–81.
McLean, S. W. 1946. Interspecific crosses involving *Datura ceratocaula* obtained by embryo dissection. Amer. Jour. Bot., 33: 630–638.
McMinn, H. E. 1942. A systematic study of the genus *Ceanothus*. In *Ceanothus*, Publ. Santa Barbara Botanic Garden, pp. 131–279.
—— 1944. The importance of field hybrids in determining species in the genus *Ceanothus*. Proc. Calif. Acad. Sci., 4th Ser., 25 (14): 323–356.
Mädler, K. 1939. Die pliozäne Flora von Frankfurt A.M. Abh. Senckenberg. Naturforsch. Ges., 446. 202 pp.
Maeda, T. 1937. Chiasma studies in *Allium fistulosum, Allium cepa,* and their F_1, F_2, and backcross hybrids. Jap. Jour. Genet., 13: 146–159.
Malte, M. O. 1934. *Antennaria* of arctic America. Rhodora, 36: 101–117.
Mangelsdorf, P. C. 1947. The origin and evolution of maize. Advances in Genetics, 1: 161–207.
Mangelsdorf, P. C., and J. W. Cameron. 1942. Western Guatemala, a secondary center of origin of cultivated maize varieties. Leaflets Bot. Mus. Harvard Univ., 10: 217–250.
Mangelsdorf, P. C., and R. G. Reeves. 1939. The origin of Indian corn and its relatives. Texas Agr. Expt. Sta., Bull. No. 574. 315 pp.
Manton, I. 1932. Introduction to the general cytology of the Cruciferae. Ann. Bot., 46: 509–556.
—— 1934. The problem of *Biscutella laevigata* L. Zeitschr. Ind. Abst. u. Vererbungsl., 67: 41–57.
—— 1935. The cytological history of watercress (*Nasturtium officinale* R. Br.). Zeitschr. Ind. Abst. u. Vererbungsl., 69: 132–157.
—— 1937. The problem of *Biscutella laevigata* L. II. The evidence from meiosis. Ann. Bot., n.s. 1: 439–462.
—— 1947. Polyploidy in *Polypodium vulgare*. Nature, 159: 136.
Marie-Victorin, Frère. 1929. Le dynamisme dans la flore du Québec. Contr. Lab. Bot. Univ. Montréal, No. 13: 1–89.
—— 1938. Phytogeographical problems of eastern Canada. Amer. Midl. Nat., 19: 489–558.
Marquardt, H. 1937. Die Meiosis von *Oenothera* I. Zeitschr. f. Zellf. u. Mikr. Anat., 27: 159–210.
Marsden-Jones, E. M. 1930. The genetics of *Geum intermedium* Willd. haud Ehrh, and its backcrosses. Jour. Genet., 23: 377–395.
Marsden-Jones, E. M., and W. B. Turrill. 1947. Researches on *Silene maritima* and *S. vulgaris*. XXVI. Kew Bull., 1946: 97–107.
Mason, H. L. 1936. The principles of geographic distribution as applied to flora analysis. Madroño, 3: 181–190.

—— 1946a. The edaphic factor in narrow endemism. I. The nature of environmental influences. Madroño, 8: 209–226.
—— 1946b. The edaphic factor in narrow endemism. II. The geographic occurrence of plants of highly restricted patterns of distribution. Madroño, 8: 241–257.
—— 1947. Evolution of certain floristic associations in western North America. Ecological Monographs, 17: 201–210.
Massart, J. 1894. La récapitulation et l'innovation en embryologie végétale. Bull. Soc. Bot. Belg., 33: 150–247.
Mather, K. 1932. Chromosome variation in *Crocus* I. Jour. Genet., 26: 129–142.
—— 1941. Variation and selection of polygenic characters. Jour. Genet., 41: 159–193.
—— 1943. Polygenic inheritance and natural selection. Biol. Rev., 18: 32–64.
—— 1947. Species crosses in *Antirrhinum*. I. Genetic isolation of the species *majus, glutinosum,* and *orontium*. Heredity, 1: 175–186.
Mather, K., and L. G. Wigan. 1942. The selection of invisible mutations. Proc. Roy. Soc. London, 131: 50–65.
Matsuura, H. 1935. On Karyo-ecotypes of *Fritillaria camschatcensis* (L.) Ker-Gawler. Jour. Fac. Sci. Hokkaido Imp. Univ., Ser. 5, III: 219–232.
Mattfeld, J. 1930. Über hybridogene Sippen der Tannen. Bibliotheca Botanica 25, No. 100: 1–84.
Matthew, W. D. 1939. Climate and evolution. 2d ed. New York Acad. Sci., Special Publ. 223 pp.
Mayr, E. 1939. The origin and history of the bird fauna of Polynesia. Proc. 6th Pac. Sci. Congr., IV: 197–216.
—— 1942. Systematics and the origin of species. New York, Columbia University Press. 334 pp.
—— 1943. The zoogeographic position of the Hawaiian Islands. The Condor, 45: 45–48.
—— 1947. Ecological factors in speciation. Evolution, 1: 263–288.
Mehlquist, G. A. L. 1945. Inheritance in the carnation. V. Tetraploid carnations from interspecific hybridization. Proc. Amer. Soc. Hort. Sci., 46: 397–406.
Melchers, G. 1939. Genetik und Evolution. Zeitschr. Ind. Abst. u. Vererbungsl., 76: 229–259.
—— 1946. Die Ursachen für die bessere Anpassungsfähigkeit der Polyploiden. Zeitschr. f. Naturforschung, 1: 160–165.
Metz, C. W. 1947. Duplication of chromosome parts as a factor in evolution. Amer. Nat., 81: 81–103.
Meurman, O., and E. Therman. 1939. Studies on the chromosome morphology and structural hybridity in the genus *Clematis*. Cytologia, 10: 1–14.
Mez, C., and H. Siegenspeck. 1926. Der Königsberger serodiagnostische Stammbaum. Bot. Archiv, 12: 163–202.

Michaelis, P. 1931. Die Bedeutung des Plasmas für die Pollenfertilität reciprok verschiedener Epilobiumbastarde. Ber. Deu. Bot. Ges., 49: 96–104.

— 1933. Entwicklungsgeschichtlich-genetische Untersuchungen an *Epilobium*. II. Die Bedeutung des Plasmas für die Pollenfertilität des *Epilobium luteum-hirsutum*-Bastardes. Zeitschr. Ind. Abst. u. Vererbungsl., 65: 1–71, 353–471.

— 1938. Über das Konstanz des Plasmons. Zeitschr. Ind. Abst. u. Vererbungsl., 74: 435–459.

— 1940. Über reziprok verschiedene Sippen-Bastarde bei *Epilobium hirsutum*. I. Die reziprok verschiedene Bastarde der *Epilobium hirsutum*-Sippe Jena. Zeitschr. Ind. Abst. u. Vererbungsl., 78: 187–222.

Michaelis, P., and E. Wertz. 1935. Entwicklungsgeschichtlich-genetische Untersuchungen an *Epilobium*. VI. Vergleichende Untersuchungen über das Plasmon von *Epilobium hirsutum, E. luteum, E. montanum,* und *E. roseum*. Zeitschr. Ind. Abst. u. Vererbungsl., 70: 138–159.

Miki, S. 1941. On the change of flora in eastern Asia since Tertiary Period. I. The clay or lignite beds flora in Japan with special reference to the *Pinus trifolia* beds in central Hondo. Jap. Jour. Bot., 11: 237–303.

Miller, A. H. 1941. Speciation in the avian genus *Junco*. Univ. of Calif. Publ. Zool., 44 (3): 173–434.

Mirov, N. T., and C. J. Kraebel. 1939. Collecting and handling seeds of wild plants. Civilian Conservation Corps Forestry Publication No. 5. 42 pp.

Miyaji, Y. 1929. Studien über die Zahlenverhältnisse der Chromosomen bei der Gattung *Viola*. Cytologia, 1: 28–58.

Modilewski, J. 1930. Neue Beträge zur Polyembryonie von *Allium odorum*. Ber. Deu. Bot. Ges., 48: 285–294.

Moffett, A. A. 1931. The chromosome constitution of the Pomoideae. Proc. Roy. Soc., B, 108: 423–446.

— 1932. Studies on the formation of multinuclear giant pollen grains in *Kniphofia*. Jour. Genet., 25: 315–336.

Molisch, H. 1938. The longevity of plants (English ed.). Transl. and publ. by E. R. Fulling, New York.

Moore, R. J. 1947. Cytotaxonomic studies in the Loganiaceae. I. Chromosome numbers and phylogeny in the Loganiaceae. Amer. Jour. Bot., 34: 527–538.

Muller, H. J. 1925. Why polyploidy is rarer in animals than in plants. Amer. Nat., 59: 346–353.

— 1930. Radiation and genetics. Amer. Nat., 64: 220–251.

— 1932. Some genetic aspects of sex. Amer. Nat., 66: 118–138.

— 1939. Reversibility in evolution, considered from the standpoint of genetics. Biol. Rev., 14: 261–280.

—— 1940. Bearings of the "Drosophila" work on systematics. In J. Huxley, ed., The New Systematics: 185–268.

—— 1942. Isolating mechanisms, evolution, and temperature. Biol. Symposia, 6: 71–125.

—— 1947. The gene. Proc. Roy. Soc., B, 134: 1–37.

Müntzing, A. 1930a. Outlines to a genetic monograph of *Galeopsis*. Hereditas, 13: 185–341.

—— 1930b. Über Chromosomenvermehrung in *Galeopsis*-Kreuzungen und ihre phylogenetische Bedeutung. Hereditas, 14: 153–172.

—— 1933. Apomictic and sexual seed formation in *Poa*. Hereditas, 17: 131–154.

—— 1936. The evolutionary significance of autopolyploidy. Hereditas, 21: 263–378.

—— 1937. The effects of chromosomal variation in *Dactylis*. Hereditas, 23: 113–235.

—— 1938. Sterility and chromosome pairing in intraspecific *Galeopsis* hybrids. Hereditas, 24: 117–188.

—— 1939. Studies on the properties and the ways of production of rye-wheat amphidiploids. Hereditas, 25: 387–430.

—— 1940. Further studies on apomixis and sexuality in *Poa*. Hereditas, 26: 115–190.

Müntzing, A., and G. Müntzing. 1941. Some new results concerning apomixis, sexuality, and polymorphism in *Potentilla*. Bot. Not. (Lund): 237–278.

Müntzing, A., and R. Prakken. 1940. The mode of chromosome pairing in *Phleum* twins with 63 chromosomes and its cytogenetic consequences. Hereditas, 26: 463–501.

Müntzing, A., O. Tedin, and G. Turesson. 1931. Field studies and experimental methods in taxonomy. Hereditas, 15: 1–12.

Munz, P. 1946. *Aquilegia*. The cultivated and wild columbines. Gentes Herbarum, 7: 1–150.

—— 1949. The *Oenothera hookeri* group. El Aliso, Publ. Rancho Santa Ana Bot. Gard., 2: 1–47.

Myers, W. M. 1943. Analysis of variance and covariance of chromosomal association and behavior during meiosis in clones of *Dactylis glomerata*. Bot. Gaz., 104: 541–552.

—— 1944. Cytological and genetic analysis of chromosomal association and behavior during meiosis in hexaploid timothy (*Phleum pratense*). Jour. Agr. Res., 68(1): 21–83.

—— 1945. Meiosis in autotetraploid *Lolium perenne* in relation to chromosomal behavior in autopolyploids. Bot. Gaz., 106: 304–316.

—— 1948. Studies on the origin of *Dactylis glomerata* L. (Abstract.) Genetics, 33: 117.

Myers, W. M., and H. D. Hill. 1942. Variations in chromosomal association and behavior during meiosis among plants from open-pollinated populations of *Dactylis glomerata*. Bot. Gaz., 104: 171–177.

Nagahuru U. See U, Nagahuru.

Nagel, L. 1939. Morphogenetic differences between *Nicotiana alata* and *N. langsdorffii* as indicated by their response to indoleacetic acid. Ann. Missouri Bot. Gard., 26: 349–368.

Nakamura, M. 1937. Cytogenetical studies in the genus *Solanum* I. Autopolyploidy of *Solanum nigrum* L. Cytologia, Fujii Jubil. Vol.: 57–68.

Nandi, H. K. 1936. The chromosome morphology, secondary association, and origin of cultivated rice. Jour. Genet., 33: 315–336.

Nannfeldt, J. A. 1937. The chromosome numbers of *Poa*, sect. *Ochlopoa* A and Gr. and their taxonomical significance. Bot. Not.: 238–254.

Nast, C. G., and I. W. Bailey. 1945. Morphology and relationships of *Trochodendron* and *Tetracentron*. II. Inflorescence, flower, and fruit. Jour. Arnold Arboretum, 26: 267–276.

—— 1946. Morphology of *Euptelea* and comparison with *Trochodendron*. Jour. Arnold Arboretum, 27: 186–191.

Natividade, J. V. 1937a. Recherches cytologiques sur quelques espèces et hybrides du genre *Quercus*. Bol. Soc. Broteriana, 12: 21–85.

—— 1937b. Investagações citologicas nalgumas espécies e hibridos do genero *Quercus*. Publ. Dir. Ger. Serv. Flor. e. Agric (Portugal), 4: 7–74.

Navashin, M. 1932. The dislocation hypothesis of evolution of chromosome numbers. Zeitschr. Ind. Abst. u. Vererbungsl., 63: 224–231.

—— 1934. Chromosome alterations caused by hybridization and their bearing upon certain general genetic problems. Cytologia, 5: 169–203.

Ness, H. 1927. Possibilities of hybrid oaks. Jour. Hered., 18: 381–386.

Newton, W. C. F., and C. Pellew. 1929. *Primula kewensis* and its derivatives. Jour. Genet., 20: 405–467.

Nielsen, E. L. 1947. Polyploidy and winter survival in *Panicum virgatum* L. Jour. Amer. Soc. Agron., 39: 822–827.

Nilsson, F. 1933. Studies in fertility and inbreeding in some herbage grasses. Hereditas, 19: 1–162.

Noggle, G. R. 1946. The physiology of polyploidy in plants. I. Review of the literature. Lloydia, 9: 153–173.

Nordenskiöld, H. 1941. Cytological studies in triploid *Phleum*. Bot. Not., 12–32.

—— 1945. Cytogenetic studies in the genus *Phleum*. Acta Agriculturae Suecanae 1(1): 1–136.

Nygren, A. 1946. The genesis of some Scandinavian species of *Calamagrostis*. Hereditas, 32: 131–262.

—— 1948. Further studies in spontaneous and synthetic *Calamagrostis purpurea*. Hereditas, 34: 113–134.

Oehler, E. 1936. Untersuchungen an einen neuen konstanten additiven *Aegilops*-Weizenbastard. Züchter, 8: 29–33.

Olmsted, C. E. 1944. Growth and development in range grasses. IV. Photoperiodic responses in twelve geographic strains of side-oats grama. Bot. Gaz., 106: 46–74.

O'Mara, J. G. 1940. Cytogenetic studies on *Triticale*. I. A method for determining the effects of individual *Secale* chromosomes on *Triticum*. Genetics, 25: 401–408.

Ono, H. 1937. Intergeneric hybridization in Cichorieae. III. Fertility and chromosome variations in F_1 and F_2 progeny of *Paraixeris denticulata* and *Crepidiastrum lanceolatum* var. *latifolium*. Cytologia, Fujii Jubil. Vol.: 535–539.

—— 1941. Intergeneric hybridization in Cichorieae. V. Variation in karyotypes and fertility of *Crepidiastrixeris denticulato-platyphylla*. Cytologia, 11 (3): 338–352.

Ono, H., and D. Sato. 1935. Intergenera Hibridigoen Cichorieae. II. Hibridoj de *Crepidiastrum lanceolatum* var. *latifolium* kaj *Paraixeris denticulata*. Jap. Jour. Genet., 11: 169–179.

Orth, R. 1939. Zur Morphologie der Primärblätter einheimischer Farne. Flora, n.s. 33: 1–55.

Osborn, A. 1941. An interesting hybrid conifer: *Cupressocyparis leylandii*. Jour. Roy. Hort. Soc., 66: 54–55.

Ostenfeld, C. H. 1910. Further studies on the apogamy and hybridization of the Hieracia. Zeitschr. Ind. Abst. u. Vererbungsl., 3: 241–285.

—— 1929. Genetic studies in *Polemonium*. II. Experiments with crosses of *P. mexicanum* Cerv. and *P. pauciflorum* Wats. Hereditas, 12: 33–40.

Owen, R. D., G. Stormont, and M. R. Irwin. 1947. An immunogenetic analysis of racial differences in dairy cattle. Genetics, 32: 64–74.

Ownbey, R. P. 1944. The liliaceous genus *Polygonatum* in North America. Ann. Missouri Bot. Gard., 31: 373–413.

Palmer, E. J. 1948. Hybrid oaks of North America. Jour. Arnold Arboretum, 29: 1–48.

Papajoannon, J. 1936. Über Artbastarde zwischen *Pinus brutia* Ten. und *Pinus halepensis* Mill. in Nordost-Chalkidiki, Griechenland. Forstwiss. Zentralbl., 58: 194–205.

Payne, F. 1920. Selection for high and low bristle number in the mutant strain "reduced." Genetics, 5: 501–542.

Peitersen, A. K. 1921. Blackberries of New England – genetic status of the plants. Vermont Agr. Expt. Sta., Bull. No. 218. 34 pp.

Perry, B. A. 1943. Chromosome number and phylogenetic relationships in the Euphorbiaceae. Amer. Jour. Bot., 30: 527–543.

Peto, F. H. 1933. The cytology of certain intergeneric hybrids between *Festuca* and *Lolium*. Jour. Genet., 28: 113–156.

—— 1936. Hybridization of *Triticum* and *Agropyron*. II. Cytology of the male parents and F_1 generation. Canad. Jour. Res., Ser. C, 14: 203–214.

—— 1938. Cytology of poplar species and natural hybrids. Canad. Jour. Res., 16: 445–455.

Pfeffer, W. 1872. Zur Blütenentwiklung der Primulaceen und Ampelideen. Pringsh. Jahrb. Wiss. Bot., 8: 194–215.

Philipson, W. R. 1934. The morphology of the lemma in grasses. New Phytologist, 33: 359–371.

Pierce, W. P. 1937. The effect of phosphorus on chromosome and nuclear volume in a violet species. Bull. Torrey Bot. Club, 64: 345–354.

Pijl, L. Van der. 1934. Über die Polyembryonie bei *Eugenia*. Rec. Trav. Bot. Neerl., 31: 113–187.

Pincher, H. C. 1937. A genetical interpretation of alternation of generations. New Phytologist, 36: 179–183.

Pirschle, K. 1942a. Quantitative Untersuchungen über Wachstum und "Ertrag" autopolyploider Pflanzen. Zeitschr. Ind. Abst. u. Vererbungsl., 80: 126–156.

—— 1942b. Weitere Untersuchungen über Wachstum und "Ertrag" von Autopolyploiden (2n, 3n, 4n) und ihren Bastarden. Zeitschr. Ind. Abst. u. Vererbungsl., 80: 247–270.

Pissarev, V. E., and N. M. Vinogradova. 1944. Hybrids between wheat and *Elymus*. Compt. Rend. (Doklady) Acad. Sci. URSS, 45: 129–132.

Pjatnitzky, S. S. 1946a. Experimental production of interspecific hybrids in the genus *Quercus*. Compt. Rend. (Doklady) Acad. Sci. URSS, 52: 343–345.

—— 1946b. New hybrid forms of oak, *Quercus komarovii* and *Quercus timjarzevii* produced experimentally. Compt. Rend. (Doklady) Acad. Sci. URSS, 52: 625–626.

Poddubnaja-Arnoldi, V. 1939a. Development of pollen and embryosac in interspecific hybrids of *Taraxacum*. Compt. Rend. (Doklady) Acad. Sci. URSS, 24: 374–377.

—— 1939b. Embryogenesis in remote hybridization in the genus *Taraxacum*. Compt. Rend. (Doklady) Acad. Sci. URSS, 24: 382–385.

Poddubnaja-Arnoldi, V., and V. Dianova. 1934. Eine zytoembryologische Untersuchung einiger Arten der Gattung *Taraxacum*. Planta, 23: 19–46.

Pontecorvo, G. 1947. Genetic systems based on heterokaryosis. Cold Spring Harbor Symp. Quant. Biol., 11: 193–201.

Pontecorvo, G., and A. R. Gemmell. 1944. Colonies of *Penicillium notatum* and other molds as models for the study of population genetics. Nature, 154: 532–534.

Poole, C. F. 1931. The interspecific hybrid *Crepis rubra* × *foetida*, and some of its derivatives. Univ. of Calif. Publ. Agr. Sci. 6: No. 6, 169–200.

Popoff, A. 1935. Über die Fortpflanzungsverhältnisse der Gattung *Potentilla*. Planta, 24: 510–522.

Powers, L. 1945. Fertilization without reduction in guayule (*Parthenium argentatum* Gray) and a hypothesis as to the evolution of apomixis and polyploidy. Genetics, 30: 323–346.

Powers, L., and R. C. Rollins. 1945. Reproduction and pollination studies on guayule, *Parthenium argentatum* Gray and *P. incanum* H.B.K. Jour. Amer. Soc. Agron., 37: 184–193.

Prat, H. 1932. L'épiderme des Graminées. Étude anatomique et systématique. Ann. Sci. Nat., Ser. 10, Bot. 14: 117–324.

— 1936. La systématique des Graminées. Ann. Sci. Nat., Ser. 10, Bot. 18: 165–258.

Propach, H. 1937. Cytogenetische Untersuchungen in der Gattung *Solanum,* sect. *Tuberarium.* Zeitschr. Ind. Abst. u. Vererbungsl., 72: 555–563.

— 1940. Cytogenetische Untersuchungen in der Gattung *Solanum,* sect. *Tuberarium.* V. Diploide Artbastarde. Zeitschr. Ind. Abst. u. Vererbungsl., 78: 115–128.

Randolph, L. F. 1935. The cytogenetics of tetraploid maize. Jour. Agr. Res., 50: 591–605.

— 1941a. An evaluation of induced polyploidy as a method of breeding crop plants. Amer. Nat., 75: 347–363.

— 1941b. Genetic characteristics of the B chromosomes in maize. Genetics, 26: 608–631.

Randolph, L. F., E. C. Abbe, and J. Einset. 1944. Comparison of shoot apex and leaf development and structure in diploid and tetraploid maize. Jour. Agr. Res., 69: 47–76.

Randolph, L. F., and H. E. Fischer. 1939. The occurrence of parthenogenetic diploids in tetraploid maize. Proc. Nat. Acad. Sci., 25: 161–164.

Raunkiaer, C. 1925. Ermitageslettens Tjorne. Isoreagentstudier. I. Kgl. Danske Videnskab. Selskab. Biol. Meddel., 5: 1–76.

— 1934. The life forms of plants and statistical plant geography. Oxford, Clarendon Press. 632 pp.

Reeder, J. R. 1946. Additional evidence of affinities between *Eragrostis* and certain Chlorideae. (Abstract.) Amer. Jour. Bot., 33: 843.

Reid, C., and E. M. Reid. 1915. The Pliocene floras of the Dutch-Prussian border, Mededeel. van de Rijksopsporing van Delfstoffen, No. 6. 178 pp.

Reid, E. M. 1920a. A comparative review of Pliocene floras, based on the study of fossil seeds. Quart. Jour. Geol. Soc. London, 76: 145–161.

— 1920b. On two preglacial floras from Castle Eden. Quart. Jour. Geol. Soc. London, 76: 104–144.

Reid, E. M., and M. E. J. Chandler. 1933. The London Clay Flora. London, British Museum. 561 pp.

Renner, O. 1917. Versuche über die gametische Konstitution der Oenotheren. Zeitschr. Ind. Abst. u. Vererbungsl., 18: 121–294.

—— 1921. Heterogamie im weiblichen Geschlecht und Embryosackentwicklung bei den Oenotheren. Zeitschr. Bot., 13: 609–621.

—— 1929. Artbastarde bei Pflanzen. Berlin, Borntraeger. 161 pp.

—— 1936. Zur Kenntnis der nichtmendelden Buntheit der Laubblätter. Flora, 30: 218–290.

Rensch, B. 1939. Über die Anwendungsmöglichkeit zoologisch-systematischer Prinzipien in der Botanik. Chon. Bot., 5: 46–49.

Resende, F. 1937. Über die Ubiquität der SAT-chromosomen bei den Blütenpflanzen. Planta, 26: 757–807.

Rhoades, M. M. 1938. Effect of the Dt gene on the mutability of the a_1 allele in maize. Genetics, 23: 377–397.

Richardson, M. M. 1935. Meiosis in Crepis. I. Jour. Genet., 31: 101–117.

—— 1936. Structural hybridity in *Lilium martagon album* × *L. hansonii.* Jour. Genet., 32: 411–450.

Richens, R. H. 1945. Forest tree breeding and genetics. Imp. Agr. Bur., Joint Publ. No. 8. Aberystwyth. 79 pp.

Rick, C. M. 1947. Partial suppression of hair development indirectly affecting fruitfulness and the proportion of cross-pollination in a tomato mutant. Amer. Nat., 81: 185–202.

Righter, F. I. 1946. New perspectives in forest tree breeding. Science, 104 (2688): 1–3.

Riley, H. P. 1938. A character analysis of colonies of *Iris fulva, Iris hexagona* var. *giganticaerulea* and natural hybrids. Amer. Jour. Bot., 25: 727–738.

—— 1939. Introgressive hybridization in a natural population of *Tradescantia.* Genetics, 24: 753–769.

—— 1942. Development and relative growth in ovaries of *Iris fulva* and *I. hexagona* var. *giganticaerulea.* Amer. Jour. Bot., 29: 323–331.

—— 1943. Cell size in developing ovaries of *Iris hexagona* var. *giganticaerulea.* Amer. Jour. Bot., 30: 356–361.

—— 1944. Relative growth of flower parts of two species of *Iris.* Bull. Torrey Bot. Club, 71: 122–133.

Robbins, W. W. 1940. Alien plants growing without cultivation in California. Univ. of Calif. Agr. Expt. Sta., Bull. No. 637. 128 pp.

Robson, G. C., and O. W. Richards. 1936. The variations of animals in nature. London, Longmans Green. 425 pp.

Rohweder, H. 1934. Beiträge zur Systematik und Phylogenie des Genus *Dianthus,* unter Berücksichtigung der karyologischen Verhältnisse. Bot. Jahrb., 66: 249–368.

—— 1937. Versuch zur Erfassung der Mengenmässigen Bedeckung des Darss und Zingst mit polyploiden Pflanzen. Ein Beitrag zur Bedeutung der Polyploidie bei der Eroberung neuer Lebensräume. Planta, 27: 500–549.

Rollins, R. 1939. The cruciferous genus *Physaria.* Rhodora, 41: 392–414.

—— 1941. Monographic study of *Arabis* in western North America. Rhodora, 43: 289–325.

—— 1944. Evidence for natural hybridity between guayule (*Parthenium argentatum*) and mariola (*Parthenium incanum*). Amer. Jour. Bot., 31: 93–99.

—— 1945a. Interspecific hybridization in *Parthenium*. I. Crosses between guayule (*P. argentatum*) and mariola (*P. incanum*). Amer. Jour. Bot., 32: 395–404.

—— 1945b. Evidence for genetic variation among apomictically produced plants of several F_1 progenies of guayule (*Parthenium argentatum*) and mariola (*P. incanum*). Amer. Jour. Bot., 32: 554–560.

—— 1946. Interspecific hybridization in *Parthenium*. II. Crosses involving *P. argentatum, P. incanum, P. stramonium, P. tomentosum,* and *P. hysterophorus*. Amer. Jour. Bot., 33: 21–30.

Rollins, R. C., D. G. Catcheside, and D. U. Gerstel. 1947. Genetics. In Final Report, Stanford Research Institute, National Rubber Research Project: 3–33. (Mimeographed.)

Rosenberg, O. 1909. Cytologische und morphologische Studien an *Drosera longifolia* × *rotundifolia*. K. Svensk. Vet. Akad. Handl., 43(11): 1–64.

—— 1917. Die Reduktionsteilung und ihre Degeneration in *Hieracium*. Svensk. Bot. Tidskr., 11: 145–206.

—— 1930. Apogamie und Parthenogenesis bei Pflanzen. Handbuch der Vererbungswiss., II. 66 pp.

Rozanova, M. 1934. Origin of new forms in the genus *Rubus*. Bot. Zhurnal, 19: 376–384.

—— 1938. On the polymorphic origin of species. Compt. Rend. Acad. Sci. URSS, 18: 677–680.

—— 1940. On genotypic differences between races of *Rubus caesius* L. Compt. Rend. (Doklady) Acad. Sci. URSS, 27: 590–593.

Rubtzov, I. A. 1945. The inequality of rates of evolution. Jour. Gen. Biol., 6: 411–441.

Ryan, F. J. 1947. Back-mutation and adaptation of nutritional mutants. Cold Spring Harbor Symp. Quant. Biol., 11: 215–227.

Ryan, F. J., and J. Lederburg. 1946. Reverse-mutation and adaptation in leucineless *Neurospora*. Proc. Nat. Acad. Sci., 32: 163–173.

Rybin, W. A. 1936. Spontane und experimentell erzeugte Bastarde zwischen Schwarzdorn und Kirschpflaume und das Abstammungsproblem der Kulturpflaume. Planta, 25: 22–58.

Sacharov, V. V., S. L. Frolova, and V. V. Mansurova. 1944. High fertility of buckwheat tetraploids obtained by means of colchicine treatment. Nature, 154: 613.

Sahni, B. 1937. Revolutions in the plant world. Proc. Nat. Acad. Sci., India, 7: 46–60.

Sakai, K. 1935. Chromosome studies in *Oryza sativa* L. I. The secondary association of the meiotic chromosomes. Jap. Jour. Genet., 11: 145–156.

Sakamura, T. 1918. Kurze Mitteilung über die Chromosomenzahlen und die Verwandschaftsverhältnisse der *Triticum*-Arten. Bot. Mag. Tokyo, 32: 151–154.

Salisbury, E. J. 1940. Ecological aspects of plant taxonomy. In J. Huxley, ed., The New Systematics: 329–340.

—— 1942. The reproductive capacity of plants. London, Bell and Sons. 244 pp.

Sando, W. J. 1935. Intergeneric hybrids of *Triticum* and *Secale* with *Haynaldia villosa*. Jour. Agr. Res., 51: 759–800.

Sansome, E. R. 1932. Segmental interchange in *Pisum*. Cytologia, 3: 200–219.

—— 1946. Heterokaryosis, mating-type factors and sexual reproduction in *Neurospora*. Bull. Torrey Bot. Club, 73: 397–409.

Sato, D. 1937. Karyotype alteration and phylogeny. I. Analysis of karyotypes in Aloinae with special reference to the SAT-chromosome. Cytologia, Fujii Jubil. Vol.: 80–95.

Saunders, A. P., and G. L. Stebbins, Jr. 1938. Cytogenetic studies in *Paeonia*. I. The compatibility of the species and the appearance of the hybrids. Genetics, 28: 65–82.

Sax, H. J. 1930. Chromosome numbers in *Quercus*. Jour. Arnold Arboretum, 11: 220–223.

Sax, K. 1922. Sterility in wheat hybrids. II. Chromosome behavior in partially sterile hybrids. Genetics, 7: 513–552.

—— 1930. Chromosome stability in the genus *Rhododendron*. Amer. Jour. Bot., 17: 247–251.

—— 1931a. Chromosome ring formation in *Rhoeo discolor*. Cytologia, 3: 36–53.

—— 1931b. The origin and relationship of the Pomoideae. Jour. Arnold Arboretum, 12: 3–22.

—— 1932. Chromosome relationships in the Pomoideae. Jour. Arnold Arboretum, 13: 363–367.

—— 1933. Species hybrids in *Platanus* and *Campsis*. Jour. Arnold Arboretum, 14: 274–278.

—— 1936. Polyploidy and geographic distribution in *Spiraea*. Jour. Arnold Arboretum, 17: 352–356.

Sax, K., and E. Anderson. 1933. Segmental interchange in chromosomes of *Tradescantia*. Genetics, 18: 53–94.

Sax, K., and J. M. Beal. 1934. Chromosomes of the Cycadales. Jour. Arnold Arboretum, 15: 255–258.

Sax, K., and H. J. Sax. 1933. Chromosome number and morphology in the conifers. Jour. Arnold Arboretum, 14: 356–375.

Schaffner, J. H. 1929. Orthogenetic series involving a diversity of morphological systems. Ohio Jour. Sci., 29: 45–60.

—— 1930. Orthogenetic series resulting from a simple progressive movement. Ohio Jour. Sci., 30: 61–71.

—— 1932. Orthogenetic evolution of degree of divergence between carpel and foliage leaf. Ohio Jour. Sci., 32: 367–378.

—— 1933. Color in various plant structures and the so-called principle of selective adaptation. Studies in determinate evolution VII. Ohio Jour. Sci., 23: 182–191.

—— 1937. The fundamental nature of the flower. Bull. Torrey Bot. Club, 64: 569–582.

Scheel, M. 1931. Karyologische Untersuchung der Gattung *Salvia*. Bot. Archiv, 32: 148–208.

Schiemann, E. 1939. Gedanken zur Genzentrentheorie Vavilovs. Naturwiss, 27: 377–383; 394–401.

Schlösser, L. A. 1936. Befruchtungs-schwierigkeiten bei Autopolyploiden und ihre Überwindung. Der Züchter, 8: 295–301.

Schmalhausen, I. I. 1946. Factors of evolution: the theory of stabilizing selection. Theodosius Dobzhansky, ed.; I. Dordick, tr. Philadelphia, Blakiston. 327 pp. Illustrated.

Schnack, B., and G. Covas, 1944. Nota sobre la validez del género *Glandularia*. Darwiniana, 6: 469–476.

—— 1945a. Un híbrido interespecífico del género *Glandularia* (*G. peruviana* × *G. megapotamica*). Rev. Arg. de Agronomía, 12(3): 224–229.

—— 1945b. Hibridación interespecífica en *Glandularia* (Verbenaceas). Darwiniana, 7: 71–79.

Schuster, J. 1931. Über das Verhältnis der systematischen Gliederung, der geographischen Verbreitung, und der paläontologischen Entwicklung der Cycadaceen. Englers Bot. Jahrb., 64: 165–260.

Schwarz, O. 1936. Über die Typologie des Eichenblattes und ihre Anwendung in der Paläobotanik. Fedde Rep. Beih., 86: 60–70.

Schwemmle, J. 1938. Genetische und zytologische Untersuchungen an Euoenotheren. I. Zeitschr. Ind. Abst. u. Vererbungsl., 75: 358–468.

Schwemmle, J., and M. Zintl. 1939. Genetische und zytologische Untersuchungen an Euoenotheren: Die Analyse der *Oenothera argentinea*. Zeitschr. Ind. Abst. u. Vererbungsl., 76: 353–410.

Scott-Moncrieff, R. 1936. A biochemical survey of some Mendelian factors for flower color. Jour. Genet., 32: 117–170.

Sears, E. R. 1941a. Amphidiploids in the seven-chromosome Triticinae. Missouri Agr. Expt. Sta., Res. Bull. 336. 46 pp.

—— 1941b. Chromosome pairing and fertility in hybrids and amphidiploids in the Triticinae. Missouri Agr. Expt. Sta., Res. Bull. 337. 20 pp.

—— 1944. Cytogenetic studies with polyploid species of wheat. II. Additional chromosomal aberrations in *Triticum vulgare*. Genetics, 29: 232–246.

—— 1947. The *sphaerococcum* gene in wheat. (Abstract.) Genetics, 32: 102–103.

Seibert, R. J. 1947. A study of *Hevea* (with its economic aspects) in the republic of Peru. Ann. Missouri Bot. Gard., 34: 261–352.

Senjaninova-Korczagina, M. 1932. Karyo-Systematical investigation of the genus *Aegilops* L. Bull. Appl. Bot., Genet., Plant Breed. 11 Ser., I: 1–90.

Senn, H. A. 1938a. Chromosome number relationships in the Leguminosae. Bibliographic Genetica, 12: 175–336.

— 1938b. Experimental data for a revision of the genus *Lathyrus*. Amer. Jour. Bot., 25: 67–78.

Setchell, W. A., T. H. Goodspeed, and R. E. Clausen. 1922. Inheritance in *Nicotiana tabacum*. I. A report on the results of crossing certain varieties. Univ. of Calif. Publ. Bot., 5: 457–582.

Seward, A. C. 1931. Plant life through the ages. Cambridge, Cambridge University Press. 601 pp.

— 1934. A phytogeographical problem, fossil plants from the Kerguelen archipelago. Ann. Bot., 48: 715–741.

Shaparenko, K. K. 1936. The nearest ancestors of Ginkgo biloba L. Acta Inst. Bot. Acad. Sci. USSR 1, v. 2: 5–32. (In Russian.)

Shapiro, N. J. 1938. The mutation process as an adaptive character of a species. Zool. Zhurnal (Moscow), 17: 592–601.

Sherman, M. 1946. Karyotype evolution: a cytogenetic study of seven species and six interspecific hybrids of *Crepis*. Univ. of Calif. Publ. Bot., 18: 369–408.

Shimotomai, N. 1933. Zur Karyogenetik der Gattung *Chrysanthemum*. Jour. Sci. Hiroshima Univ., Ser. B, Div. 2, Vol. 2: 1–100.

Silow, R. A. 1944. The genetics of species development in the Old World cottons. Jour. Genet., 46: 62–77.

Simonet, M. 1934. Nouvelles recherches cytologiques et génétiques chez les *Iris*. Ann. Sci. Nat. Bot., Ser. 10, 16: 229–383.

— 1935. Observations sur quelques espèces et hybrides d'*Agropyrum*. I. Revision de l'*Agropyrum junceum* (L.) P. B. et de l'*A. elongatum* (Host.) P. B. d'après l'étude cytologique. Bull. Soc. Bot. France, 82: 624–632.

Simpson, G. C. 1940. Possible causes of change in climate and their limitations. Proc. Linn. Soc. London, 152: 190–219.

Simpson, G. G. 1944. Tempo and mode in evolution. New York, Columbia University Press. 237 pp.

Sinnott, E. W. 1914. Investigations on the phylogeny of the angiosperms. I. The anatomy of the node as an aid in the classification of angiosperms. Amer. Jour. Bot., 1: 303–322.

— 1916. Comparative rapidity of evolution in various plant types. Amer. Nat., 50: 466–478.

— 1936. A developmental analysis of inherited shape differences in Cucurbit fruits. Amer. Nat., 70: 245–254.

— 1937. The genetic control of developmental relationships. Amer. Nat., 71: 113–119.

— 1939. A developmental analysis of the relation between cell size and fruit size in Cucurbits. Amer. Jour. Bot., 26: 179–189.

Sinnott, E. W., and S. Kaiser. 1934. Two types of genetic control over the development of shape. Bull. Torrey Bot. Club, 61: 1–7.

Sinskaia, E. N. 1928. The oleiferous plants and root crops of the family Cruciferae. Bull. Appl. Bot., Genet., Plant Breed., 19 (3): 1–619.

— 1931a. The study of species in their dynamics and interrelation with different types of vegetation. Bull. Appl. Bot., Gen., Plant Breed., 25 (2): 1–97.

Sinskaia, E. N., and A. A. Beztuzheva. 1931. The forms of *Camelina sativa* in connection with climate, flax, and man. Bull. Appl. Bot., Genet., Plant Breed., 25: 98–200.

Skalinska, M. 1928a. Sur les causes d'une disjunction non typique des hybrides du genre *Aquilegia*. Acta Soc. Bot. Polon., 5: 141–173.

— 1928b. Études sur la sterilité partielle des hybrides du genre *Aquilegia*. Zeitschr. Ind. Abst. u. Vererbungsl., Suppl.: 1343–1372.

— 1929. Das Problem des Nichterscheinens des väterlichen Typus in der Spaltung der partiell sterilen *Aquilegia*-Species-Bastarde. Acta Soc. Bot. Polon., 6: 138–164.

— 1931. A new case of unlike reciprocal hybrids in *Aquilegia*. Proc. 5th Int. Bot. Cong., Cambridge: 250.

— 1935. Cytogenetic investigations of an allotetraploid *Aquilegia*. Bull. Acad. Pol. Sci. et Lettres, Ser. B: 33–63.

Skirm, G. W. 1942. Bivalent pairing in an induced tetraploid of *Tradescantia*. Genetics, 27: 635–640.

Skottsberg, C. 1925. Juan Fernandez and Hawaii, a phytogeographical discussion. B. P. Bishop Mus., Bull. 16: 1–47.

— 1936. Antarctic plants in Polynesia. Essays in Geobotany in honor of W. A. Setchell, Univ. of California Press: 291–311.

— 1938. Geographical isolation as a factor in species formation, and its relation to certain insular floras. Proc. Linn. Soc. London, 150: 286–293.

— 1939. Remarks on the Hawaiian flora. Proc. Linn. Soc. London, 151: 181–186.

Skovsted, A. 1934. Cytological studies in cotton. II. Two interspecific hybrids between Asiatic and New World cottons. Jour. Genet., 28: 407–424.

— 1943. Successive mutations in *Nadsonia richteri* Kostka. Compt. Rend. Carlsberg Lab., Ser. Phys. 23(20): 409–457.

Small, J. 1937a. Quantitative evolution. II. Compositae D_p-ages in relation to time. Proc. Roy. Soc. Edinburgh, 57: 215–220.

— 1937b. Quantitative evolution. III. D_p-ages of Gramineae. Proc. Roy. Soc. Edinburgh, 57: 221–227.

— 1945. Tables to illustrate the geological history of species-number in diatoms. Proc. Roy. Irish Acad., 50, Sec. B: 295–309.

— 1946. Quantitative evolution. VIII. Numerical analysis of tables to illustrate the geological history of species number in Diatoms; an introductory summary. Proc. Roy. Irish Acad., 51, Sec. B: 53–80.

Small, J., and I. K. Johnston. 1937. Quantitative evolution in Compositae. Proc. Roy. Soc. Edinburgh, 57: 26–54.

Smith, A. C. 1945. Geographical distribution of the Winteraceae. Jour. Arnold Arboretum, 26: 48–59.

—— 1946. A taxonomic review of *Euptelea*. Jour. Arnold Arboretum, 27: 175–185.

Smith, D. C. 1944. Pollination and seed formation in grasses. Jour. Agr. Res., 68: 79–95.

Smith, E. C. 1941. Chromosome behavior in *Catalpa hybrida* Spaeth. Jour. Arnold Arboretum, 22: 219–221.

Smith, E. C., and C. Nichols, Jr. 1941. Species hybrids in forest trees. Jour. Arnold Arboretum, 22: 443–454.

Smith, H. E. 1946. *Sedum pulchellum*: a physiological and morphological comparison of diploid, tetraploid, and hexaploid races. Bull. Torrey Bot. Club, 73: 495–541.

Smith, J. P. 1919. Climatic relations of the Tertiary and Quaternary faunas of the California region. Proc. Calif. Acad. Sci., Ser. 4, 9: 123–173.

Smith, L. 1936. Cytogenetic studies in *Triticum monococcum* L. and *T. aegilopoides* Bal. Univ. of Missouri Agr. Expt. Sta., Res. Bull., 248: 1–38.

—— 1946. Haploidy in einkorn. Jour. Agr. Res., 73: 291–301.

Sokolovskaya, A. P., and O. S. Strelkova. 1940. Karyological investigation of the alpine flora on the main Caucasus range and the problem of geographical distribution of polyploids. Compt. Rend. Acad. Sci. URSS, 29: 415–418.

Sörensen, T., and G. Gudjónsson. 1946. Spontaneous chromosome-aberrants in apomictic Taraxaca. Kongel. Danske Vidensk. Selsk., Biol. Skr. 4(2): 1–48.

Sparrow, A. H., M. L. Ruttle, and B. R. Nebel. 1942. Comparative cytology of sterile intra- and fertile intervarietal hybrids of *Antirrhinum majus* L. Amer. Jour. Bot., 29: 711–715.

Spencer, W. P. 1947. Genetic drift in a population of *Drosophila immigrans*. Evolution, 1: 103–110.

Spiegelman, S., and C. C. Lindegren. 1944. A comparison of the kinetics of enzymatic adaptation in genetically homogeneous and heterogeneous populations of yeast. Ann. Missouri Bot. Gard., 31: 219–233.

Sprecher, M. A. 1919. Étude sur la sémence et la germination du *Garcinia mangostana*. L. Rev. Gén. Bot., 31: 513–531; 611–634.

Springer, E. 1935. Über apogame (vegetative entstandene) Sporogone an der bivalenten Rasse des laubmooses *Phascum cuspidatum*. Zeitschr. Ind. Abst. u. Vererbungsl., 69: 249–262.

Stadler, L. J. 1932. On the genetic nature of induced mutations in plants. Proc. 6th Int. Congr. Genet., 1: 274–294.

—— 1942. Some observations on gene variability and spontaneous gene mutation. Spragg Memorial Lectures on Plant Breed., 3d Ser., 15 pp. Michigan State College, East Lansing.

Stadler, L. J., and G. F. Sprague. 1936. Genetic effects of ultraviolet radiation in maize, I, II, and III. Proc. Nat. Acad. Sci., 22: 572–591.

Stakman, E. C., M. F. Kernkamp, T. H. King, and W. J. Martin. 1943. Genetic factors for mutability and mutant characters in *Ustilago zeae*. Amer. Jour. Bot., 30: 37–48.

Stebbins, G. L., Jr. 1932a. Cytology of *Antennaria*. I. Normal species. Bot. Gaz., 94: 134–151.

—— 1932b. Cytology of *Antennaria*. II. Parthenogenetic species. Bot. Gaz., 94: 322–345.

—— 1935. A new species of *Antennaria* from the Appalachian region. Rhodora, 37: 229–237.

—— 1937. Critical notes on the genus *Ixeris*. Jour. Bot., 1937: 43–51.

—— 1938a. Cytogenetic studies in *Paeonia*. II. The cytology of the diploid species and hybrids. Genetics, 23: 83–110.

—— 1938b. Cytological characteristics associated with the different growth habits in the dicotyledons. Amer. Jour. Bot., 25: 189–198.

—— 1939. Notes on the systematic relationships of the Old World species and of some horticultural forms of *Paeonia*. Univ. of Calif. Publ. Bot., 19: 245–266.

—— 1940a. The significance of polyploidy in plant evolution. Amer. Nat., 74: 54–66.

—— 1940b. Studies in the Cichorieae: *Dubyaea* and *Soroseris*, endemics of the Sino-Himalayan region. Mem. Torrey Bot. Club, 19: 1–76.

—— 1941a. Apomixis in the angiosperms. Bot. Rev., 7: 507–542.

—— 1941b. Comparative growth rates of diploid and autotetraploid *Stipa lepida*. (Abstract.) Amer. Jour. Bot., 28, Suppl: 6s.

—— 1941c. Additional evidence for a holarctic dispersal of flowering plants in the Mesozoic era. Proc. 6th Pac. Sci. Congr.: 649–660.

—— 1942a. The role of isolation in the differentiation of plant species. Biol. Symposia, 6: 217–233.

——1942b. The genetic approach to problems of rare and endemic species. Madroño, 6: 240–258.

—— 1942c. Polyploid complexes in relation to ecology and the history of floras. Amer. Nat., 76: 36–45.

—— 1945. The cytological analysis of species hybrids. Bot. Rev., 11: 463–486.

—— 1947a. Types of polyploids: their classification and significance. Advances in Genetics, I: 403–429.

—— 1947b. Evidence on rates of evolution from the distribution of existing and fossil plant species. Ecological Monographs, 17: 149–158.

—— 1947c. The origin of the complex of *Bromus carinatus* and its phytogeographic implications. Contr. Gray. Herb., 165: 42–55.

—— 1948. Review of "A Study of the Genus *Paeonia*," by F. C. Stern. Madroño, 9: 193–199.

—— 1949a. Rates of evolution in plants. Genetics, Paleontology, and Evolution, Princeton University Press: 229–242.

—— 1949b. The evolutionary significance of natural and artificial polyploids in the family Gramineae. Proc. 8th Int. Congr. Genetics: 461–485.

Stebbins, G. L., Jr., and E. B. Babcock. 1939. The effect of polyploidy and apomixis on the evolution of species in *Crepis*. Jour. Hered., 30: 519–530.

Stebbins, G. L., Jr., and S. Ellerton. 1939. Structural hybridity in *Paeonia californica* and *P. brownii*. Jour. Genet., 38: 1–36.

Stebbins, G. L., Jr., and J. A. Jenkins. 1939. Aposporic development in the North American species of *Crepis*. Genetica, 21: 1–34.

Stebbins, G. L., Jr., and M. Kodani. 1944. Chromosomal variation in guayule and mariola. Jour. Hered., 35: 163–172.

Stebbins, G. L., Jr., and R. M. Love. 1944. A cytological study of California forage grasses. Amer. Jour. Bot., 28: 371–382.

Stebbins, G. L., Jr., E. B. Matzke, and C. Epling. 1947. Hybridization in a population of *Quercus marilandica* and *Quercus ilicifolia*. Evolution, 1: 79–88.

Stebbins, G. L., Jr., and H. A. Tobgy. 1944. The cytogenetics of hybrids in *Bromus*. I. Hybrids within the section *Ceratochloa*. Amer. Jour. Bot., 31: 1–11.

Stebbins, G. L., Jr., H. A. Tobgy, and J. R. Harlan. 1944. The cytogenetics of hybrids in Bromus. II. *Bromus carinatus* and *Bromus arizonicus*. Proc. Calif. Acad. Sci., 25: 307–322.

Stebbins, G. L., Jr., J. I. Valencia, and R. M. Valencia. 1946a. Artificial and natural hybrids in the Gramineae, tribe Hordeae. I. *Elymus, Sitanion,* and *Agropyron*. Amer. Jour. Bot., 33: 338–351.

—— 1946b. Artificial and natural hybrids in the Gramineae, tribe Hordeae. II. *Agropyron, Elymus,* and *Hordeum*. Amer. Jour. Bot., 33: 579–586.

Stephens, S. G. 1947. Cytogenetics of *Gossypium* and the problem of the New World cottons. Advances in Genetics, I: 431–442.

Stern, F. C. 1946. A study of the genus *Paeonia*. London, Royal Horticultural Society. viii + 155 pp.

—— 1947. Plant distribution in the Northern Hemisphere. Geogr. Jour., 108: 24–40.

Steyermark, C. S. 1939. Distribution and hybridization of *Vernonia* in Missouri. Bot. Gaz., 100: 548–562.

Stockwell, P., and F. I. Righter. 1946. *Pinus*: the fertile species hybrid between knobcone and Monterey pines. Madroño, 8: 157–160.

Stout, A. B., and C. Chandler. 1941. Change from self-incompatibility to self-compatibility accompanying change from diploidy to tetraploidy. Science, 94: 118.

Stout, A. B., R. H. McKee, and E. I. Schreiner. 1927. The breeding of forest trees for pulp wood. Jour. N. Y. Bot. Gard., 28: 49–63.

Stout, A. B., and E. J. Schreiner. 1934. Hybrids between the necklace cottonwood and the large-leafed aspen. Jour. N. Y. Bot. Gard., 35: 140–143.

Straub, J. 1940. Quantitative und qualitative Verschiedenheiten innerhalb von polyploiden Pflanzenreihen. Biol. Zentralbl., 60: 659–669.

Stubbe, H., and F. von Wettstein. 1941. Über die Bedeutung von Klein- und Grossmutationen in der Evolution. Biol. Zentralbl., 61: 265–297.

Sturtevant, A. H. 1937. Essays on evolution. I. On the effects of selection on mutation rate. Quart. Rev. Biol., 12: 464–467.

—— 1938. Essays on evolution. III. On the origin of interspecific sterility. Quart. Rev. Biol., 13: 333–335.

Sturtevant, A. H., and G. W. Beadle. 1939. An introduction to genetics. Philadelphia, Saunders. 391 pp.

Sugiura, T. 1936. Studies on the chromosome numbers in higher plants, with special reference to cytokinesis. Cytologia, 7: 544–595.

Sukatschew, W. 1928. Einige experimentelle Untersuchungen über den Kampf ums Dasein zwischen Biotypen derselben Art. Zeitschr. Ind. Abst. u. Vererbungsl., 47: 54–74.

Sule, E. P. 1946. On the question of the production of perennial forms of wheat (Russian). Trudy Zonal. Inst. Zern. Khoz. Nechern. Polocy USSR, 13: 32–49.

Sumner, F. B. 1942. Where does adaptation come in? Amer. Nat., 76: 433–444.

Suneson, C. A., and G. A. Wiebe. 1942. Survival of barley and wheat varieties in mixtures. Jour. Amer. Soc. Agron., 34(11): 1052–1056.

Suomalainen, Eeva. 1947. On the cytology of the genus *Polygonatum*, group *Alternifolia*. Ann. Acad. Sci. Fenn., Ser. A, IV(13): 1–66.

Suomalainen, Esko. 1940a. Beiträge zur Zytologie der parthenogenetischen Insekten. I. Coleoptera. Ann. Acad. Sci. Fenn., Ser. A, 54(7): 1–143.

—— 1940b. Polyploidy in parthenogenetic Curculionidae. Hereditas, 26: 51–64.

—— 1945. Zu den Chromosomenverhältnissen und dem Artbildungsproblem bei parthenogenetischen Tieren. Sitzungsber. Finn. Akad. Wiss., 181–201.

—— 1947. Parthenogenese und Polyploidie bei Rüsselkäfern (Curculionidae). Hereditas, 33: 425–456.

Svärdson, G. 1945. Chromosome studies on Salmonidae. Rep. Swedish State Inst. Fresh-Water Fishery Res., Drottningholm, No. 23: 1–151.

Svedelius, N. 1929. An evaluation of the structural evidences for genetic relationships in plants: Algae. Proc. Int. Congr. Plant Sci., Ithaca, N.Y., 1: 457–471.

Sveshnikova, I. N. 1927. Die Genese des Kerns im Genus *Vicia*. Verh. 5th Int. Kongr. Vererb.: 1415–1421.

—— 1936. Translocations in hybrids as an indicator of karyotype evolution. Biol. Zhurn., 5: 303–326.

Swamy, B. G. L. 1949. Further contributions to the morphology of the Degeneriaceae. Jour. Arnold Arboretum, 30: 10–38.

Sweadner, W. R. 1937. Hybridization and the phylogeny of the genus *Platysamia*. Ann. Carnegie Mus., 25: 163–242.

Sweet, E. D. 1938. Chiasmata, crossing over, and mutation in *Oenothera* hybrids. Jour. Genet., 35: 397–419.

Swingle, W. T. 1932. Recapitulation of seedling characters by nucellar buds developing in the embryo-sac of *Citrus*. Proc. 6th Internat. Cong. of Genetics, 2: 196–197.

Sylvén, N. 1937. The influence of climatic conditions on type composition. Imp. Bureau Plant Genetics, Herbage Bull. No. 21. 8 pp.

Syrach-Larsen, C. 1937. The employment of species, types, and individuals in forestry. Roy. Vet. Agr. Coll. Yearbook, Copenhagen. 69–222.

Täckholm, G. 1922. Zytologische Studien über die Gattung *Rosa*. Acta Hort. Berg., 7: 97–381.

Tahara, M. 1915. Cytological studies on *Chrysanthemum*. II. Bot. Mag. Tokyo, 29: 48–50.

Takhtadzhian, A. 1943. Comparative ontogeny and phylogeny in the higher plants. Trans. Molotov State University, Erevan 22: 71–176. (In Russian, with English summary.)

Talbot, M. W., H. H. Biswell, and A. L. Hormay. 1939. Fluctuations in the annual vegetation of California. Ecology, 20: 394–402.

Tanaka, N. 1937. Chromosome studies in Cyperaceae. I. Cytologia, Fujii Jubil. Vol.: 814–821.

—— 1939. Chromosome studies in Cyperaceae. IV. Chromosome number of *Carex* species. Cytologia, 10: 51–58.

Tarnavschi, T. 1939. Karyologische Untersuchungen an Halophyten aus Rumanien im Lichte zyto-ökologischer und zyto-geographischer Forschung. Bul. Fac. Stiinte din Cernauti, 12: 68–106.

Tatebe, T. 1936. Genetic and cytological studies on the F_1 hybrid of scarlet or tomato eggplant (*Solanum integrifolium* Poir.) $\times$ eggplant (*Solanum melongena* L.). Bot. Mag. Tokyo, 50: 457–462.

Taylor, H. 1945. Cytotaxonomy and phylogeny of the Oleaceae. Brittonia, 5: 337–367.

Tedin, O. 1925. Vererbung, Variation, und Systematik in der Gattung *Camelina*. Hereditas, 6: 275–386.

Terao, H., and U. Midusima. 1939. Some considerations on the classification of *Oryza sativa* L. into two subspecies, so-called "japonica" and "indica." Jap. Jour. Bot., 10: 213–258.

Thomas, H. H. 1925. The Caytoniales, a new group of angiospermous plants from the Jurassic rocks of Yorkshire. Phil. Trans. Roy. Soc., B, 213: 299–313.

Thomas, P. T. 1940a. Reproductive versatility in *Rubus*. II. The chromosomes and development. Jour. Genet., 40: 119–128.

—— 1940b. The origin of new forms in *Rubus*. III. The chromosome constitution of *R. loganobaccus* Bailey, its parents and derivatives. Jour. Genet., 40: 141–156.

Thompson, J. M. 1933. Studies in advancing sterility. VI. The theory of scitaminean flowering. Publ. Hartley Bot. Lab., Liverpool, No. 11. 114 pp.

—— 1934. Studies in advancing sterility. VII. The state of flowering known as angiospermy. Publ. Hartley Bot. Lab., Liverpool, No. 12. 47 pp.

Thompson, W. P., E. J. Britten, and J. C. Harding. 1943. The artificial synthesis of a 42-chromosome species resembling common wheat. Canad. Jour. Res., C, 21: 134–144.

Timofeeff-Ressovsky, N. W. 1940. Mutations and geographical variation. In J. Huxley, ed., The New Systematics: 73–136.

Tinney, F. W., and G. S. Aamodt. 1940. The progeny test as a measure of the types of seed development in *Poa pratensis* L. Jour. Hered., 31: 457–464.

Tischler, G. 1931. Pflanzliche Chromosomenzahlen. Tab. Biolog. Period. 7: 109–226.

—— 1935. Die Bedeutung der Polyploidie für die Verbreitung der Angiospermen, erlautert an den Arten Schleswig Holsteins, mit Ausblicken auf andere Florengebiete. Bot. Jahrb., 67 (1): 1–36.

—— 1936. Pflanzliche Chromosomenzahlen. II. Tab. Biol. Per. 11: 281–304; 12: 57–115.

—— 1937. Die Halligen flora der Nordsee im Lichte cytologischer Forschung. Cytologia, Fujii Jubil. Vol.: 162–169.

—— 1938. Pflanzliche Chromosomenzahlen. IV. Tab. Biol. Per. 16: 162–218.

—— 1946. Über die Siedlungsfähigkeit von Polyploiden. Zeitschr. f. Naturforschung, 1: 157–159.

Tobgy, H. A. 1943. A cytological study of *Crepis fuliginosa, C. neglecta,* and their F_1 hybrid, and its bearing on the mechanism of phylogenetic reduction in chromosome number. Jour. Genet., 45: 67–111.

Tobler, M. 1931. Experimentelle Analyse der Genom- und Plasmonwirkung bei Moosen. IV. Zur Variabilität des Zellvolumens einer Sippenkreuzung von *Funaria hygrometrica* und deren bivalenten Rassen. Zeitschr. Ind. Abst. u. Vererbungsl., 60: 39–62.

Tomé, G. A., and I. J. Johnson. 1945. Self- and cross-fertility relationships in *Lotus corniculatus* L. and *Lotus tenuis* Wald. and Kit. Jour. Amer. Soc. Agron., 37: 1011–1023.

Tometorp, G. 1939. Cytological studies on haploid *Hordeum distichum*. Hereditas, 25: 241–254.

Trelease, W. F. 1924. The American Oaks. Mem. Nat. Acad. Sci., 20: 1–255.

Turesson, G. 1922a. The species and the variety as ecological units. Hereditas, 3: 100–113.
—— 1922b. The genotypical response of the plant species to the habitat. Hereditas, 3: 211–350.
—— 1925. The plant species in relation to habitat and climate. Contributions to the knowledge of genecological units. Hereditas, 6: 147–236.
—— 1927. Contributions to the genecology of glacial relics. Hereditas, 9: 81–101.
—— 1929. Zur Natur und Begrenzung der Arteinheiten. Hereditas, 12: 323–333.
—— 1930. Studien über *Festuca ovina*. II. Chromosomenzahl und Viviparie. Hereditas, 13: 177–184.
—— 1931a. The selective effect of climate upon the plant species. Hereditas, 14: 99–152.
—— 1931b. Studien über *Festuca ovina*. III. Weitere beiträge zur kenntnis der Chromosomenzahlen viviparen formen. Hereditas, 15: 13–16.
—— 1932. Die Genenzentrumtheorie und das Entwicklungszentrum der Pflanzenart. K. Fys. Sallsk. Lund Forhandl., 2(6): 1–11.
—— 1936. Rassenökologie und Pflanzengeographie. Bot. Not.: 420.
—— 1943. Variation in the apomictic microspecies of *Alchemilla vulgaris* L. Bot. Not. (Lund): 413–427.
Turrill, W. B. 1936a. Contacts between plant classification and experimental botany. Nature, 137: 563–566.
—— 1936b. Natural selection and the distribution of plants. Proc. Roy. Soc., B, 121: 49–52.
—— 1938a. The expansion of taxonomy, with special reference to the Spermatophyta. Biol. Rev., 13: 342–373.
—— 1938b. Taxonomy and genetics. Jour. Bot. London, 76: 33–39.
—— 1938c. Problems of British Taraxaca. Proc. Linn. Soc., 150: 120–124.
—— 1938d. Material for a study of taxonomic problems in *Taraxacum*. Rep. Bot. Exchange Club, 11: 570–589.
—— 1940. Experimental and synthetic plant taxonomy. In J. Huxley, ed., The New Systematics: 47–71.
—— 1942a. Differences in the systematics of plants and animals and their dependence on differences in structure, function, and behavior in the two groups. Proc. Linn. Soc. London, 153: 272–277.
—— 1942b. Experimental attacks on species problems. Chronica Botanica, 7: 281–283.
—— 1946. The ecotype concept; a consideration with appreciation and criticism, especially of recent trends. New Phytologist, 45: 34–43.

U, Nagahuru. 1935. Genome analysis in *Brassica* with special reference to the experimental formation of *B. Napus* and peculiar mode of fertilization. Jap. Jour. Bot., 7: 389–452.

Upcott, M. 1936. The parents and progeny of *Aesculus carnea*. Jour. Genet., 33: 135–149.

—— 1939. The nature of tetraploidy in *Primula kewensis*. Jour. Genet., 39: 79–100.

Vaarama, A. 1939. Cytological studies on some Finnish species and hybrids of the genus *Rubus* L. Jour. Sci. Agr. Soc. Finland, 11: 1–13.

Valentine, D. H. 1948. Studies in British Primulas. II. Ecology and taxonomy of primrose and oxlip (*Primula vulgaris* Huds. and *P. elatior* Schreb.) New Phytologist, 47: 111–130.

Vandel, A. 1938. Chromosome number, polyploidy and sex in the animal kingdom. Proc. Zool. Soc. London, A, 107: 519–541.

Vassilchenko, I. T. 1939. *Camelina,* in Komarov. V. Flora USSR, 8: 596–602.

Vavilov, N. I. 1926. Studies on the origin of cultivated plants. Bull. Appl. Bot. Plant Breed. (Leningrad), 16(2): 1–248.

Viosca, P. 1935. The irises of southeastern Louisiana. Bull. Amer. Iris Soc., April, No. 57. 56 pp.

Vries, H. de. 1923. Über die Entstehung von *Oenothera lamarckiana* mut. *velutina*. Biol. Zentralbl., 43: 213–224.

Wahl, H. A. 1940. Chromosome numbers and meiosis in the genus *Carex*. Amer. Jour. Bot., 27: 458–470.

Wakar, B. A. 1935a. Zytologische Untersuchung über F_1 der Weizen-Queckengrasbastarde. Cytologia, 7: 293–312.

—— 1935b. Cytologische Untersuchung der ersten Generation der Weizen-Queckengrasbastarde. Der Züchter, 7: 199–207.

—— 1937. Cytologische Untersuchung der selbstfertilen ersten Generation der Weizen-Queckengrasbastarde. Cytologia, 8: 67–90.

Walters, J. L. 1942. Distribution of structural hybrids in *Paeonia californica*. Amer. Jour. Bot., 29: 270–275.

Warmke, H. E. 1945. Experimental polyploidy and rubber content in *Taraxacum kok-saghyz*. Bot. Gaz., 106: 316–324.

Warmke, H. E., and A. F. Blakeslee. 1940. The establishment of a dioecious race in *Melandrium*. Amer. Jour. Bot., 27: 751–762.

Watkins, A. E. 1932. Hybrid sterility and incompatibility. Jour. Genet., 30: 257–266.

Webber, H. J. 1931. The economic importance of apogamy in *Citrus* and *Mangifera*. Proc. Amer. Soc. Hort. Sci., 28: 57–61.

Webber, J. M. 1935. Interspecific hybridization in *Gossypium* and the meiotic behavior of F_1 plants. Jour. Agr. Res., 51: 1047–1070.

Weber, W. A. 1946. A taxonomic and cytological study of the genus *Wyethia,* family Compositae with notes on the related genus *Balsamorhiza*. Amer. Midl. Nat., 35: 400–452.

Westergaard, M. 1940. Studies on cytology and sex determination in polyploid forms of *Melandrium album*. Dansk. Bot. Arkiv., 10: 1–131.

— 1948. The aspects of polyploidy in the genus *Solanum*. III. Seed production in autopolyploid and allopolyploid Solanums. Kgl. Danske Vidensk. Selsk. Biol. Meddel., 18, No. 3: 1–18.

Wetmore, R. H., and A. L. Delisle. 1939. Studies in the genetics and cytology of two species in the genus *Aster* and their polymorphy in nature. Amer. Jour. Bot., 26: 1–12.

Wettstein, F. von. 1924. Kreuzungsversuche mit multiploiden Moosrassen. II. Biol. Zentralbl., 44: 145–168.

— 1927. Die Erscheinung der Heteroploidie, besonders im Pflanzenreich. Ergebn. der Biologie, 2: 311–356.

— 1937. Experimentelle Untersuchungen zum Artbildungsproblem. I. Zellgrössenregulation und Fertilwerden einer polyploiden *Bryum*-Sippe. Zeitschr. Ind. Abst. u. Vererbungsl., 74: 34–53.

Wettstein, F., and J. Straub. 1942. Experimentelle Untersuchungen zum Artbildungsproblem. III. Weitere Beobachtungen an polyploiden *Bryum*.-Sippen. Zeitschr. Ind. Abst. u. Vererbungsl., 80: 271–280.

Wettstein, W. von. 1933. Die Kreuzungsmethode und die Beschreibung von F_1-Bastarden bei *Populus*. Zeitschr. f. Züchtung A, 18: 597–626.

— 1937. Leistungssteigerung durch Herkunftskreuzung bei *Populus tremula*. Naturwissenschaften, 25: 434–436.

Wherry, Edgar T. 1939. A provisional key to the Polemoniaceae. Bartonia: 13–17.

Whitaker, T. W. 1933. Chromosome number and relationship in the Magnoliales. Jour. Arnold Arboretum, 14: 376–385.

— 1934a. Chromosome constitution in certain monocotyledons. Jour. Arnold Arboretum, 15: 135–143.

— 1934b. The occurrence of tumors on certain *Nicotiana* hybrids. Jour. Arnold Arboretum, 15: 144–153.

— 1936. Fragmentation in *Tradescantia*. Amer. Jour. Bot., 23: 517–519.

— 1944. The inheritance of certain characters in a cross of two American species of *Lactuca*. Bull. Torrey Bot. Club, 71: 347–355.

Whitaker, T. W., and R. C. Thompson. 1941. Cytological studies in *Lactuca*. Bull. Torrey Bot. Club, 68: 388–394.

White, M. J. D. 1945. Animal cytology and evolution. Cambridge, Cambridge University Press. 375 pp.

White, W. J. 1940. Intergeneric crosses between *Triticum* and *Agropyron*. Scient. Agr., 21: 198–232.

Wiegand, K. M. 1935. A taxonomist's experience with hybrids in the wild. Science, 81: 161–166.

Willis, J. C. 1922. Age and area. Cambridge, Cambridge Univ. Press. 259 pp.

—— 1940. The course of evolution by differentiation or divergent mutation rather than by selection. Cambridge, Cambridge University Press. 207 pp.

Wilson, G. B. 1946. Cytological studies in the Musae. I. Meiosis in some triploid clones. Genetics, 31: 241–258.

Winge, O. 1917. The chromosomes. Their numbers and general importance. Compt. Rend. Trav. Lab. Carlsberg, 13: 131–275.

—— 1933. A case of amphidiploidy within the collective species *Erophila verna*. Hereditas, 18: 181–191.

—— 1938. Inheritance of species characters in *Tragopogon*. A cytogenetic investigation. Compt. Rend. Trav. Lab. Carlsberg, Sér. Phys. 22: 155–193.

—— 1940. Taxonomic and evolutionary studies in *Erophila* based on cytogenetic investigations. Compt. Rend. Lab. Carlsberg, Sér. Phys. 23: 17–39.

—— 1944. On segregation and mutation in yeast. Compt. Rend. Trav. Lab. Carlsberg, Sér. Phys. 24: 79–96.

Winkler, H. 1916. Über die experimentelle Erzeugung von Pflanzen mit abweichenden chromosomenzahlen. Zeitschr. Bot., 8: 417–531.

—— 1920. Verbreitung und Ursache der Parthenogenesis im Pflanzen- und Tierreiche. Jena, G. Fischer.

—— 1934. Fortpflanzung der Gewächse 7. Apomixis. Handw. d. Naturwiss., 4: 451–461. Jena, G. Fischer.

Winter, F. L. 1929. The mean and variability as affected by continuous selection for composition in corn. Jour. Agr. Res., 39: 451–476.

Wodehouse, R. P. 1935. Pollen grains. New York, McGraw-Hill. 574 pp.

Wolf, C. B. 1938. California plant notes. II. Rancho Santa Ana Bot. Gard., Occas. Papers, 1: 44–90.

—— 1944. The Gander oak, a new hybrid oak from San Diego County, California. Proc. Calif. Acad. Sci., 25: 177–188.

Woodson, R. E., Jr. 1941. The North American Asclepiadaceae. I. Perspective of the genera. Ann. Missouri Bot. Gard., 28: 193–248.

—— 1947. Some dynamics of leaf variation in *Asclepias tuberosa*. Ann. Missouri Bot. Gard., 34: 353–432.

Wright, S. 1931. Evolution in Mendelian populations. Genetics, 16: 97–159.

—— 1940a. The statistical consequences of Mendelian heredity in relation to speciation. In J. Huxley, ed., The New Systematics: 161–183.

—— 1940b. Breeding structure of populations in relation to speciation. Amer. Nat., 74: 232–248.

—— 1941a. The material basis of evolution. Scient. Monthly, 53: 165–170.

—— 1941b. The "age and area" concept extended. Ecology, 22: 345–347.

—— 1943. An analysis of local variability of flower color in *Linanthus parryae*. Genetics, 28: 139–156.
—— 1946. Isolation by distance under diverse systems of mating. Genetics, 31: 39–59.
Wright, S., and Th. Dobzhansky. 1946. Genetics of natural populations. XII. Experimental reproduction of some of the changes caused by natural selection in certain populations of *Drosophila pseudoobscura*. Genetics, 31: 125–156.
Wulff, E. V. 1943. An introduction to historical plant geography. Engl. trans., based on the 2d ed. (1936) revised in 1939. Waltham, Mass., Chronica Botanica. 223 pp.
Wulff, H. D. 1937a. Karyologische Untersuchungen an der Halophyten-flora Schleswig-Holsteins. Jahrb. Wiss. Bot., 841: 812–840.
—— 1937b. Chromosomenstudien an der schleswig-holsteinischen Angiospermen-Flora. I. Ber. Deu. Bot. Ges., 55: 262–269.
Wynne-Edwards, V. C. 1937. Isolated arctic-alpine floras in eastern North America: a discussion of their glacial and recent history. Trans. Roy. Soc. Canada, III (5) 31: 33–58.

Yakovleva, S. V. 1933. Karyological investigation of some *Salvia* species. Bull. Appl. Bot., Ser. 2, 5: 207–213.
Yarnell, S. H. 1933. Inheritance in an oak species hybrid. Jour. Arnold Arboretum, 14: 68–75.
—— 1936. Chromosome behavior in blackberry-raspberry hybrids. Jour. Agr. Res., 52: 385–396.

Zakharjevsky, A. 1941. Control of sterility and the breeding of hybrids of *T. durum* × *T. timopheevi*. Proc. Acad. Agr. Sci. USSR, 6(2): 5–9. (In Russian.)
Zhebrak, A. 1944a. Synthesis of new species of wheats. Nature, 153(3888): 549–551.
—— 1944b. Synthesis of new species of wheats. Acta Timiriazev Agr. Acad., 6 (Genet. div.): 5–54. (In Russian with English summary.)
Zhebrak, A. R. 1946. New amphidiploid species of wheat and their significance for selection and evolution. Amer. Nat., 80: 271–279.
Zimmerman, E. C. 1942. Distribution and origin of some eastern Oceanic insects. Amer. Nat., 76: 280–307.
—— 1948. Insects of Hawaii. Vol. I. Introduction. Honolulu, University of Hawaii Press. 206 pp.
Zimmermann, W. 1930. Die Phylogenie der Pflanzen. Jena, G. Fischer. 452 pp.
Zinger, H. B. 1909. On the species of *Camelina* and *Spergularia* occurring as weeds in sowings of flax and their origin. Trudy Bot. Muz. Akad. Nauk. USSR, 6: 1–303. (In Russian.)

Index

Bei Fragen zur Produktsicherheit wenden Sie sich bitte an:
If you have any questions regarding product safety,
please contact:

Walter de Gruyter GmbH
Genthiner Straße 13
10785 Berlin
productsafety@degruyterbrill.com